Handbook of Natural Computing

Main Editor
Grzegorz Rozenberg

Editors
Thomas Bäck
Joost N. Kok

Handbook of Natural Computing

Volume 2

With 734 Figures and 75 Tables

 Springer

Editors
Grzegorz Rozenberg
LIACS
Leiden University
Leiden, The Netherlands
and
Computer Science Department
University of Colorado
Boulder, USA

Thomas Bäck
LIACS
Leiden University
Leiden, The Netherlands

Joost N. Kok
LIACS
Leiden University
Leiden, The Netherlands

ISBN 978-3-540-92909-3   ISBN 978-3-540-92910-9 (eBook)
ISBN 978-3-540-92911-6 (print and electronic bundle)
DOI 10.1007/978-3-540-92910-9
Springer Heidelberg Dordrecht London New York

Library of Congress Control Number: 2010933716

Printed on acid-free paper

Springer is part of Springer Science+Business Media (www.springer.com)

Preface

Natural Computing is the field of research that investigates human-designed computing inspired by nature as well as computing taking place in nature, that is, it investigates models and computational techniques inspired by nature, and also it investigates, in terms of information processing, phenomena taking place in nature.

Examples of the first strand of research include neural computation inspired by the functioning of the brain; evolutionary computation inspired by Darwinian evolution of species; cellular automata inspired by intercellular communication; swarm intelligence inspired by the behavior of groups of organisms; artificial immune systems inspired by the natural immune system; artificial life systems inspired by the properties of natural life in general; membrane computing inspired by the compartmentalized ways in which cells process information; and amorphous computing inspired by morphogenesis. Other examples of natural-computing paradigms are quantum computing and molecular computing, where the goal is to replace traditional electronic hardware, by, for example, bioware in molecular computing. In quantum computing, one uses systems small enough to exploit quantum-mechanical phenomena to perform computations and to perform secure communications more efficiently than classical physics, and, hence, traditional hardware allows. In molecular computing, data are encoded as biomolecules and then tools of molecular biology are used to transform the data, thus performing computations.

The second strand of research, computation taking place in nature, is represented by investigations into, among others, the computational nature of self-assembly, which lies at the core of the nanosciences; the computational nature of developmental processes; the computational nature of biochemical reactions; the computational nature of bacterial communication; the computational nature of brain processes; and the systems biology approach to bionetworks where cellular processes are treated in terms of communication and interaction, and, hence, in terms of computation.

Research in natural computing is genuinely interdisciplinary and forms a bridge between the natural sciences and computer science. This bridge connects the two, both at the level of information technology and at the level of fundamental research. Because of its interdisciplinary character, research in natural computing covers a whole spectrum of research methodologies ranging from pure theoretical research, algorithms, and software applications to experimental laboratory research in biology, chemistry, and physics.

Computer Science and Natural Computing

A preponderance of research in natural computing is centered in computer science. The spectacular progress in Information and Communication Technology (ICT) is highly supported by the evolution of computer science, which designs and develops the instruments needed for this progress: computers, computer networks, software methodologies, etc. As ICT has such a tremendous impact on our everyday lives, so does computer science.

However, there is much more to computer science than ICT: it is the science of information processing and, as such, a fundamental science for other disciplines. On one hand, the only common denominator for research done in such diverse areas of computer science is investigating various aspects of information processing. On the other hand, the adoption of Information and Information Processing as central notions and thinking habit has been an important development in many disciplines, biology and physics being prime examples. For these scientific disciplines, computer science provides not only instruments but also a way of thinking.

We are now witnessing exciting interactions between computer science and the natural sciences. While the natural sciences are rapidly absorbing notions, techniques, and methodologies intrinsic to information processing, computer science is adapting and extending its traditional notion of computation, and computational techniques, to account for computation taking place in nature around us. Natural Computing is an important catalyst for this two-way interaction, and this handbook constitutes a significant record of this development.

The Structure of the Handbook

Natural Computing is both a well-established research field with a number of classical areas, and a very dynamic field with many more recent, novel research areas. The field is vast, and so it is quite usual that a researcher in a specific area does not have sufficient insight into other areas of Natural Computing. Also, because of its dynamic development and popularity, the field constantly attracts more and more scientists who either join the active research or actively follow research developments.

Therefore, the goal of this handbook is two-fold:

(i) to provide an authoritative reference for a significant and representative part of the research in Natural Computing, and
(ii) to provide a convenient gateway to Natural Computing for motivated newcomers to this field.

The implementation of this goal was a challenge because this field and its literature are vast — almost all of its research areas have an extensive scientific literature, including specialized journals, book series, and even handbooks. This implies that the coverage of the whole field in reasonable detail and within a reasonable number of pages/volumes is practically impossible.

Thus, we decided to divide the presented material into six areas. These areas are by no means disjoint, but this division is convenient for the purpose of providing a representative picture of the field — representative with respect to the covered research topics and with respect to a good balance between classical and emerging research trends.

Each area consists of individual chapters, each of which covers a specific research theme. They provide necessary technical details of the described research, however they are self-contained and of an expository character, which makes them accessible for a broader audience. They also provide a general perspective, which, together with given references, makes the chapters valuable entries into given research themes.

This handbook is a result of the joint effort of the handbook editors, area editors, chapter authors, and the Advisory Board. The choice of the six areas by the handbook editors in consultation with the Advisory Board, the expertise of the area editors in their respective areas, the choice by the area editors of well-known researchers as chapter writers, and the peer-review for individual chapters were all important factors in producing a representative and reliable picture of the field. Moreover, the facts that the Advisory Board consists of 68 eminent scientists from 20 countries and that there are 105 contributing authors from 21 countries provide genuine assurance for the reader that this handbook is an authoritative and up-to-date reference, with a high level of significance and accuracy.

Handbook Areas

The material presented in the handbook is organized into six areas: Cellular Automata, Neural Computation, Evolutionary Computation, Molecular Computation, Quantum Computation, and Broader Perspective.

Cellular Automata

Cellular automata are among the oldest models of computation, dating back over half a century. The first cellular automata studies by John von Neumann in the late 1940s were biologically motivated, related to self-replication in universal systems. Since then, cellular automata gained popularity in physics as discrete models of physical systems, in computer science as models of massively parallel computation, and in mathematics as discrete-time dynamical systems. Cellular automata are a natural choice to model real-world phenomena since they possess several fundamental properties of the physical world: they are massively parallel, homogeneous, and all interactions are local. Other important physical constraints such as reversibility and conservation laws can be added as needed, by properly choosing the local update rule. Computational universality is common in cellular automata, and even starkly simple automata are capable of performing arbitrary computation tasks. Because cellular automata have the advantage of parallelism while obeying natural constraints such as locality and uniformity, they provide a framework for investigating realistic computation in massively parallel systems. Computational power and the limitations of such systems are most naturally investigated by time- and space-constrained computations in cellular automata. In mathematics — in terms of symbolic dynamics — cellular automata are viewed as endomorphisms of the full shift, that is, transformations that are translation invariant and continuous in the product topology. Interesting questions on chaotic dynamics have been studied in this context.

Neural Computation

Artificial neural networks are computer programs, loosely modeled after the functioning of the human nervous system. There are neural networks that aim to gain understanding of biological neural systems, and those that solve problems in artificial intelligence without necessarily creating a model of a real biological system. The more biologically oriented neural networks model the real nervous system in increasing detail at all relevant levels of information processing: from synapses to neurons to interactions between modules of interconnected neurons. One of the major challenges is to build artificial brains. By reverse-engineering the mammalian brain in silicon, the aim is to better understand the functioning of the (human) brain through detailed simulations. Neural networks that are more application-oriented tend to drift further apart from real biological systems. They come in many different flavors, solving problems in regression analysis and time-series forecasting, classification, and pattern recognition, as well as clustering and compression. Good old multilayered perceptrons and self-organizing maps are still pertinent, but attention in research is shifting toward more recent developments, such as kernel methods (including support vector machines) and Bayesian techniques. Both approaches aim to incorporate domain knowledge in the learning process in order to improve prediction performance, e.g., through the construction of a proper kernel function or distance measure or the choice of an appropriate prior distribution over the parameters of the neural network. Considerable effort is devoted to making neural networks efficient so that large models can be learned from huge databases in a reasonable amount of time. Application areas include, among many others, system dynamics and control, finance, bioinformatics, and image analysis.

Evolutionary Computation

The field of evolutionary computation deals with algorithms gleaned from models of organic evolution. The general aim of evolutionary computation is to use the principles of nature's processes of natural selection and genotypic variation to derive computer algorithms for solving hard search and optimization tasks. A wide variety of instances of evolutionary algorithms have been derived during the past fifty years based on the initial algorithms, and we are now witnessing astounding successes in the application of these algorithms: their fundamental understanding in terms of theoretical results; understanding algorithmic principles of their construction; combination with other techniques; and understanding their working principles in terms of organic evolution. The key algorithmic variations (such as genetic algorithms, evolution strategies, evolutionary programming, and genetic programming) have undergone significant developments over recent decades, and have also resulted in very powerful variations and recombinations of these algorithms. Today, there is a sound understanding of how all of these algorithms are instances of the generic concept of an evolutionary search approach. Hence the generic term "evolutionary algorithm" is nowadays being used to describe the generic algorithm, and the term "evolutionary computation" is used for the field as a whole. Thus, we have observed over the past fifty years how the field has integrated the various independently developed initial algorithms into one common principle. Moreover, modern evolutionary algorithms benefit from their ability to adapt and self-adapt their strategy parameters (such as mutation rates, step sizes, and search distributions) to the needs of the task at hand. In this way, they are robust and flexible metaheuristics for problem-solving even without requiring too much special expertise from their users. The feature of self-adaptation illustrates the ability of evolutionary principles to work on different levels at the same time, and therefore provides a nice demonstration of the universality of the evolutionary principle for search and optimization tasks. The widespread use of evolutionary computation reflects these capabilities.

Molecular Computation

Molecular computing is an emergent interdisciplinary field concerned with programming molecules so that they perform a desired computation, or fabricate a desired object, or control the functioning of a specific molecular system. The central idea behind molecular computing is that data can be encoded as (bio)molecules, e.g., DNA strands, and tools of molecular science can be used to transform these data. In a nutshell, a molecular program is just a collection of molecules which, when placed in a suitable substrate, will perform a specific function (execute the program that this collection represents). The birth of molecular computing is often associated with the 1994 breakthrough experiment by Leonard Adleman, who solved a small instance of a hard computational problem solely by manipulating DNA strands in test tubes. Although initially the main effort of the area was focused on trying to obtain a breakthrough in the complexity of solving hard computational problems, this field has evolved enormously since then. Among the most significant achievements of molecular computing have been contributions to understanding some of the fundamental issues of the nanosciences. One notable example among them is the contribution to the understanding of self-assembly, a central concept of the nanosciences. The techniques of molecular programming were successfully applied in experimentally constructing all kinds of molecular-scale objects or devices with prescribed functionalities. Well-known examples here are self-assembly of Sierpinski triangles, cubes, octahedra, DNA-based logic circuits, DNA "walkers" that move along a track, and autonomous molecular motors. A complementary approach to understanding bioinformation and computation is through studying the information-processing capabilities of cellular organisms. Indeed, cells and nature "compute" by "reading" and "rewriting" DNA through processes that modify DNA (or RNA) sequences. Research into the computational abilities of cellular organisms has the potential to uncover the laws governing biological information, and to enable us to harness the computational power of cells.

Quantum Computation

Quantum computing has been discussed for almost thirty years. The theory of quantum computing and quantum information processing is simply the theory of information processing with a classical notion of information replaced by its quantum counterpart. Research in quantum computing is concerned with understanding the fundamentals of information processing on the level of physical systems that realize/implement the information. In fact, quantum computing can be seen as a quest to understand the fundamental limits of information processing set by nature itself. The mathematical description of quantum information is more complicated than that of classical information — it involves the structure of Hilbert spaces. When describing the structure behind known quantum algorithms, this reduces to linear algebra over complex numbers. The history of quantum algorithms spans the last fifteen years, and some of these algorithms are extremely interesting, and even groundbreaking — the most remarkable are Shor's factorization in polynomial time and Grover's search algorithm. The nature of quantum information has also led to the invention of novel cryptosystems, whose security is not based on the complexity of computing functions, but rather on the physical properties of quantum information. Quantum computing is now a well-established discipline, however implementation of a large-scale quantum computer continues to be extremely challenging, even though quantum information processing primitives, including those allowing secure cryptography, have been demonstrated to be practically realizable.

Broader Perspective

In contrast to the first five areas focusing on more-established themes of natural computing, this area encompasses a perspective that is broader in several ways. First, the reader will find here treatments of certain well-established and specific techniques inspired by nature (e.g., simulated annealing) not covered in the other five areas. Second, the reader will also find application-centered chapters (such as natural computing in finance), each covering, in one chapter, a collection of natural computing methods, thus capturing the impact of natural computing as a whole in various fields of science or industry. Third, some chapters are full treatments of several established research fields (such as artificial life, computational systems biology, evolvable hardware, and artificial immune systems), presenting alternative perspectives and cutting across some of the other areas of the handbook, while introducing much new material. Other elements of this area are fresh, emerging, and novel techniques or perspectives (such as collision-based computing, nonclassical computation), representing the leading edge of theories and technologies that are shaping possible futures for both natural computing and computing in general. The contents of this area naturally cluster into two kinds (sections), determined by the essential nature of the techniques involved. These are "Nature-Inspired Algorithms" and "Alternative Models of Computation". In the first section, "Nature-Inspired Algorithms", the focus is on algorithms inspired by natural processes realized either through software or hardware or both, as additions to the armory of existing tools we have for dealing with well-known practical problems. In this section, we therefore find application-centered chapters, as well as chapters focusing on particular techniques, not otherwise dealt with in other areas of the handbook, which have clear and proven applicability. In the second section, "Alternative Models of Computation", the emphasis changes, moving away from specific applications or application areas, toward more far-reaching ideas. These range from developing computational approaches and "computational thinking" as fundamental tools for the new science of systems biology to ideas that take inspiration from nature as a platform for suggesting entirely novel possibilities of computing.

Handbook Chapters

In the remainder of this preface we will briefly describe the contents of the individual chapters. These chapter descriptions are grouped according to the handbook areas where they belong and given in the order that they appear in the handbook. This section provides the reader with a better insight into the contents, allowing one to design a personal roadmap for using this handbook.

Cellular Automata

This area is covered by nine chapters.

The first chapter, "Basic Concepts of Cellular Automata", by Jarkko J. Kari, reviews some classical results from the theory of cellular automata, relations between various concepts of injectivity and surjectivity, and some basic dynamical system concepts related to chaos in cellular automata. The classical results discussed include the celebrated Garden-of-Eden and Curtis–Hedlund–Lyndon theorems, as well as the balance property of surjective cellular automata. All these theorems date back to the 1960s. The results are provided together with examples that illustrate proof ideas. Different variants of sensitivity to initial conditions and mixing properties are introduced and related to each other. Also undecidability results concerning cellular automata are briefly discussed.

A popular mathematical approach is to view cellular automata as dynamical systems in the context of symbolic dynamics. Several interesting results in this area were reported as early as 1969 in the seminal paper by G.A. Hedlund, and still today this research direction is among the most fruitful sources of theoretical problems and new results. The chapter "Cellular Automata Dynamical Systems", by Alberto Dennunzio, Enrico Formenti, and Petr Kůrka, reviews some recent developments in this field. Recent research directions considered here include subshifts attractors and signal subshifts, particle weight functions, and the slicing construction. The first two concern one-dimensional cellular automata and give precise descriptions of the limit behavior of large classes of automata. The third one allows one to view two-dimensional cellular automata as one-dimensional systems. In this way combinatorial complexity is decreased and new results can be proved.

Programming cellular automata for particular tasks requires special techniques. The chapter "Algorithmic Tools on Cellular Automata", by Marianne Delorme and Jacques Mazoyer, covers classical algorithmic tools based on signals. Linear signals as well as signals of nonlinear slope are discussed, and basic transformations of signals are addressed. The chapter provides results on using signals to construct functions in cellular automata and to implement arithmetic operations on segments. The methods of folding the space–time, freezing, and clipping are also introduced.

The time-complexity advantage gained from parallelism under the locality and uniformity constraints of cellular automata can be precisely analyzed in terms of language recognition. The chapter "Language Recognition by Cellular Automata", by Véronique Terrier, presents results and questions about cellular automata complexity classes and their relationships to other models of computations. Attention is mainly directed to real-time and linear-time complexity classes, because significant benefits over sequential computation may be obtained at these low time complexities. Both parallel and sequential input modes are considered. Separate complexity classes are given also for cellular automata with one-way communications and two-way communications.

The chapter "Computations on Cellular Automata", by Jacques Mazoyer and Jean-Baptiste Yunès, continues with the topic of algorithmic techniques in cellular automata. This chapter uses the basic tools, such as signals and grids, to build natural implementations of common algorithms in cellular automata. Examples of implementations include real-time multiplication of integers and the prime number sieve. Both parallel and sequential input and output modes are discussed, as well as composition of functions and recursion.

The chapter "Universalities in Cellular Automata", by Nicolas Ollinger, is concerned with computational universalities. Concepts of universality include Turing universality (the ability to compute any recursive function) and intrinsic universality (the ability to simulate any other cellular automaton). Simulations of Boolean circuits in the two-dimensional case are explained in detail in order to achieve both kinds of universality. The more difficult one-dimensional case is also discussed, and seminal universal cellular automata and encoding techniques are presented in both dimensions. A detailed chronology of important papers on universalities in cellular automata is also provided.

A cellular automaton is reversible if every configuration has only one previous configuration, and hence its evolution process can be traced backward uniquely. This naturally corresponds to the fundamental time-reversibility of the microscopic laws of physics. The chapter "Reversible Cellular Automata", by Kenichi Morita, discusses how reversible cellular automata are defined, as well as their properties, how they are designed, and their computing abilities. After providing the definitions, the chapter surveys basic properties of injectivity and surjectivity. Three design methods of reversible cellular automata are provided: block rules, partitioned, and second-order cellular automata. Then the computational power of reversible cellular automata is discussed. In particular, simulation methods of irreversible cellular automata, reversible Turing machines, and some other universal systems are given to clarify universality of reversible cellular automata. In spite of the strong constraint of reversibility, it is shown that reversible cellular automata possess rich information processing capabilities, and even very simple ones are computationally universal.

A conservation law in a cellular automaton is a statement of the invariance of a local and additive energy-like quantity. The chapter "Conservation Laws in Cellular Automata", by Siamak Taati, reviews the basic theory of conservation laws. A general mathematical framework for formulating conservation laws in cellular automata is presented and several characterizations are summarized. Computational problems regarding conservation laws (verification and existence problems) are discussed. Microscopic explanations of the dynamics of the conserved quantities in terms of flows and particle flows are explored. The related concept of dissipating energy-like quantities is also discussed.

The chapter "Cellular Automata and Lattice Boltzmann Modeling of Physical Systems", by Bastien Chopard, considers the use of cellular automata and related lattice Boltzmann methods as a natural modeling framework to describe and study many physical systems composed of interacting components. The theoretical basis of the approach is introduced and its potential is illustrated for several applications in physics, biophysics, environmental science, traffic models, and multiscale modeling. The success of the technique can be explained by the close relationship between these methods and a mesoscopic abstraction of many natural phenomena.

Neural Computation

This area is covered by ten chapters.

Spiking neural networks are inspired by recent advances in neuroscience. In contrast to classical neural network models, they take into account not just the neuron's firing rate, but also the time moment of spike firing. The chapter "Computing with Spiking Neuron Networks", by Hélène Paugam-Moisy and Sander Bohte, gives an overview of existing approaches to modeling spiking neural neurons and synaptic plasticity, and discusses their computational power and the challenge of deriving efficient learning procedures.

Image quality assessment aims to provide computational models to predict the perceptual quality of images. The chapter "Image Quality Assessment — A Multiscale Geometric Analysis-Based Framework and Examples", by Xinbo Gao, Wen Lu, Dacheng Tao, and Xuelong Li, introduces the fundamentals and describes the state of the art in image quality assessment. It further proposes a new model, which mimics the human visual system by incorporating concepts such as multiscale analysis, contrast sensitivity, and just-noticeable differences. Empirical results clearly demonstrate that this model resembles subjective perception values and reflects the visual quality of images.

Neurofuzzy networks have the important advantage that they are easy to interpret. When applied to control problems, insight about the process characteristics at different operating regions can be easily obtained. Furthermore, nonlinear model predictive controllers can be developed as a nonlinear combination of several local linear model predictive controllers that have analytical solutions. Through several applications, the chapter "Nonlinear Process Modelling and Control Using Neurofuzzy Networks", by Jie Zhang, demonstrates that neurofuzzy networks are very effective in the modeling and control of nonlinear processes.

Similar to principal component and factor analysis, independent component analysis is a computational method for separating a multivariate signal into additive subcomponents. Independent component analysis is more powerful: the latent variables corresponding to the subcomponents need not be Gaussian and the basis vectors are typically nonorthogonal. The chapter "Independent Component Analysis", by Seungjin Choi, explains the theoretical foundations and describes various algorithms based on those principles.

Neural networks has become an important method for modeling and forecasting time series. The chapter "Neural Networks for Time-Series Forecasting", by G. Peter Zhang, reviews some recent developments (including seasonal time-series modeling, multiperiod forecasting, and ensemble methods), explains when and why they are to be preferred over traditional forecasting models, and also discusses several practical data and modeling issues.

Support vector machines have been extensively studied and applied in many domains within the last decade. Through the so-called kernel trick, support vector machines can efficiently learn nonlinear functions. By maximizing the margin, they implement the principle of structural risk minimization, which typically leads to high generalization performance. The chapter "SVM Tutorial — Classification, Regression and Ranking", by Hwanjo Yu and Sungchul Kim, describes these underlying principles and discusses support vector machines for different learning tasks: classification, regression, and ranking.

It is well known that single-hidden-layer feedforward networks can approximate any continuous target function. This still holds when the hidden nodes are automatically and randomly generated, independent of the training data. This observation opened up many possibilities for easy construction of a broad class of single-hidden-layer neural networks. The chapter "Fast Construction of Single-Hidden-Layer Feedforward Networks", by Kang Li, Guang-Bin Huang, and Shuzhi Sam Ge, discusses new ideas that yield a more compact network architecture and reduce the overall computational complexity.

Many recent experimental studies demonstrate the remarkable efficiency of biological neural systems to encode, process, and learn from information. To better understand the experimentally observed phenomena, theoreticians are developing new mathematical approaches and tools to model biological neural networks. The chapter "Modeling Biological Neural Networks", by Joaquin J. Torres and Pablo Varona, reviews some of the most popular models of neurons and neural networks. These not only help to understand how living systems perform information processing, but may also lead to novel bioinspired paradigms of artificial intelligence and robotics.

The size and complexity of biological data, such as DNA/RNA sequences and protein sequences and structures, makes them suitable for advanced computational tools, such as neural networks. Computational analysis of such databases aims at exposing hidden information that provides insights that help in understanding the underlying biological principles. The chapter "Neural Networks in Bioinformatics", by Ke Chen and Lukasz A. Kurgan, focuses on proteins. In particular it discusses prediction of protein secondary structure, solvent accessibility, and binding residues.

Self-organizing maps is a prime example of an artificial neural network model that both relates to the actual (topological) organization within the mammalian brain and at the same time has many practical applications. Self-organizing maps go back to the seminal work of Teuvo Kohonen. The chapter "Self-organizing Maps", by Marc M. Van Hulle, describes the state of the art with a special emphasis on learning algorithms that aim to optimize a predefined criterion.

Evolutionary Computation

This area is covered by thirteen chapters.

The first chapter, "Generalized Evolutionary Algorithms", by Kenneth De Jong, describes the general concept of evolutionary algorithms. As a generic introduction to the field, this chapter facilitates an understanding of specific instances of evolutionary algorithms as instantiations of a generic evolutionary algorithm. For the instantiations, certain choices need to be made, such as representation, variation operators, and the selection operator, which then yield particular instances of evolutionary algorithms, such as genetic algorithms and evolution strategies, to name just a few.

The chapter "Genetic Algorithms — A Survey of Models and Methods", by Darrell Whitley and Andrew M. Sutton, introduces and discusses (including criticism) the standard genetic algorithm based on the classical binary representation of solution candidates and a theoretical interpretation based on the so-called schema theorem. Variations of genetic algorithms with respect to solution representations, mutation operators, recombination operators, and selection mechanisms are also explained and discussed, as well as theoretical models of genetic algorithms based on infinite and finite population size assumptions and Markov chain theory concepts. The authors also critically investigate genetic algorithms from the perspective of identifying their limitations and the differences between theory and practice when working with genetic algorithms. To illustrate this further, the authors also give a practical example of the application of genetic algorithms to resource scheduling problems.

The chapter "Evolutionary Strategies", by Günter Rudolph, describes a class of evolutionary algorithms which have often been associated with numerical function optimization and continuous variables, but can also be applied to binary and integer domains. Variations of evolutionary strategies, such as the $(\mu+\lambda)$-strategy and the (μ,λ)-strategy, are introduced and discussed within a common algorithmic framework. The fundamental idea of self-adaptation of strategy parameters (variances and covariances of the multivariate normal distribution used for mutation) is introduced and explained in detail, since this is a key differentiating property of evolutionary strategies.

The chapter "Evolutionary Programming", by Gary B. Fogel, discusses a historical branch of evolutionary computation. It gives a historical perspective on evolutionary programming by describing some of the original experiments using evolutionary programming to evolve finite state machines to serve as sequence predictors. Starting from this canonical evolutionary programming approach, the chapter also presents extensions of evolutionary programming into continuous domains, where an attempt towards self-adaptation of mutation step sizes has been introduced which is similar to the one considered in evolutionary strategies. Finally, an overview of some recent applications of evolutionary programming is given.

The chapter "Genetic Programming — Introduction, Applications, Theory and Open Issues", by Leonardo Vanneschi and Riccardo Poli, describes a branch of evolutionary algorithms derived by extending genetic algorithms to allow exploration of the space of computer programs. To make evolutionary search in the domain of computer programs possible, genetic programming is based on LISP S-expression represented by syntax trees, so that genetic programming extends evolutionary algorithms to tree-based representations. The chapter gives an overview of the corresponding representation, search operators, and technical details of genetic programming, as well as existing applications to real-world problems. In addition, it discusses theoretical approaches toward analyzing genetic programming, some of the open issues, as well as research trends in the field.

The subsequent three chapters are related to the theoretical analysis of evolutionary algorithms, giving a broad overview of the state of the art in our theoretical understanding. These chapters demonstrate that there is a sound theoretical understanding of capabilities and limitations of evolutionary algorithms. The approaches can be roughly split into convergence velocity or progress analysis, computational complexity investigations, and global convergence results.

The convergence velocity viewpoint is represented in the chapter "The Dynamical Systems Approach — Progress Measures and Convergence Properties", by Silja Meyer-Nieberg and Hans-Georg Beyer. It demonstrates how the dynamical systems approach can be used to analyze the behavior of evolutionary algorithms quantitatively with respect to their progress rate. It also provides a complete overview of results in the continuous domain, i.e., for all types of evolution strategies on certain objective functions (such as sphere, ridge, etc.). The chapter presents results for undisturbed as well as for noisy variants of these objective functions, and extends the approach to dynamical objective functions where the goal turns into optimum tracking. All results are presented by means of comparative tables, so the reader gets a complete overview of the key findings at a glance.

The chapter "Computational Complexity of Evolutionary Algorithms", by Thomas Jansen, deals with the question of optimization time (i.e., the first point in time during the run of an evolutionary algorithm when the global optimum is sampled) and an investigation of upper bounds, lower bounds, and the average time needed to hit the optimum. This chapter presents specific results for certain classes of objective functions, most of them defined over binary search spaces, as well as fundamental limitations of evolutionary search and related results on the "no free lunch" theorem and black box complexity. The chapter also discusses the corresponding techniques for analyses, such as drift analysis and the expected multiplicative distance decrease.

Concluding the set of theoretical chapters, the chapter "Stochastic Convergence", by Günter Rudolph, addresses theoretical results about the properties of evolutionary algorithms concerned with finding a globally optimal solution in the asymptotic limit. Such results exist for certain variants of evolutionary algorithms and under certain assumptions, and this chapter summarizes the existing results and integrates them into a common framework. This type of analysis is essential in qualifying evolutionary algorithms as global search algorithms and for understanding the algorithmic conditions for global convergence.

The remaining chapters in the area of evolutionary computation report some of the major current trends.

To start with, the chapter "Evolutionary Multiobjective Optimization", by Eckart Zitzler, focuses on the application of evolutionary algorithms to tasks that are characterized by multiple, conflicting objective functions. In this case, decision-making becomes a task of identifying good compromises between the conflicting criteria. This chapter introduces the concept and a variety of state-of-the-art algorithmic concepts to use evolutionary algorithms for approximating the so-called Pareto front of solutions which cannot be improved in one objective without compromising another. This contribution presents all of the required formal concepts, examples, and the algorithmic variations introduced into evolutionary computation to handle such types of problems and to generate good approximations of the Pareto front.

The term "memetic algorithms" is used to characterize hybridizations between evolutionary algorithms and more classical, local search methods (and agent-based systems). This is a general concept of broad scope, and in order to illustrate and characterize all possible instantiations, the chapter "Memetic Algorithms", by Natalio Krasnogor, presents an algorithmic engineering approach which allows one to describe these algorithms as instances of generic patterns. In addition to explaining some of the application areas, the chapter presents some theoretical remarks, various different ways to define memetic algorithms, and also an outlook into the future.

The chapter "Genetics-Based Machine Learning", by Tim Kovacs, extends the idea of evolutionary optimization to algorithmic concepts in machine learning and data mining, involving applications such as learning classifier systems, evolving neural networks, and genetic fuzzy systems, to mention just a few. Here, the application task is typically a data classification, data prediction, or nonlinear regression task — and the quality of solution candidates is evaluated by means of some model quality measure. The chapter covers a wide range of techniques for applying evolutionary computation to machine learning tasks, by interpreting them as optimization problems.

The chapter "Coevolutionary Principles", by Elena Popovici, Anthony Bucci, R. Paul Wiegand, and Edwin D. de Jong, deals with a concept modeled after biological evolution in which an explicit fitness function is not available, but solutions are evaluated by running them against each other. A solution is evaluated in the context of the other solutions, in the actual population or in another. Therefore, these algorithms develop their own dynamics, because the point of comparison is not stable, but coevolving with the actual population. The chapter provides a fundamental understanding of coevolutionary principles and highlights theoretical concepts, algorithms, and applications.

Finally, the chapter "Niching in Evolutionary Algorithms", by Ofer M. Shir, describes the biological principle of niching in nature as a concept for using a single population to find, occupy, and keep multiple local minima in a population. The motivation for this approach is to find alternative solutions within a single population and run of evolutionary algorithms, and this chapter discusses approaches for niching, and the application in the context of genetic algorithms as well as evolutionary strategies.

Molecular Computation

This area is covered by eight chapters.

The chapter "DNA Computing — Foundations and Implications", by Lila Kari, Shinnosuke Seki, and Petr Sosík, has a dual purpose. The first part outlines basic molecular biology notions necessary for understanding DNA computing, recounts the first experimental demonstration of DNA computing by Leonard Adleman in 1994, and recaps the 2001 milestone wet laboratory experiment that solved a 20-variable instance of 3-SAT and thus first demonstrated the potential of DNA computing to outperform the computational ability of an unaided human. The second part describes how the properties of DNA-based information, and in particular the Watson–Crick complementary of DNA single strands, have influenced areas of theoretical computer science such as formal language theory, coding theory, automata theory, and combinatorics on words. More precisely, it explores several notions and results in formal language theory and coding theory that arose from the problem of the design of optimal encodings for DNA computing experiments (hairpin-free languages, bond-free languages), and more generally from the way information is encoded on DNA strands (sticker systems, Watson–Crick automata). Lastly, it describes the influence that properties of DNA-based information have had on research in combinatorics on words, by presenting several natural generalizations of classical concepts (pseudopalindromes, pseudoperiodicity, Watson–Crick conjugate and commutative words, involutively bordered words, pseudoknot bordered words), and outlining natural extensions in this context of two of the most fundamental results in combinatorics of words, namely the Fine and Wilf theorem and the Lyndon–Schützenberger result.

The chapter "Molecular Computing Machineries — Computing Models and Wet Implementations", by Masami Hagiya, Satoshi Kobayashi, Ken Komiya, Fumiaki Tanaka, and Takashi Yokomori, explores novel computing devices inspired by the biochemical properties of biomolecules. The theoretical results section describes a variety of molecular computing models for finite automata, as well as molecular computing models for Turing machines based on formal grammars, equality sets, Post systems, and logical formulae. It then presents molecular computing models that use structured molecules such as hairpins and tree structures. The section on wet implementations of molecular computing models, related issues, and applications includes: an enzyme-based DNA automaton and its applications to drug delivery, logic gates and circuits using DNAzymes and DNA tiles, reaction graphs for representing various dynamics of DNA assembly pathways, DNA whiplash machines implementing finite automata, and a hairpin-based implementation of a SAT engine for solving the 3-SAT problem.

The chapter "DNA Computing by Splicing and by Insertion–Deletion", by Gheorghe Păun, is devoted to two of the most developed computing models inspired by DNA biochemistry: computing by splicing, and computing by insertion and deletion. DNA computing by splicing was defined by Tom Head already in 1987 and is based on the so-called splicing operation. The splicing operation models the recombination of DNA molecules that results from cutting them with restriction enzymes and then pasting DNA molecules with compatible ends by ligase enzymes. This chapter explores the computational power of the splicing operation showing that, for example, extended splicing systems starting from a finite language and using finitely many splicing rules can generate only the family of regular languages, while extended splicing systems starting from a finite language and using a regular set of rules can generate all recursively enumerable languages. Ways in which to avoid the impractical notion of a regular infinite set of rules, while maintaining the maximum computational power, are presented. They include using multisets and adding restrictions on the use of rules such as permitting contexts, forbidding contexts, programmed splicing systems, target languages, and double splicing. The second model presented, the insertion–deletion system, is based on a finite set of axioms and a finite set of contextual insertion rules and contextual deletion rules. Computational power results described here include the fact that insertion–deletion systems with context-free insertion rules of words of length at most one and context-free deletion rules of words of unbounded length can generate only regular languages. In contrast, for example, the family of insertion–deletion systems where the insertion contexts, deletion contexts, and the words to be inserted/deleted are all of length at most one, equals the family of recursively enumerable languages.

The chapter "Bacterial Computing and Molecular Communication", by Yasubumi Sakakibara and Satoshi Hiyama, investigates attempts to create autonomous cell-based Turing machines, as well as novel communication paradigms that use molecules as communication media. The first part reports experimental research on constructing *in vivo* logic circuits as well as efforts towards building *in vitro* and *in vivo* automata in the framework of DNA computing. Also, a novel framework is presented to develop a programmable and autonomous *in vivo* computer in a bacterium. The first experiment in this direction uses DNA circular strands (plasmids) together with the cell's protein-synthesis mechanism to execute a finite state automaton in *E. coli*. Molecular communication is a new communication paradigm that proposes the use of molecules as the information medium, instead of the traditional electromagnetic waves. Other distinctive features of molecular communication include its stochastic nature, its low energy consumption, the use of an aqueous transmission medium, and its high compatibility with biological systems. A molecular communication system starts with a sender (e.g., a genetically modified or an artificial cell) that generates molecules, encodes information onto the molecules (called information molecules), and emits the information molecules into a propagation environment (e.g., aqueous solution within and between cells). A molecular propagation system (e.g., lipid bilayer vesicles encapsulating the information molecules) actively transports the information molecules to an appropriate receiver. A receiver (e.g., a genetically modified or an artificial cell) selectively receives the transported information molecules, and biochemically reacts to the received information molecules, thus "decoding" the information. The chapter describes detailed examples of molecular communication system designs, experimental results, and research trends.

The chapter "Computational Nature of Gene Assembly in Ciliates", by Robert Brijder, Mark Daley, Tero Harju, Nataša Jonoska, Ion Petre, and Grzegorz Rozenberg, reviews several approaches and results in the computational study of gene assembly in ciliates. Ciliated protozoa contain two functionally different types of nuclei, the macronucleus and the micronucleus. The macronucleus contains the functional genes, while the genes of the micronucleus are not functional due to the presence of many interspersing noncoding DNA segments. In addition, in some ciliates, the coding segments of the genes are present in a permuted order compared to their order in the functional macronuclear genes. During the sexual process of conjugation, when two ciliates exchange genetic micronuclear information and form two new micronuclei, each of the ciliates has to "decrypt" the information contained in its new micronucleus to form its new functional macronucleus. This process is called gene assembly and involves deleting the noncoding DNA segments, as well as rearranging the coding segments in the correct order. The chapter describes two models of gene assembly, the intermolecular model based on the operations of circular insertion and deletion, and the intramolecular model based on the three operations of "loop, direct-repeat excision", "hairpin, inverted-repeat excision", and "double-loop, alternating repeat excision". A discussion follows of the mathematical properties of these models, such as the Turing machine computational power of contextual circular insertions and deletions, and properties of the gene assembly process called invariants, which hold independently of the molecular model and assembling strategy. Finally, the template-based recombination model is described, offering a plausible hypothesis (supported already by some experimental data) about the "bioware" that implements the gene assembly.

The chapter "DNA Memory", by Masanori Arita, Masami Hagiya, Masahiro Takinoue, and Fumiaki Tanaka, summarizes the efforts that have been made towards realizing Eric Baum's dream of building a DNA memory with a storage capacity vastly larger than the brain. The chapter first describes the research into strategies for DNA sequence design, i.e., for finding DNA sequences that satisfy DNA computing constraints such as uniform melting temperature, avoidance of undesirable Watson–Crick bonding between sequences, preventing secondary structures, avoidance of base repeats, and absence of forbidden sequences. Various implementations of memory operations, such as access, read, and write, are described. For example, the "access" to a memory word in Baum's associative memory model, where a memory word consists of a single-stranded portion representing the address and a double-stranded portion representing the data, can be implemented by using the Watson–Crick complement of the address fixed to a solid support. In the Nested Primer Molecular Memory, where the double-stranded data is flanked on both sides by address sequences, the data can be retrieved by Polymerase Chain Reaction (PCR) using the addresses as primer pairs. In the multiple hairpins DNA memory, the address is a catenation of hairpins and the data can be accessed only if the hairpins are opened in the correct order by a process called DNA branch migration. After describing implementations of writable and erasable hairpin memories either in solution or immobilized on surfaces, the topic of *in vivo* DNA memory is explored. As an example, the chapter describes how representing the digit 0 by regular codons, and the digit 1 by wobbled codons, was used to encode a word into an essential gene of *Bacillus subtilis*.

The chapter "Engineering Natural Computation by Autonomous DNA-Based Biomolecular Devices", by John H. Reif and Thomas H. LaBean, overviews DNA-based biomolecular devices that are autonomous (execute steps with no external control after starting) and programmable (the tasks executed can be modified without entirely redesigning the DNA nanostructures). Special attention is given to DNA tiles, roughly square-shaped DNA nanostructures that have four "sticky-ends" (DNA single strands) that can specifically bind them to other tiles via Watson–Crick complementarity, and thus lead to the self-assembly of larger and more complex structures. Such tiles have been used to execute sequential Boolean computation via linear DNA self-assembly or to obtain patterned 2D DNA lattices and Sierpinski triangles. Issues such as error correction and self-repair of DNA tiling are also addressed. Other described methods include the implementation of a DNA-based finite automaton via disassembly of a double-stranded DNA nanostructure effected by an enzyme, and the technique of whiplash PCR. Whiplash PCR is a method that can achieve state transitions by encoding both transitions and the current state of the computation on the same DNA single strand: The free end of the strand (encoding the current state) sticks to the appropriate transition rule on the strand forming a hairpin, is then extended by PCR to a new state, and finally is detached from the strand, this time with the new state encoded at its end. The technique of DNA origami is also described, whereby a scaffold strand (a long single DNA strand, such as from the sequence of a virus) together with many specially designed staple strands (short single DNA strands) self-assemble by folding the scaffold strand — with the aid of the staples — in a raster pattern that can create given arbitrary planar DNA nanostructures. DNA-based molecular machines are then described such as autonomous DNA walkers and programmable DNA nanobots (programmable autonomous DNA walker devices). A restriction-enzyme-based DNA walker consists of a DNA helix with two sticky-ends ("feet") that moves stepwise along a "road" (a DNA nanostructure with protruding "steps", i.e., single DNA strands).

The chapter "Membrane Computing", by Gheorghe Păun, describes theoretical results and applications of membrane computing, a branch of natural computing inspired by the architecture and functioning of living cells, as well as from the organization of cells in tissues, organs, or other higher-order structures. The cell is a hierarchical structure of compartments, defined by membranes, that selectively communicate with each other. The computing model that abstracts this structure is a membrane system (or P system, from the name of its inventor, Gheorghe Păun) whose main components are: the membrane structure, the multisets of objects placed in the compartments enveloped by the membranes, and the rules for processing the objects and the membranes. The rules are used to modify the objects in the compartments, to transport objects from one compartment to another, to dissolve membranes, and to create new membranes. The rules in each region of a P system are used in a maximally parallel manner, nondeterministically choosing the applicable rules and the objects to which they apply. A computation consists in repeatedly applying rules to an initial configuration of the P system, until no rule can be applied anymore, in which case the objects in a priori specified regions are considered the output of the computation. Several variants of P systems are described, including P systems with symport/antiport rules, P systems with active membranes, splicing P systems, P systems with objects on membranes, tissue-like P systems, and spiking neural P systems. Many classes of P systems are able to simulate Turing machines, hence they are computationally universal. For example, P systems with symport/antiport rules using only three objects and three membranes are computationally universal. In addition, several types of P systems have been used to solve NP-complete problems in polynomial time, by a space–time trade-off. Applications of P systems include, among others, modeling in biology, computer graphics, linguistics, economics, and cryptography.

Quantum Computation

This area is covered by six chapters.

The chapter "Mathematics for Quantum Information Processing", by Mika Hirvensalo, contains the standard Hilbert space formulation of finite-level quantum systems. This is the language and notational system allowing us to speak, describe, and make predictions about the objects of quantum physics. The chapter introduces the notion of quantum states as unit-trace, self-adjoint, positive mappings, and the vector state formalism is presented as a special case. The physical observables are introduced as complete collections of mutually orthogonal projections, and then it is discussed how this leads to the traditional representation of observables as self-adjoint mappings. The minimal interpretation, which is the postulate connecting the mathematical objects to the physical world, is presented. The treatment of compound quantum systems is based mostly on operative grounds. To provide enough tools for considering the dynamics needed in quantum computing, the formalism of treating state transformations as completely positive mappings is also presented. The chapter concludes by explaining how quantum versions of finite automata, Turing machines, and Boolean circuits fit into the Hilbert space formalism.

The chapter "Bell's Inequalities — Foundations and Quantum Communication", by Časlav Brukner and Marek Żukowski, is concerned with the nature of quantum mechanics. It presents the evidence that excludes two types of hypothetical deterministic theories: neither a nonlocal nor a noncontextual theory can explain quantum mechanics. This helps to build a true picture of quantum mechanics, and is therefore essential from the philosophical point of view. The Bell inequalities show that nonlocal deterministic theories cannot explain the quantum mechanism, and the Kochen–Specker theorem shows that noncontextual theories are not possible as underlying theories either. The traditional Bell theorem and its variants, GHZ and CHSH among them, are presented, and the Kochen–Specker theorem is discussed. In this chapter, the communication complexity is also treated by showing how the violations of classical locality and noncontextuality can be used as a resource for communication protocols. Stronger-than quantum violations of the CHSH inequality are also discussed. They are interesting, since it has been shown that if the violation of CHSH inequality is strong enough, then the communication complexity collapses into one bit (hence the communication complexity of the true physical world seems to settle somewhere between classical and stronger-than quantum).

The chapter "Algorithms for Quantum Computers", by Jamie Smith and Michele Mosca, introduces the most remarkable known methods that utilize the special features of quantum physics in order to gain advantage over classical computing. The importance of these methods is that they form the core of designing discrete quantum algorithms. The methods presented and discussed here are the quantum Fourier transform, amplitude amplification, and quantum walks. Then, as specific examples, Shor's factoring algorithm (quantum Fourier transform), Grover search (amplitude amplification), and element distinctness algorithms (quantum random walks) are presented. The chapter not only involves traditional methods, but it also contains discussion of continuous-time quantum random walks and, more importantly, an extensive presentation of an important recent development in quantum algorithms, viz., tensor network evaluation algorithms. Then, as an example, the approximate evaluation of Tutte polynomials is presented.

The chapter "Physical Implementation of Large-Scale Quantum Computation", by Kalle-Antti Suominen, discusses the potential ways of physically implementing quantum computers. First, the DiVincenzo criteria (requirements for building a successful quantum computer) are presented, and then quantum error correction is discussed. The history, physical properties, potentials, and obstacles of various possible physical implementations of quantum computers are covered. They involve: cavity QED, trapped ions, neutral atoms and single electrons, liquid-form molecular spin, nuclear and electron spins in silicon, nitrogen vacancies in diamond, solid-state qubits with quantum dots, superconducting charge, flux and phase quantum bits, and optical quantum computing.

The chapter "Quantum Cryptography", by Takeshi Koshiba, is concerned with quantum cryptography, which will most likely play an important role in future when quantum computers make the current public-key cryptosystems unreliable. It gives an overview of classical cryptosystems, discusses classical cryptographic protocols, and then introduces the quantum key distribution protocols BB84, B92, and BBM92. Also protocol OTU00, not known to be vulnerable under Shor's algorithm, is presented. In future, when quantum computers are available, cryptography will most probably be based on quantum protocols. The chapter presents candidates for such quantum protocols: KKNY05 and GC01 (for digital signatures). It concludes with a discussion of quantum commitment, oblivious transfer, and quantum zero-knowledge proofs.

The complexity class BQP is the quantum counterpart of the classical class BPP. Intuitively, BQP can be described as the class of problems solvable in "reasonable" time, and, hence, from the application-oriented point of view, it will likely become the most important complexity class in future, when quantum computers are available. The chapter "BQP-Complete Problems", by Shengyu Zhang, introduces the computational problems that capture the full hardness of BQP. In the very fundamental sense, no BQP-complete problems are known, but the promise problems (the probability distribution of outputs is restricted by promise) bring us as close as possible to the "hardest" problems in BQP, known as BQP-complete promise problems. The chapter discusses known BQP-complete promise problems. In particular, it is shown how to establish the BQP-completeness of the Local Hamiltonian Eigenvalue

Sampling problem and the Local Unitary Phase Sampling problem. The chapter concludes with an extensive study showing that the Jones Polynomial Approximation problem is a BQP-complete promise problem.

Broader Perspective

This area consists of two sections, "Nature-Inspired Algorithms" and "Alternative Models of Computation".

Nature-Inspired Algorithms

This section is covered by six chapters.

The chapter "An Introduction to Artificial Immune Systems", by Mark Read, Paul S. Andrews, and Jon Timmis, provides a general introduction to the field. It discusses the major research issues relating to the field of Artificial Immune Systems (AIS), exploring the underlying immunology that has led to the development of immune-inspired algorithms, and focuses on the four main algorithms that have been developed in recent years: clonal selection, immune network, negative selection, and dendritic cell algorithms; their use in terms of applications is highlighted. The chapter also covers evaluation of current AIS technology, and details some new frameworks and methodologies that are being developed towards more principled AIS research. As a counterpoint to the focus on applications, the chapter also gives a brief outline of how AIS research is being employed to help further the understanding of immunology.

The chapter on "Swarm Intelligence", by David W. Corne, Alan Reynolds, and Eric Bonabeau, attempts to demystify the term Swarm Intelligence (SI), outlining the particular collections of natural phenomena that SI most often refers to and the specific classes of computational algorithms that come under its definition. The early parts of the chapter focus on the natural inspiration side, with discussion of social insects and stigmergy, foraging behavior, and natural flocking behavior. Then the chapter moves on to outline the most successful of the computational algorithms that have emerged from these natural inspirations, namely ant colony optimization methods and particle swarm optimization, with also some discussion of different and emerging such algorithms. The chapter concludes with a brief account of current research trends in the field.

The chapter "Simulated Annealing", by Kathryn A. Dowsland and Jonathan M. Thompson, provides an overview of Simulated Annealing (SA), emphasizing its practical use. The chapter explains its inspiration from the field of statistical thermodynamics, and then overviews the theory, with an emphasis again on those aspects that are important for practical applications. The chapter then covers some of the main ways in which the basic SA algorithm has been modified by various researchers, leading to improved performance for a variety of problems. The chapter briefly surveys application areas, and ends with several useful pointers to associated resources, including freely available code.

The chapter "Evolvable Hardware", by Lukáš Sekanina, surveys this growing field. Starting with a brief overview of the reconfigurable devices used in this field, the elementary principles and open problems are introduced, and then the chapter considers, in turn, three main areas: extrinsic evolution (evolving hardware using simulators), intrinsic evolution (where the evolution is conducted within FPGAs, FPTAs, and so forth), and adaptive hardware (in which real-world adaptive hardware systems are presented). The chapter finishes with an overview of major achievements in the field.

The first of two application-centered chapters, "Natural Computing in Finance — A Review", by Anthony Brabazon, Jing Dang, Ian Dempsey, Michael O'Neill, and David Edelman, provides a rather comprehensive account of natural computing applications in what is, at the time of writing (and undoubtedly beyond), one of the hottest topics of the day. This chapter introduces us to the wide range of different financial problems to which natural computing methods have been applied, including forecasting, trading, arbitrage, portfolio management, asset allocation, credit risk assessment, and more. The natural computing areas that feature in this chapter are largely evolutionary computing, neural computing, and also agent-based modeling, swarm intelligence, and immune-inspired methods. The chapter ends with a discussion of promising future directions.

Finally, the chapter "Selected Aspects of Natural Computing", by David W. Corne, Kalyanmoy Deb, Joshua Knowles, and Xin Yao, provides detailed accounts of a collection of example natural computing applications, each of which is remarkable or particularly interesting in some way. The thrust of this chapter is to provide, via such examples, an idea of both the significant impact that natural computing has already had, as well as its continuing significant promise for future applications in all areas of science and industry. While presenting this eclectic collection of marvels, the chapter also aims at clarity and demystification, providing much detail that helps see how the natural computing methods in question were applied to achieve the stated results. Applications covered include Blondie24 (the evolutionary neural network application that achieves master-level skill at the game of checkers), the design of novel antennas using evolutionary computation in conjunction with developmental computing, and the classic application of learning classifier systems that led to novel fighter-plane maneuvers for the USAF.

Alternative Models of Computation

This section is covered by seven chapters.

The chapter "Artificial Life", by Wolfgang Banzhaf and Barry McMullin, traces the roots, raises key questions, discusses the major methodological tools, and reviews the main applications of this exciting and maturing area of computing. The chapter starts with a historical overview, and presents the fundamental questions and issues that Artificial Life is concerned with. Thus the chapter surveys discussions and viewpoints about the very nature of the differences between living and nonliving systems, and goes on to consider issues such as hierarchical design, self-construction, and self-maintenance, and the emergence of complexity. This part of the chapter ends with a discussion of "Coreworld" experiments, in which a number of systems have been studied that allow spontaneous evolution of computer programs. The chapter moves on to survey the main theory and formalisms used in Artificial Life, including cellular automata and rewriting systems. The chapter concludes with a review and restatement of the main objectives of Artificial Life research, categorizing them respectively into questions about the origin and nature of life, the potential and limitations of living systems, and the relationships between life and intelligence, culture, and other human constructs.

The chapter "Algorithmic Systems Biology — Computer Science Propels Systems Biology", by Corrado Priami, takes the standpoint of computing as providing a philosophical foundation for systems biology, with at least the same importance as mathematics, chemistry, or physics. The chapter highlights the value of algorithmic approaches in modeling, simula-tion, and analysis of biological systems. It starts with a high-level view of how models and experiments can be tightly integrated within an algorithmic systems biology vision, and then deals in turn with modeling languages, simulations of models, and finally the postprocessing of results from biological models and how these lead to new hypotheses that can then re-enter the modeling/simulation cycle.

The chapter "Process Calculi, Systems Biology and Artificial Chemistry", by Pierpaolo Degano and Andrea Bracciali, concentrates on the use of process calculi and related techniques for systems-level modeling of biological phenomena. This chapter echoes the broad viewpoint of the previous chapter, but its focus takes us towards a much deeper understanding of the potential mappings between formal systems in computer science and systems interpretation of biological processes. It starts by surveying the basics of process calculi, setting out their obvious credentials for modeling concurrent, distributed systems of interacting parts, and mapping these onto a "cells as computers" view. After a process calculi treatment of systems biology, the chapter goes on to examine process calculi as a route towards artificial chemistry. After considering the formal properties of the models discussed, the chapter ends with notes on some case studies showing the value of process calculi in modeling biological phenomena; these include investigating the concept of a "minimal gene set" prokaryote, modeling the nitric oxide-cGMP pathway (central to many signal transduction mechanisms), and modeling the calyx of Held (a large synapse structure in the mammalian auditory central nervous system).

The chapter on "Reaction—Diffusion Computing", by Andrew Adamatzky and Benjamin De Lacy Costello, introduces the reader to the concept of a reaction—diffusion computer. This is a spatially extended chemical system, which processes information via transforming an input profile of ingredients (in terms of different concentrations of constituent ingredients) into an output profile of ingredients. The chapter takes us through the elements of this field via case studies, and it shows how selected tasks in computational geometry, robotics, and logic can be addressed by chemical implementations of reaction—diffusion computers. After introducing the field and providing a treatment of its origins and main achievements, a classical view of reaction—diffusion computers is then described. The chapter moves on to discuss varieties of reaction—diffusion processors and their chemical constituents, covering applications to the aforementioned tasks. The chapter ends with the authors' thoughts on future developments in this field.

The chapter "Rough—Fuzzy Computing", by Andrzej Skowron, shifts our context towards addressing a persistent area of immense difficulty for classical computing, which is the fact that real-world reasoning is usually done in the face of inaccurate, incomplete, and often inconsistent evidence. In essence, concepts in the real world are vague, and computation needs ways to address this. We are hence treated, in this chapter, to an overarching view of rough set theory, fuzzy set theory, their hybridization, and applications. Rough and fuzzy computing are broadly complementary approaches to handling vagueness, focusing respectively on capturing the level of distinction between separate objects and the level of membership of an object in a set. After presenting the basic concepts of rough computing and fuzzy computing in turn, in each case going into some detail on the main theoretical results and practical considerations, the chapter goes on to discuss how they can be, and have been, fruitfully combined. The chapter ends with an overview of the emerging field of "Wisdom Technology" (Wistech) as a paradigm for developing modern intelligent systems.

The chapter "Collision-Based Computing", by Andrew Adamatzky and Jérôme Durand-Lose, presents and discusses the computations performed as a result of spatial localizations in systems that exhibit dynamic spatial patterns over time. For example, a collision may be between two gliders in a cellular automaton, or two separate wave fragments within an excitable chemical system. This chapter introduces us to the basics of collision-based computing and overviews collision-based computing schemes in 1D and 2D cellular automata as well as continuous excitable media. Then, after some theoretical foundations relating to 1D cellular automata, the chapter presents a collision-based implementation for a 1D Turing machine and for cyclic tag systems. The chapter ends with discussion and presentation of "Abstract Geometrical Computation", which can be seen as collision-based computation in a medium that is the continuous counterpart of cellular automata.

The chapter "Nonclassical Computation — A Dynamical Systems Perspective", by Susan Stepney, takes a uniform view of computation, in which inspiration from a dynamical systems perspective provides a convenient way to consider, in one framework, both classical discrete systems and systems performing nonclassical computation. In particular, this viewpoint presents a way towards computational interpretation of physical embodied systems that exploit their natural dynamics. The chapter starts by discussing "closed" dynamical systems, those whose dynamics involve no inputs from an external environment, examining their computational abilities from a dynamical systems perspective. Then it discusses continuous dynamical systems and shows how these too can be interpreted computationally, indicating how material embodiment can give computation "for free", without the need to explicitly implement the dynamics. The outlook then broadens to consider open systems, where the dynamics are affected by external inputs. The chapter ends by looking at constructive, or developmental, dynamical systems, whose state spaces change during computation. These latter discussions approach the arena of biological and other natural systems, casting them as computational, open, developmental, dynamical systems.

Acknowledgements

This handbook resulted from a highly collaborative effort. The handbook and area editors are grateful to the chapter writers for their efforts in writing chapters and delivering them on time, and for their participation in the refereeing process.

We are indebted to the members of the Advisory Board for their valuable advice and fruitful interactions. Additionally, we want to acknowledge David Fogel, Pekka Lahti, Robert LaRue, Jason Lohn, Michael Main, David Prescott, Arto Salomaa, Kai Salomaa, Shinnosuke Seki, and Rob Smith, for their help and advice in various stages of production of this handbook. Last, but not least, we are thankful to Springer, especially to Ronan Nugent, for intense and constructive cooperation in bringing this project from its inception to its successful conclusion.

Leiden; Edinburgh; Nijmegen;   Grzegorz Rozenberg (Main Handbook Editor)
Turku; London, Ontario   Thomas Bäck (Handbook Editor and Area Editor)
October 2010   Joost N. Kok (Handbook Editor and Area Editor)
                                                      David W. Corne (Area Editor)
                                                         Tom Heskes (Area Editor)
                                                    Mika Hirvensalo (Area Editor)
                                                      Jarkko J. Kari (Area Editor)
                                                           Lila Kari (Area Editor)

Editor Biographies

Prof. Dr. Grzegorz Rozenberg

Prof. Rozenberg was awarded his Ph.D. in mathematics from the Polish Academy of Sciences, Warsaw, and he has since held full-time positions at Utrecht University, the State University of New York at Buffalo, and the University of Antwerp. Since 1979 he has been a professor at the Department of Computer Science of Leiden University, and an adjunct professor at the Department of Computer Science of the University of Colorado at Boulder, USA. He is the founding director of the Leiden Center for Natural Computing.

Among key editorial responsibilities over the last 30 years, he is the founding Editor of the book series Texts in Theoretical Computer Science (Springer) and Monographs in Theoretical Computer Science (Springer), founding Editor of the book series Natural Computing (Springer), founding Editor-in-Chief of the journal Natural Computing (Springer), and founding Editor of Part C (Theory of Natural Computing) of the journal Theoretical Computer Science (Elsevier). Altogether he's on the Editorial Board of around 20 scientific journals.

He has authored more than 500 papers and 6 books, and coedited more than 90 books. He coedited the "Handbook of Formal Languages" (Springer), he was Managing Editor of the "Handbook of Graph Grammars and Computing by Graph Transformation" (World Scientific), he coedited "Current Trends in Theoretical Computer Science" (World Scientific), and he coedited "The Oxford Handbook of Membrane Computing" (Oxford University Press).

He is Past President of the European Association for Theoretical Computer Science (EATCS), and he received the Distinguished Achievements Award of the EATCS "in recognition of his outstanding scientific contributions to theoretical computer science". Also he is a cofounder and Past President of the International Society for Nanoscale Science, Computation, and Engineering (ISNSCE).

He has served as a program committee member for most major conferences on theoretical computer science in Europe, and among the events he has founded or helped to establish are the International Conference on Developments in Language Theory (DLT), the International

Conference on Graph Transformation (ICGT), the International Conference on Unconventional Computation (UC), the International Conference on Application and Theory of Petri Nets and Concurrency (ICATPN), and the DNA Computing and Molecular Programming Conference.

In recent years his research has focused on natural computing, including molecular computing, computation in living cells, self-assembly, and the theory of biochemical reactions. His other research areas include the theory of concurrent systems, the theory of graph transformations, formal languages and automata theory, and mathematical structures in computer science.

Prof. Rozenberg is a Foreign Member of the Finnish Academy of Sciences and Letters, a member of the Academia Europaea, and an honorary member of the World Innovation Foundation. He has been awarded honorary doctorates by the University of Turku, the Technical University of Berlin, and the University of Bologna. He is an ISI Highly Cited Researcher.

He is a performing magician, and a devoted student of and expert on the paintings of Hieronymus Bosch.

Prof. Dr. Thomas Bäck

Prof. Bäck was awarded his Ph.D. in Computer Science from Dortmund University in 1994, for which he received the Best Dissertation Award from the Gesellschaft für Informatik (GI). He has been at Leiden University since 1996, where he is currently full Professor for Natural Computing and the head of the Natural Computing Research Group at the Leiden Institute of Advanced Computer Science (LIACS).

He has authored more than 150 publications on natural computing technologies. He wrote a book on evolutionary algorithms, "Evolutionary Algorithms in Theory and Practice" (Oxford University Press), and he coedited the "Handbook of Evolutionary Computation" (IOP/Oxford University Press).

Prof. Bäck is an Editor of the book series Natural Computing (Springer), an Associate Editor of the journal Natural Computing (Springer), an Editor of the journal Theoretical Computer Science (Sect. C, Theory of Natural Computing; Elsevier), and an Advisory Board member of the journal Evolutionary Computation (MIT Press). He has served as program chair for all major conferences in evolutionary computation, and is an Elected Fellow of the International Society for Genetic and Evolutionary Computation for his contributions to the field.

His main research interests are the theory of evolutionary algorithms, cellular automata and data-driven modelling, and applications of these methods in medicinal chemistry, pharmacology and engineering.

Prof. Dr. Joost N. Kok

Prof. Kok was awarded his Ph.D. in Computer Science from the Free University in Amsterdam in 1989, and he has worked at the Centre for Mathematics and Computer Science in Amsterdam, at Utrecht University, and at the Åbo Akademi University in Finland. Since 1995 he has been a professor in computer science, and since 2005 also a professor in medicine at Leiden University. He is the Scientific Director of the Leiden Institute of Advanced Computer Science, and leads the research clusters Algorithms and Foundations of Software Technology.

He serves as a chair, member of the management team, member of the board, or member of the scientific committee of the Faculty of Mathematics and Natural Sciences of Leiden University, the ICT and Education Committee of Leiden University, the Dutch Theoretical Computer Science Association, the Netherlands Bioinformatics Centre, the Centre for Mathematics and Computer Science Amsterdam, the Netherlands Organisation for Scientific Research, the Research Foundation Flanders (Belgium), the European Educational Forum, and the International Federation for Information Processing (IFIP) Technical Committee 12 (Artificial Intelligence).

Prof. Kok is on the steering, scientific or advisory committees of the following events: the Mining and Learning with Graphs Conference, the Intelligent Data Analysis Conference, the Institute for Programming and Algorithms Research School, the Biotechnological Sciences Delft–Leiden Research School, and the European Conference on Machine Learning and Principles and Practice of Knowledge Discovery in Databases. And he has been a program committee member for more than 100 international conferences, workshops or summer schools on data mining, data analysis, and knowledge discovery; neural networks; artificial intelligence; machine learning; computational life science; evolutionary computing, natural computing and genetic algorithms; Web intelligence and intelligent agents; and software engineering.

He is an Editor of the book series Natural Computing (Springer), an Associate Editor of the journal Natural Computing (Springer), an Editor of the journal Theoretical Computer Science (Sect. C, Theory of Natural Computing; Elsevier), an Editor of the Journal of Universal

Computer Science, an Associate Editor of the journal Computational Intelligence (Wiley), and a Series Editor of the book series Frontiers in Artificial Intelligence and Applications (IOS Press).

His academic research is concentrated around the themes of scientific data management, data mining, bioinformatics, and algorithms, and he has collaborated with more than 20 industrial partners.

Advisory Board

Area Editors

Table of Contents

Volume 1

Volume 2

Volume 3

Volume 4

List of Contributors

Andrew Adamatzky
Department of Computer Science
University of the West of England
Bristol
UK
andrew.adamatzky@uwe.ac.uk

Paul S. Andrews
Department of Computer Science
University of York
UK
psa@cs.york.ac.uk

Masanori Arita
Department of Computational Biology,
Graduate School of Frontier Sciences
The University of Tokyo
Kashiwa
Japan
arita@k.u-tokyo.ac.jp

Wolfgang Banzhaf
Department of Computer Science
Memorial University of Newfoundland
St. John's, NL
Canada
banzhaf@cs.mun.ca

Hans-Georg Beyer
Department of Computer Science
Fachhochschule Vorarlberg
Dornbirn
Austria
hans-georg.beyer@fhv.at

Sander Bohte
Research Group Life Sciences
CWI
Amsterdam
The Netherlands
s.m.bohte@cwi.nl

Eric Bonabeau
Icosystem Corporation
Cambridge, MA
USA
eric@icosystem.com

Anthony Brabazon
Natural Computing Research and
Applications Group
University College Dublin
Ireland
anthony.brabazon@ucd.ie

Andrea Bracciali
Department of Computing Science and
Mathematics
University of Stirling
UK
braccia@cs.stir.ac.uk

Robert Brijder
Leiden Institute of Advanced Computer
Science
Universiteit Leiden
The Netherlands
robert.brijder@uhasselt.be

Časlav Brukner
Faculty of Physics
University of Vienna
Vienna
Austria
caslav.brukner@univie.ac.at

Anthony Bucci
Icosystem Corporation
Cambridge, MA
USA
anthony@icosystem.com

Ke Chen
Department of Electrical and Computer
Engineering
University of Alberta
Edmonton, AB
Canada
kchen1@ece.ualberta.ca

Seungjin Choi
Pohang University of Science and
Technology
Pohang
South Korea
seungjin@postech.ac.kr

Bastien Chopard
Scientific and Parallel Computing Group
University of Geneva
Switzerland
bastien.chopard@unige.ch

David W. Corne
School of Mathematical and Computer
Sciences
Heriot-Watt University
Edinburgh
UK
dwcorne@macs.hw.ac.uk

Mark Daley
Departments of Computer Science and
Biology
University of Western Ontario
London, Ontario
Canada
daley@csd.uwo.ca

Jing Dang
Natural Computing Research and
Applications Group
University College Dublin
Ireland
jing.dang@ucd.ie

Edwin D. de Jong
Institute of Information and Computing
Sciences
Utrecht University
The Netherlands
dejong@cs.uu.nl

Kenneth De Jong
Department of Computer Science
George Mason University
Fairfax, VA
USA
kdejong@gmu.edu

Benjamin De Lacy Costello
Centre for Research in Analytical, Material
and Sensor Sciences, Faculty of Applied
Sciences
University of the West of England
Bristol
UK
ben.delacycostello@uwe.ac.uk

Kalyanmoy Deb
Department of Mechanical Engineering
Indian Institute of Technology
Kanpur
India
deb@iitk.ac.in

Pierpaolo Degano
Dipartimento di Informatica
Università di Pisa
Italy
degano@di.unipi.it

Marianne Delorme
Laboratoire d'Informatique Fondamentale
de Marseille (LIF)
Aix-Marseille Université and CNRS
Marseille
France
delorme.marianne@orange.fr

Ian Dempsey
Pipeline Financial Group, Inc.
New York, NY
USA
ian.dempsey@pipelinefinancial.com

Alberto Dennunzio
Dipartimento di Informatica
Sistemistica e Comunicazione, Università
degli Studi di Milano-Bicocca
Italy
dennunzio@disco.unimib.it

Kathryn A. Dowsland
Gower Optimal Algorithms, Ltd.
Swansea
UK
k.a.dowsland@btconnect.com

Jérôme Durand-Lose
LIFO
Université d'Orléans
France
jerome.durand-lose@univ-orleans.fr

David Edelman
School of Business
UCD Michael Smurfit Graduate Business
School
Dublin
Ireland
david.edelman@ucd.ie

Gary B. Fogel
Natural Selection, Inc.
San Diego, CA
USA
gfogel@natural-selection.com

Enrico Formenti
Département d'Informatique
Université de Nice-Sophia Antipolis
France
enrico.formenti@unice.fr

Xinbo Gao
Video and Image Processing System Lab,
School of Electronic Engineering
Xidian University
China
xbgao@ieee.org

Shuzhi Sam Ge
Social Robotics Lab
Interactive Digital Media Institute, The
National University of Singapore
Singapore
elegesz@nus.edu.sg

Masami Hagiya
Department of Computer Science
Graduate School of Information Science
and Technology
The University of Tokyo
Tokyo
Japan
hagiya@is.s.u-tokyo.ac.jp

Tero Harju
Department of Mathematics
University of Turku
Finland
harju@utu.fi

Mika Hirvensalo
Department of Mathematics
University of Turku
Finland
mikhirve@utu.fi

Satoshi Hiyama
Research Laboratories
NTT DOCOMO, Inc.
Yokosuka
Japan
hiyama@nttdocomo.co.jp

Guang-Bin Huang
School of Electrical and Electronic
Engineering
Nanyang Technological University
Singapore
egbhuang@ntu.edu.sg

Thomas Jansen
Department of Computer Science
University College Cork
Ireland
t.jansen@cs.ucc.ie

Nataša Jonoska
Department of Mathematics
University of South Florida
Tampa, FL
USA
jonoska@math.usf.edu

Jarkko J. Kari
Department of Mathematics
University of Turku
Turku
Finland
jkari@utu.fi

Lila Kari
Department of Computer Science
University of Western Ontario
London
Canada
lila@csd.uwo.ca

Sungchul Kim
Data Mining Lab, Department of Computer
Science and Engineering
Pohang University of Science and
Technology
Pohang
South Korea
subright@postech.ac.kr

Joshua Knowles
School of Computer Science and
Manchester Interdisciplinary
Biocentre (MIB)
University of Manchester
UK
j.knowles@manchester.ac.uk

Satoshi Kobayashi
Department of Computer Science
University of Electro-Communications
Tokyo
Japan
satoshi@cs.uec.ac.jp

Ken Komiya
Interdisciplinary Graduate School of
Science and Engineering
Tokyo Institute of Technology
Yokohama
Japan
komiya@dis.titech.ac.jp

Takeshi Koshiba
Graduate School of Science and
Engineering
Saitama University
Japan
koshiba@mail.saitama-u.ac.jp

Tim Kovacs
Department of Computer Science
University of Bristol
UK
kovacs@cs.bris.ac.uk

Natalio Krasnogor
Interdisciplinary Optimisation Laboratory,
The Automated Scheduling, Optimisation
and Planning Research Group, School of
Computer Science
University of Nottingham
UK
natalio.krasnogor@nottingham.ac.uk

Lukasz A. Kurgan
Department of Electrical and Computer
Engineering
University of Alberta
Edmonton, AB
Canada
lkurgan@ece.ualberta.ca

Petr Kůrka
Center for Theoretical Studies
Academy of Sciences and Charles
University in Prague
Czechia
kurka@cts.cuni.cz

Thomas H. LaBean
Department of Computer Science and
Department of Chemistry and Department
of Biomedical Engineering
Duke University
Durham, NC
USA
thomas.labean@duke.edu

Kang Li
School of Electronics, Electrical Engineering
and Computer Science
Queen's University
Belfast
UK
k.li@ee.qub.ac.uk

Xuelong Li
Center for OPTical IMagery Analysis and
Learning (OPTIMAL), State Key Laboratory
of Transient Optics and Photonics
Xi'an Institute of Optics and Precision
Mechanics, Chinese Academy of Sciences
Xi'an, Shaanxi
China
xuelong_li@opt.ac.cn

Wen Lu
Video and Image Processing System Lab,
School of Electronic Engineering
Xidian University
China
luwen@mail.xidian.edu.cn

Jacques Mazoyer
Laboratoire d'Informatique Fondamentale
de Marseille (LIF)
Aix-Marseille Université and CNRS
Marseille
France
mazoyerj2@orange.fr

Barry McMullin
Artificial Life Lab, School of Electronic
Engineering
Dublin City University
Ireland
barry.mcmullin@dcu.ie

Silja Meyer-Nieberg
Fakultät für Informatik
Universität der Bundeswehr München
Neubiberg
Germany
silja.meyer-nieberg@unibw.de

Kenichi Morita
Department of Information Engineering,
Graduate School of Engineering
Hiroshima University
Japan
morita@iec.hiroshima-u.ac.jp

Michele Mosca
Institute for Quantum Computing and
Department of Combinatorics &
Optimization
University of Waterloo and St. Jerome's
University and Perimeter Institute for
Theoretical Physics
Waterloo
Canada
mmosca@iqc.ca

Nicolas Ollinger
Laboratoire d'informatique fondamentale
de Marseille (LIF)
Aix-Marseille Université, CNRS
Marseille
France
nicolas.ollinger@lif.univ-mrs.fr

Michael O'Neill
Natural Computing Research and
Applications Group
University College Dublin
Ireland
m.oneill@ucd.ie

Hélène Paugam-Moisy
Laboratoire LIRIS – CNRS
Université Lumière Lyon 2
Lyon
France
and
INRIA Saclay – Ile-de-France
Université Paris-Sud
Orsay
France
helene.paugam-moisy@univ-lyon2.fr
hpaugam@lri.fr

Gheorghe Păun
Institute of Mathematics of the Romanian
Academy
Bucharest
Romania
and
Department of Computer Science and
Artificial Intelligence
University of Seville
Spain
gpaun@us.es
george.paun@imar.ro

Ion Petre
Department of Information Technologies
Åbo Akademi University
Turku
Finland
ipetre@abo.fi

Riccardo Poli
Department of Computing and Electronic
Systems
University of Essex
Colchester
UK
rpoli@essex.ac.uk

Elena Popovici
Icosystem Corporation
Cambridge, MA
USA
elena@icosystem.com

Corrado Priami
Microsoft Research
University of Trento Centre for
Computational and Systems Biology
(CoSBi)
Trento
Italy
and
DISI
University of Trento
Trento
Italy
priami@cosbi.eu

Mark Read
Department of Computer Science
University of York
UK
markread@cs.york.ac.uk

John H. Reif
Department of Computer Science
Duke University
Durham, NC
USA
reif@cs.duke.edu

Alan Reynolds
School of Mathematical and Computer
Sciences
Heriot-Watt University
Edinburgh
UK
a.reynolds@hw.ac.uk

Grzegorz Rozenberg
Leiden Institute of Advanced Computer
Science
Universiteit Leiden
The Netherlands
and
Department of Computer Science
University of Colorado
Boulder, CO
USA
rozenber@liacs.nl

Günter Rudolph
Department of Computer Science
TU Dortmund
Dortmund
Germany
guenter.rudolph@tu-dortmund.de

Yasubumi Sakakibara
Department of Biosciences and Informatics
Keio University
Yokohama
Japan
yasu@bio.keio.ac.jp

Lukáš Sekanina
Faculty of Information Technology
Brno University of Technology
Brno
Czech Republic
sekanina@fit.vutbr.cz

Shinnosuke Seki
Department of Computer Science
University of Western Ontario
London
Canada
sseki@csd.uwo.ca

Ofer M. Shir
Department of Chemistry
Princeton University
NJ
USA
oshir@princeton.edu

Andrzej Skowron
Institute of Mathematics
Warsaw University
Poland
skowron@mimuw.edu.pl

Jamie Smith
Institute for Quantum Computing and
Department of Combinatorics &
Optimization
University of Waterloo
Canada
ja5smith@iqc.ca

Petr Sosík
Institute of Computer Science
Silesian University in Opava
Czech Republic
and
Departamento de Inteligencia Artificial
Universidad Politécnica de Madrid
Spain
petr.sosik@fpf.slu.cz

Susan Stepney
Department of Computer Science
University of York
UK
susan.stepney@cs.york.ac.uk

Kalle-Antti Suominen
Department of Physics and Astronomy
University of Turku
Finland
kalle-antti.suominen@utu.fi

Andrew M. Sutton
Department of Computer Science
Colorado State University
Fort Collins, CO
USA
sutton@cs.colostate.edu

Siamak Taati
Department of Mathematics
University of Turku
Finland
siamak.taati@gmail.com

Masahiro Takinoue
Department of Physics
Kyoto University
Kyoto
Japan
takinoue@chem.scphys.kyoto-u.ac.jp

Fumiaki Tanaka
Department of Computer Science
Graduate School of Information Science
and Technology
The University of Tokyo
Tokyo
Japan
fumi95@is.s.u-tokyo.ac.jp

Dacheng Tao
School of Computer Engineering
Nanyang Technological University
Singapore
dacheng.tao@gmail.com

Véronique Terrier
GREYC, UMR CNRS 6072
Université de Caen
France
veroniqu@info.unicaen.fr

Jonathan M. Thompson
School of Mathematics
Cardiff University
UK
thompsonjm1@cardiff.ac.uk

Jon Timmis
Department of Computer Science and
Department of Electronics
University of York
UK
jtimmis@cs.york.ac.uk

Joaquin J. Torres
Institute "Carlos I" for Theoretical and
Computational Physics and Department of
Electromagnetism and Matter Physics,
Facultad de Ciencias
Universidad de Granada
Spain
jtorres@ugr.es

Marc M. Van Hulle
Laboratorium voor Neurofysiologie
K.U. Leuven
Leuven
Belgium
marc@neuro.kuleuven.be

Leonardo Vanneschi
Department of Informatics, Systems and
Communication
University of Milano-Bicocca
Italy
vanneschi@disco.unimib.it

Pablo Varona
Departamento de Ingeniería Informática
Universidad Autónoma de Madrid
Spain
pablo.varona@uam.es

Darrell Whitley
Department of Computer Science
Colorado State University
Fort Collins, CO
USA
whitley@cs.colostate.edu

R. Paul Wiegand
Institute for Simulation and Training
University of Central Florida
Orlando, FL
USA
wiegand@ist.ucf.edu

Xin Yao
Natural Computation Group, School of
Computer Science
University of Birmingham
UK
x.yao@cs.bham.ac.uk

Takashi Yokomori
Department of Mathematics, Faculty of
Education and Integrated Arts and
Sciences
Waseda University
Tokyo
Japan
yokomori@waseda.jp

Hwanjo Yu
Data Mining Lab, Department of Computer
Science and Engineering
Pohang University of Science and
Technology
Pohang
South Korea
hwanjoyu@postech.ac.kr

Jean-Baptiste Yunès
Laboratoire LIAFA
Université Paris 7 (Diderot)
France
jean-baptiste.yunes@liafa.jussieu.fr

G. Peter Zhang
Department of Managerial Sciences
Georgia State University
Atlanta, GA
USA
gpzhang@gsu.edu

Jie Zhang
School of Chemical Engineering and
Advanced Materials
Newcastle University
Newcastle upon Tyne
UK
jie.zhang@newcastle.ac.uk

Shengyu Zhang
Department of Computer Science and
Engineering
The Chinese University of Hong Kong
Hong Kong S.A.R.
China
syzhang@cse.cuhk.edu.hk

Eckart Zitzler
PHBern – University of Teacher Education,
Institute for Continuing Professional
Education
Bern
Switzerland
eckart.zitzler@phbern.ch
eckart.zitzler@tik.ee.ethz.ch

Marek Żukowski
Institute of Theoretical Physics and
Astrophysics
University of Gdansk
Poland
marek.zukowski@univie.ac.at
fizmz@univ.gda.pl

Evolutionary Computation

Thomas Bäck

20 Generalized Evolutionary Algorithms

Kenneth De Jong
Department of Computer Science, George Mason University,
Fairfax, VA, USA
kdejong@gmu.edu

G. Rozenberg et al. (eds.), *Handbook of Natural Computing*, DOI 10.1007/978-3-540-92910-9_20,
© Springer-Verlag Berlin Heidelberg 2012

Abstract

People have been inventing and tinkering with various forms of evolutionary algorithms (EAs) since the 1950s when digital computers became more readily available to scientists and engineers. Today we see a wide variety of EAs and an impressive array of applications. This diversity is both a blessing and a curse. It serves as strong evidence for the usefulness of these techniques, but makes it difficult to see "the big picture" and make decisions regarding which EAs are best suited for new application areas. The purpose of this chapter is to provide a broader "generalized" perspective.

1   Introduction

Some of the earliest examples of evolutionary algorithms (EAs) appeared in the literature in the 1950s (see, e.g., Box 1957 or Friedberg 1959). Readers interested in more details of this early history are encouraged to see the excellent summary in Fogel (1998). However, three developments that took place in the 1960s have had the most direct and lasting impact on the field today. At the Technical University of Berlin, Ingo Rechenberg and Hans-Paul Schwefel developed an algorithm they called an evolution strategy (ES) in order to solve difficult real-valued parameter optimization problems. At the same time, Larry Fogel and his colleagues at UCLA developed a technique they called evolutionary programming (EP) to automate the process of constructing finite state machines that controlled the behavior of intelligent agents. Independently, John Holland at the University of Michigan developed a class of techniques called genetic algorithms (GAs) for designing robust adaptive systems. Readers interested in more details of these important historical developments are encouraged to consult one or more of the readily available books such as Fogel (1995), Schwefel (1995), and Holland (1975).

Historically, there was little interaction among these groups until the early 1990s. By then each group had developed their own notation and nomenclature. As the interactions became more frequent, it became clear that there was a need for a more general perspective in which GAs, EP, and ESs were seen as important EA instances of a unified "evolutionary computation" (EC) framework (see, e.g., Bäck 1996).

The influence of these historical developments cannot be understated. Many of the EAs used today are variants of one of the historical forms (ES, EP, or GA). At the same time, we are the beneficiaries today of 50 years of exploring, understanding, and extending these ideas. The result is a pretty clear picture of "simple" EAs and a sense of how best to extend and apply them, as indicated in the following sections.

2   Simple Evolutionary Algorithms

The word *evolution* has several meanings. It is used in a general sense to describe a process of incremental change such as "evolving" medical procedures. When we use the term *evolutionary algorithm*, we have a more specific meaning in mind, namely, a Darwinian-like evolutionary process that involves a population of individuals dynamically changing over time with some sort of "survival-of-the-fittest" selection pressure. One might imagine algorithms of this sort whose purpose is to model an existing evolutionary system and explore possible perturbations to such systems. In this case, there is a need for model fidelity to capture the details of particular evolutionary systems. However, as noted in ❷ Sect. 1, the historical motivation for developing EAs was to harness their intrinsic adaptive capabilities to solve difficult computational problems. As such, these EAs are *inspired* by the features of natural systems, but not constrained by them. These algorithms solve problems in the following way:

- They maintain a population of individuals that represent possible solutions to a problem.
- Individuals have a notion of "fitness" associated with them representing the quality of the solution they represent. That fitness is used to bias the population dynamics (i.e., the birth/death cycle).
- Individuals are selected to be parents and produce "offspring," that is, new solutions that are similar but not identical to the parents.
- Individuals are selected to die (i.e., be deleted from the population).

If one of these EAs is run for several generations (i.e., several birth/death cycles), the fitness-biased selection processes produce a steady increase in the fitness of the population, conveying a strong sense of fitness optimization. This observation leads to the following general template for a simple evolutionary optimization algorithm:

```
Randomly generate an initial population of size M.
Do until some stopping criterion is met:
  Select some parents to produce some offspring.
  Select some individuals to die.
End Do
Return as the problem solution, the individual with the highest observed
fitness.
```

Of course, this general template is not computationally specific enough to be coded and executed. In order to instantiate this EA template for a particular class of problems, we must answer the following questions:

- How are individuals (problem solutions) represented in the population?
- How big should the population be?
- How are parents selected?
- How are offspring produced from selected parents?
- How many offspring should be produced?
- How are the survivors chosen?

The fact that there is more than one answer to each of these questions was already evident in the 1960s when ES, EP, and GA variants were being developed. As an example, consider parameter optimization problems. A traditional ES approach would be to represent solutions internally as n-dimensional vectors of parameter values and maintain a small parent population (e.g., $M < 10$). Each member of the parent population produces one (or more) offspring by cloning the parent and "mutating" some of the offspring's parameter values via a Gaussian perturbation operator. The combined parent and offspring populations are sorted by fitness and only the top M individuals survive to the next generation.

By contrast, a traditional GA approach would convert the n-dimensional parameter vectors into an internal binary string representation and maintain a population of moderate size (e.g., $M \in [50, 100]$). Pairs of parents are selected stochastically using a fitness-proportional selection mechanism to produce offspring that inherit combinations of binary substrings from their parents and a small number of mutations (bit flips). Exactly M offspring are produced. They all survive and the parents all die.

Both of these traditional models were inspired by natural systems, but clearly behave differently as computational procedures for solving search and optimization problems. It is known from various "no-free-lunch" theorems that efficient algorithms incorporate a bias that allows them to exploit the structure inherent in a particular class of problems, and that the same bias is likely to be a hindrance on other problem classes with different structures. This perspective suggests a general strategy for answering the questions listed above: make the choices so as to tune the EA bias to the particular class of problems of interest. That is, view the EA template above as a "meta-heuristic" to be instantiated in problem-specific ways.

This, in turn, requires an understanding of how these EA design decisions affect its problem-solving bias. These issues are explored in the following sections.

2.1   Internal Representation

Choices for the internal representation of problem solutions fall into two biologically inspired categories: genotypic representations and phenotypic representations. Genotypic representations encode problem solutions internally in a manner analogous to the biological genetic code, that is, solutions are represented as strings formed from a universal alphabet (e.g., binary strings in traditional GAs). Reproductive operators are defined to manipulate the universal encodings, significantly increasing the portability of the EA code from one application to the next. Applying such EAs to a new problem class simply requires writing the encoding/decoding procedures.

By contrast, phenotypic representations represent problem solutions directly with no intermediate encoding (e.g., parameter vectors in ESs). Reproductive operators are defined to manipulate the solutions directly. This provides the opportunity to introduce problem-specific biases into the EA, but does imply significant redesign and re-implementation of the reproductive operators.

To illustrate these issues, consider how one might apply an EA to traveling salesperson problems. A GA practitioner would focus on designing a mapping of permutation spaces to/from binary strings, leaving the reproductive operators unchanged. An ES practitioner would focus on replacing the standard Gaussian mutation operator with something more appropriate for permutation spaces.

2.1.1   Reproductive Operators

The most important property of reproductive operators is their ability to use existing solutions as a springboard for creating new and interesting solutions. From a biological perspective, this is the notion of heritability and is achieved in two rather distinct ways: asexual and sexual reproduction. Asexual reproduction involves cloning single parents and then applying a mutation operator to provide some variability. This is the strategy used in traditional ES and EP approaches. Sexual reproduction involves combining elements of more than one parent (generally with a small dose of mutation) to produce offspring that inherit some features from each parent. This is the strategy used in traditional GAs in which offspring are produced from two parents (as is usual in natural systems). However, there is no *a priori* computational reason to limit this to two parents (see, for example, Eiben et al. 1994).

From a search perspective, these two reproductive strategies differ significantly. Asexual reproduction tends to be more of a local search operator producing offspring in a nearby neighborhood of their parents. A key element in the effectiveness of asexual reproduction is the ability to dynamically control the expected distance between parents and offspring, tuning this step size to the properties of the fitness landscape (e.g., the "1:5" step size adaptation rule in traditional ESs (Schwefel 1995), or more recent "covariance matrix" approaches (Hansen and Ostermeier 2001)).

By contrast, sexual reproduction tends to be more of a global search operator producing offspring that are frequently quite far from their parents. A key element in the effectiveness of sexual reproduction is the ability to recombine modular subcomponents in useful ways. For example, in parameter optimization problems, inheriting coupled (epistatic) values is highly desirable, but generally not known *a priori*. An interesting property of many of the recombination operators in use is that they intrinsically adapt to such couplings, in that, interactions that improve fitness spread throughout the population and thus are more likely to be inherited.

Both sexual and asexual reproductive operators complement each other nicely producing a blend of local and global search. As a consequence, many EAs today use a combination of both.

2.2 Selection Mechanisms

In simple EAs, there are two opportunities for selection: when choosing parents and when determining survivors. Both are opportunities to use fitness information to bias the search. As a first thought, one might consider selecting only the best individuals. But, one must exercise care here since this introduces a certain amount of "greediness" into the search process. In computer science, greedy algorithms are a well-studied class of procedures that have the general property of rapid convergence, but not necessarily to the best solution. For some problem domains and fitness landscapes, rapid convergence to a nearby local optimum is more than adequate. For others, a slower, more diffuse search is required to avoid getting immediately trapped in a local optimum.

So, a key aspect to EA design is determining the right amount of selection pressure (greediness) for a particular application. This is accomplished by making several interrelated observations. The first is that having fitness bias for *both* parent selection and survival selection invariably produces too much greediness and results in rapid convergence to suboptimal solutions. As a consequence, most EAs use fitness-biased selection for only one of the two selection steps, leaving the other one unbiased. To introduce fitness bias, we have a well-studied collection of selection mechanisms to choose from, which is listed here in order of decreasing selection pressure:

- Truncation selection: choose only the N most fit individuals.
- Tournament selection: (repeatedly) choose K individuals at random and keep only the best of the K selected.
- Rank-proportional selection: choose individuals proportional to their fitness rank in the population.
- Fitness-proportional selection: choose individuals proportional to the value of their fitness.

Tournament and fitness-proportional selection have analogs in naturally evolving biological systems. Truncation and rank selection are similar to strategies used in breeding farms. All have been formally analyzed regarding the degree of selection pressure they induce (see, for example, De Jong 2006).

One can generally choose whether to implement these selection mechanisms as deterministic or stochastic procedures. There is a growing amount of evidence that the stochastic versions converge a bit more slowly but are less likely to get stuck in suboptimal areas (De Jong 2006).

A related observation has to do with whether or not parents and offspring compete with each other for survival. Mathematical models of evolutionary systems frequently adopt a "nonoverlapping" generation assumption: only the offspring survive. This is done primarily to make the mathematical analysis simpler. From a computer science perspective, the choice affects the greediness of an EA. Overlapping generations produce greedier search behavior than nonoverlapping populations. This is why traditional ES procedures come in two forms: "plus" (overlapping) and "comma" (nonoverlapping) strategies (Schwefel 1995), and why there are both nonoverlapping and "steady state" GAs (De Jong 2006).

The final observation that is important to keep in mind when choosing a selection mechanism is the need to achieve an effective balance between exploration and exploitation. In simple EAs, exploration is encouraged by having weaker selection pressure and more reproductive variation, and the converse for exploitation. Effective balance can be achieved in a variety of ways as illustrated by the traditional EAs. Traditional ESs typically have strong survival selection (overlapping generations with truncation selection) and balance that off with a fairly aggressive mutation operator. By contrast, traditional GAs have much weaker selection pressure (fitness-proportional parent selection with nonoverlapping generations) and much milder forms of reproductive variation.

2.3 Population Sizing

In natural systems, the size of a population can grow and shrink based on a variety of environmental factors. To date, EA developers have found little computational advantage for allowing the population size to vary from one generation to the next. Since fixed-size population models are also easier to implement, most EAs maintain a fixed-size population of size M. However, as we have seen, internal to a generation some number K of offspring are produced, and by the end of the generation a survival selection mechanism is used to reduce this intermediate number of $M + K$ individuals back to M.

So, an EA designer has two decisions to make: choosing a value for M and for K. Traditionally, the choices have varied widely for the following reasons. Choices for M are related to the degree of parallel search required. Highly complex, rugged, multi-peaked landscapes need a much higher degree of parallelism (larger M) than, say, a convex quadratic surface. Values for K are associated with feedback delays – the amount of exploration done with the current population before updating it with new information. Larger values of K are needed for more complex, rugged, multi-peaked landscapes. As a consequence, we see effective ESs with $M = 1$ and $K = 1$, effective EPs with $M = 10$ and $K = 10$, and effective GAs with $M = 50$ and $K = 50$. More recently, EAs being used to search complex program spaces have $M = 1,000$ or more.

Since EAs are frequently used on problems with fitness landscapes of unknown (*a priori*) properties, the values of M and K are generally chosen via some preliminary experimentation.

2.4   Fitness Landscapes

For many problems the fitness landscape is defined by the objective function that measures the quality of a solution. Understanding how EAs use fitness feedback to bias, the search can often suggest ways to improve an objective function to make it more amenable to evolutionary search. A classic example is that of solving constraint-satisfaction problems. In this case, the formal fitness landscape is defined as 1 (true) if all constraints are satisfied and 0 (false) otherwise, creating a "needle-in-the-haystack" search problem that is lacking any intermediate feedback suggesting areas to look in. The problem becomes significantly more EA friendly if the objective function is modified to given nonuniform feedback on partial solutions (e.g., the number of constraints satisfied).

3   Example Application Areas

At the most general level, these simple EAs can be viewed as parallel adaptive search procedures with a very natural sense of fitness optimization. It is no surprise, then, that the majority of applications involve solving some sort of optimization problem. However, as we will see in this section, the breadth and variety of these applications goes well beyond the scope of traditional mathematical optimization areas.

3.1   Parameter Optimization

Perhaps the most pervasive and best understood EA application area is that of parameter optimization. This is partly because linear vectors of parameters map quite nicely onto our (admittedly simplistic) notion of biological genetic material consisting of linear strands of genes, the values of which affect an organism's fitness. By analogy, recombining and mutating parameter values provide an interesting and frequently effective strategy for finding optimal combinations of parameter values.

Techniques for solving parameter optimization problems have been developed for many years by the mathematical optimization community. The key difference is that simple EAs make far fewer assumptions about the characteristics of the objective function to be optimized. They do not *a priori* require continuity, differentiability, convexity, etc. Rather, they exhibit a slower, but more robust search behavior even in the presence of noise. At the same time, domain-specific knowledge can be incorporated into an EA via specialized representations and reproductive operators when appropriate and desirable. As a consequence, optimization-oriented EAs continue to provide a nice complement to the existing and more traditional optimization techniques.

3.2   Nonlinear and Variable-Size Structures

Somewhat more challenging are optimization problems in which solutions are most naturally represented as nonlinear structures possibly varying in size. One example of that has already been discussed – the early EP work involving the evolution of finite state machines. To that, we can easily add a variety of useful structures including decision trees, artificial neural networks, and job-shop schedules. Traditional approaches to such problems focus on developing very problem-specific techniques. The EA approach is more meta-heuristic in nature in the sense that the standard EA framework gets instantiated by choosing an internal representation for the structures and designing effective mutation and recombination operators capable of generating useful offspring. So, for example, one might choose an adjacency list representation for finite state machines and then experiment with various recombination and mutation operators to find ones that produce offspring finite state machines that are interesting variants of their parents.

3.3 Executable Objects

Perhaps the most challenging problems are those in which the structures to be optimized are executable objects. This challenge is actually quite close to our modern view of biological genetic material as a complex program being executed within each cell, involving complex regulatory control mechanisms, and generating phenotypic traits in highly nonlinear ways. In the EA community, this challenge has given rise to entire subgroups focused on how best to evolve executable objects. An interesting open question is the extent to which more traditional programming languages are evolution friendly. As one might expect, languages like C and Java are much better suited for humans than evolution. However, languages like Lisp and rule-oriented languages have produced many striking positive results (see, for example, Koza 1992 or Grefenstette et al. 1990).

3.4 Other Than Optimization Problems

As noted earlier, standard EAs are best described as parallel adaptive search procedures. As such they can be usefully applied whenever one is faced with difficult search problems. For example, many constraint-satisfaction problems are really about finding *any* solution that satisfies all the constraints. Many aspects of data mining involve searching for interesting patterns and/or unusual events. Machine learning problems are generally characterized as searching hypothesis spaces for descriptions of concepts or behaviors. In each case, EAs have been shown to provide effective search strategies that complement more domain-specific approaches.

4 More Complex Evolutionary Algorithms

The EAs discussed in ❷ Sect. 2 were characterized as "simple" in the sense that they can easily be described in a few lines of pseudocode. They are also simple from a biological perspective, abstracting away many standard features of natural evolutionary systems. Yet, as we have seen, these simple EAs produce surprisingly powerful and robust search heuristics. However, one might wonder why obvious features of natural evolutionary systems are missing, such as a notion of age, gender, multistranded chromosomes with dominance relationships among gene values, etc. The short answer is that many of these features have been experimented with but appear to provide no useful computational benefit when solving static optimization problems.

On the other hand, the successful application of simple EAs has inspired their application to problem areas that are significantly more complex than standard optimization problems. As a result, there has been a corresponding increase in the complexity of the EAs developed to address these problems. In this section, a few of the more important "complexifications" are described.

4.1 New Parallel EA Models

One of the important features of EAs is their natural parallelism which can be exploited using the increasing array of inexpensive parallel hardware architectures (e.g., multi-core machines, closely coupled multi-cpu clusters, loosely coupled clusters, etc.). In many of the more challenging applications, evaluating the fitness of an individual is the most cpu-intensive EA component. A good example of this is the area of "evolutionary robotics" in which evaluating the fitness of a robot controller is accomplished via a cpu-intensive simulation. In such cases, a simple master–slave configuration provides the ability to evaluate individuals in parallel on multiple loosely coupled machines.

In some EA applications, it is useful to impose a spatial topology on the population, replacing random mating and replacement with local neighborhood interactions (see, for example, Sarma 1998). In this case, computer languages supporting multi-threaded implementations provide a means for implementing efficient parallel local activities on multiprocessor machines.

At the other extreme, "island models" have been shown to be quite useful for solving large and difficult optimization problems (see, e.g., Skolicki and De Jong 2004). These island models represent a coarsely grained parallel framework in which multiple EAs are running on separate "islands" with occasional "migrations" of individuals from one island to another. These models map easily onto loosely coupled networks of machines, allowing a simple means for exploiting the collective computing power of a cluster of machines.

4.2 Multi-population EA Models

In addition to island models, there are a variety of multi-population models that are inspired by the biological notion of coevolving populations in which the fitness of individuals is dependent in part on the individuals in other populations. Classic examples of this in nature are competitive relationships such as predator–prey models or more cooperative relationships such as host–parasite models. As an example, competitive coevolutionary EAs have been used to evolve better game-playing programs (e.g., Rosin and Belew 1997), while cooperative coevolutionary EAs have been successfully applied to difficult design problems that can be broken down into a collection of subproblems (e.g., Potter and De Jong 2000).

4.3 Multi-objective Optimization

Most "real life" design problems involve trying to simultaneously satisfy multiple objectives. Trying to optimize a design for one objective often reduces the desirability of the design with respect to another objective (e.g., power versus fuel efficiency), requiring a designer to identify acceptable trade-off points. This is often done by finding a set of Pareto-optimal points, that is, points in design space for which it is impossible to improve on one objective without making another worse, and then choosing from among them.

The ability to compute a set of Pareto-optimal points is, in general, a computationally difficult task. Some of the best algorithms to date for doing so are suitably modified EAs (see, for example, Deb 2001 or Coello et al. 2002). The success in this area has led to the organization of conferences focused just on this topic.

4.4   Handling Dynamic Environments

Historically the EC field has focused primarily on solving "static" optimization problems, that is, problems which do not change during the optimization process. However, there are many interesting and difficult optimization problems associated with dynamic environments (e.g., traffic control problems, stock market trading, etc.). As one might expect, standard EAs do not handle dynamic problems very well. However, suitable variations do quite well (see, for example, Branke 2002 or Morrison 2004).

4.5   Exploiting Morphogenesis

One of the most striking differences between simple EAs and natural systems is the fact that in simple EAs there is no concept of infancy, growth, and maturation. Offspring come into existence as mature adults whose fitness is immediately determinable and can be immediately selected to be a parent. Part of the reason for this is that, from a computational point of view, this additional complexity adds little or no value when solving standard static parameter optimization problems. However, as we extend the range of EA applications to problems like evolving complex engineering designs or complex agent behaviors, it becomes increasingly difficult to design effective representations and reproductive operators. If we reflect on how this is accomplished in natural systems, we note that the reproductive operators are manipulating the plans for growing complex objects. That is, there is a clear distinction between the genotype of an individual and its phenotype. A process of growth (morphogenesis) produces the phenotype from the phenotype during a maturation process, and fitness is associated with the resulting phenotype.

These observations have led to considerable interest of EAs in "generative representations" capable of producing complex phenotypes from much simpler evolving phenotypes (see, for example, Kicinger et al. 2004). Independently, this notion of generative representation has been extensively explored in other areas of science as well, including biology (e.g., Lindenmayer 1968) and complex systems (e.g., Wolfram 1994).

5   Summary and Conclusions

Although evolutionary computation was one of the early fields to develop around the notion of "inspiration from nature," it is certainly not the only one. Other familiar examples are "simulated annealing" techniques and "artificial neural networks." More recently, there has been considerable interest in nature-inspired "ant colony" optimization and "particle swarm" optimization. This has led to the broader concept of "natural computation" as documented in this handbook.

As emphasized in this chapter, adopting a generalized view of EAs leads naturally to the perception of EAs as a powerful "meta-heuristic," a template for constructing problem-specific heuristics. This is an important perspective that continues to expand the range of EA applications.

Included in this chapter are a sprinkling of references to more detailed material. Far more information is available from the proceedings of the many EC-related conferences such as GECCO, CEC, PPSN, and FOGA, and from the established journals of the field such as *Evolutionary Computation* (MIT Press), *Transactions on Evolutionary Computation* (IEEE Press), and *Genetic Programming and Evolvable Machines* (Springer). In addition, there is a wealth of information and open-source code available on the Internet.

References

Bäck T (1996) Evolutionary algorithms in theory and practice. Oxford University Press, New York

Box G (1957) Evolutionary operation: a method for increasing industrial productivity. Appl Stat 6(2):81–101

Branke J (2002) Evolutionary optimization in dynamic environments. Kluwer, Boston, MA

Coello C, Veldhuizen D, Lamont G (2002) Evolutionary algorithms for solving multi-objective problems. Kluwer, Boston, MA

De Jong K (2006) Evolutionary computation: a unified approach. MIT Press, Cambridge, MA

Deb K (2001) Multi-objective optimization using evolutionary algorithms. Wiley, New York

Eiben A, Raué P, Ruttkay Z (1994) Genetic algorithms with multi-parent recombination. In: Davidor Y, Schwefel H-P, Männer R (eds) Proceedings of the parallel problem solving from nature conference (PPSN III). Springer, Berlin, pp 78–87

Fogel D (1995) Evolutionary computation: toward a new philosophy of machine intelligence. IEEE Press, Piscataway, NJ

Fogel D (1998) Evolutionary computation: the fossil record. IEEE Press, Piscataway, NJ

Friedberg R (1959) A learning machine: Part 1. IBM Res J 3(7):282–287

Grefenstette J, Ramsey CL, Schultz AC (1990) Learning sequential decision rules using simulation models and competition. Mach Learn 5(4):355–381

Hansen N, Ostermeier A (2001) Completely derandomized self-adaptation in evolution strategies. Evolut Comput 9(2):159–195

Holland J (1975) Adaptation in natural and artificial systems. University of Michigan Press, Ann Arbor, MI

Kicinger R, Arciszewski T, De Jong K (2004) Morphogenesis and structural design: cellular automata representations of steel structures in tall buildings. In: Proceedings of the congress on evolutionary computation, Portland, IEEE Press, Piscataway, NJ pp 411–418

Koza J (1992) Genetic programming. MIT Press, Cambridge, MA

Lindenmayer A (1968) Mathematical models for cellular interaction in development. J Theor Biol 18:280–315

Morrison R (2004) Designing evolutionary algorithms for dynamic environments. Springer, Berlin

Potter M, De Jong K (2000) Cooperative coevolution: an architecture for evolving coadapted subcomponents. Evolut Comput 8(1):1–29

Rosin C, Belew R (1997) New methods for competitive coevolution. Evolut Comput 5(1):1–29

Sarma J (1998) An analysis of decentralized and spatially distributed genetic algorithms. Ph.D. thesis, George Mason University

Schwefel H-P (1995) Evolution and optimum seeking. Wiley, New York

Skolicki Z, De Jong K (2004) Improving evolutionary algorithms with multi-representation island models. In: Yao X, Burke EK, Lozano JA, Smith J, Merelo Guervos JJ, Bullinaria JA, Rowe JE, Tino P, Kaban A, Schwefel H-P (eds) Proceedings of the parallel problem solving from nature conference (PPSN VIII). Springer, Berlin, pp 420–429

Wolfram S (1994) Cellular automata and complexity. Addison-Wesley, Reading, MA

21 Genetic Algorithms — A Survey of Models and Methods

Darrell Whitley[1] · *Andrew M. Sutton*[2]
[1]Department of Computer Science, Colorado State University, Fort Collins, CO, USA
whitley@cs.colostate.edu
[2]Department of Computer Science, Colorado State University, Fort Collins, CO, USA
sutton@cs.colostate.edu

G. Rozenberg et al. (eds.), *Handbook of Natural Computing*, DOI 10.1007/978-3-540-92910-9_21,
© Springer-Verlag Berlin Heidelberg 2012

Abstract

This chapter first reviews the simple genetic algorithm. Mathematical models of the genetic algorithm are also reviewed, including the schema theorem, exact infinite population models, and exact Markov models for finite populations. The use of bit representations, including Gray encodings and binary encodings, is discussed. Selection, including roulette wheel selection, rank-based selection, and tournament selection, is also described. This chapter then reviews other forms of genetic algorithms, including the steady-state Genitor algorithm and the CHC (cross-generational elitist selection, heterogenous recombination, and cataclysmic mutation) algorithm. Finally, landscape structures that can cause genetic algorithms to fail are looked at, and an application of genetic algorithms in the domain of resource scheduling, where genetic algorithms have been highly successful, is also presented.

1 The Basics of Genetic Algorithms

Genetic algorithms were the first form of evolutionary algorithms to be widely accepted across a diverse set of disciplines ranging from operations research to artificial intelligence. Today, genetic algorithms and other evolutionary algorithms are routinely used as search and optimization tools for engineering and scientific applications.

Genetic algorithms were largely developed by John Holland and his students in the 1960s, 1970s, and 1980s. The term "genetic algorithms" came into common usage with the publication of Ken De Jong's 1975 PhD dissertation. Holland's classic 1975 book, *Adaptation in Natural and Artificial Systems* (Holland 1975), used the term "genetic plans" rather than "genetic algorithms." In the mid-1980s genetic algorithms started to reach other research communities. An explosion of research in genetic algorithms came soon after a similar explosion of research in artificial neural networks. Both areas of research draw inspiration from biological systems as a computational model.

There are several forms of evolutionary algorithms that use simulated evolution as a mechanism to solve problems where the problems can be expressed as a search or optimization problem. Other areas of evolutionary computation, such as evolutionary programming, evolution strategies, and genetic programming, are discussed in other chapters. Compared to other evolutionary algorithms, genetic algorithms put a great deal of emphasis on the combined interactions of selection, recombination, and mutation acting on a genotype. In most early forms of genetic algorithms, recombination was emphasized over mutation. This emphasis still endures in some forms of the genetic algorithm. However, hybridization of genetic algorithms with local search is also very common.

Genetic algorithms emphasize the use of a "genotype" that is decoded and evaluated. These genotypes are often simple data structures. In most early applications, the genotypes (which are sometimes thought of as artificial chromosomes) are bit strings which can be recombined in a simplified form of "sexual reproduction" and can be mutated by bit flips. Because of the bit encoding, it is sometimes common to think of genetic algorithms as function optimizers. However, this does not mean that they yield globally optimal solutions. Instead, Holland (in the introduction to the 1992 edition of his book *Adaptation in Natural and Artificial Systems* Holland (1992)) and De Jong (1993) have both emphasized that these algorithms find competitive solutions rather than optimal solutions, but both also suggest that it is probably best to view genetic algorithms not as an optimization process, but rather as adaptive systems.

At the risk of overemphasizing optimization, an example application from function optimization provides a useful vehicle for explaining certain aspects of these algorithms. For example, consider a control or production process which must be optimized with respect to some evaluation criterion. The input domain to the problem, $\mathcal{X}$, is the set of all possible configurations to the system. Given an evaluation function f, one seeks to maximize (or minimize) $f(x)$, $x \in \mathcal{X}$. The input domain can be characterized in many ways. For instance, if $\mathcal{X} = \{0, 1, \ldots, 2^L - 1\}$ is the search space then our input domain might be the set of binary strings of length L. If we elect to use a "real-valued" encoding, a floating point representation might be used, but in any discrete computer implementation, the input domain is still a finite set.

In the evolutionary algorithm community, it has become common to refer to the evaluation function as a *fitness* function. Technically, one might argue that the objective function $f(x)$ is the evaluation function, and that the fitness function is a second function more closely linked with selection. For example, we might assign fitness to an artificial chromosome based on its rank in the population, assigning a "fitness" of 2.0 to the best member of the population and a fitness of 1.0 to the median member of the population. Thus, the fitness function and evaluation function are not always the same. Nevertheless in common usage, the term fitness function has become a surrogate name for the evaluation function.

Returning to our simple example of an optimization problem, assume that there is a system with three parameters we can control: temperature, pressure, and duration. These are in effect three inputs to a black box optimization problem where inputs from the domain of the function are fed into the black box and a value from the co-domain of the function is produced as an output. The system's response could be a *quality* measure dependent on the temperature, pressure, and duration parameters.

One could represent the three parameters using a vector of three real-valued parameters, such as

$$\langle 32.56, 18.21, 9.83 \rangle$$

or the three parameters could be represented as bit strings, such as

$$\langle 000111010100, 110100101101, 001001101011 \rangle.$$

Using an explicit bit encoding automatically raises the question as to what precision should be used, and what should be the mapping between bit strings and real values. Picking the right precision can have a large impact on performance. Historically, genetic algorithms have typically been implemented using low precision; using 10 bits per parameter is common in many test functions. Using high precision (e.g., 32 bits per parameters) generally results in poor performance, and real-valued representations should be considered if high precision is required.

Recombination is central to genetic algorithms. For now, assume that the artificial chromosomes are bit strings and that 1-point crossover will be used. Consider the string 1101001100101101 and another binary string, yxyyxyxxyyyxyxxy, in which the values 0 and 1 are denoted by x and y. Using a single randomly chosen crossover point, a 1-point recombination might occur as follows.

```
11010 \/ 01100101101

yxyyx /\ yxxyyyxyxxy
```

■ **Fig. 1**

One generation is broken down into a selection phase and a recombination phase. This figure shows strings being assigned into adjacent slots during selection. It is assumed that selection randomizes the location of strings in the intermediate population. Mutation can be applied before or after crossover.

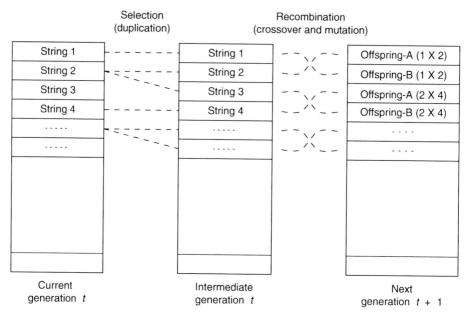

Swapping the fragments between the two parents produces the following two offspring.

    `11010yxxyyyxyxxy`   and   `yxyyx01100101101`

Parameter boundaries are ignored. We can also apply a mutation operator. For each bit in the population, mutate with some low probability p_m. For bit strings of length L a mutation rate of $1/L$ is often suggested; for separable problems, this rate of mutation can be shown to be particularly effective.

In addition to mutation and recombination operators, the other key component to a genetic algorithm (or any other evolutionary algorithm) is the selection mechanism. For a genetic algorithm, it is instructive to view the mechanism by which a standard genetic algorithm moves from one generation to the next as a two-stage process. Selection is applied to the current population to create an *intermediate generation*, as shown in ❯ *Fig. 1*. Then, recombination and mutation are applied to the intermediate population to create the *next generation*. The process of going from the current population to the next population constitutes one generation in the execution of a genetic algorithm.

We will assume that our selection mechanism is *tournament selection*. This is not the mechanism used in early genetic algorithms, but it is commonly used today. A simple version of tournament selection randomly samples two strings out of the initial population, then "selects" the best of the two strings to insert into the intermediate generation. If the population size is N, then this must be done N times to construct the intermediate population. We can then apply mutation and crossover to the intermediate population.

1.1 The Canonical Holland-Style Genetic Algorithm

In a Holland-style genetic algorithm, tournament selection is not used; instead *fitness proportional selection* is used. Proportional fitness is defined by $f_i/\bar{f}$, where f_i is the evaluation associated with string i and $\bar{f}$ is the average evaluation of all the strings in the population. Under fitness proportional selection, fitness is usually being maximized. All early forms of genetic algorithms used fitness proportional selection. Theoretical models also often assume fitness proportional selection because it is easy to model mathematically.

The fitness value f_i may be the direct output of an evaluation function, or it may be scaled in some way. After calculating $f_i/\bar{f}$ for all the strings in the current population, selection is carried out. In the canonical genetic algorithm, the probability that strings in the current population are copied (i.e., duplicated) and placed in the intermediate generation that is proportional to their fitness.

For a maximization problem, if $f_i/\bar{f}$ is used as a measure of fitness for string i, then strings where $f_i/\bar{f}$ is greater than 1.0 have above average fitness and strings where $f_i/\bar{f}$ is less than 1.0 have below average fitness. We would like to allocate more chances to reproduce to those strings that are above average. One way to do this is to directly duplicate those strings that are above average. To do so, f_i is broken into an integer part, x_i, and a remainder, r_i. We subsequently place x_i duplicates of string i directly into the intermediate population and place 1 additional copy with probability r_i.

This is efficiently implemented using *stochastic universal sampling*. Assume that the population is laid out in random order as a number line where each individual is assigned space on the number line in proportion to fitness. Now generate a random number between 0 and 1 denoted by k. Next, consider the position of the number $i+k$ for all integers i from 1 to N where N is the population size. Each number $i+k$ will fall on the number line in some space corresponding to a member of the population. The position of the N numbers $i+k$ for $i=1$ to N in effect selects the members of the intermediate population. This is illustrated in ❯ *Fig. 2*.

This method is also known as roulette wheel selection: the space assigned to the 25 members of the population in ❯ *Fig. 2* could be laid out along the circumference of a circle representing the roulette wheel. The choice of k in effect "spins" the roulette wheel (since only the fractional part of the spins effects the outcome) and determines the position of the evenly spaced pointers, thereby simultaneously picking all N members of the intermediate population. The resulting selection is unbiased (Baker 1987).

After selection has been executed, the construction of the intermediate population is complete. The *next generation* of the population is created from the intermediate population.

◻ **Fig. 2**
Stochastic Universal Sampling. The fitnesses of the population can be seen as being laid out on a number line in random order as shown at the bottom of the figure. A single random value, $0 \leq k \leq 1$, shifts the uniformly spaced "pointers" which now select the member of the next intermediate population.

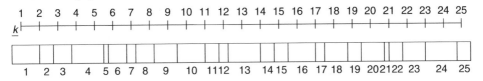

Crossover is applied to randomly paired strings with a probability denoted p_c. The offsprings created by "recombination" go into the next generation (thus replacing the parents). If no recombination occurs, the parents can pass directly into the next generation. However, parents that do not undergo recombination may still be changed due to the application of a mutation operator.

After the process of selection, recombination, and mutation is complete, the next generation of the population can be evaluated. The process of evaluation, selection, recombination, and mutation forms one generation in the execution of a genetic algorithm.

One way to characterize *selective pressure* is to calculate the selection bias toward the best individual in the population compared to the average fitness or the median fitness of the population. This is more useful for "rank-based" than fitness proportional, but it still serves to highlight the fact that fitness proportional selection can result in a number of problems. First, selective pressure can be too strong: too many duplicates are sometimes allocated to the best individual(s) in the population in the early generations of the search. Second, in the later stages of search, as the individuals in the population improve over time, there tends to be less variation in fitness, with more individuals (including the best individual(s)) being close to the population average. As the population average fitness increases, the fitness variance decreases and the corresponding uniformity in fitness values causes selective pressure to go down. When this happens, the search stagnates.

Because of these problems, few implementations of genetic algorithms use simple fitness proportionate reproduction. At the very least, the fitness may be rescaled before fitness proportionate reproduction. Fitness scaling mechanisms are discussed in detail in Goldberg's textbook *Genetic Algorithm in Search, Optimization and Machine Learning* (Goldberg 1989b). In recent years, the use of rank-based selection and in particular tournament selection has become common.

1.2   Rank-Assigned Tournament Selection

Selection can be based on fitness assigned according to rank, or relative fitness. This can be done explicitly. Assume the population is sorted by the evaluation function. A linear ranking mechanism with selective pressure Z (where $1 < Z \leq 2$) allocates a fitness bias of Z to the top-ranked individual, $2 - Z$ to the bottom-ranked individual, and a fitness bias of 1.0 to the median individual. Note that the difference in selective pressure between the best and the worst member of the population is constant and independent of how many generations have passed. This has the effect of making selective pressure more constant and controlled. Code for linear ranking is given by Whitley (1989).

The use of rank-based selection also transforms the search into what is known as an *ordinal optimization method*. The exact evaluation of sample points from the search space is no longer important. All that is important is the relative value (or relative fitness) of the strings representing sample points from the search space. Note that the fitness function can be an ordinal measure, but the evaluation function itself is not ordinal.

Ordinal optimization may also sometimes relax computation requirements (Ho et al. 1992; Ho 1994). It may be easier to determine $f(x_1) < f(x_2)$ than to exactly compute $f(x_1)$ and $f(x_2)$. An approximate evaluation or even noisy evaluation may still provide enough information to determine the relative ranking, or to determine the relative ranking with high probability.

This can be particularly important when the evaluation is a simulation rather than an exact mathematical objective function.

A fast and easy way to implement ranking is *tournament selection* (Goldberg 1990; Goldberg and Deb 1991). We have already seen a simple form of tournament selection which selects two strings at random and places the best in the intermediate population. In expectation, every string is sampled twice. The best string wins both tournaments and gets two copies in the intermediate population. The median string wins one and loses one and gets one copy in the intermediate population. The worst string loses both tournaments and does not reproduce. In expectation, this produces a linear ranking with a selective pressure of 2.0 toward the best individual. If the winner of the tournament is placed in the intermediate population with probability $0.5 < p < 1.0$, then the selective pressure is less than 2.0. If a tournament size larger than 2 is used and the winner is chosen deterministically, then selective pressure is greater than 2.0.

We can first generalize tournament selection to generate a linear selection bias less than 2, or a nonlinear bias greater than 2. For stronger selection, pick $t > 2$ individuals uniformly at random and select the best one of those individuals for reproduction. The process is repeated the number of times necessary to achieve the desired size of the intermediate population. The *tournament size*, t, has a direct correspondence with selective pressure because the expected number of times we sample the fittest individual in the intermediate population is t.

The original motivation behind the modern versions of tournament selection is that the algorithm is embarrassingly parallel because each tournament is independent (Suh and Gucht 1987). If one has the same number of processors as the population size, all N tournaments can be run simultaneously in parallel and constructing an intermediate population would require the same amount of time as executing a single tournament. Each processor samples the population t times and selects the best of those individuals.

Poli (2005) has noted that there are two factors that lead to the loss of diversity in regular tournament selection. Due to the randomness of tournaments, some individuals might not get *sampled* to participate in a tournament at all. Other individuals might be sampled but not be *selected* for the intermediate population because the individual loses the tournament.

One way to think of tournament selection (where $t = 2$) is a comparison of strings based on two vectors of random numbers.

```
vector A:  5 4 2 6 2 2 8 3
vector B:  5 3 6 5 4 3 4 6
```

The integer stored at $A(i)$ identifies a member of the population. We will assume that the lower integers correspond to better evaluations. During the ith tournament, $A(i)$ is compared to $B(i)$ to determine the winner of the tournament. Note that in this example, if the integers 1–8 represent the population members, then individuals 1 and 7 are not *sampled*. Poli points out that the number of individuals neglected in the first decision if two offsprings are produced by recombination is $P(1 - 1/P)^{TP}$ where P is the population size and T is the tournament size. Expressed another way, what Poli calculates is the expected number of population members that fail to appear in vector A or B if these were used to keep track of tournament selection during one generation (or the amount of time needed to sample P new points in the search space). It is also possible to under-sample. If the tournament size is 2, in expectation, all members of the population would be sampled twice if tournament selection were used in combination with a generational genetic algorithm. However, some members of the population might be sampled only once.

Other forms of selection do not have this problem. Universal stochastic sampling guarantees that *all* individuals with above-average fitness get to reproduce, and all individuals below average get a chance to reproduce with a probability proportional to their fitness. Note that fitness in this case could be based on the evaluation function or it could still be a rank-based assignment of space on the roulette wheel in ❯ *Fig. 2*.

Unbiased tournament selection (Sokolov and Whitley 2005) eliminates loss of diversity related to the failure to sample. Unbiased tournament selection also reduces the variance associated with how often an individual is sampled in one generation (variance reduction tournament selection might have been a more accurate name, but is somewhat more cumbersome). Unbiased tournament selection operates much like regular tournament selection except that permutations are used in place of random sampling during tournament construction. Rather than randomly sampling the population (or using random vectors of numbers to sample the population), t random permutations are generated. The ith element in each permutation indexes to a population member: the t population members pointed to by the ith element of each permutation form a tournament. An effective improvement is to use only $t-1$ permutations; the indices from 1 to P in sorted order can serve as one permutation (e.g., as generated in a for-loop). Other permutations can be constructed by sequentially sampling the population without replacement.

An example of unbiased tournament selection for tournament size $t = 3$ can be seen in ❯ *Fig. 3*. Assume that $f(x) = x$ and that lower numbers correspond to better individuals; then the last row, labeled "Winners," presents the resulting intermediate population as chosen by unbiased tournament selection. Note that permutation 1 is in sorted order. Recall that the selective pressure is partially controlled through the number of permutations.

Unbiased stochastic tournament selection can also be employed for selective pressure $S < 2$. We align two permutations, perform a pairwise comparison of elements, but select the best with $0.5 < p_s < 1.0$. For selective pressure less than or equal to 2 we also enforce the constraint that the ith element of the second permutation is not equal to i. This provides a guarantee that every individual will participate in a tournament twice. This is easily enforced: when generating the ith element of a permutation, we withhold the ith individual from consideration. There are not enough degrees of freedom to guarantee that this constraint holds for the last element of the permutation, but a violation can be fixed by swapping the last element with its immediate predecessor. The constraint is not enforced for $t>2$ as too much overhead is involved.

To illustrate the congruence between "formula-based" ranking and tournament selection, a simulation was performed to compare three methods of linear ranking: (1) by random sampling using an explicit mathematical formula, (2) using regular tournament selection, and (3) using unbiased tournament selection. The number of times an individual with each rank was selected was measured as the search progressed. ❯ *Figure 4* presents these measurements

◼ **Fig. 3**

Unbiased tournament selection with tournament size $t = 3$. In this case, each column represents a slot in the population; the best individual in each column wins the tournament for that slot.

```
Permutation 1: 1  2  3  4  5  6  7  8
Permutation 2: 6  2  1  5  8  4  3  7
Permutation 3: 2  8  1  4  3  6  7  5
      Winners: 1  2  1  4  3  4  3  5
```

◘ Fig. 4

The number of times an individual with a particular rank was selected for recombination. The results are presented for rank-based, regular, and unbiased tournament selection schemes. *Solid line* is the mean computed across 50 runs with 100 generations/run. *Dashed lines* lie one standard deviation away from the mean.

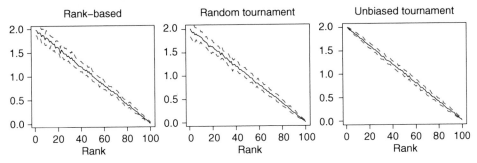

for a population of size 100 averaged over 50 runs with 100 generations each. A selective pressure of 2.0 was used. The solid line represented the mean and the two dashed lines were one standard deviation away from it. Notice that linear ranking behavior is problem independent because the only measure used in the calculations is the relative goodness of solutions.

As can be seen, formula-based ranking and regular tournament selection are essentially the same. On the other hand, unbiased tournament selection controls for variance by insuring the actual sample in each case is within one integer value of its expected value.

1.3 Different Forms of Representation, Crossover, and Mutation

One of the long-standing debates in the field of evolutionary algorithms involves the use of binary versus real-valued encodings for parameter optimization problems. The genetic algorithms community has largely emphasized bit representations. The main argument for bit encodings is that this representation decomposes the problem into the largest number of smallest possible building blocks and that a genetic algorithm works by processing these building blocks. This viewpoint, which was widely accepted 10 years ago, is now considered to be controversial. On the other hand, the evolution strategies community (Schwefel 1981, 1995; Bäck 1996) has emphasized the use of real-valued encodings for parameter optimization problems. Application-oriented researchers were also among the first in the genetic algorithms community to experiment with real-valued encodings (Davis 1991; Schaffer and Eshelman 1993).

Real-valued encodings can be recombined using simple crossover. The assignment of parameter values can be directly inherited from one parent or the other. But other forms of recombination are also possible, such as blending crossover that averages the real-valued parameter values. This idea might be generalized by considering the two parents as points p_1 and p_2 in a multidimensional space. These points might be combined as a weighted average (using some bias $0 \leq \alpha \leq 1$) such that an offspring might be produced by computing a new point p_0 in the following way:

$$p_0 = \alpha p_1 + (1 - \alpha)p_2$$

One can also compute a centroid associated with multiple parents.

Because real-valued encodings are commonly used in the evolution strategy community, the reader is encouraged to consult the chapter on evolution strategies. This chapter will focus on binary versus Gray coded bit representations.

1.3.1 Binary Versus Gray Representations

A related issue that has long been debated in the evolutionary algorithms community is the relative merit of Gray codes versus standard binary representations for parameter optimization problems. Generally, "Gray code" refers to the *standard binary reflected Gray code* (Bitner et al. 1976); but there are exponentially many possible Gray codes for a particular Hamming space. A Gray code is a bit encoding where adjacent integers are also Hamming distance 1 neighbors in Hamming space.

There are at least two basic ways to compute a Gray code. In general, a Gray matrix can be used that acts as a transform of a standard binary string. The standard binary reflected Gray code encoding matrix has 1s along the diagonal and 1s along the upper minor diagonal and 0s everywhere else. The matrix for decoding the standard binary reflected Gray code has 1s along the diagonal and all of the upper triangle of the matrix is composed of 1 bits. The lower triangle of the matrix is composed of 0 bits. The following is an example of the Graying matrix G (on the left) and the deGraying matrix D (on the right).

$$G = \begin{pmatrix} 1 & 1 & 0 & 0 \\ 0 & 1 & 1 & 0 \\ 0 & 0 & 1 & 1 \\ 0 & 0 & 0 & 1 \end{pmatrix} \qquad D = \begin{pmatrix} 1 & 1 & 1 & 1 \\ 0 & 1 & 1 & 1 \\ 0 & 0 & 1 & 1 \\ 0 & 0 & 0 & 1 \end{pmatrix}$$

Given a binary string (vector) b, a Graying matrix G, and a deGraying matrix D, bG produces a binary string (vector), which is the binary reflected Gray code of string b. Using the deGraying matrix, $(bG)D = b$.

A faster way to produce the standard binary reflected Gray code is to shift vector b by 1 bit, which will be denoted by the function $s(b)$. Using exclusive-or (denoted by $\oplus$) over the bits in b and $s(b)$,

$$b \oplus s(b) = bG$$

This produces a string with $L+1$ bits; the last bit is deleted. In implementation, no shift is necessary and one can have exclusive-or pairs of adjacent bits.

Assume that bit strings are assigned to the corner of a hypercube so that the strings assigned to the adjacent points in the hypercube differ by one bit. In general, a Gray code is a permutation of the corners of the hypercube, such that the resulting ordering forms a path that visits all of the points in the space and every point along the path differs from the point before and after it by a single bit flip.

There is no known closed form for computing the number of Gray codes for strings of length L. Given any Gray code (such as the standard binary reflected Gray code), one can (1) permute the order of the bits themselves to create $L!$ different Gray codes, and for each of these Gray codes one can (2) shift the integer assignment to the strings by using modular addition of a constant and still maintain a Gray code. This can be done in 2^L different ways. If all the Gray codes produced by these permutation and shift operations were unique, there would be at least $2^L(L!)$ Gray codes. Different Gray codes have different neighborhood structures. In practice, it is easy to use the trick of shifting the Gray code by a constant to explore different Gray codes with different neighborhoods. But one can also prove that the neighborhood structure repeats after a shift of $2^L/4$.

The "reflected" part of the standard binary reflected Gray code derives from the following observation. Assume that we have a sequence of 2^L strings and we wish to extend that sequence to 2^{L+1} strings. Let the sequence of 2^L strings be denoted by

$$a\ b\ c\ \dots\ x\ y\ z$$

Without loss of generality, assume the strings can be decoded to correspond to integers such that $a = 0, b = 1, c = 2, \dots, y = (2^L - 2), z = (2^L - 1)$. In standard binary representations, we extend the string by duplicating the sequence and appending a 0 to each string in the first half of the sequence and a 1 to each string in the second half of the sequence. Thus, if we start with a standard binary sequence of strings, then the following is also a standard binary encoding:

$$0a\ 0b\ 0c\ \dots 0x\ 0y\ 0z\ 1a\ 1b\ 1c\ \dots 1x\ 1y\ 1z$$

However, in a reflected Gray code, we extend the string by duplicating and reversing the sequence; then a 0 is appended to all the strings in the original sequence and a 1 appended to the string in the reflected or reversed sequence.

$$0a\ 0b\ 0c\ \dots 0x\ 0y\ 0z\ 1z\ 1y\ 1x\ \dots 1c\ 1b\ 1a$$

It is easy to verify that if the original sequence is a Gray code, then the reflected expansion is also a Gray code.

Over all possible discrete functions that can be mapped onto bit strings, the space of any and all Gray code representations and the space of binary representations must be identical. This is another example of what has come to be known as a "no-free-lunch" result (Wolpert and Macready 1995; Radcliffe and Surry 1995; Whitley and Rowe 2008). The empirical evidence suggests, however, that Gray codes are often superior to binary encodings. It has long been known that Gray codes remove Hamming cliffs, where adjacent integers are represented by complementary bit strings: for example, 7 and 8 encoded as 0111 and 1000. Whitley et al. (1996) first made the rather simple observation that every Gray code must preserve the connectivity of the original real-valued functions and that this impacts the number of local optima that exists under Gray and binary representation. This is illustrated in ❯ *Fig. 5*.

A consequence of the connectivity of the Gray code representation is that for every 1-dimensional parameter optimization problem, the number of optima in the Gray coded space must be less than or equal to the number of optima in the original real-valued function. Binary encodings offer no such guarantee. Binary encodings destroy half of the connectivity of the original real-valued function; thus, given a large basin of attraction with a globally competitive local optimum, many of the points near the optimum that are not optima become new local optima under a binary representation. The theoretical and empirical evidence suggests (Whitley 1999) that for parameter optimization problems with a bounded number of optima, Gray codes are better than binary in the sense that Gray codes induce fewer optima.

As for the debate over whether Gray bit encodings are better or worse than real-coded representations, the evidence is very unclear. If high precision is required, real-valued encodings are usually better. In other cases, a lower precision Gray code outperforms a real-valued encoding. This also means that an accurate comparison is difficult. The same parameter optimization problem encoded with high precision real-valued representation versus a low

☐ **Fig. 5**

Adjacency in 4-bit Hamming space for Gray and standard binary encodings. The binary representation destroys half of the connectivity of the original function.

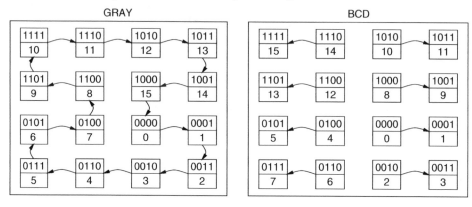

precision bit encoding results in two different search spaces. There are no clear theoretical or empirical answers to this question.

1.3.2 *N*-Point and Uniform Recombination

Early mathematical models of genetic algorithms often employed 1-point crossover in conjunction with binary strings. This bias again seems to be connected to the fact that 1-point crossover is easy to model mathematically. It is easy to see that 2-point crossover has advantages over 1-point crossover. In 1-point crossover, bits at opposite ends of the encoding are virtually always separated by recombination. Thus, there is a bias against inheriting bits from the same parents if they are at opposite ends of the encoding. However, with 2-point crossover, there is no longer any bias with regard to the end of the encodings. In 1-point crossover, the "end" of the encoding is a kind of explicit crossover point. 2-Point crossover treats the encoding as if it were circular so that the choice of both of the two crossover points are randomized.

Spears and De Jong (1991) argue for a variable *N*-point recombination. By controlling the number of recombination points, one can also control the likelihood of the two bits (or real-values) which are close together on one parent string and which are likely to be inherited together. With a small number of crossover points, recombination is less disruptive in terms of assorting which bits are inherited from which parents. With a large number of crossover points, recombination is more disruptive.

Is disruptive crossover good or bad? The early literature on genetic algorithms generally argued disruption was bad. On the other hand, disruption is also a source of new exploration. The most disruptive form of crossover is *uniform crossover*. Syswerda (1989) was one of the first application-oriented researchers to champion the use of uniform crossover. When uniform crossover is applied to a binary encoding, every bit is inherited randomly from one of the two parents. This means that uniform crossover on bit strings can actually be viewed as a blend of crossover and mutation. Any bits that are shared by the two parents must be passed along to offspring, because no matter which parent the bit comes from, the bit is the same. But if the two parents differ in a particular bit position, then the bit that is inherited is randomly chosen. Another way of looking at uniform crossover is that all bit assignments that are shared by the two parents are automatically inherited. All bits that do not share the same value are set to a random value, because the inheritance is random. In some sense, the bits that are not set to the same value in the parents are determined by mutation: each of their values has an equal probability of being either one or zero.

Does it really matter if bits from a single parent that are close together on the encoding are inherited together or not? The early dogma of genetic algorithms held that inheritance of bits from a single parent that are close to each other on the chromosome was very important; very disruptive recombination, and uniform crossover in particular, should be bad. This in some sense was the central dogma at the time. This bias had a biological source.

This idea was borrowed in part from the biological concept of *coadapted alleles*. A widely held tenant of biological evolution is that distinct fragments of genetic information, or alleles, that work well *together* to enhance survivability of an organism should be inherited together. This view had a very strong influence on Holland's theory regarding the computational power of genetic algorithms: a theory based on the concept of hyperplane sampling.

1.4　Schemata and Hyperplanes

In his 1975 book, *Adaptation in Natural and Artificial Systems* (Holland 1975), Holland develops the concepts of schemata and hyperplane sampling to explain how a genetic algorithm can yield a robust search by implicitly sampling subpartitions of a search space. The idea that genetic algorithms derive their primary search power by hyperplane sampling is now controversial.

The set of strings of length L over a binary alphabet correspond to the vertices of an L-dimensional hypercube. A *hyperplane* is simply a subset of such vertices. These are defined in terms of the bit values they share in common. The concept of schemata is used to describe hyperplanes containing strings that contain particular shared bit patterns. Bits that are shared are represented in the schema; bits that are not necessarily shared are replaced by the * symbol. We will say that a bit string *matches* a particular schema if that bit string can be constructed from the schema by replacing a * symbol with the appropriate bit value. Thus, a 10-bit schema such as 1********* defines a subset that contains half the points in the search space, namely, all the strings that begin with a 1 bit in the search space. All bit strings that match a particular schema are contained in the hyperplane partition represented by that particular schema. The string of all * symbols corresponds to the space itself and is not counted as a partition of the space. There are $3^L - 1$ possible schemata since there are L positions in the bit string and each position can be a 0,1, or * symbol and we do not count the string with no zeros or ones. A *low order* hyperplane is represented by a schema that has few bits, but many * symbols. The number of points contained in a hyperplane is 2^k where k is the number of * symbols in the corresponding schema.

The notion of a population-based search is critical to the schema-based theory of the search power of genetic algorithms. A population of sample points provides information about numerous hyperplanes; furthermore, low order hyperplanes should be sampled by numerous points in the population. Holland introduced the concept of *intrinsic* or *implicit parallelism* to describe a situation where many hyperplanes are sampled when a population of strings is evaluated; it has been argued that far more hyperplanes are sampled than the number of strings contained in the population.

Holland's theory suggests that schemata representing competing hyperplanes increase or decrease their representation in the population according to the relative fitness of the strings that lie in those hyperplane partitions. By doing this, more trials are allocated to regions of the search space that have been shown to contain above average solutions.

1.5 An Illustration of Hyperplane Sampling

Holland (1975) suggested the following view of hyperplane sampling. In ❷ *Fig.* 6, a function over a single variable is plotted as a one-dimensional space. The function is to be maximized. Assume the encoding uses 8 bits. The hyperplane 0******* spans the first half of the space and 1******* spans the second half of the space. Since the strings in the 0******* partition are on average better than those in the 1******* partition, we would like the search to be proportionally biased toward this partition. In the middle graph of ❷ *Fig.* 6, the portion of the space corresponding to **1***** is shaded, which also highlights the intersection of 0******* and **1*****, namely, 0*1****. Finally, in the bottom graph, 0*10***** is highlighted.

One of the assumptions behind the illustration in ❷ *Fig.* 6 is that the sampling of hyperplane partitions is not affected to a significant degree by local minima. At the same time, increasing the sampling rate of regions of the search space (as represented by hyperplanes) that are above average compared to other competing regions does not guarantee convergence to a global optimum. The global optimum could be a relatively isolated peak that might never be sampled, for example. A small randomly placed peak that is not sampled is basically invisible; all search algorithms are equally blind to such peaks if there is no information to guide search.

Nevertheless, good solutions that are globally competitive might be found by such a strategy. This is particularly true if the search space is structured in such a way that pieces of good solutions can be recombined to find better solutions. The notion that hyperplane sampling is a useful way to guide search should be viewed as heuristic. In general, even having perfect knowledge of schema averages up to some fixed order provides little guarantee as to the quality of the resulting search. This is discussed in detail in ❷ Sect. 2.

1.6 The Schema Theorem

Holland (1975) developed the *schema theorem* to provide a lower bound on the change in the sampling rate for a single hyperplane from generation t to generation $t+1$. By developing the theorem as a lower bound, Holland was able to make the schema theorem hold independently for every schema/hyperplane. At the same time, as a lower bound, the schema theorem is inexact, and the bounds hold for only one generation into the future. After one generation, the bounds are no longer guaranteed to hold. This weakness is just one of the many reasons that the concept of "hyperplane sampling" is controversial.

Let $P(H, t)$ be the proportion of the population that samples a hyperplane H at time t. Let $P(H, t + \text{intermediate})$ be the proportion of the population that samples hyperplane H after fitness proportionate selection but before crossover or mutation. Let $f(H, t)$ be the average fitness of the strings sampling hyperplane H at time t and denote the population average by $\bar{f}$. Note that $\bar{f}$ should also have a time index, but this is often not denoted explicitly. This is

◘ Fig. 6

A function and various partitions of hyperspace. Fitness is scaled to a 0 to 1 range in this diagram.

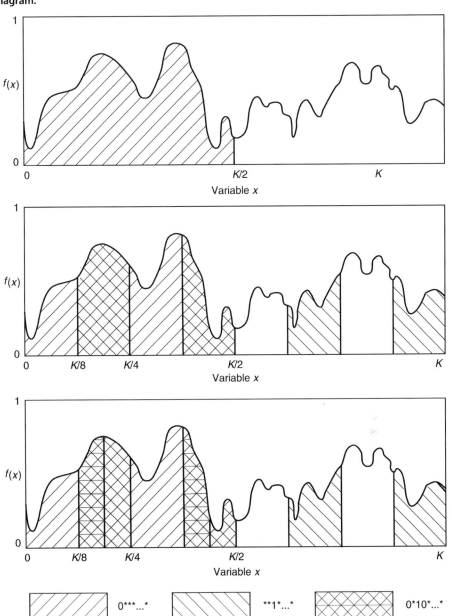

important because the average fitness of the population is *not* constant. Assuming that selection is carried out using fitness proportional selection:

$$P(H, t + \text{intermediate}) = P(H, t)\frac{f(H, t)}{\bar{f}}.$$

Thus, ignoring crossover and mutation, under just selection, the sampling rate of hyperplanes changes according to their average fitness. Put another way, selection "focuses" the search in what appears to be promising regions where the strings sampled so far have above-average fitness compared to the remainder of the search space. Some of the controversy related to "hyperplane sampling" begins immediately with this characterization of selection. The equation accurately describes the focusing effects of selection; the concern, however, is that the focusing effect of selection is not limited to the $3^L - 1$ hyperplanes that Holland considered to be relevant. Selection acts exactly the same way on any arbitrarily chosen subset of the search space. Thus, it acts in exactly the same way on the $2^{(2^L)}$ members of the power set over the set of all strings. While there appears to be nothing special about the sampling rate of hyperplanes under selection, all subsets of strings are not acted on in the same way by crossover and mutation. Some subsets of bit patterns corresponding to schemata are more likely to survive and be inherited in the population under crossover and mutation. For example, the sampling rate of order 1 hyperplanes is not disrupted by crossover and, in general, lower order hyperplanes are less affected by crossover than higher order hyperplanes.

Laying this issue aside for a moment, it is possible to write an exact version of the schema theorem that considers selection, crossover, and mutation. What we want to compute is $P(H, t+1)$, the proportion of the population that samples hyperplane H at the next generation as indexed by $t+1$. First just consider selection and crossover.

$$P(H, t+1) = (1 - p_c)P(H, t)\frac{f(H, t)}{\bar{f}} + p_c\left[P(H, t)\frac{f(H, t)}{\bar{f}}(1 - \text{losses}) + \text{gains}\right]$$

where p_c is the probability of performing a crossover operation. When crossover does not occur (which happens with probability $(1 - p_c)$), only selection changes the sampling rate. However, when crossover does occur (with probability p_c) then we have to consider how crossover and mutation can destroy hyperplane samples (denoted by losses) and how crossover can create new samples of hyperplanes (denoted by gains).

For example, assume we are interested in the schema 11*****. If a string such as 1110101 were recombined between the first two bits with a string such as 1000000 or 0100000, no disruption would occur in hyperplane 11***** since one of the offspring would still reside in this partition. Also, if 1000000 and 0100000 were recombined exactly between the first and second bit, a new independent offspring would sample 11*****; this is the source of *gains* that is referred to in the above calculation.

We will return to an exact computation, but, for now, instead of computing *losses* and *gains*, what if we compute a bound on them instead? To simplify things, *gains* are ignored and the conservative assumption is made that crossover falling in the significant portion of a schema always leads to disruption. Thus, we now have a bound on the sampling rate of schemata rather than an exact characterization:

$$P(H, t+1) \geq (1 - p_c)P(H, t)\frac{f(H, t)}{\bar{f}} + p_c\left[P(H, t)\frac{f(H, t)}{\bar{f}}(1 - \text{disruptions})\right].$$

The *defining length* of a schema is based on the distance between the first and last bits in the schema with value either 0 or 1 (i.e., not a * symbol). Given that each position in a schema can be 0, 1, or *, scanning left to right, if I_x is the index of the position of the rightmost occurrence of either a 0 or a 1 and I_y is the index of the leftmost occurrence of either a 0 or a 1, then the defining length is merely $I_x - I_y$. The defining length of a schema representing a hyperplane H

is denoted here by $\Delta(H)$. If 1-point is used, then the defining length can be used to also calculate a bound on disruption:

$$\frac{\Delta(H)}{L-1}(1 - P(H,t))$$

and including this term (and applying simple algebra) yields:

$$P(H,t+1) \geq P(H,t)\frac{f(H,t)}{\bar{f}}\left[1 - p_c\frac{\Delta(H)}{L-1}(1 - P(H,t))\right]$$

We now have a useful version of the schema theorem (although it does not yet consider mutation). This version assumes that selection for the first parent string is fitness-based and the second parent is chosen randomly. But typically both parents are chosen based on fitness. This can be added to the schema theorem by merely indicating the alternative parent chosen from the intermediate population after selection (Schaffer 1987).

$$P(H,t+1) \geq P(H,t)\frac{f(H,t)}{\bar{f}}\left[1 - p_c\frac{\Delta(H)}{L-1}\left(1 - P(H,t)\frac{f(H,t)}{\bar{f}}\right)\right]$$

Finally, mutation is included. Let $o(H)$ be a function that returns the order of the hyperplane H. The order of H exactly corresponds to a count of the number of bits in the schema representing H that have value 0 or 1. Let the mutation probability be p_m where mutation always flips the bit. Thus, the probability that mutation does not affect the schema representing H is $(1 - p_m)^{o(H)}$. This leads to the following expression of the schema theorem.

$$P(H,t+1) \geq P(H,t)\frac{f(H,t)}{\bar{f}}\left[1 - p_c\frac{\Delta(H)}{L-1}\left(1 - P(H,t)\frac{f(H,t)}{\bar{f}}\right)\right](1 - p_m)^{o(H)}$$

2   Interpretations and Criticisms of the Schema Theorem

For many years, the schema theorem was central to the theory of how genetic algorithms are able to effectively find good solutions in complex search spaces. Groups of bits that are close together on the encoding are less likely to be disrupted by crossover. Therefore, if groups of bits that are close together define a (hyperplane) subregion of the subspace that contains good solutions, selection should increase the representation of these bits in the population, and crossover and mutation should not "interfere" with selection since the probability of disruption should be low. In effect, such groups of bits act as coadapted sets of alleles; these are so important that they have been termed the "building blocks" of genetic search. As different complexes of coadaptive alleles (or bits) emerge in a population, these building blocks are put together by recombination to create even better individuals. The most aggressive interpretation of the schema theorem is that a genetic algorithm would allocate nearly optimal trials to sample different partitions of the search space in order to achieve a near-optimal global search strategy.

There are many different criticisms of the schema theorem. The schema theorem is not incorrect, but arguments have been made that go beyond what is actually proven by the schema theorem. First of all, the schema theorem is an inequality, and it only applies to one generation into the future. So while the bound provided by the schema theorem absolutely holds for one generation into the future, it provides no guarantees about how strings or hyperplanes will be sampled in future generations.

It is true that the schema theorem does hold true independently for all possible hyperplanes for one generation. However, over multiple generations, the interactions between different subpartitions of the search space (as represented by hyperplanes and schemata) are extremely important. For example, in some search space of size 2^8 suppose that the schemata 11^{******} and $^*00^{*****}$ are both "above average" in the current generation of a population in some run of a genetic algorithm. Assume the schema theorem indicates that both will have increasing representation in the next generation. But trials allocated to schemata 11^{******} and $^*00^{*****}$ are in conflict because they disagree about the value of the second bit. Over multiple generations, both regions cannot receive increasing trials. These schema are *inconsistent* about what bit value is to be preferred in the second position. The schema theorem does not predict how such inconsistencies will be sorted out.

Whitley et al. (1995b) and Heckendorn et al. (1996) have shown that problems can have varying degrees of consistency in terms of which hyperplanes appear to be promising. For problems that display higher *consistency*, the "most fit" schemata tend to agree about what the values of particular bits should be. In a problem where there is a great deal of consistency, genetic search is usually effective. But other problems can be highly inconsistent, so that the most fit individuals (and sets of individuals as represented by schemata) display a large degree of conflict in terms of what bit values are preferred in different positions. It seems reasonable to assume that a genetic algorithm should do better on problems that display greater consistency, since inconsistency means that the search is being guided by conflicting information (this is very much related to the notion of "deception" but the concept of deception is controversial and much misunderstood).

One criticism of pragmatic significance is that users of the standard or canonical genetic algorithm often use very small populations. The number of positions containing 0 or 1 is referred to as the order of a schema. Thus, $^{**}1^{*****}$ is an order 1 schema, $^{***}0^{***}1$ is an order 2 schema, and $^*1^{**}0^*1^*$ is an order 3 schema. Many users employ a population size of 100 or smaller. In a population of size 100, we would expect 50 samples of any order 1 schema, 25 samples of any order 2 schema, 12.5 samples of any order 3 schema, and exponentially decaying numbers of samples to higher order schema. Thus, if we suppose that the genetic algorithm is implicitly attempting to allocate trials to different regions of the search space based on schema averages, a small population (e.g., 100) is inadequate unless we only care about very low order schemata. Therefore, even if hyperplane sampling is a robust form of heuristic search, the user destroys this potential by using small population sizes. Small populations require that the search rely more on hill-climbing. But in some cases, hill-climbing is very productive.

What if we had perfect schema information? What if we could compute schema information exactly in polynomial time? Rana et al. (1998) have shown that schema information up to any fixed order can be computed in polynomial time for some nondeterministic polynomial time (NP)-Complete problems. This includes Maximum Satisfiability (MAXSAT) problems and NK-Landscapes. This is very surprising. One theoretical consequence of this is captured by the following theorem:

▶ If $P \neq NP$ then, in the general case, exact knowledge of the static schema fitness averages up to some fixed order cannot provide information that can be used to guarantee finding a global optimum, or even an above average solution, in polynomial time. (For proofs, see Heckendorn et al. 1999a,b).

This seems like a very negative result. But it is dangerous to overinterpret either positive or negative results. In practice, random MAXSAT problems are characterized by highly inconsistent schema information, so there is really little or no information that can be exploited to guide the search (Heckendorn et al. 1999b). And in practice, genetic algorithms perform very poorly on MAXSAT problems (Rana et al. 1998) unless they are aided by additional hill-climbing techniques. On the other hand, genetic algorithms are known to work well in many other domains. Again, the notion of using schema information to guide search is heuristic.

There are many other criticisms of the schema theorem. In the early literature, too much was claimed about schema and hyperplane processing that was not backed up by solid proofs. It is no longer accepted that genetic algorithms allocate trials in an "optimal way" and it is certainly not the case that the genetic algorithm is guaranteed to yield optimal or even near-optimal solutions. In fact, there are good counterexamples to these claims. On the other hand, some researchers have attacked the entire notion of schema processing as invalid or false. Yet, the schema theorem itself is clearly a valid bound; and, experimentally, in problems where there are clearly defined regions that are above average, the genetic algorithm does quickly allocate more trials to such regions as long as these regions are relatively large.

There is still a great deal of work to be done to understand the role that hyperplane sampling plays in genetic search. Historically, the role of hyperplane sampling has been exaggerated, and the role played by hill-climbing has been underestimated. At the same time, many of the empirical results that call into question how genetic algorithms really work have been carried out on a highly biased sample of test problems. These test problems tend to be *separable*. A separable optimization problem is one in which the parameters are independent in terms of interaction. These problems are inherently easy to solve. For example, using a fixed precision (e.g., 32 bits per parameter), a separable problem can be solved exactly in a time that is a polynomial in the number of parameters by searching each parameter separately. On such problems, hill-climbing is a highly effective search strategy. Perhaps because of these simple test problems, the effectiveness (and speed) of simple hill-climbing has been overestimated.

In the next section, exact models of Holland's simple genetic algorithm will be introduced.

3 Infinite Population Models of Simple Genetic Algorithms

Goldberg (1987, 1989b) and Bridges and Goldberg (1987) were the first to model critical details of how the genetic algorithm processes infinitely large populations under recombination. These models were independently derived in 1990 by Vose and Whitley in a more precise and generalized form. Vose and Liepins (1991) and Vose (1993) extended and generalized this model while Whitley et al. (1992) and Whitley (1993) introduced another version of the infinite population model that connects the work of Goldberg and Vose. In this section, the Vose model is reviewed and it is shown how the effects of various operators fit into this model.

The models presented here all use fitness proportionate reproduction because it is simpler to model mathematically. Vose (1999) also presents models for rank-based selection.

The vector $p^t \in \mathbb{R}$ is such that the kth component of the vector is equal to the proportional representation of string k at generation t. It is assumed that $n = 2^L$ is the number of points in the search space defined over strings of length L and that the vector p is indexed 0 to $n-1$. The vector $s^t \in \mathbb{R}$ represents the tth generation of the genetic algorithm after selection and the ith component of s^t is the proportional representation of string i in the population after selection, but before any operators (e.g., recombination, mutation, and local search) are applied. Likewise, p_i^t represents the proportional representation of string i at generation t *before* selection occurs.

The function $r_{i,j}(k)$ yields the probability that string k results from the recombination of strings i and j. (For now, assume that r only yields the results for recombination; the effect of other operators could also be included in r.) Now, using $\mathcal{E}$ to denote expectation,

$$\mathcal{E}\{p_k^{t+1}\} = \sum_{i,j} s_i^t \, s_j^t \, r_{i,j}(k) \tag{1}$$

To begin the construction of a general model, we first consider how to calculate the proportional representation of string 0 (i.e., the string composed of all zeros) at generation $t+1$; in other words, we compute p_0^{t+1}. A mixing matrix M is constructed where the (i,j)th entry $m_{i,j} = r_{i,j}(0)$. Here M is built by assuming that each recombination generates a single offspring. The calculation of the change in representation for string $k = 0$ is now given by

$$\mathcal{E}\{p_0^{t+1}\} = \sum_{i,j} s_i^t s_j^t r_{i,j}(0) = s^T M s \tag{2}$$

where T denotes transpose. Note that this computation gives the expected representation of a single string, 0, in the next genetic population.

It is simple to see that the exact model for the infinite population genetic algorithm has the same structure as the model on which the schema theorem is based. For example, if we accept that the string of all zeros (denoted here simply by the integer 0) is a special case of a hyperplane, then when $H = 0$, the following equivalence holds:

$$s^T M s = P(H, t+1) = (1 - p_c) P(H, t) \frac{f(H, t)}{\bar{f}}$$
$$+ p_c \left[P(H, t) \frac{f(H, t)}{\bar{f}} (1 - \text{losses}) + \text{gains} \right] \tag{3}$$

The information about losses and gains is contained in the matrix M. To keep matters simple, assume the probability of crossover is $p_c = 1$. Then, in the first row and column of matrix M, the calculation is really the probability of *retaining* a copy of the string, which is $(1 - \text{losses})$ (except at the intersection of the first row and column); the probabilities elsewhere in M are the gains in the above equation. To include the probability of crossover p_c in the model one must include the crossover probability in the construction of the M matrix.

The point of ❷ Eq. 3 is that the exact Vose/Liepins model and the model on which the schema theorem is based is really the same. Whitley (1993) presents the calculations of losses and gains so as to make the congruent aspects of the infinite population model and the schema theorem more obvious.

The equations that have been looked at so far tell us how to compute the expected representation of one point in the search space one generation into the future. The key is to generalize this calculation to all points in the search space. Vose and Liepins formalized the notion that bitwise exclusive-or can be used to access various probabilities from the recombination function r. Specifically,

$$r_{i,j}(k) = r_{i,j}(k \oplus 0) = r_{i \oplus k, j \oplus k}(0). \tag{4}$$

This implies that the mixing matrix M, which was defined such that entry $m_{i,j} = r_{i,j}(0)$, can provide mixing information for any string k just by changing how M is accessed. By reorganizing the components of the vector, s, the mixing matrix M can yield information about the probability $r_{i,j}(k)$. A permutation function, σ, is defined as follows:

$$\sigma_j \langle s_0, \ldots, s_{n-1} \rangle^T = \langle s_{j \oplus 0}, \ldots, s_{j \oplus (n-1)} \rangle^T \tag{5}$$

where the vectors are treated as columns and n is the size of the search space. The computation

$$(\sigma_q \, s^t)^T M (\sigma_q \, s^t) = p_q^{t+1} \tag{6}$$

thus reorganizes s with respect to string q and produces the expected representation of string q at generation $t+1$. A general operator $\mathcal{M}$ can now be defined over s, which remaps $s^T M s$ to cover all strings in the search space.

$$\mathcal{M}(s) = \langle (\sigma_0 \, s)^T M (\sigma_0 \, s), \ldots, (\sigma_{n-1} \, s)^T M (\sigma_{n-1} \, s) \rangle^T \tag{7}$$

This model has not yet addressed how to generate the vector s^t given p^t. A fitness matrix F is defined such that fitness information is stored along the diagonal; the (i,i)th element is given by $f(i)$ where f is the fitness function. Following Vose and Wright (1997),

$$s^t = F p^t / 1^T F p^t \tag{8}$$

since $F p^t = \langle f_0 p_0^t, f_1 p_1^t, \ldots, f_{n-1}, p_{n-1}^t \rangle$ and the population average is given by $1^T F p^t$.

Vose (1999) refers to this complete model as the $\mathcal{G}$ function. Given any population distribution p, $\mathcal{G}(p)$ can be interpreted in two ways. On one hand, if the population is infinitely large, then $\mathcal{G}(p)$ is the exact distribution of the next population. On the other hand, given any *finite* population p, if strings in the next population are chosen one at a time, then $\mathcal{G}(p)$ also defines a vector such that element i is chosen for the next generation with probability $\mathcal{G}(p)_i$. This is because $\mathcal{G}(p)$ defines the exact sampling distribution for the next generation. This is very useful when constructing finite Markov models of the simple genetic algorithm.

Next, we look at how mutation can be added to this model in a modular fashion. This could be built directly into the construction of the M matrix. But looking at mutation as an additional process is instructive.

Recall that M is the mixing matrix, which we initially defined to cover only crossover. Define Q as the mutation matrix. Assuming mutation is independently applied to each bit with the same probability, Q can be constructed by defining a mutation vector Γ such that component Γ_i is the probability of mutating string i and producing string 0. The vector Γ is the first column of the mutation matrix and in general Γ can be reordered to yield column j of the matrix by reordering such that element $q_{i,j} = \Gamma_{i \oplus j}$.

Having defined a mutation matrix, mutation can now be applied after recombination in the following fashion:

$$p^{t+1,m} = (p^{t+1})^T Q,$$

where $p^{t+1,m}$ is just the p vector at time $t+1$ after mutation has occurred. Mutation also can be done before crossover; the effect of mutation on the vector s immediately after selection produces the following change: $s^T Q$, or equivalently, $Q^T s$.

Now we drop the $p^{t+1,m}$ notation and assume that the original p^{t+1} vector is defined to include mutation, such that

$$p_0{}^{t+1} = (Q^T s)^T M(Q^T s)$$
$$p_0{}^{t+1} = s^T (QMQ^T) s$$

and we can therefore define a new matrix M_2 such that

$$p_0{}^{t+1} = s^T M_2 s \quad \text{where} \quad M_2 = (QMQ^T)$$

As long as the mutation rate is independently applied to each bit in the string, it makes no difference whether mutation is applied before or after recombination. Also, this view of mutation makes it clear how the general mixing matrix can be built by combining matrices for mutation and crossover.

4   The Markov Model for Finite Populations

Nix and Vose (1992) show how to structure the finite population for a simple genetic algorithm. Briefly, the Markov model is an $N \times N$ transition matrix Q, where N is the number of finite populations of K strings and $Q_{i,j}$ is the probability that the kth generation will be population $\mathcal{P}_j$, given that the $(k-1)$th population is $\mathcal{P}_i$.

Let

$$\langle Z_{0,j}, Z_{1,j}, Z_{2,j}, \ldots, Z_{r-1,j} \rangle$$

represent a population, where $Z_{x,j}$ represents the number of copies of string x in population j, and $r=2^L$. The population is built incrementally. The number of ways to place the $Z_{0,j}$ copies of string 0 in the population is:

$$\binom{K}{Z_{0,j}}$$

The number of ways $Z_{1,j}$ strings can be placed in the population is:

$$\binom{K - Z_{0,j}}{Z_{1,j}}$$

Continuing for all strings,

$$\binom{K}{Z_{0,j}} \binom{K - Z_{0,j}}{Z_{1,j}} \binom{K - Z_{0,j} - Z_{1,j}}{Z_{2,j}} \ldots \binom{K - Z_{0,j} - Z_{1,j} - \cdots - Z_{r-2,j}}{Z_{r-1,j}}$$

which yields

$$\frac{K!}{(K - Z_{0,j})! Z_{0,j}!} \frac{(K - Z_{0,j})!}{(K - Z_{0,j} - Z_{1,j})! Z_{1,j}!} \frac{(K - Z_{0,j} - Z_{1,j})!}{(K - Z_{0,j} - Z_{1,j} - Z_{2,j})! Z_{2,j}!}$$
$$\ldots \frac{(K - Z_{0,j} - Z_{1,j} - \cdots - Z_{r-2,j})!}{Z_{r-1,j}!}$$

which in turn reduces to

$$\frac{K!}{Z_{0,j}! Z_{1,j}! Z_{2,j}! \ldots Z_{r-1,j}!}$$

Let $C_i(y)$ be the probability of generating string y from the finite population P_i. Then

$$Q_{i,j} = \frac{K!}{Z_{0,j}! Z_{1,j}! Z_{2,j}! \ldots Z_{r-1,j}!} \prod_{y=0}^{r-1} C_i(y)^{Z_{y,j}}$$

and

$$Q_{i,j} = K! \prod_{y=0}^{r-1} \frac{C_i(y)^{Z_{y,j}}}{Z_{y,j}!}$$

So, how do we compute $C_i(y)$? Note that the finite population P_i can be described by a vector p. Also note that the sampling distribution from which P_j is constructed is given by the infinite population model $\mathcal{G}(p)$ (Vose 1999). Thus, replacing $C_i(y)$ by $\mathcal{G}(p)_y$ yields

$$Q_{i,j} = K! \prod_{y=0}^{r-1} \frac{(\mathcal{G}(p)_y)^{Z_{y,j}}}{Z_{y,j}!}$$

5 Theory Versus Practice

As the use of evolutionary algorithms became more widespread, a number of alternative genetic algorithm implementations have also come into common use. Some of the evolutionary algorithms in common use are based on evolution strategies. Select algorithms closest to traditional genetic algorithms will be reviewed.

The widespread use of alternative forms of genetic algorithms also means there is a fundamental tension between (at least some part of) the theory community and the application community. There are beautiful mathematical models and some quite interesting results for Holland's original genetic algorithm. But there are few results for the alternative forms of genetic algorithms that are used by many practitioners.

5.1 Steady-State and Island Model Genetic Algorithms

Genitor (Whitley and Kauth 1988; Whitley 1989) was the first of what was later termed "steady-state" genetic algorithms by Syswerda (1989). The name "steady-state" is somewhat unfortunate and the term "monotonic" genetic algorithm has been suggested by Alden Wright: these algorithms keep the best solutions found so far, and thus the population average monotonically improves over time. The distinction between steady-state genetic algorithms and regular generational genetic algorithms was also foreshadowed by the evolution strategy community. The Genitor algorithm, for example, can also be seen as a variant of a $(\mu+1)$-Evolution strategy in terms of its selection mechanism. In contrast, Holland's generational genetic algorithm is an example of a (μ,λ)-Evolution Strategy where $\mu = \lambda$. The genetic algorithms community also used "monotonic" selection for classifier systems in the early 1980s.

Reproduction occurs one individual at a time in the Genitor algorithm. Two parents are selected for reproduction and produce an offspring that is immediately placed back into the population. The worst individual in the population is deleted. Ignoring the worst member of the population, the remainder of the population monotonically improves.

Another major difference between Genitor and other forms of genetic algorithms is that fitness is assigned according to rank rather than by fitness proportionate reproduction. In the original Genitor algorithm, the population is maintained in a sorted data structure. Fitness is assigned according to the position of the individual in the sorted population. This also allows one to prevent duplicates from being introduced into the population. This selection schema also means that the best $N-1$ solutions are always preserved in a population of size N. Goldberg and Deb (1991) have shown that by replacing the worst member of the population, Genitor generates much higher selective pressure than the canonical genetic algorithm.

Steady-state genetic algorithms retain the flavor of a traditional genetic algorithm. But the differences are significant. Besides the additional selective pressure, keeping the best strings seen so far means that the resulting search is more focused. This can be good or bad. The resulting search has a strong hill-climbing flavor, and the algorithm will continue to perturb the (best) strings in the population by crossover and mutation looking for an improved solution. If a population contains adequate building blocks to reach a solution in reasonable time, this focus can pay off. However, if the population does not contain adequate building blocks to generate further improvements, the search will be stuck. A traditional genetic algorithm or a (μ, λ)-evolution strategy allows a certain degree of drift and has the ability to continue moving in the space of possible populations. Thus, the greedy focus of a steady-state genetic algorithm sometimes pays off, and sometimes it does not.

To combat stagnation, it is sometimes necessary to use larger populations with steady-state genetic algorithms, or to use more aggressive mutation. Generally, larger populations result in slower progress, but improved solutions in the long run. Another way to combat stagnation is to use an *island model genetic algorithm*. Steady-state genetic algorithms seem to work better in conjunction with the island model paradigm than generational genetic algorithms. This may be due to the fact that steady-state genetic algorithms are more prone to stagnation.

Instead of using one large population, the population can be broken into several smaller populations. Thus, instead of running one population of size 1,000, one might run five populations of size 200. While the smaller population will tend to converge faster, diversity can be maintained by allowing migration between the subpopulations. For example, a small number of individuals (e.g., 1–5) might migrate from one subpopulation to another every 5–10 generations. It is important that migration be limited in size and frequency, otherwise the set of subpopulation becomes homogeneous too quickly (Starkweather et al. 1990).

In practice, steady-state genetic algorithms such as Genitor are often better optimizers than the canonical generational genetic algorithm. This has especially been true for scheduling problems such as the traveling salesman problem.

Current implementations of steady-state genetic algorithms are likely to use tournament selection instead of explicit rank-based selection. The use of tournament selection makes it unnecessary to keep the population in sorted order for the purposes of selection. But the population must still be sorted to keep track of the worst member of the population over time. Of course, once the population is sorted, inserting a new offspring takes $O(N)$ time, where N is the population size. In practice, poor offspring that are not inserted incur no insertion cost, and by starting from the bottom of the population, the insertion cost is proportional to the fitness of the offspring.

Some researchers have experimented with other ways of deciding which individuals should be replaced after new offspring are generated. Tournament selection can be used in reverse to decide if a new offspring should be allowed into the population and which member of the current population should be replaced. When tournament selection is used in reverse, the tournament sizes are typically larger. One can also use an aspiration level, so that new offspring are not allowed to compete for entry into the population unless this aspiration level is met.

5.2 CHC

The CHC (Cross generational elitist selection, Heterogeneous recombination and Cataclysmic mutation) (Eshelman 1991; Eshelman and Schaffer 1991) algorithm was created by Larry Eshelman with the explicit idea of borrowing from both the genetic algorithm and the evolution strategy community. CHC explicitly borrows the $(\mu+\lambda)$ strategy of evolution strategies. After recombination, the N best unique individuals are drawn from the parent population and offspring population to create the next generation. This also implies that duplicates are removed from the population. This form of selection is referred to as *truncation selection*. From the genetic algorithm community, CHC builds on the idea that recombination should be the dominant search operator. A bit representation is typically used for parameter optimization problems. In fact, CHC goes so far as to use *only* recombination in the main search algorithm. However, it restarts the search where progress is no longer being made by employing what Eshelman refers to as *cataclysmic mutation*.

Since truncation selection is used, parents can be paired randomly for recombination. However, the CHC algorithm also employs a *heterogeneous recombination* restriction as a method of "incest prevention" (Eshelman 1991). This is accomplished by only mating those string pairs which differ from each other by some number of bits; Eshelman refers to this as a mating threshold. The initial mating threshold is set at $L/4$, where L is the length of the string. If a generation occurs, in which no offspring are inserted into the new population, then the threshold is reduced by 1.

The crossover operator in CHC performs uniform crossover; bits are randomly and independently exchanged, but exactly half of the bits that differ are swapped. This operator, called HUX (Half Uniform Crossover) ensures that offspring are equidistant between the two parents. This serves as a diversity preserving mechanism. An offspring that is *closer* to a parent in terms of Hamming distance will have a tendency to be more similar to that parent in terms of evaluation. Even if this tendency is weak, it can contribute to loss of diversity. If the offspring and parent that are closest in Hamming space tend to be selected together in the next generation, diversity is reduced. By requiring that the offspring be exactly halfway in between the two parents in terms of Hamming distance, the crossover operator attempts to slow down loss of genetic diversity. One could also argue that the HUX operator attempts to maximize the distribution of new samples in new regions of the search space by placing the offspring as far as possible from the two parents while still retaining the bits which the two parents share in common.

No mutation is applied during the regular search phase of the CHC algorithm. When no offspring can be inserted into the population of a succeeding generation and the mating threshold has reached a value of 0, CHC infuses new diversity into the population via a form of restart. Cataclysmic mutation uses the best individual in the population as a template to re-initialize the population. The new population includes one copy of the template string; the remainder of the population is generated by mutating some percentage of bits (e.g., 35%) in the template string.

Bringing this all together, CHC stands for *cross generational elitist selection, heterogeneous recombination* (by incest prevention) and *cataclysmic mutation*, which is used to restart the search when the population starts to converge.

The rationale behind CHC is to have a very aggressive search by using truncation selection which guarantees the survival of the best strings, but to offset the aggressiveness of the search by using highly disruptive uniform crossover. Because of these mechanisms, CHC is able to use a relatively small population size. It generally works well with a population size of 50. Eshelman and Schaffer have reported quite good results using CHC on a wide variety of test problems (Eshelman 1991; Eshelman and Schaffer 1991). Other experiments (c.f. Mathias and Whitley 1994; Whitley et al. 1995) have shown that it is one of the most effective evolutionary algorithms for parameter optimization on many common test problems. Given the small population size and the operators, it seems unreasonable to think of an algorithm such as CHC as a "hyperplane sampling" genetic algorithm. Rather, it can be viewed as an aggressive population-based hill-climber that also uses restarts to quickly explore new regions of the search space. This is also why the small population size is important to the performance of CHC. If the population size is increased, there is some additional potential for increased performance before a restart is triggered. But as with steady-state genetic algorithms, an increase in population size generally results in a small increase in performance after a significant increase in time to convergence/stagnation. The superior performance of CHC using a population of 50 suggests that more restarts have a better payoff than increasing the population size.

6 Where Genetic Algorithms Fail

Do genetic algorithms have a particular mode of failure? There are probably various modes of failure, some of which are common to many search algorithms. If functions are too random, too noisy, or fundamentally unstructured, then search is inherently difficult. "Deception" in the form of misleading hyperplane samples is also a mode of failure that has been studied (Whitley 1991; Goldberg 1989a; Grefenstette 1993).

There is another fundamental mode of failure such that there exists a well-defined set of problems where the performance of genetic algorithms is likely to be poor. Experiments show that various genetic algorithms and local search methods are more or less blind to "ridges" in the search space of parameter optimization problems. In two dimensions, the *ridge problem* is essentially this: a method that searches parallel to the x and y axes cannot detect improving moves that are oriented at a 45° angle to these axes.

A simplified representation of a ridge problem appears in ❯ *Fig. 7*. Changing one variable at a time will move local search to the diagonal. However, looking in either the x-dimension or the y-dimension, every point along the diagonal appears to be a local optimum. There is actually gradient information if one looks *along* the diagonal; however, this requires either (1) changing both variables at once or (2) transforming the coordinate system of the search space so as to "expose" the gradient information.

The ridge problem is relatively well documented in the mathematical literature on derivative free minimization algorithms (Rosenbrock 1960; Brent 1973). However, until recently, there has been little discussion of this problem in the evolutionary algorithm literature.

Let $\Omega = \{0, 1, \ldots, 2^{\ell} - 1\}$ be the search space which can be mapped onto a hypercube. Elements $x, z \in \Omega$ are *neighbors* when (x, z) is an edge in the hypercube. Bit climbing search algorithms terminate at a *local optimum*, denoted by $x \in \Omega$, such that none of the points in the neighborhood $N(x)$ improve upon x when evaluated by some objective function. Of course, the neighborhood structure of a problem depends upon the coding scheme used. Gray codes are often used for bit representations because, by definition, adjacent integers are adjacent neighbors.

◻ Fig. 7

Local search moves only in the horizontal and vertical directions. It therefore "finds" the diagonal, but gets stuck there. Every point on the diagonal is locally optimal. Local search is blind to the fact that there is gradient information moving along the diagonal.

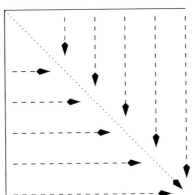

Suppose the objective function is defined on the unit interval $0 \leq x < 1$. To optimize this function, the interval is discretized by selecting n points. The *natural encoding* is then a map from Ω to the graph that has edges between points x and $x+1$ for all $x = 0, 1, \ldots, n-2$.

Under a Gray encoding, adjacent integers have bit representations that are neighbors at Hamming distance 1 (e.g., $3 = 010$, $4 = 110$). Thus a Gray encoding has the following nice property:

▶ A function $f : \Omega \to \mathbb{R}$ cannot have more local optima under a Gray encoding than it does under the natural encoding.

A proof first appears in Rana and Whitley (1997); this theorem states that when using a Gray code, local optima of the objective function considered as a function on the unit interval can be destroyed, but no new local optima can be created (❷ *Fig. 8*).

But are there unimodal functions where the natural encoding is multimodal? If the function is one dimensional, the answer is no. However, if the function is not one dimensional, the answer is yes. False local optima are induced on ridges since there are points along the ridge where improving moves become invisible to search.

This limitation is not unique to local search, and it is not absolute for genetic algorithms. Early population sampling can potentially allow the search to avoid being trapped by "ridges." It is also well known that genetic algorithms quickly lose diversity and then the search must use mutation or otherwise random jumps to move along the ridge. Any method that tends to search one dimension at a time (or to find improvements by changing one dimension at a time via mutation) has the same limitation, including local search and simple "line search" methods.

Salomon (1960) showed that most benchmarks become much more difficult when the problems are *rotated*. Searching a simple two-dimensional elliptical bowl is optimally solved by one iteration of line search when the ellipse is oriented with the x and y axis. However, when the space is rotated $45°$, the bowl becomes a ridge and the search problem is more difficult for many search algorithms.

☐ **Fig. 8**

The two leftmost images show how local search moves on a 2D ridge using different step sizes. Smaller step sizes results in more progress on the ridge, but more steps to get there. The rightmost figure shows points at which local search becomes stuck on a real-world application.

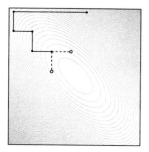 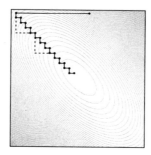 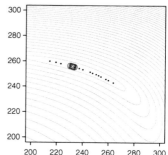

Modern evolution strategies are invariant under rotation. In particular, the covariance matrix adaptation (CMA) evolution strategy uses a form of principal component analysis (PCA) to rotate the search space as additional information is obtained about the local landscape (Hansen 2006). Efforts are currently underway to generalize the concept of a rotationally invariant representation that could be used by different search algorithms (Hansen 2008). Until this work advances, current versions of genetic algorithms still have problems with some ridge structures. Crossover operators such as interval crossover (Schaffer and Eshelman 1993) attempt to deal with the ridge problem but are not as powerful as CMA.

7 An Example of Genetic Algorithms for Resource Scheduling

Resource scheduling problems involve the allocation of time and/or a resource to a finite set of requests. When demand for a resource becomes greater than the supply, conflicting requests require some form of arbitration and the scheduling problem can be posed as an optimization problem over a combinatorial domain. Thus, these problems are combinatorial in nature and are quite distinct from parameter optimization problems. Indeed, genetic algorithms have been very successful on scheduling applications of this nature.

A schedule may attempt to maximize the total number of requests that are filled, or the aggregate value (or priority) of requests filled, or to optimize some other metric of resource utilization. In some cases, a request needs to be assigned a time window on some appropriate resource with suitable capabilities and capacity. In other cases, the request does not correspond to a particular time window, but rather just a quantity of an oversubscribed resource.

One example of resource scheduling is the scheduling of flight simulators (Syswerda 1991). Assume a company has three flight simulators, and 100 people who want to use them. In this case, the simulators are a limited resource. A user might request to use a simulator from 9:00 a.m. to 10:00 a.m. on Tuesday. But if six users want a simulator at this time, then not all requests can be fulfilled. Some users may only be able to utilize the simulators at certain times or on certain days. How does one satisfy as many requests as possible? Or which requests are most important to satisfy? Does it make sense to schedule a user for less time than they requested?

One way to attempt to generate an approximate solution to this problem is to have users indicate a number of prioritized choices. If a user's first choice cannot be filled, then perhaps their second choice of times is available. A hypothetical evaluation function is to have a schedule evaluator that awards ten points for every user that gets their first choice, and five points for those that get their second choice and three points for those that get their third choice, one point for those scheduled (but not in their first, second, or third prioritized requests) and -10 points for user requests that cannot be scheduled.

Is this the best evaluation function for this problem? This raises an interesting issue. In real-world applications, sometimes the evaluation function is not strictly determined and developers must work with users to define a reasonable evaluation function. Sometimes what appears to be an "obvious" evaluation function turns out to be the wrong evaluation function.

Reasonably good solutions to a resource scheduling problem can usually be obtained by using a greedy scheduler which allocates to each request the best available resource at the best available time on a first-come, first-served basis.

The problem with a simple greedy strategy is that requests are not independent – when one request is assigned a slot on a resource, that resource is no longer available (or less available). Thus, placing a single request at its optimal position may preclude the optimal placement of multiple other requests. This is the standard problem with all greedy methods.

A significant improvement on a greedy first-come, first-served strategy is to explore the space of possible permutations of the requests where the permutation defines a priority ordering of the requests to be placed in the schedule. A genetic algorithm can then be applied to search the space of permutations. The fitness function is some measure of cost or quality of the resulting schedule.

The use of a permutation-based representation also requires the construction of a "schedule builder" which maps the permutation of requests to an actual schedule. In a sense, the schedule builder is also a greedy scheduling algorithm. Greedy scheduling on its own can be a relatively good strategy. However, by exploring the space of different permutations, one changes the order in which requests get access to resources; this can also be thought of as changing the order in which requests arrive. Thus, exploring the space of permutations provides the opportunity to improve on the simple greedy scheduler.

The separation of "permutation space" and "schedule space" allows for the use of two levels of optimization. At a lower level, a greedy scheduler converts each permutation into a schedule; at a higher level, a genetic algorithm is used to search the space of permutations. This approach thus creates a strong separation between the problem representation and the actual details of the particular scheduling application. This allows the use of relatively generic "genetic recombination" operators or other local search operators. A more direct representation of the scheduling problem would require search operators that are customized to the application. When using a permutation-based representation, this customization is hidden inside the schedule builder. Changes from one application to another only require that a new schedule builder be constructed for that particular application. This makes the use of permutation-based representations rather flexible.

Whitley et al. (1989) first used a strict permutation-based representation in conjunction with genetic algorithms for real-world applications. However, Davis (1985b) had previously used "an intermediary, encoded representation of schedules that is amenable to crossover operations, while employing a decoder that always yields legal solutions to the problem." This is also a strategy later adopted by Syswerda (1991) and Syswerda and Palmucci (1991), which he enhanced by refining the set of available recombination operators.

Typically, simple genetic algorithms encode solutions using bit-strings, which enable the use of "standard" crossover operators such as 1-point and 2-point (Goldberg 1989b). However, when solutions for scheduling problems are encoded as permutations, a special crossover operator is required to ensure that the recombination of two parent permutations results in a child that (1) inherits good characteristics of both parents and (2) is still a legal permutation. Numerous crossover operators have been proposed for permutations representing scheduling problems. For instance, Syswerda's (1991) *order* and *position* crossover are methods for producing legal permutations that inherit various ordering or positional elements from parents.

Syswerda's order crossover and position crossover differ from other permutation crossover operators such as Goldberg's Partially Mapped Crossover (PMX) operator (Goldberg and Lingle 1985) or Davis' order crossover (Davis 1985a) in that no contiguous block is directly passed to the offspring. Instead, several elements are randomly selected by absolute position. These operators are largely used for scheduling applications (e.g., Syswerda 1991; Watson et al. 1999; Syswerda and Palmucci 1991 for Syswerda's operator) and are distinct from the permutation recombination operators that have been developed for the traveling salesman problem (Nagata and Kobayashi 1997; Whitley et al. 1989). Operators that work well for scheduling applications do not work well for the traveling salesman problem, and operators that work well for the traveling salesman problem do not work well for scheduling.

Syswerda's order crossover operator can be seen as a generalization of Davis' order crossover (Davis 1991) that also borrows from the concept of uniform crossover for bit strings. Syswerda's order crossover operator starts by selecting K uniform-random positions in Parent 2. The corresponding elements from Parent 2 are then located in Parent 1 and reordered, so that they appear in the same relative order as they appear in Parent 2. Elements in Parent 1 that do not correspond to selected elements in Parent 2 are passed directly to the offspring.

```
        Parent 1:   "A B C D E F G"
        Parent 2:   "C F E B A D G"
Selected Elements:     * * *
```

The selected elements in Parent 2 are F B and A in that order. A remapping operator reorders the relevant elements in Parent 1 in the same order found in Parent 2.

```
"A B _ _ _ F _"  remaps to  "F B _ _ _ A _"
```

The other elements in Parent 1 are untouched, thus yielding

```
"F B C D E A G"
```

Syswerda also defined a "position crossover." Whitley and Yoo (1995) prove that Syswerda's order crossover and position crossover are identical in expectation when order crossover selects K positions and position crossover selects $L - K$ positions over permutations of length L.

7.1   The Coors Warehouse Scheduling Problem

The Coors production facility (circa 1990) consists of 16 production lines, a number of loading docks, and a warehouse for product inventory. At the time this research was originally carried out (Starkweather et al. 1991), each production line could manufacture approximately 500 distinct products. Orders could be filled directly from the production lines or from inventory.

A solution is a priority ordering of customer orders. While the ultimate resource is the product that is being ordered, another limiting resource are the loading docks. Once a customer order is selected to be filled, it is assigned a loading dock and (generally) remains at the dock until it is completely filled, at which point the dock becomes empty and available for another order. All orders compete for product from either the production line or inventory. Note that scheduling also has secondary effects on how much product is loaded from the production line onto a truck (or train) and how much must temporarily be placed in inventory.

The problem representation used here is permutation of customer orders; the permutation queues up the customer orders which then wait for a vacant loading dock. When a dock becomes free, an order is removed from the queue and assigned to the dock.

A simulation is used to compute the evaluation function. Assume we wish to schedule one 24 h period. The simulation determines how long it takes to fill each order and how many orders can be filled in 24 h. The simulation must also track which product is drawn out of inventory, how much product is directly loaded off the production line, and how much product must first go into inventory until it is needed.

❯ *Figure 9* illustrates how a permutation is mapped to a schedule. Customer orders are assigned a dock based on the order in which they appear in the permutation; the permutation in effect acts as a customer priority queue. In the right-hand side of the illustration, note that initially customer orders **A–I** get first access to the docks (in a left to right order). **C** finishes first, and the next order, **J**, replaces **C** at the dock. Order **A** finishes next and is replaced by **K**. **G** finishes next and is replaced by **L**.

For the Coors warehouse scheduling problem, one is interested in producing schedules that simultaneously achieve two goals. One of these is to minimize the mean time that customer orders to remain at dock. Let N be the number of customer orders. Let M_i be the time that the truck or rail car holding customer order i spends at dock. Mean time at dock, $\mathcal{M}$, is then given by

$$\mathcal{M} = \frac{1}{N} \sum_{i=0}^{N} M_i.$$

The other goal is to minimize the running average inventory. Let F be the makespan of the schedule. Let $\mathcal{J}_t$ be inventory at time t. The running average inventory, I, is given by

$$I = \frac{1}{F} \sum_{t=0}^{F} \mathcal{J}_t.$$

Technically this is a multi-objective problem. This problem was transformed into a single-objective problem using a linear combination of the individual objectives:

$$\text{obj} = \frac{(\mathcal{M} - \mu_{\mathcal{M}})}{\sigma_{\mathcal{M}}} + \frac{(I - \mu_I)}{\sigma_I} \tag{9}$$

where I represents running average inventory, $\mathcal{M}$ represents order mean time at dock, while μ and σ represent the respective means and standard deviations over a set of solutions.

In ❯ *Table 1*, results are given for the Genitor steady-state genetic algorithm compared to a stochastic hill-climber (Watson et al. 1999). Mean time at dock and average inventory are reported, also with performance and standard deviations over 30 runs. The move operator for the hill-climber was an "exchange operator." This operator selects two random customers and then swaps their position in the permutation. All of the algorithms reported here

◘ Fig. 9
The warehouse model includes production lines, inventory, and docks. The columns in the schedule represent different docks. Customer orders are assigned to a dock left to right and product is drawn from inventory and the production lines.

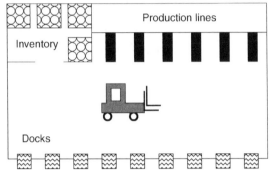

Customer priority queue: A, B, C, D, E, F, G, H, I, . . ., Z

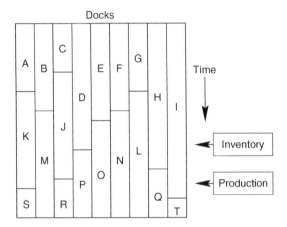

◘ Table 1
Performance results. The final column indicates a human-generated solution used by Coors

	Genetic algorithm	Hill climber	Coors
Mean time-at-dock			
μ	392.49	400.14	437.55
σ	0.2746	4.7493	n.a.
Average inventory			
μ	364,080	370,458	549,817
σ	1,715	20,674	n.a.

used 100,000 function evaluations. The Genitor algorithm used a population size of 500 and a selective pressure of 1.1.

For our test data, we have an actual customer order sequence developed and used by Coors personnel to fill customer orders. This solution produced an average inventory of 549,817.25 product units and an order mean time of 437.55 min at dock.

Genitor was able to improve the mean time at dock by approximately 9%. The big change, however, is in average inventory. Both Genitor and the Hill Climber show a dramatic reduction in average inventory, meaning more product came out of inventory and that the schedule did a better job of directly loading product off the line. Watson et al. (1999) provide a more detailed discussion of the Coors warehouse scheduling application.

8   Conclusions

This paper has presented a broad survey of both theoretical and practical work related to genetic algorithms. For the reader who may be interested in using genetic algorithms to solve a particular problem, two comments might prove to be useful.

First, the details matter. A small change in representation or a slight modification in how an algorithm is implemented can change algorithm performance. The literature is full of papers that claim one algorithm is better than the other. What is often not obvious is how hard the researchers had to work to get good performance and how sensitive the results are to tuning the search algorithms? Are the benchmarks representative of real-world problems, and do the results generalize?

Second, genetic algorithms are a general purpose approach. An application-specific solution will almost always be better than a general purpose solution. Of course, in virtually every real-world application, application-specific knowledge gets integrated into the evaluation function, the operators, and the representation. Doing this well takes a good deal of experience and intuition about how to solve optimization and search problems which makes the practice of search algorithm design somewhat of a specialized art. This observation leads back to the first point: the details matter, and a successful implementation usually involves keen insight into the problem.

Acknowledgments

This research was partially supported by a grant from the Air Force Office of Scientific Research, Air Force Materiel Command, USAF, under grant number FA9550-08-1-0422. The U.S. Government is authorized to reproduce and distribute reprints for Governmental purposes, notwithstanding any copyright notation thereon. Funding was also provided by the Coors Brewing Company, Golden, Colorado.

References

Bäck T (1996) Evolutionary algorithms in theory and practice. Oxford University Press, Oxford

Baker J (1987) Reducing bias and inefficiency in the selection algorithm. In: Grefenstette J (ed) GAs and their applications: 2nd international conference, Erlbaum, Hilsdale, NJ, pp 14–21

Bitner JR, Ehrlich G, Reingold EM (1976) Efficient generation of the binary reflected gray code and its applications. Commun ACM 19 (9):517–521

Brent R (1973) Algorithms for minization with derivatives. Dover, Mineola, NY

Bridges C, Goldberg D (1987) An analysis of reproduction and crossover in a binary coded genetic algorithm. In: Grefenstette J (ed) GAs and their applications: 2nd international conference, Erlbaum, Cambridge, MA

Davis L (1985a) Applying adaptive algorithms to epistatic domains. In: Proceedings of the IJCAI-85, Los Angeles, CA

Davis L (1985b) Job shop scheduling with genetic algorithms. In: Grefenstette J (ed) International conference on GAs and their applications. Pittsburgh, PA, pp 136–140

Davis L (1991) Handbook of genetic algorithms. Van Nostrand Reinhold, New York

DeJong K (1993) Genetic algorithms are NOT function optimizers. In: Whitley LD (ed) FOGA – 2, Morgan Kaufmann, Los Altos, CA, pp 5–17

Eshelman L (1991) The CHC adaptive search algorithm: how to have safe search when engaging in nontraditional genetic recombination. In: Rawlins G (ed) FOGA – 1, Morgan Kaufmann, Los Altos, CA, pp 265–283

Eshelman L, Schaffer D (1991) Preventing premature convergence in genetic algorithms by preventing incest. In: Booker L, Belew R (eds) Proceedings of the 4th international conference on GAs. Morgan Kaufmann, San Diego, CA

Goldberg D (1987) Simple genetic algorithms and the minimal, deceptive problem. In: Davis L (ed) Genetic algorithms and simulated annealing. Pitman/ Morgan Kaufmann, London, UK, chap 6

Goldberg D (1989a) Genetic algorithms and Walsh functions: Part II, deception and its analysis. Complex Syst 3:153–171

Goldberg D (1989b) Genetic algorithms in search, optimization and machine learning. Addison-Wesley, Reading, MA

Goldberg D (1990) A note on Boltzmann tournament selection for genetic algorithms and population-oriented simulated annealing. Tech. Rep. Nb. 90003. Department of Engineering Mechanics, University of Alabama, Tuscaloosa, AL

Goldberg D, Deb K (1991) A comparative analysis of selection schemes used in genetic algorithms. In: Rawlins G (ed) FOGA – 1, Morgan Kaufmann, San Mateo, CA, pp 69–93

Goldberg D, Lingle R (1985) Alleles, loci, and the traveling salesman problem. In: Grefenstette J (ed) International conference on GAs and their applications. London, UK, pp 154–159

Grefenstette J (1993) Deception considered harmful. In: Whitley LD (ed) FOGA – 2, Morgan Kaufmann, Vail, CO, pp 75–91

Hansen N (2006) The CMA evolution strategy: a comparing review. In: Toward a new evolutionary computation: advances on estimation of distribution algorithms. Springer, Heidelberg, Germany, pp 75–102

Hansen N (2008) Adaptive encoding: how to render search coordinate system invariant. In: Proceedings of 10th international conference on parallel problem solving from nature. Springer, Dortmund, Germany, pp 205–214

Heckendorn R, Rana S, Whitley D (1999a) Polynomial time summary statistics for a generalization of MAXSAT. In: GECCO-99, Morgan Kaufmann, San Francisco, CA, pp 281–288

Heckendorn R, Rana S, Whitley D (1999b) Test function generators as embedded landscapes. In: Foundations of genetic algorithms FOGA – 5, Morgan Kaufmann, Los Atlos, CA

Heckendorn RB, Whitley LD, Rana S (1996) Nonlinearity, Walsh coefficients, hyperplane ranking and the simple genetic algorithm. In: FOGA – 4, San Diego, CA

Ho Y (1994) Heuristics, rules of thumb, and the 80/20 proposition. IEEE Trans Automat Cont 39(5): 1025–1027

Ho Y, Sreenivas RS, Vakili P (1992) Ordinal optimization of discrete event dynamic systems. Discrete Event Dyn Syst 2(1):1573–7594

Holland J (1975) Adaptation in natural and artificial systems. University of Michigan Press, Ann Arbor, MI

Holland JH (1992) Adaptation in natural and artificial systems, 2nd edn. MIT Press, Cambridge, MA

Schaffer JD, Eshelman L (1993) Real-coded genetic algorithms and interval schemata. In: Whitley LD (ed) FOGA – 2, Morgan Kaufmann, Los Atlos, CA

Mathias KE, Whitley LD (1994) Changing representations during search: a comparative study of delta coding. J Evolut Comput 2(3):249–278

Nagata Y, Kobayashi S (1997) Edge assembly crossover: a high-power genetic algorithm for the traveling salesman problem. In: Bäck T (ed) Proceedings of the 7th international conference on GAs, Morgan Kaufmann, California, pp 450–457

Nix A, Vose M (1992) Modelling genetic algorithms with Markov chains. Ann Math Artif Intell 5:79–88

Poli R (2005) Tournament selection, iterated coupon-collection problem, and backward-chaining evolutionary algorithms. In: Foundations of genetic algorithms, Springer, Berlin, Germany, pp 132–155

Radcliffe N, Surry P (1995) Fundamental limitations on search algorithms: evolutionary computing in perspective. In: van Leeuwen J (ed) Lecture notes in computer science, vol 1000, Springer, Berlin, Germany

Rana S, Whitley D (1997) Representations, search and local optima. In: Proceedings of the 14th national conference on artificial intelligence AAAI-97. MIT Press, Cambridge, MA, pp 497–502

Rana S, Heckendorn R, Whitley D (1998) A tractable Walsh analysis of SAT and its implications for genetic algorithms. In: AAAI98, MIT Press, Cambridge, MA, pp 392–397

Rosenbrock H (1960) An automatic method for finding the greatest or least value of a function. Comput J 3:175–184

Salomon R (1960) Reevaluating genetic algorithm performance under coordinate rotation of benchmark functions. Biosystems 39(3):263–278

Schaffer JD (1987) Some effects of selection procedures on hyperplane sampling by genetic algorithms. In: Davis L (ed) Genetic algorithms and simulated annealing. Morgan Kaufmann, San Francisco, CA, pp 89–130

Schwefel HP (1981) Numerical optimization of computer models. Wiley, New York

Schwefel HP (1995) Evolution and optimum seeking. Wiley, New York

Sokolov A, Whitley D (2005) Unbiased tournament selection. In: Proceedings of the 7th genetic and evolutionary computation conference. The Netherlands, pp 1131–1138

Spears W, Jong KD (1991) An analysis of multi-point crossover. In: Rawlins G (ed) FOGA – 1, Morgan Kaufmann, Los Altos, CA, pp 301–315

Starkweather T, Whitley LD, Mathias KE (1990) Optimization using distributed genetic algorithms. In: Schwefel H, Männer R (eds) Parallel problem solving from nature. Springer, London, UK, pp 176–185

Starkweather T, McDaniel S, Mathias K, Whitley D, Whitley C (1991) A comparison of genetic sequencing operators. In: Booker L, Belew R (eds) Proceedings of the 4th international conference on GAs. Morgan Kaufmann, San Mateo, pp 69–76

Suh J, Gucht DV (1987) Distributed genetic algorithms. Tech. rep., Indiana University, Bloomington, IN

Syswerda G (1989) Uniform crossover in genetic algorithms. In: Schaffer JD (ed) Proceedings of the 3rd international conference on GAs, Morgan Kaufmann, San Mateo, CA

Syswerda G (1991) Schedule optimization using genetic algorithms. In: Davis L (ed) Handbook of genetic algorithms, Van Nostrand Reinhold, New York, chap 21

Syswerda G, Palmucci J (1991) The application of genetic algorithms to resource scheduling. In: Booker L, Belew R (eds) Proceedings of the 4th international conference on GAs, Morgan Kaufmann, San Mateo, CA

Vose M (1993) Modeling simple genetic algorithms. In: Whitley LD (ed) FOGA – 2, Morgan Kaufmann, San Mateo, CA, pp 63–73

Vose M (1999) The simple genetic algorithm. MIT Press, Cambridge, MA

Vose M, Liepins G (1991) Punctuated equilibria in genetic search. Complex Syst 5:31–44

Vose M, Wright A (1997) Simple genetic algorithms with linear fitness. Evolut Comput 2(4):347–368

Watson JP, Rana S, Whitley D, Howe A (1999) The impact of approximate evaluation on the performance of search algorithms for warehouse scheduling. J Scheduling 2(2):79–98

Whitley D (1999) A free lunch proof for gray versus binary encodings. In: GECCO-99, Morgan Kaufmann, Orlando, FL, pp 726–733

Whitley D, Kauth J (1988) GENITOR: A different genetic algorithm. In: Proceedings of the 1988 Rocky Mountain conference on artificial intelligence, Denver, CO

Whitley D, Rowe J (2008) Focused no free lunch theorems. In: GECCO-08, ACM Press, New York

Whitley D, Yoo NW (1995) Modeling permutation encodings in simple genetic algorithm. In: Whitley D, Vose M (eds) FOGA – 3, Morgan Kaufmann, San Mateo, CA

Whitley D, Starkweather T, Fuquay D (1989) Scheduling problems and traveling salesmen: the genetic edge recombination operator. In: Schaffer JD (ed) Proceedings of the 3rd international conference on GAs. Morgan Kaufmann, San Francisco, CA

Whitley D, Das R, Crabb C (1992) Tracking primary hyperplane competitors during genetic search. Ann Math Artif Intell 6:367–388

Whitley D, Beveridge R, Mathias K, Graves C (1995a) Test driving three 1995 genetic algorithms. J Heuristics 1:77–104

Whitley D, Mathias K, Pyeatt L (1995b) Hyperplane ranking in simple genetic algorithms. In: Eshelman L (ed) Proceedings of the 6th international conference on GAs. Morgan Kaufmann, San Francisco, CA

Whitley D, Mathias K, Rana S, Dzubera J (1996) Evaluating evolutionary algorithms. Artif Intell J 85:1–32

Whitley LD (1989) The GENITOR algorithm and selective pressure: why rank based allocation of reproductive trials is best. In: Schaffer JD (ed) Proceedings of the 3rd international conference on GAs. Morgan Kaufmann, San Francisco, CA, pp 116–121

Whitley LD (1991) Fundamental principles of deception in genetic search. In: Rawlins G (ed) FOGA – 1, Morgan Kaufmann, San Francisco, CA, pp 221–241

Whitley LD (1993) An executable model of the simple genetic algorithm. In: Whitley LD (ed) FOGA – 2, Morgan Kaufmann, Vail, CO, pp 45–62

Wolpert DH, Macready WG (1995) No free lunch theorems for search. Tech. Rep. SFI-TR-95-02-010, Santa Fe Institute, Santa Fe, NM

22 Evolutionary Strategies

Günter Rudolph
Department of Computer Science, TU Dortmund, Dortmund, Germany
guenter.rudolph@tu-dortmund.de

G. Rozenberg et al. (eds.), *Handbook of Natural Computing*, DOI 10.1007/978-3-540-92910-9_22,
© Springer-Verlag Berlin Heidelberg 2012

Abstract

Evolutionary strategies (ES) are part of the all-embracing class of evolutionary algorithms (EA). This chapter presents a compact tour through the developments regarding ES from the early beginning up to the recent past before advanced techniques and the state-of-the-art techniques are explicated in detail. The characterizing features that distinguish ES from other subclasses of EA are discussed as well.

Prelude

This chapter on evolutionary strategies (ES) has a number of predecessors (Rechenberg 1978; Bäck et al. 1991; Schwefel and Rudolph 1995; Beyer and Schwefel 2002), each of them is still worth reading as they reveal which techniques have been developed in the course of time and which methods were considered as the state of the art at that particular time. Here, the aim is to update the comprehensive overview provided in Beyer and Schwefel (2002) to the present state of the art. To this end, a condensed version of Beyer and Schwefel (2002) is presented in ❷ Sects. 1–3 before advanced techniques and the current standard are introduced in ❷ Sect. 4. Finally, ❷ Sect. 5 reports ongoing developments that may become part of a subsequent state-of-the-art survey on evolutionary strategies in the near future.

1 Historical Roots

In the 1960s, difficult multimodal and noisy optimization problems from engineering sciences awaited their solution at the Technical University of Berlin, Germany. At that time, mathematical models or numerical simulations of these particular problems were not available. As a consequence, the optimization had to be done experimentally with the real object at the hardware level. Three students, Peter Bienert, Ingo Rechenberg, and Hans-Paul Schwefel, envisioned an automated cybernetic system that alters the object parameters mechanically or electrically, runs the experiment, measures the outcome of the experiment, and uses this information in the context of some optimization strategy for the decision on how to alter the object parameters of the real object for the experiment in the subsequent iteration. Obvious candidates for the optimization methods in the framework of this early "hardware-in-the-loop" approach were all kinds of gradient-like descent methods for minimization tasks. But these methods failed. Inspired by lectures on biological evolution, they tried a randomized method that may be seen as the simplest algorithm driven by mutation and selection – a method nowadays known as $(1 + 1)$-ES. This approach was successful (Rechenberg 1965) and it was extended in various directions including first theoretical results (Schwefel 1964, 1977; Rechenberg 1973) during the next decade. After some years of only few novelties, the research activities gained momentum again in the late 1980s making evolutionary strategies an active field of research since that time.

Since early theoretical publications mainly analyzed simple ES without recombination, somehow the myth arose that ES put more emphasis on mutation than on recombination. This is a fatal misconception! Recombination has been an important ingredient of ES from the very beginning and this is still valid today.

2   Algorithmic Skeleton: The ES Metaheuristic

A metaheuristic may be defined as an algorithmic framework of iterative nature that specifies the sequence of abstract operations. Only if the abstract operations are instantiated by concrete operations the metaheuristic turns into a real algorithm. As a consequence, many variants of instantiated metaheuristics are possible. And this holds true also for the metaheuristic termed an *evolutionary algorithm* that works as follows:

Algorithm 1   Metaheuristic: Evolutionary Algorithm

initialize population of individuals
evaluate all individuals by fitness function
repeat
    select individuals (parents) for reproduction
    vary selected individuals in randomized manner to obtain new individuals (offspring)
    evaluate offspring by fitness function
    select individuals for survival from offspring and possibly parents according to fitness
until stopping criterion fulfilled

The metaheuristic known as *evolutionary strategies* narrows the degrees of freedom in the instantiation of the EA metaheuristic. These characterizing features are

1. Selection for reproduction is unbiased.
2. Selection for survival is ordinal and deterministic.
3. Variation operators are parametrized and hence controllable.
4. Individuals consist of candidate solution and control parameters.

This leads to the ES metaheuristic:

Algorithm 2   Metaheuristic: Evolutionary Strategy

initialize population of individuals
evaluate all individuals by fitness function
repeat
    select parents for reproduction uniformly at random
    vary selected individuals in randomized manner to obtain offspring
    evaluate offspring by fitness function
    rank offspring and possibly parents according to fitness
    select individuals with best ranks for survival
until stopping criterion fulfilled

Next, the ES metaheuristic is narrowed further and specified in a more formal manner by following Schwefel and Rudolph (1995) in essence. A population at generation $t \geq 0$ is denoted by $P^{(t)}$. An individual $p \in P^{(t)}$ is a pair $p = (x, \Psi)$ where $x \in X$ is an element of the feasible set X and Ψ is a finite set of strategy parameters of arbitrary kind. The objective function $f: X \to \mathbb{R}$ is termed fitness function and it maps an element from the feasible set X to a value in $\mathbb{R}$. Now we are in the position to specify the $(\mu \,/\, \rho, \kappa, \lambda)$-ES, where μ denotes the number of parents, λ the number of offspring, ρ the number of parents participating in

```
Algorithm 3  (μ/ρ, κ, λ)-ES

initialize population P^(0) with μ individuals
∀p ∈ P^(0): set p.Ψ.age = 1
set t = 0
repeat
    Q^(t) = { }
    for i = 1 to λ do
        select ρ parents p_1, ..., p_ρ ∈ P^(t) uniformly at random
        q = variation (p_1,..., p_ρ) with q.Ψ.age = 0
        Q^(t) = Q^(t) ∪ {q}
    end for
    P^(t+1) = selection of μ best individuals from Q^(t) ∪ {p ∈ P^(t): p.Ψ.age < κ}
    ∀p ∈ P^(t+1): increment p.Ψ.age
    increment t
until stopping criterion fulfilled
```

◘ **Table 1**

Default parameterization associated with original shorthand notation

Algorithm	Parents	Mates	Offspring	Age
$(1 + 1)$	$\mu = 1$	$\rho = 1$	$\lambda = 1$	$\kappa = \infty$
$(1 + \lambda)$	$\mu = 1$	$\rho = 1$	$\lambda \geq 1$	$\kappa = \infty$
$(1, \lambda)$	$\mu = 1$	$\rho = 1$	$\lambda \geq 2$	$\kappa = 1$
$(\mu + 1)$	$\mu \geq 2$	$\rho = 2$	$\lambda = 1$	$\kappa = \infty$
$(\mu + \lambda)$	$\mu \geq 2$	$\rho = 2$	$\lambda \geq 2$	$\kappa = \infty$
(μ, λ)	$\mu \geq 2$	$\rho = 2$	$\lambda \geq \mu$	$\kappa = 1$

producing an offspring, and κ the maximum age of an individual. Thus, we have $\mu, \rho, \lambda \in \mathbb{N}$ and $\kappa \in \mathbb{N} \cup \{\infty\}$ with $\rho \leq \mu$ and, if setting $\kappa < \infty$, we require $\lambda > \mu$.

Prior to 1995, the age parameter was not part of an individual. Only the special cases $\kappa = 1$ (the "comma"-strategy) and $\kappa = \infty$ (the "plus" strategy) were in deployment. ❷ *Table 1* contains the default parameterization associated with the original shorthand notation. For example, $(\mu + \lambda)$-ES usually stands for a $(\mu / 2, \infty, \lambda)$-ES.

3 Design Principles: Instantiation of the Metaheuristic

In the course of instantiating the $(\mu / \rho, \kappa, \lambda)$-ES to a real optimization algorithm, three tasks have to be settled:

1. Choice of an appropriate problem representation
2. Choice (and possibly design) of variation operators acting on the problem representation
3. Choice of strategy parameters (includes initialization)

Subsequently, these tasks will be elucidated in more detail.

3.1   Representation

In the ES community, there is no doctrine concerning the usage of a special representation. Rather, the credo is to use the most natural representation for the problem under consideration. For example, if the decision variables only can attain the value 0 and 1 then a bit string is used as representation; and if all the variables can attain real values then a real vector is used as representation. If some variables are binary, some integer-valued and some real-valued then the representation will be a compound or composite representation, that is, a tuple from $\mathbb{B}^{n_1} \times \mathbb{Z}^{n_2} \times \mathbb{R}^{n_3}$. This flexibility in the representation requires a larger number of variation operators than the fixed-representation case.

3.2   Variation

Suppose the representation has been chosen, that is, the candidate solution x of an individual (x, Ψ) is an element of some n-fold Cartesian product X. Long-time practical experiences within the ES community in designing variation operators has led to three design guidelines:

1. Reachability
   Every solution $x \in X$ should be reachable from an arbitrary solution $x_0 \in X$ after a finite number of repeated applications of the variation operator with positive probability bounded from zero.
2. Unbiasedness
   Unless knowledge about the problem has been gathered, the variation operator should not favor a particular subset of solutions. Formally, this can be achieved by choosing the maximum entropy distribution that obeys knowledge about the problem as constraints. As can be seen from some examples given later, this automatically leads to distributions obeying the principle of strong causality, that is, small variations are more likely than large variations.
3. Control
   The variation operator should have parameters that affect the shape of the distribution. As known from theory, when approaching the optimal solution, the strength of variation must be weakened steadily.

These design guidelines will now be exemplified for binary, integer, and real search spaces.

3.2.1   Binary Search Space

An evolution strategy with $X = \mathbb{B}^n$ was used by Rechenberg (1973) to minimize the linear pseudoboolean function $f(x) = \|x - x^*\|_1$ for some given pattern $x^* \in \mathbb{B}^n$. The variation operators are recombination and mutation.

Let $x, y \in \mathbb{B}^n$ and $k \in \{1, 2, \ldots, n-1\}$ uniformly at random. An offspring $z \in \mathbb{B}^n$ is said to be created/recombined by *1-point crossover* if $z_i = x_i$ for $i \leq k$ and $z_i = y_i$ for $i > k$. Offspring z is generated by *uniform crossover* if for each i holds $z_i = x_i$ or $z_i = y_i$ with equal probability. The standard method for mutation flips every bit independently with some mutation probability $p_m \in (0, 1)$.

Do these operators obey the design guidelines? As for reachability: regardless of the result of the recombination operator, the mutation operator can move from every $x \in \mathbb{B}^n$ to an

22

arbitrary $y \in \mathbb{B}^n$ in a single step with positive probability $p^{H(x,y)} (1-p)^{n-H(x,y)} > 0$. Verifying the unbiasedness requires the definition of a maximum entropy distribution.

Definition 1 Let X be a discrete random variable with $p_k = P\{X = x_k\}$ for some index set K. The quantity

$$H(X) = -\sum_{k \in K} p_k \log p_k$$

is called the *entropy* of the distribution of X. If X is a continuous random variable with density $f_X(\cdot)$, the entropy is given by

$$H(X) = -\int_{-\infty}^{\infty} f_X(x) \log f_X(x)$$

The distribution of a random variable X for which $H(X)$ attains its maximum value is termed the *maximum entropy distribution*. $\square$

Example 1 Let $p_k = P\{X = k\}$ for $k = 1, \ldots, n$. We are seeking values p_k such that $H(X)$ becomes maximal under the constraint that the sum of the p_k must be 1. This leads to the Lagrangian function

$$L(p, a) = -\sum_{k=1}^{n} p_k \log p_k + a \left(\sum_{k=1}^{n} p_k - 1 \right)$$

with partial derivatives

$$\frac{\partial L(p, a)}{\partial p_k} = -1 - \log p_k + a \overset{!}{=} 0 \tag{1}$$

and

$$\frac{\partial L(p, a)}{\partial a} = \sum_{k=1}^{n} p_k - 1 \overset{!}{=} 0 \tag{2}$$

Rearrangement of ❯ Eq. 1 yields $p_k \overset{!}{=} e^{a-1}$, which can be substituted in ❯ Eq. 2 leading to

$$\sum_{k=1}^{n} p_k = \sum_{k=1}^{n} e^{a-1} = n e^{a-1} \overset{!}{=} 1$$

and finally to $p_k \overset{!}{=} e^{a-1} \overset{!}{=} \frac{1}{n}$. Thus we have shown that the discrete uniform distribution has maximum entropy. $\square$

The cutpoint k in case of 1-point crossover was drawn uniformly at random among all possible locations. This is the maximum entropy distribution and no cut point has been given any preference. In case of uniform crossover, each bit is selected from the two parents with equal probability. Again, this is the maximum entropy distribution.

Mutations are realized by flipping each bit independently with some probability p_m. This is the maximum entropy distribution for given p_m. But if the distribution of the number K of bit flips is considered, the situation changes. Notice that $K \sim B(n, p_m)$ has binomial distribution with expectation $E[K] = n\, p_m$ and variance $V[X] = n\, p_m (1 - p_m)$. The maximum entropy distribution for a given expectation would be a Boltzmann distribution. If the variance is also given, then one obtains an apparently nameless distribution.

Finally, control of the mutation operator is ensured as we can change the value of p_m. Since recombination and mutation together form the variation operation, the design guideline of providing a control parameter is fulfilled.

3.2.2 Integer Search Space

The first ES for integer search space $\mathbb{Z}^n$ was developed by Schwefel (1964) where mutations were binomially distributed. Here, we shall develop our a mutation operator according to the design guidelines (Rudolph 1994).

Notice that 1-point and uniform crossover can be defined analogous to the binary case. Reachability can be ensured if the support of our mutation distribution is $\mathbb{Z}^n$. If the random integer vector is to be generated by drawing each component independently with the same distribution, then it suffices to find the maximum entropy distribution with support $\mathbb{Z}$.

Thus, the question addressed here is: Which discrete distribution with support $\mathbb{Z}$ is symmetric with respect to 0, has a mean deviation $m > 0$ and possesses maximum entropy? The answer requires the solution of the following nonlinear optimization problem:

$$- \sum_{k=-\infty}^{\infty} p_k \log p_k \rightarrow \max !$$

subject to

$$p_k = p_{-k} \quad \forall k \in \mathbb{Z} \tag{3}$$

$$\sum_{k=-\infty}^{\infty} p_k = 1 \tag{4}$$

$$\sum_{k=-\infty}^{\infty} |k| p_k = m \tag{5}$$

$$p_k \geq 0 \quad \forall k \in \mathbb{Z} \tag{6}$$

We may neglect condition (❯ 6), if the solution of the surrogate problem fulfills these inequalities. We may therefore differentiate the Lagrangian

$$L(p, a, b) = - \sum_{k=-\infty}^{\infty} p_k \log p_k + a \cdot \left(\sum_{k=-\infty}^{\infty} p_k - 1 \right) + b \cdot \left(\sum_{k=-\infty}^{\infty} |k| p_k - m \right)$$

to obtain the necessary condition (❯ 4) and (❯ 5) and $-1 - \log p_k + a + b|k| = 0$ or alternatively

$$p_k = e^{a-1} \cdot \left(e^b \right)^{|k|} \quad \forall k \in \mathbb{Z} \tag{7}$$

Exploitation of the symmetry condition (❯ 3) and substitution of ❯ Eq. 7 in ❯ Eq. 4 leads to

$$\sum_{k=-\infty}^{\infty} p_k = p_0 + 2 \cdot \sum_{k=1}^{\infty} p_k = e^{a-1} \cdot \left(1 + 2 \cdot \sum_{k=1}^{\infty} \left(e^b \right)^{|k|} \right) = 1 \tag{8}$$

Let $q = e^b < 1$ so that $S(q) := \sum_{k=1}^{\infty} q^{|k|} < \infty$ and $q \cdot S'(q) = \sum_{k=1}^{\infty} |k| q^{|k|}$. Then condition (❯ 8) reads

$$e^{1-a} = 1 + 2 \cdot S(q) = \frac{1+q}{1-q} \tag{9}$$

Substitution of ❯ Eq. 7 in ❯ Eq. 5 yields

$$\sum_{k=-\infty}^{\infty} |k| p_k = 2 \cdot \sum_{k=1}^{\infty} |k| \, p_k = 2e^{a-1} \cdot \sum_{k=1}^{\infty} |k| q^{|k|} = 2e^{a-1} \cdot q \cdot S'(q) = m$$

so that with substitution of (❯ Eq. 9) one obtains

$$p = 1 - \frac{m}{(1+m^2)^{1/2} + 1} \tag{10}$$

with $q = 1 - p$. Substitution of ❯ Eq. 9 in ❯ Eq. 7 yields

$$p_k = \frac{p}{2-p} (1-p)^{|k|}$$

which is the maximum entropy distribution. It can be shown (Rudolph 1994) that a random number with this distribution can be generated by the difference $G_1 - G_2$ of two independent geometrically distributed random numbers G_1 and G_2 with parameter p. Thus, random numbers according to the maximum entropy distribution are easy to generate and the variability of distribution can be controlled by choosing the mean deviation $m > 0$ in ❯ Eq. 10.

3.2.3 Continuous Search Space

The continuous search space $\mathbb{R}^n$ imposed the necessity on ES researchers to develop highly sophisticated variation operators. These will be presented in detail in ❯ Sect. 4. Here, focus is on the basics only.

As for recombination, 1-point crossover or uniform crossover (also called discrete or dominant recombination) can be implemented analogous to the binary or integer case. Another common recombination operator is termed *intermediary recombination*: Two parents $x, y \in \mathbb{R}^n$ generate offspring $z = \frac{1}{2}(x + y)$ by averaging. This can be generalized to $\rho \geq 2$ parents easily. A randomized version of intermediary recombination would draw a random number u uniformly distributed in $[0, 1]$ so that the offspring $z = u x + (1 - u) y$ will be located somewhere on the line between the two parents. Notice that this is a maximum entropy distribution.

Before the choice of mutation distribution is discussed, some terminology has to be introduced.

Definition 2 Let $X = (X_1, X_2, \ldots, X_n)'$ be an n-dimensional random vector. The vector

$$\mathsf{E}[X] = (\mathsf{E}[X_1], \mathsf{E}[X_2], \ldots, \mathsf{E}[X_n])'$$

is termed its *expectation vector* whereas the $n \times n$-matrix

$$\mathsf{Cov}[X] = \mathsf{E}[X - \mathsf{E}[X]]\mathsf{E}[X - \mathsf{E}[X]]'$$

is said to be its *covariance matrix*. □

Matrices are extensively used not only for compact notation (Abadir and Magnus 2005).

Definition 3   A square matrix $A \in \mathbb{R}^{n,n}$ is *symmetric* if $A = A'$, *orthogonal* if $A'A = AA' = I_n$ (where I_n is the unit matrix), *positive definite* (p.d.) if $x'Ax > 0$ for all $x \in \mathbb{R}^n \backslash \{0\}$, *positive semidefinite* (p.s.d.) if $x'Ax \geq 0$ for all $x \in \mathbb{R}^n$, *regular* if $\det(A) \neq 0$ and *singular* if $\det(A) = 0$. $\square$

Usually, the mutation distribution of ES is the multivariate normal distribution.

Definition 4   (see Rao 1973, p. 518 or Tong 1990, p. 29) An n-dimensional random vector X with $\mathsf{E}[X] = v$ and $\mathsf{Cov}[X] = \Sigma$ is *multivariate normal* if and only if $c'X$ is univariate normal with $\mathsf{E}[c'X] = c' \, \mathsf{E}[X]$ and $\mathsf{Cov}[c'X] = c'\Sigma c$ for all $c \in \mathbb{R}^n \backslash \{0\}$. $\square$

This definition includes also the case if the covariance matrix is only positive semidefinite. If Σ is positive definite, the distribution has a probability density function

$$f_X(x) = \frac{1}{(2\,\pi)^{n/2} \det(\Sigma)^{1/2}} \exp\left(-\frac{1}{2}(x - v)'\Sigma^{-1}(x - v)\right) \tag{11}$$

for $x \in \mathbb{R}^n$ (as p.d. matrices are regular which in turn can be inverted). Since the support of the distribution is $\mathbb{R}^n$, every point in $\mathbb{R}^n$ can be reached in principle in a single mutation. Moreover, since the multivariate normal distribution has maximum entropy for given expectation vector and covariance matrix (Rao 1973), the requirement of unbiasedness can be fulfilled. But as long as nothing is known about the problem, the covariance matrix should be the unit matrix I_n. The distribution can be controlled by parameter $\sigma > 0$ if a mutation of the current individual $X^{(t)} \in \mathbb{R}^n$ at generation $t \geq 0$ is done as follows: $X^{(t+1)} = X^{(t)} + \sigma Z$ where Z is drawn from a multivariate normal distribution with zero mean and covariance matrix I_n, denoted by $Z \sim N(0, I_n)$. Mechanisms for controlling the mutation distribution are presented in ❯ Sect. 4.

3.3   Parameterization

The parameterization requires not only the fixation of μ, λ, ρ, and κ but also other parameters that may be specific to the variation operators deployed. This task demands skill, experience, intuition, and instinct. Fortunately, there are methods that facilitate this task (Bartz-Beielstein 2006; Smit and Eiben 2009). Moreover, we have to decide how to initialize our population: This may happen uniformly at random in a certain search region, by stratified or Latin hypercube sampling (McKay et al. 1979) or in accordance with standard NLP methods at/around some given start position if no range can be specified in advance.

4   Advanced Adaptation Techniques in $\mathbb{R}^n$

All control or adaptation mechanisms tacitly presume that the operational reach of the ES is within a vicinity of its current position $\tilde{x} \in \mathbb{R}^n$ where the fitness function can be appropriately approximated by a second order Taylor expansion if the problem was differentiable, that is,

$$f(x) \approx f(\tilde{x}) + (x - \tilde{x})'\nabla f(\tilde{x}) + \frac{1}{2}(x - \tilde{x})'\nabla^2 f(\tilde{x})(x - \tilde{x})$$

where $\nabla f(\tilde{x})$ is the gradient and $\nabla^2 f(\tilde{x})$ the Hessian matrix, which is symmetric and positive definite. If the fitness function is quadratic, then the Taylor expansion is exact and the Hessian matrix contains information about the scaling and orientation of the iso-hyperellipsoids that specify positions in the search space with the same fitness value. If the Taylor expansion is not exact, then the Hessian matrix provides at least a reasonable approximation.

Suppose without loss of generality that $f(x) = \frac{1}{2} x'Ax$ with p.d. and symmetric square matrix A. Owing to part (a) of Theorem 1 below, we can decompose matrix A via $A = T D T'$.

Theorem 1 (see Abadir and Magnus 2005) *Let $A \in \mathbb{R}^{n,n}$ be a p.d. matrix.*

(a) *There exist an orthogonal matrix $T \in \mathbb{R}^{n,n}$ and a diagonal matrix $D \in \mathbb{R}^{n,n}$ with positive diagonal entries such that $A = T D T'$ (spectral decomposition), where the columns of T are the eigenvectors of A and the diagonal entries in D are the associated eigenvalues. Moreover, the decomposition can be written as*

$$A = \sum_{i=1}^{n} d_i t_i t_i'$$

where d_i and t_i are the eigenvalues in D and eigenvectors in T, respectively.
(b) *There exists a lower triangular matrix R with positive diagonal entries such that $A = R R'$ (Cholesky decomposition).* □

With this decomposition, the fitness function can be formulated as follows

$$f(x) = \frac{1}{2} x'Ax = \frac{1}{2} x'T DT'x = \frac{1}{2} (T'x)'D(T'x) = \frac{1}{2} y'Dy = \frac{1}{2} \sum_{i=1}^{n} d_i y_i^2 \qquad (12)$$

where $y = T'x$ describes the rotation of the original coordinate system by orthogonal matrix T'. Now recall that the surface of the unit hyperellipsoid $\mathscr{E}_n$ centered in the origin of $\mathbb{R}^n$ is defined by the set

$$\mathscr{E}_n = \left\{ y \in \mathbb{R}^n : \sum_{i=1}^{n} \frac{y_i^2}{a_i^2} = 1 \right\}$$

where the $a_i > 0$ are the lengths of the semi-principal axes from the origin to the intersection with the associated axis of the coordinate system. Inspection of ❷ Eq. 12 reveals that the eigenvalues of A, that is, the diagonal entries in D, determine the lengths of the semi-principal axes of the ellipsoid: $a_i = \sqrt{2/d_i}$ for $i = 1, \ldots, n$. Thus, the smallest eigenvalue of A is associated with the longest semi-principal axis and hence to the strongest curvature of the ellipsoid.

Not surprisingly, the curvature will have some impact on the performance and hence on the control mechanism of the mutation distribution of the ES. This fact will be discussed in detail in this section. The following general result reveals how we can generate arbitrarily shaped multivariate normal distributions.

Theorem 2 (see Rao 1973, p. 522 or Tong 1990, p. 32)

If $X \sim N_n(v, \Sigma)$ and $Y = AX + b$ where $A \in \mathbb{R}^{m,n}$ and $b \in \mathbb{R}^m$, then

$$Y \sim N_m(Av + b, A\Sigma A')$$ □

For our purposes it suffices to consider the special case with $m = n$, $v = b = 0$ and $\Sigma = I_n$.

Corollary 1   *If $X \sim N_n(0, I_n)$ and $Y = AX$ with $A \in \mathbb{R}^{n,n}$ then $Y \sim N_n(0, AA')$.*   □

But as long as nothing is known about the problem, one should choose $A = I_n$, which leads to a spherical distribution, that is, points with the same probability are located on the surface of a hypersphere. If A is a diagonal matrix with nonidentical diagonal entries, then the resulting distribution is elliptical and the principal-axes of the ellipsoid with same probabilities are parallel to the axes of the coordinate system. In the general case, the ellipsoid with same probabilities is also rotated.

Apart from the shape of the mutation distribution, its variance has to be controlled. The next result reveals how the mutations' variance is related to a random step length of the ES:

Theorem 3   (see Fang et al. 1990) *If $X \sim N_n(0, \sigma^2 I_n)$ then $X \stackrel{d}{=} r \cdot u_n$ where $r \stackrel{d}{=} \|X\|$ is a scalar random variable with $\chi_n(\sigma)$-distribution and u_n is an n-dimensional random vector that is uniformly distributed on the surface of the unit hypersphere, i.e., $\|u_n\| = 1$ with probability 1. Moreover, r and u_n are independent.*   □

Thus, a mutation according to a multivariate normal distribution can be interpreted as a step in random direction u_n with random step length r. It can be shown (see, e.g., Rudolph 1997, p. 21) that

$$\mathsf{E}[\|X\|] = \sigma \, \frac{\Gamma\left(\frac{n+1}{2}\right)}{\Gamma\left(\frac{n}{2}\right)} \, \sqrt{2} \approx \sigma \sqrt{n - \frac{1}{2} + \frac{1}{16\,n}} \quad \text{and} \quad \mathsf{V}[\|X\|] \approx \frac{\sigma^2}{2}\left(1 - \frac{1}{8\,n}\right)$$

indicating that the random step length actually does not vary strongly. Thus, controlling the step sizes in ES means controlling the variances of the mutation distribution and vice versa.

The subsequent sections present different techniques of controlling step sizes and shape of the mutation distributions. Some are of historical interest but, in principle, any technique can sometimes beat another technique on a set of problems as predicted by the NFL theorem (Wolpert and Macready 1997; Auger and Teytaud 2007). Nevertheless, there is a clear recommendation currently: First try the CMA-ES (❷ Sect. 4.3.4) and switch to other methods if not satisfied with the results.

4.1   Control of Step Sizes

4.1.1   Adaptation of Step Sizes Based on Success Frequency

Early experiments (Schwefel 1964) have led to the observation that the step size must not be fixed during the search to achieve good performance. Intuitively, it is easy to accept the argument that the step sizes should become smaller the closer the ES approaches the optimum — otherwise most of the mutations would not lead to improvements. Actually, there exists an optimal step size depending on the current position and the distance to the optimum. Since the position of the optimum is not known, a clever mechanism is required for getting a hint whether the current step size is too large or too small.

Rechenberg (1973) analyzed two extremely different objective functions (called "sphere model" and "corridor model") theoretically with the result that the largest progress toward the optimum is achieved for a step size that leads to a relative frequency of successful mutations

---

Algorithm 4 (1 + 1)-ES

choose $X^{(0)} \in \mathbb{R}^n, \sigma^{(0)} > 0$, set $s = 0, t = 0$
repeat
    draw $Z \sim N(0, I_n)$
    $Y^{(t)} = X^{(t)} + \sigma^{(t)} Z$
    if $f(Y^{(t)}) \leq f(X^{(t)})$ **then**
        $s{+}{+}$
        $X^{(t+1)} = Y^{(t)}$
    else
        $X^{(t+1)} = X^{(t)}$
    end if
    if $t \bmod p \neq 0$ **then**
        $\sigma^{(t+1)} = \sigma^{(t)}$
    else
        if $\frac{s}{p} < \frac{1}{5}$ **then**
            $\sigma^{(t+1)} = \sigma^{(t)}/\gamma$
        else
            $\sigma^{(t+1)} = \sigma^{(t)} \cdot \gamma$
        end if
        $s = 0$
    end if
    increment t
until stopping criterion fulfilled

(i.e., mutations that lead to a better fitness value) of about 0.270 for the sphere model and about 0.184 for the corridor model. This finding led to the conjecture that for an arbitrary fitness function, the step size should be adjusted to a length that leads to a success frequency between the two extreme values. Choosing $0.200 = \frac{1}{5}$ as a compromise between the extreme cases, Rechenberg's analysis led to the first step size control mechanism in ES, the so-called $\frac{1}{5}$-*success rule* that is implemented as follows: Count the number of successful mutations s within a certain period of time p. If s/p is less than $1/5$, then the step size is too large and it should be decreased by a factor $1/\gamma < 1$. Otherwise, the step size is too small and it should be increased by a factor of $\gamma > 1$. Clearly, p should be a multiple of 5 and $\gamma = \left(\frac{20}{17}\right)^{\frac{1}{n}}$ is recommended in Schwefel (1981).

4.1.2 Mutative Self-Adaptation of Step Sizes

Since the step sizes are altered only every p iterations when using the $\frac{1}{5}$-rule, the adaptation happens staircase-like and sometimes it may be necessary to alter the step sizes at a quicker rate to achieve optimum performance. Therefore, Schwefel (1977) proposed a mutative step-size adaptation that leads to smoother adjustment of the step sizes (see ❷ *Algorithm 5*).

An individual now consists of its position $x \in \mathbb{R}^n$ and a single step size $\sigma > 0$. As can be seen in ❷ *Algorithm 5*, the step size $\sigma^{(t)}$ is multiplicatively mutated with a lognormally distributed random number L, denoted $L \sim LN(0, \tau^2)$ where $\tau = 1/\sqrt{n}$. Random variable L is generated as follows: If $N \sim N(0, \tau^2)$ then $L = \exp(N) \sim LN(0, \tau^2)$. The adaptation process works quite well for $\lambda \geq 5$ but larger values are recommended.

Algorithm 5   (1, λ)-ES

choose $X^{(0)} \in \mathbb{R}^n$, set $t = 0$
repeat
    for $i = 1$ to λ **do**
        draw $Z_i \sim N(0, I_n)$, $L_i \sim LN(0, \tau^2)$
        $Y^{(i)} = X^{(i)} + \sigma^{(t)} L_i Z_i$
    end for
    let $b \in \{1,\dots,\lambda\}$ such that $f\left(Y_b^{(t)}\right) = \min\{f\left(Y_i^{(t)}\right): i = 1,\dots,\lambda\}$
    $X^{(t+1)} = Y_b^{(t)}$
    $\sigma^{(t+1)} = L_b \cdot \sigma^{(t)}$
    increment t
until stopping criterion fulfilled

The mutation distribution used so far was a spherically shaped normal distribution, that is, the covariance matrix was the unit matrix. An extension (Bäck et al. 1991) of the method above is a variant with scaled mutations (❷ *Algorithm 6*). Now an individual additionally has n scaling parameters $s \in \mathbb{R}_+^n$ that are object to mutations as well. The parameters change to $\tau = 1/\sqrt{2\sqrt{n}}$ and $\eta = 1/\sqrt{2n}$. But notice that experiments have revealed (Schwefel 1987) that a scaling parallel to the coordinate axes only is achieved if $\mu > 1$ and multiparent recombination is deployed. Moreover, it is quite useless to learn this kind of scaling since it is helpful for additively separable problems only. There exist specialized methods optimizing additively separable problems much more efficiently than evolutionary algorithms. In short, ❷ *Algorithm 6* with $\mu = 1$ cannot be recommended for general problems.

Algorithm 6   (1, λ)-ES with Mutative Scaling

choose $X^{(0)} \in \mathbb{R}^n, \sigma^{(0)} > 0$, set $s^{(0)} = (1,\dots,1)', t = 0$
repeat
    for $i = 1,\dots,\lambda$ **do**
        draw $Z_i \sim N(0, I_n), L_i \sim LN(0, \tau^2), S_i \sim LN(0, \eta^2 I_n)$
        $Y^{(i)} = X^{(i)} + \sigma^{(t)} \cdot L_i \cdot \mathrm{diag}(s^{(t)}) \cdot \mathrm{diag}(S_i) \cdot Z_i$
    end for
    let $b \in \{1,\dots,\lambda\}$ such that $f\left(Y_b^{(t)}\right) = \min\{f\left(Y_i^{(t)}\right): i = 1,\dots,\lambda\}$
    $X^{(t+1)} = Y_b^{(t)}$
    $\sigma^{(t+1)} = L_b \cdot \sigma^{(t)}$
    $s^{(t+1)} = \mathrm{diag}(S_b) \cdot s^{(t)}$
    increment t
until stopping criterion fulfilled

4.1.3   Mutative Self-Adaptation of Step Sizes with Momentum Adaptation

Ostermeier conjectured that (Ostermeier 1992) mutative scaling is not efficient especially for large dimensional problems since all mutations are drawn independently. Therefore, he proposed to replace the mutative scaling by a deterministic mechanism that additionally extrapolates good moves in the past to the future (called *momentum adaptation*).

Let vector $m^{(t)} = (m_1^{(t)}, \ldots, m_n^{(t)})'$ with $m^{(0)} = 0 \in \mathbb{R}^n$ induce the diagonal matrix $M^{(t)} = \text{diag}(|m_1^{(t)}|, \ldots, |m_n^{(t)}|)$ where $|\cdot|$ denotes the absolute value. The random variables $Z_i \sim N(0, I_n)$ and S_i with $P\{S_i = 1\} = P\{S_i = 2/3\} = P\{S_i = 3/2\} = 1/3$ are mutually independent for all $i = 1, \ldots, \lambda$. Notice that the global step size σ is now mutated by a discrete random variable S instead of the continuous lognormal random variable L used previously.

The parameter $\gamma \in [0, 1]$ is a learning rate. No recommendations are given. If $\gamma = 0$ then this method reduces to the original $(1, \lambda)$-ES with (discrete) mutative step size control.

Algorithm 7 **(1, λ)-ES with Momentum Adaptation**

choose $X^{(0)} \in \mathbb{R}^n$, set $t = 0$
repeat
  for $i = 1$ to λ **do**
    draw Z_i and S_i
    $Y_i^{(t)} = X^{(t)} + \sigma^{(t)} \cdot S_i \cdot \frac{1}{\sqrt{n}} \cdot [m^{(t)} + (I_n + M^{(t)})Z_i]$
  end for
  let $b \in \{1, \ldots, \lambda\}$ such that $f\left(Y_b^{(t)}\right) = \min\{f\left(Y_i^{(t)}\right) : i = 1, \ldots, \lambda\}$
  $X^{(t+1)} = Y_b^{(t)}$
  $\sigma^{(t+1)} = S_b \cdot \sigma^{(t)}$
  $m^{(t+1)} = m^{(t)} + \gamma \cdot (I_n + M^{(t)})Z_b$
  increment t
until stopping criterion fulfilled

4.1.4 Mutative Self-Adaptation of Step Sizes with Derandomized Scaling

Ostermeier et al. (1994a) followed the idea of Ostermeier (1992) and proposed another deterministic method for adapting the scaling of the mutation distribution. Hints about

---

Algorithm 8 **(1, λ)-ES with Derandomized Scaling**

choose $X^{(0)} \in \mathbb{R}^n$; set $\sigma^{(0)} > 0, S^{(0)} = I_n, t = 0$
repeat
  for $i = 1$ to λ **do**
    draw $Z_i \sim N(0, I_n)$ and G_i
    $Y_i^{(t)} = X^{(t)} + \sigma^{(t)} \cdot G_i \cdot S^{(t)} \cdot Z_i$
  end for
  let $b \in \{1, \ldots, \lambda\}$ such that $f\left(Y_b^{(t)}\right) = \min\{f\left(Y_i^{(t)}\right) : i = 1, \ldots, \lambda\}$
  $X^{(t+1)} = Y_b^{(t)}$
  $\sigma^{(t+1)} = \sigma^{(t)} G_b$
  for $j = 1$ to n **do**
    $s_j^{(t+1)} = s_j^{(t)} \exp\left(|Z_{b,i}| - \sqrt{\frac{2}{\pi}}\right)$
  end for
  $S^{(t+1)} = \text{diag}\left(s_1^{(t+1)}, \ldots, s_n^{(t+1)}\right)$
  increment t
until stopping criterion fulfilled

an appropriate scaling is now extracted from the best mutation vector Z_b. This base concept turned out to be fruitful in future.

The global step size σ is multiplicatively mutated by the discrete random variable G with $P\{G = g\} = P\{G = 1/g\} = \frac{1}{2}$ for $g = 1.4$.

Algorithm 9 $(1, \lambda)$-CSA-ES

choose $X^{(0)} \in \mathbb{R}^n$; set $\sigma^{(0)} > 0, S^{(0)} = I_n, t = 0$
repeat
  for $i = 1$ to λ **do**
    draw Z_i
    $Y_i^{(t)} = X^{(t)} + \sigma^{(t)} \cdot S^{(t)} \cdot Z_i$
  end for
  let $b \in \{1, \dots, \lambda\}$ such that $f\left(Y_b^{(t)}\right) = \min\{f\left(Y_i^{(t)}\right) : i = 1, \dots, \lambda\}$
  $X^{(t+1)} = Y_b^{(t)}$
  $p^{(t+1)} = (1 - \alpha)p^{(t)} + \alpha Z_b$
  $\sigma^{(t+1)} = \sigma^{(t)} \exp\left(\|p^{(t+1)}\|\sqrt{\frac{2-c}{nc}} - 1 + \frac{1}{5n}\right)^{1/\sqrt{n}}$
  for $j = 1$ to n **do**
    $s_j^{(t+1)} = s_j^{(t)} \left[|p_j^{(t+1)}| \cdot \sqrt{\frac{2-c}{c}} + \frac{7}{20}\right]^{1/n}$
  end for
  $S^{(t+1)} = \text{diag}\left(s_1^{(t+1)}, \dots, s_n^{(t+1)}\right)$
  increment t
until stopping criterion fulfilled

4.1.5   Cumulative Step-Size Adaptation (CSA)

The concept of extracting information from the best mutations in the previous method was extended by the same authors (Ostermeier et al. 1994b) by extracting information out of the entire sequence of best mutations. This leads to the method of CSA, which has been adopted for many other algorithms. The best mutation vectors Z_b of each iteration are accumulated in a weighted manner in an evolution path p that represents a good estimator of how to mutate in future. The step size control as well as the scaling of the mutation distribution now depends on the evolution path; they are no longer subject to a mutative adaptation. The parameters should be set to $\alpha = c = \sqrt{1/n}$.

4.2   Control of Descent Direction Without Covariances

The methods considered so far generated mutation distributions where the covariance matrix is a diagonal matrix. If most general distributions are required then one needs not only n scaling parameters but additionally $n(n - 1)/2$ further parameters to fully describe a general covariance matrix. Of course, less general distributions with fewer parameters are imaginable. Some authors refrained from maintaining a full covariance matrix due to increased space and computation requirements. Therefore, they proposed methods that can direct mutations toward good descent directions without using a full covariance matrix.

These arguments concerning space and time restrictions appear obsolete nowadays. The main memory of contemporary computers can store huge matrices and if the evaluation of the fitness function requires a run of a complex simulation, then the time required for matrix computations is neglectable.

4.2.1 Mutative Self-Adaptation of Step Sizes with Correlated Momentum Adaptation

Ostermeier (1992) extended his method with momentum adaptation by estimating correlations from the best mutation Z_b. Random variable S is distributed as in ❯ Sect. 4.1.3. Again, $\gamma \in [0, 1)$ is a learning parameter and the experiments reported in Ostermeier (1992) have been made in the range $0.001 \le \gamma \le 0.3$. Finally, the range $\gamma \in \left[\frac{1}{n}, \frac{3}{n}\right]$ is recommended for $n \ge 3$.

Algorithm 10 (1, λ)-ES with Momentum and Correlations

choose $X^{(0)} \in \mathbb{R}^n$, set $t = 0$
repeat
    for $i = 1$ to λ **do**
        draw $Z_i \sim N(0, I_n)$, $K_i \sim N(0, \frac{1}{4})$ and S_i
        $Y_i^{(t)} = X^{(t)} + \sigma^{(t)} \cdot S_i \cdot \frac{1}{\sqrt{n}} \cdot \left[(1 + K_i)m^{(t)} + \left(I_n + \frac{1}{c^{(t)}}M^{(t)}\right) \cdot Z_i\right]$
    end for
    let $b \in \{1, \dots, \lambda\}$ such that $f\left(Y_b^{(t)}\right) = \min\{f\left(Y_i^{(t)}\right) : i = 1, \dots, \lambda\}$
    $X^{(t+1)} = Y_b^{(t)}$
    $\sigma^{(t+1)} = S_b \cdot \sigma^{(t)}$
    $m^{(t+1)} = m^{(t)} + \gamma \cdot \left[K_b \cdot m^{(t)} + \left(I_n + \frac{1}{c^{(t)}}M^{(t)}\right)\right]Z_b$
    $c^{(t+1)} = \max\{c^{(t)} \cdot (1 + \gamma)^{K_b - 1/20}, 1\}$
    increment t
until stopping criterion fulfilled

4.2.2 Main Vector Adaptation (MVA-ES)

Poland and Zell (2001) came up with the idea to adapt a vector that represents the main descent direction. Actually, they generate mildly correlated mutations. It is evident from the algorithm that they have made heavy use of the concept of the evolution path from the CSA-ES for adapting the main vector.

The parameters are set as follows: $\alpha = \beta = 4/(n + 4)$ and $\gamma = 2/(n + \sqrt{n})^2$. For conceptual reasons, the weighting function is defined as $w(v) = 1 + 2/\|v\|$ but the experiments reveal that $w(v) \equiv 3$ yields best results!

4.2.3 Evolutionary Gradient Search (EGS)

Salomon (1998) observed that sophisticated ES and stochastic gradient methods are quite similar. As a result, he proposed an evolutionary gradient search that approximates the (negative) gradient in a randomized manner by a central differences schema. Notice that $2\lambda + 2$ fitness function evaluations are necessary per iteration. The parameters should obey $\kappa \ge 1, \gamma_1 > 1$ and $\gamma_2 \in (0, 1)$. For example, $\gamma_1 = 1.4$ and $\gamma_2 = 1/\gamma_1$. As for η, no recommendation is given, but $\eta = 2$ seems reasonable, especially in the presence of noise.

Algorithm 11 **(1, λ)-MVA-ES**

choose $X^{(0)} \in \mathbb{R}^n$; set $\sigma^{(0)} > 0, t = 0$
repeat
  for $i = 1$ to λ **do**
    draw $Z_i \sim N(0, I_n)$ and $N_i \sim N(0, 1)$
    $Y_i^{(t)} = X^{(t)} + \sigma^{(t)} \cdot (Z_i + N_i \cdot w(v^{(t)}) \cdot v^{(t)})$
  end for
  let $b \in \{1, \ldots, \lambda$ such that $f\left(Y_b^{(t)}\right) = \min\{f\left(Y_i^{(t)}\right): i = 1, \ldots, \lambda\}$
  $X^{(t+1)} = Y_b^{(t)}$
  $s^{(t+1)} = (1 - \alpha)s^{(t)} + \sqrt{\alpha(2 - \alpha)}Z_b$
  $\sigma^{(t+1)} = \sigma^{(t)} \cdot \exp\left(\frac{\|s^{(t+1)}\| - \sqrt{n - \frac{1}{2}}}{(1 - \beta^{-1})\sqrt{n - \frac{1}{2}}}\right)$
  $p^{(t+1)} = (1 - \beta)p^{(t)} + \sqrt{\beta(2 - \beta)}(Z_b + N_b \cdot w(v^{(t)}) \cdot v^{(t)})$
  $v^{(t+1)} = (1 - \gamma)\,\mathrm{sgn}((v^{(t)})'p^{(t)}) \cdot v^{(t)} + \gamma p^{(t)}$
  increment t
until stopping criterion fulfilled

4.3 Control of Covariances

The first ES that used and adapted a full covariance matrix was the ES with correlated mutations (named CORR-ES) proposed by Schwefel (1981). He conjectured that it would be optimal to align the hyperellipsoid of equal probabilities of the mutation distribution to the hyperellipsoid of equal fitness values if the problem is adequately described by a second order Taylor expansion. If this is true, then it would be optimal (Rudolph 1992) to set the covariance matrix C proportional to the inverse Hessian matrix of the Taylor expansion. This is easily seen from the probability density function (❷ 11) of the multivariate normal distribution where the isolines of equal probabilities are basically determined by the exponential expression $\exp(-\frac{1}{2}xC^{-1}x)$. Thus, if A is the Hessian matrix of the Taylor approximation, then $C^{-1} = A$ is needed to obtain equally aligned hyperellipsoids, or, equivalently, $C = A^{-1}$.

But it is not obvious that this is always the best choice. If the Hessian matrix is known, then we can perform a coordinate transformation to reduce the original problem

Algorithm 12 **(1, (2 λ + 2))-Evolutionary Gradient Search**

choose $X^{(0)} \in \mathbb{R}^n$; set $\sigma^{(0)} > 0, t = 0$
repeat
  for $i = 1$ to λ **do**
    draw $Z_i \sim N(0, I_n)$
  end for
  $\bar{Z} = \sum_{i=1}^{\lambda}\left[f(x^{(t)} - \sigma^{(t)} \cdot Z_i) - f(x^{(t)} + \sigma^{(t)} \cdot Z_i)\right] \cdot Z_i$
  $Y_1^{(t)} = X^{(t)} + \gamma_1 \cdot \frac{\sqrt{n}}{\eta} \cdot \frac{\bar{Z}}{\|\bar{Z}\|}$
  $Y_2^{(t)} = X^{(t)} + \gamma_2 \cdot \frac{\sqrt{n}}{\eta} \cdot \frac{\bar{Z}}{\|\bar{Z}\|}$
  let $b \in \{1, 2\}$ such that $f\left(Y_b^{(t)}\right) = \min\{f\left(Y_1^{(t)}\right), f\left(Y_2^{(t)}\right)\}$
  $X^{(t+1)} = Y_b^{(t)}$
  $\sigma^{(t+1)} = \sigma^{(t)} \cdot \gamma_b$
  increment t
until stopping criterion fulfilled

$f(x) = \frac{1}{2} x' A x + b' x + c$, where the Hessian matrix A is p.d. and symmetric, to a simpler problem for which a single global step size is sufficient, that is, the covariance matrix is the unit matrix

$$
\begin{aligned}
f(Qx) &= \frac{1}{2} (Qx)' A (Qx) + b'(Qx) + c \\
&= \frac{1}{2} x' Q' A Q x + b' Q x + c & (13) \\
&= \frac{1}{2} x' Q' B' B Q x + b' Q x + c \\
\\
&= \frac{1}{2} x' (B^{-1})' B' B B^{-1} x + b' B^{-1} x + c \\
&= \frac{1}{2} x' x + b' B^{-1} x + c & (14)
\end{aligned}
$$

using $A = B'B$ in ❷ Eq. 13 and choosing $Q = B^{-1}$ in ❷ Eq. 14. Thus, if the Hessian A is known, it can be decomposed to $A = B'B$ owing to part (b) of ❷ Theorem 1. It remains to invert matrix B such that the elliptical problem reduces to a spherical problem.

The bad news, however, is that this is never done. Actually, none of the ES methods perform such a coordinate transformation. Rather the mutation vector is scaled and rotated – the problem itself remains unaltered! Using the stochastic decomposition $\sigma Q Z \stackrel{d}{=} r Q u$ of ❷ Theorem 3 and noting that $\nabla f(x) = Ax + b$ what happens is easily seen:

$$
\begin{aligned}
f(x + rQu) &= \frac{1}{2} (x + rQu)' A (x + rQu) + b'(x + rQu) + c \\
&= \frac{1}{2} (x' A x + 2 r x' A Q u + r^2 u' Q' A Q u) + b' x + r b' Q u + c \\
&= f(x) + r x' A Q u + r b' Q u + \frac{1}{2} r^2 u' Q' A Q u \\
&= f(x) + r(Ax + b + \frac{r}{2} A Q u)' Q u & (15) \\
&= f(x) + r(\nabla f(x) + \frac{r}{2} A Q u)' Q u \\
&= f(x) + r \nabla f(x)' Q u + \frac{r^2}{2} u' Q' A Q u \\
&= f(x) + r \nabla f(x)' Q u + \frac{r^2}{2}
\end{aligned}
$$

if $A = B'B$ and $Q = B^{-1}$ because $u' Q' A Q u = u' Q' B' B Q u = u' u = \|u\|^2 = 1$ with probability 1. Evidently, $\nabla f(x)' Q u < 0$ is necessary for an improvement, that is, Qu must be a direction of descent. If we assume that Qu and r in ❷ Eq. 15 are deterministic, then we know that $Qu = -\nabla f(x)$ yields the steepest descent and the optimal step length becomes $r^* = -\nabla f(x)' Q u = \|\nabla f(x)\|^2$.

In any case, the optimal alignment of the longest principal semiaxis of the mutation hyperellipsoid should point toward the negative gradient for optimal performance and not always toward the longest principal semiaxis of the Hessian matrix. But luckily, the ES drives itself in a situation in which negative gradient and the longest principal semiaxis of the Hessian matrix coincide: Regardless of the starting point, the ES runs roughly in negative gradient direction by construction of the method (try many directions and choose those with steepest descent). Sooner or later, but inevitably, it will reach a position in the search space that is close to the longest principal semiaxis of the Hessian matrix. Now, the direction of steepest descent and the longest principal semiaxis of the Hessian matrix coincide: Let $f(x) = \frac{1}{2} x'Ax + b'x + c$ with $x \in \mathbb{R}^n$, A p.d. and symmetric. If $\xi > 0$ is an eigenvalue of A and v its associated eigenvector, then the gradient direction in $\tilde{x} \in \mathbb{R}^n$ is parallel to the eigenvector provided the current position $\tilde{x} \in \{x \in \mathbb{R}^n : x = -A^{-1}b + s \cdot v, s \in \mathbb{R}\}$ is on the line spanned by the eigenvector: $\nabla f(\tilde{x}) = A\tilde{x} + b = A(-A^{-1}b + s \cdot v) + b = -b + sAv + b = sAv = s\xi \cdot v$.

From now on, the best points to be sampled are close to the longest principal semiaxis of the Hessian matrix so that the situation remains basically unaltered during the subsequent optimization phase and the adaptation process for the covariance matrix can stabilize. Therefore, it is actually optimal to use the inverse Hessian matrix as covariance matrix as soon as the longest principal semiaxis of the Hessian matrix is reached.

4.3.1 Mutative Self-Adaptation of Covariances (MSC)

Schwefel (1981) devised the first method to adapt arbitrary covariance matrices. The method uses the fact that every orthogonal matrix can be decomposed in the product of $n(n-1)/2$ elementary rotation matrices. An elementary rotation matrix $R_{jk}(\omega)$ is an unit matrix except that for the following entries holds: $r_{jj} = r_{kk} = \cos \omega$ and $r_{jk} = -r_{kj} = -\sin \omega$. Owing to ❯ Theorem 1 every positive definite matrix can be decomposed via $A = T'S'ST$ where T is orthogonal and S is diagonal matrices with positive entries. As a consequence, every p.d. matrix can be generated by adapting $n(n-1)/2$ angles (for building T) and n scale parameters (for building S) (Rudolph 1992). Arbitrary elliptical mutation vectors can then be generated as follows: If $Z \sim N(0, I_n)$ then $TSZ \sim N(0, (TS)(TS)')$. As a consequence, an individual $(x, \sigma, \omega) \in \mathbb{R}^n \times \mathbb{R}_+^n \times (-\pi, \pi]^{n(n-1)/2}$ consists of its current position $x \in \mathbb{R}^n$, n positive step sizes (or scaling parameters) $\sigma \in \mathbb{R}_+^n$ and $n(n-1)/2$ angles for the elementary rotations. For the sake of notational clarity, ❯ Algorithm 13 is described for the case $\mu = 1$. Please note that the CORR-ES has never been used in this manner! Rather, in practice, choose $\mu > 1$ and deploy recombination on object parameters x and strategy parameters σ and ω.

Algorithm 13 (1, λ)-CORR-ES or (1, λ)-MSC-ES

choose $X^{(0)} \in \mathbb{R}^n$; set $\sigma^{(0)} > 0, t = 0$
repeat
    for $i = 1$ to λ **do**
        draw $Z_i \sim N(0, I_n), S_i \sim LN(0, \eta^2 I_n), L_i \sim LN(0, \tau^2), W_i \sim N\left(0, \left(\frac{5\pi}{180}\right)^2\right)$
        $\sigma_i^{(t+1)} = L_i \cdot \text{diag}(S_i) \cdot \sigma_i^{(t+1)}$
        $\omega_i^{(t+1)} = \left(\omega_i^{(t)} + W_i + (\pi, \ldots \pi)'\right) \mod 2(\pi, \ldots \pi)' - (\pi, \ldots \pi)'$
        $Y^{(t)} = X^{(t)} + \prod_{j=1}^{n-1} \prod_{k=j+1}^{n} R_{jk}\left(\omega_{i,j,k}^{(t+1)}\right) \cdot \text{diag}\left(\sigma_i^{(t+1)}\right) Z_i$
    end for
    let $b \in \{1, \ldots \lambda\}$ such that $f\left(Y_b^{(t)}\right) = \min\{f\left(Y_i^{(t)}\right): i = 1, \ldots, \lambda\}$
    $X^{(t+1)} = Y_b^{(t)}$
    $\sigma^{(t+1)} = \sigma_b^{(t+1)}$
    $\omega^{(t+1)} = \omega_b^{(t+1)}$
    increment t
until stopping criterion fulfilled

4.3.2   Covariance Matrix Estimation by Least Squares (LS-CME-ES)

Rudolph (1992) has proposed to estimate the covariance matrix by least squares estimation from the already evaluated individuals. This method was rediscovered by Auger et al. (2004). The idea is as follows: Assume that the problem is appropriately determined by a second order Taylor expansion: $f(x) = \frac{1}{2} x'Ax + b'x + c$ where $A \in \mathbb{R}^{n,n}, b \in \mathbb{R}^n$ and $c \in \mathbb{R}$. Whenever an individual was evaluated, we have a pair $(x, f(x))$ with known values. But the constants in A, b, and c are unknown. Thus, the above equation has $n(n+1)/2 + n + 1 = (n^2 + 3n + 2)/2 = N$ unknown values if a known pair $(x, f(x))$ was fed in. In this case, this equation is a linear equation where the unknown values for A, b, and c are the variables. To estimate N by the least squares method, at least $N + 1$ equations are needed, that is, at least $N + 1$ evaluated individuals. The estimation of the N parameters requires $O(N^2) = O(n^6)$ operations. The values for b and c are not needed. Finally, a Cholesky decomposition of $A = BB'$ (see ❯ Theorem 1) delivers B and subsequent inversion yields the matrix $Q = B^{-1}$ required for generating the desired elliptically distributed mutation vector. Although the computational effort of $O(n^6)$ seems daunting initially, this approach may be useful if the dimension is not too large and a single fitness function evaluation requires much time.

4.3.3   Generating Set Adaptation (GSA)

Hansen et al. (1995) deployed a different method to generate elliptically distributed normal random vectors with arbitrary covariance matrix. To understand the process, some results should be recalled. If $n = 1$, $m \geq 2$, $b = 0$, $\Sigma = 1$ and $C \in \mathbb{R}^{m,1}$ then ❯ Theorem 2 immediately proves the following result:

Corollary 2   Let $X \sim N_1(0, 1)$ and $Y = X \cdot c$ for some $c \in \mathbb{R}^m$. Then $Y \sim N_m(0, c\,c')$ with $\det(c\,c') = 0$.

Thus, the product of a standard normal random variable and a deterministic vector leads to a multivariate normal random vector with singular covariance matrix. What happens if several random vectors generated in this manner are added?

Theorem 4   (see Rao 1973, p. 524 or Tong 1990, p. 33) If $X_i \sim N_n(v_i, \Sigma_i)$ independent for $i = 1,\ldots,m$ and $a \in \mathbb{R}^m$, then

$$X = \sum_{i=1}^{m} a_i X_i \sim N_n \left(\sum_{i=1}^{m} a_i v_i, \sum_{i=1}^{m} a_i^2 \Sigma_i \right) \qquad \square$$

Notice that there are no restrictions imposed on the covariance matrices, they may be singular as well as regular. Thus, the sum of multivariate normal random vectors with singular covariance matrices is a multivariate normal random vector whose covariance matrix is the sum of the singular covariance matrices. The resulting covariance matrix is regular if the deterministic vectors used are a basis for the vector space $\mathbb{R}^n$.

Theorem 5   Let $X \sim N(0, I_n)$, $B = (b_1, b_2, \ldots, b_n) \in \mathbb{R}^{n,n}$ where column vectors $b_1, \ldots, b_n$ are a basis of $\mathbb{R}^n$, and let e_i be the ith unit vector. Then

$$\sum_{i=1}^{n} X_i b_i \sim N(BB')$$

where BB' is regular.

Proof Thanks to Theorem 2 we know that $BX \sim N(0, BB')$. Since B contains a basis of $\mathbb{R}^n$ its rank is n so that B is regular. It follows that BB' is regular as it has rank n (Abadir and Magnus 2005, p. 81). Notice that $X = \sum_{i=1}^n X_i e_i$ and hence

$$BX = B \sum_{i=1}^n X_i e_i = \sum_{i=1}^n X_i B e_i = \sum_{i=1}^n X_i b_i \sim N(0, BB')$$

which proves the result. □

What happens if additional singular random vectors are added although the distribution is regular already?

Lemma 1 *Let $A \in \mathbb{R}^{n,n}$ be positive semidefinite. If $B \in \mathbb{R}^{n,n}$ is p.s.d., then $A + B$ is positive semidefinite; if $B \in \mathbb{R}^{n,n}$ is p.d., then $A + B$ is positive definite.*

Proof Only the second part is shown: By definition, $x'Ax \geq 0$ and $x'Bx > 0$. Consequently, $x'(A+B)x = x'Ax + x'Bx > 0$ which proves that $A + B$ is positive definite. □

Thus, adding singular normal vectors to a regular normal vector yields a regular normal vector since the covariance matrix stays positive definite. If more than n vectors are used to span $\mathbb{R}^n$ and the set contains a basis then this set of vectors is termed a *generating set* of $\mathbb{R}^n$.

After these preparations, the process of adapting the covariance matrix becomes transparent. Adapt more than n vectors $v_1, \ldots, v_m$ with $m > n$ and generate the random vector via

$$\sum_{i=1}^m v_i Z_i$$

where $Z \sim N(0, I_m)$. Arbitrary covariance matrices can be constructed in this manner.

Algorithm 14 (1, λ)-GSA-ES

choose $X^{(0)} \in \mathbb{R}^n$; set $\sigma^{(0)} > 0, t = 0$
define a generating set $\left\{ v_1^{(0)}, \ldots, v_m^{(0)} \right\}$ of $m \geq n$ vectors spanning the $\mathbb{R}^n$
repeat
　for $i = 1$ to λ **do**
　　draw $Z_i \sim N(0, I_m)$ and G_i
　　$\tilde{Z}_i^{(t)} = \gamma \sum_{j=1}^m Z_{ij} \cdot b_j^{(t)}$
　　$Y_i^{(t)} = X^{(t)} + \sigma^{(t)} \cdot G_i \cdot \tilde{Z}_i^{(t)}$
　end for
　let $b \in \{1, \ldots \lambda\}$ such that $f\left(Y_b^{(t)}\right) = \min\left\{ f\left(Y_i^{(t)}\right) : i = 1, \ldots, \lambda \right\}$
　$X^{(t+1)} = Y_b^{(t)}$
　$\sigma^{(t+1)} = \sigma^{(t)} \cdot G_b$
　$v_1^{(t+1)} = (1 - \alpha)v_1^{(t)} + \sqrt{\alpha(2 - \alpha)}G_b \tilde{Z}_b^{(t)}$
　$v_{j+1}^{(t+1)} = v_j^{(t)}$ for $j = 1, \ldots, m - 1$.
　increment t
until stopping criterion fulfilled

In the beginning, it is recommended to set $v_1^{(0)} = 0 \in \mathbb{R}^n$ and $v_j^{(0)} \sim N\left(0, \frac{1}{n}I_n\right)$ for $j = 2$ to m. The size of the generating set should be in the range $m \in \{n^2, \ldots, 2n^2\}$, guaranteeing that the initial mutation vector has a regular covariance matrix. Notice that the step-size adaptation is mutative again with the random variable G_i distributed as

$$P\left\{ G_i^\beta = \left(\frac{3}{2}\right)^\beta \right\} = P\left\{ G_i^\beta = \left(\frac{2}{3}\right)^\beta \right\} = \frac{1}{2}$$

where $\beta = 1/\sqrt{n}$. Moreover, set $\alpha = 1/\sqrt{n}$ and $\gamma = (1 + 1/m)/\sqrt{m}$.

4.3.4 Covariance Matrix Adaptation (CMA)

Hansen and Ostermeier (1996, 2001) and Hansen (2009) finally introduced a covariance matrix adaptation method that received the status of a standard method. Now, the covariance matrix is continuously updated with information gathered from evaluated points.

Algorithm 15 (μ, λ)-CMA-ES

choose $X^{(0)} \in \mathbb{R}^n$; set $\sigma^{(0)} > 0, C^{(0)} = T^{(0)} = S^{(0)} = I_n, p_c^{(0)} = p_\sigma^{(0)} = 0 \in \mathbb{R}^n, t = 0$

repeat

  for $i = 1$ to λ **do**

    draw $Z_i \sim N(0, I_n)$

    $Y_i^{(t)} = X^{(t)} + \sigma_i^{(t)} \cdot T^{(t)} \cdots S^{(t)} Z_i^{(t)}$

  end for

  let $b_1, b_2, \ldots, b_\lambda \in \{1, \ldots, \lambda\}$ such that $f\left(Y_{b_1}^{(t)}\right) \le f\left(Y_{b_2}^{(t)}\right) \le \cdots \le f\left(Y_{b_\lambda}^{(t)}\right)$

  $X^{(t+1)} = \sum_{i=1}^{\mu} w_{b_i} \cdot Y_{b_i}^{(t)}$

  $p_\sigma^{(t+1)} = (1 - \alpha)p_\sigma^{(t)} + \sqrt{\alpha(2 - \alpha)\mu_{\text{eff}}} \sum_{i=1}^{\mu} w_{b_i} \cdot Z_{b_i}$

  $\sigma^{(t+1)} = \sigma^{(t)} \cdot \exp\left(\frac{\alpha}{\delta} \frac{\|p_\sigma^{(t+1)}\| - \chi_n}{\chi_n}\right)$

  $p_c^{(t+1)} = (1 - \beta)p_c^{(t)} + \eta\sqrt{\beta(1 - \beta)\mu_{\text{eff}}} T^{(t)} \cdot S^{(t)} \sum_{i=1}^{\mu} w_{b_i} \cdot Z_{b_i}$

  $C^{(t+1)} = (1 - \gamma)C^{(t)} + \frac{\gamma}{\nu}p_c^{(t+1)}\left(p_c^{(t+1)}\right)' + \gamma(1 - \nu^{-1}) \sum_{i=1}^{\mu} w_{b_i} T^{(t)} S^{(t)} Z_{b_i} \left(T^{(t)} S^{(t)} Z_{b_i}\right)'$

  get $T^{(t+1)}$ and $S^{(t+1)}$ by spectral decomposition of $C^{(t+1)}$

  increment t

until stopping criterion fulfilled

The parameters are recommended as follows:

$$\lambda = 4 + \lfloor 3 \log n \rfloor$$

$$\mu = \left\lceil \frac{\lambda + 1}{2} \right\rceil$$

$$w_i = \frac{\log\left(\frac{\lambda+1}{2i}\right)}{\mu \log\left(\frac{\lambda+1}{2}\right) - \log(\mu!)}$$

$$\mu_{\text{eff}} = \frac{1}{w'w}$$

$$\nu = \mu_{\text{eff}}$$

$$\alpha = \frac{\mu_{\text{eff}} + 2}{n + \mu_{\text{eff}} + 3}$$

$$\beta = \frac{4}{n+4}$$

$$\gamma = \frac{1}{v}\frac{2}{(n+\sqrt{2})^2} + \left(1-\frac{1}{v}\right)\min\left\{1,\frac{2v-1}{(n+2)^2+v}\right\}$$

$$\delta = 1 + 2\max\left\{0,\sqrt{\frac{\mu_{\text{eff}}-1}{n+1}}-1\right\} + \alpha$$

$$\chi_n = \sqrt{n-\frac{1}{2}}$$

Jastrebski and Arnold (2006) proposed to exploit the information from the worst mutation vectors also, a lesson learned from the analysis of optimally weighted recombination (Rudolph 1997; Arnold 2006): If w is the worst out of λ mutations then it points in a direction of steep ascent. Therefore, $-w$ should point in the direction of steep descent.

4.3.5 Modified Evolutionary Gradient Search (CMA-EGS)

The method presented in ❷ Sect. 4.2.3 was modified with covariance adaptation by Arnold and Salomon (2007). It also requires spectral decomposition of the covariance matrix. As this is a costly operation, they recommend to perform it only every $n/10$ iterations.

Algorithm 16 $(1, \lambda)$-CMA-EGS

choose $X^{(0)} \in \mathbb{R}^n$; set $C^{(0)} = I_n, \sigma^{(0)} > 0, t = 0$
repeat
  for $i = 1$ to λ **do**
    draw $Z_i \sim N(0, C^{(t)})$
  end for
  $\bar{Z} = \sum_{i=1}^{\lambda}\left[f(x^{(t)} - \sigma^{(t)}Z_i) - f(x^{(t)} + \sigma^{(t)}Z_i)\right] \cdot Z_i$
  $\hat{Z} = \frac{\sqrt{n}}{\eta} \cdot \frac{\bar{Z}}{\|\bar{Z}\|}$
  let $T^{(t)}$ and $S^{(t)}$ such that $C^{(t)} = T^{(t)}S^{(t)}\left(T^{(t)}S^{(t)}\right)'$ (see text for details)
  $X^{(t+1)} = X^{(t)} + \sigma^{(t)} \cdot T^{(t)}S^{(t)} \cdot \hat{Z}$
  $s_c^{(t+1)} = (1-\alpha)s_c^{(t)} + \eta\sqrt{\alpha(2-\alpha)}T^{(t)}S^{(t)} \cdot \hat{Z}$
  $s_\sigma^{(t+1)} = (1-\beta)s_\sigma^{(t)} + \eta\sqrt{\beta(2-\beta)}T^{(t)} \cdot \hat{Z}$
  $C^{(t+1)} = (1-\gamma)C^{(t)} + \gamma s_c^{(t+1)}\left(s_c^{(t+1)}\right)'$
  $\sigma^{(t+1)} = \sigma^{(t)} \cdot \exp\left(\frac{\|s_\sigma^{(t)}\|^2-n}{2n(1+\beta^{-1})}\right)$
  increment t
until stopping criterion fulfilled

There is no clear recommendation for $\eta \geq 1$, but according to their experiments $\eta = 2$ seems to be reasonable (especially in the case of noise). The remaining parameters are recommended to be set as follows:

$$\alpha = \beta = \frac{4}{n+4}, \qquad \gamma = \frac{2}{(n+\sqrt{2})^2}$$

4.3.6 Covariance Matrix Self-Adaptation (CMSA)

Beyer and Sendhoff (2008) returned to a mutative self-adaptive step length control but adapted the basic mechanism of the covariance update rule. This opens the door for replacing the costly spectral decomposition by the simpler, quicker, and more stable Cholesky decomposition.

Algorithm 17 (μ, λ)-CMSA-ES

choose $\mu > 1, X^{(0)} \in \mathbb{R}^n$; set $\sigma^{(0)} > 0, C^{(0)} = R^{(0)} = I_n, t = 0$
repeat
  for $i = 1$ to λ **do**
    draw $Z_i \sim N(0, I_n)$ and $L_i \sim LN(0, \tau^2)$
    $Y_i^{(t)} = X^{(t)} + \sigma_i^{(t)} \cdot L_i \cdot R^{(t)} Z_i^{(t)}$
  end for
  let $b_1, b_2, \ldots, b_\lambda \in \{1, \ldots \lambda\}$ such that $f\left(Y_{b_1}^{(t)}\right) \leq f\left(Y_{b_2}^{(t)}\right) \leq \cdots \leq f\left(Y_{b_\lambda}^{(t)}\right)$
  $X^{(t+1)} = \frac{1}{\mu} \sum_{i=1}^{\mu} Y_{b_i}^{(t)}$
  $\sigma^{(t+1)} = \sigma^{(t)} \cdot \frac{1}{\mu} \sum_{i=1}^{\mu} L_{b_i}$
  $C^{(t+1)} = (1 - \gamma)C^{(t)} + \gamma \sum_{i=1}^{\mu} R^{(t)} Z_{b_i} \left(R^{(t)} Z_{b_i}\right)'$
  get $R^{(t+1)}$ by Cholesky decomposition of $C^{(t+1)}$
  increment t
until stopping criterion fulfilled

Only two parameters are used:

$$\tau \in \left[\frac{1}{2\sqrt{2n}}, \frac{1}{\sqrt{2n}}\right] \quad \text{and} \quad \gamma = \frac{2\mu}{2\mu + n(n+1)}$$

The authors reported results comparable to the CMA-ES for small population sizes and considerable better results for large population sizes that are necessary in the presence of noise. If this behavior can be confirmed in future applications, then this method may become a serious competitor for the CMA-ES.

5   Further Reading

In this chapter, only the basic facts and methods have been presented. Many aspects of ES have been left for further reading. For example, Auger and Hansen (2005) recommend to start the CMA-ES as usual but as soon as a termination criterion is fulfilled, the method should be restarted with increased population size. The results reported are impressive. This technique could be useful for other variants as well.

The CMA-ES has been extended to work in the case of multi-objective optimization (Igel et al. 2006) whereas the constraint handling (Kramer and Schwefel 2006) is fairly unexplored. Model-assisted ES have been developed (Emmerich et al. 2002; Ulmer et al. 2004; Hoffmann and Hölemann 2006) as well as methods that work in the presence of noise (Arnold and Beyer 2006; Beyer and Meyer-Nieberg 2006). Also, the technique of niching (Preuss et al. 2005, 2006) for exploring several promising regions simultaneously has been deployed for CMA-ES (Shir et al. 2007). A survey on parallel ES is given in Rudolph (2005).

As already mentioned, there are many other techniques that deserve further reading. Some ideas toward an integration of currently unexplored principles from biological evolution may be found in Rudolph and Schwefel (2008).

References

Abadir K, Magnus J (2005) Matrix algebra. Cambridge University Press, New York

Arnold D (2006) Weighted multirecombination evolution strategies. Theor Comput Sci 361(1):18–37

Arnold D, Beyer HG (2006) A general noise model and its effects on evolution strategy performance. IEEE Trans Evolut Comput 10(4):380–391

Arnold D, Salomon R (2007) Evolutionary gradient search revisited. IEEE Trans Evolut Comput 11(4): 480–495

Auger A, Hansen N (2005) A restart CMA evolution strategy with increasing population size. In: Proceedings of the 2005 IEEE congress on evolutionary computation (CEC 2005), vol 2. IEEE Press, Piscataway, NJ, pp 1769–1776

Auger A, Teytaud O (2007) Continuous lunches are free! In: Thierens D et al. (eds) Proceedings of the genetic and evolutionary computation conference (GECCO 2007). ACM Press, New York, pp 916–922

Auger A, Schoenauer M, Vanhaecke N (2004) LS-CMA-ES: a second-order algorithm for covariance matrix adaptation. In: Yao X et al. (eds) Proceedings of the 8th international conference on parallel problem solving from nature (PPSN VIII). Springer, Berlin, Germany, pp 182–191

Bäck T, Hoffmeister F, Schwefel HP (1991) A survey of evolution strategies. In: Belew RK, Booker LB (eds) Proceedings of the fourth international conference on genetic algorithms (ICGA'91). Morgan Kaufmann, San Mateo CA, pp 2–9

Bartz-Beielstein T (2006) Experimental research in evolutionary computation. The new experimentalism. Springer, Heidelberg

Beyer HG, Meyer-Nieberg S (2006) Self-adaptation of evolution strategies under noisy fitness evaluations. Genet Programming Evolvable Mach 7(4):295–328

Beyer HG, Schwefel HP (2002) Evolution strategies – a comprehensive introduction. Nat Comput 1(1): 3–52

Beyer HG, Sendhoff B (2008) Covariance matrix adaptation revisited – the CMSA evolution strategy. In: Rudolph G et al. (eds) Proceedings of the 10th international conference on parallel problem solving from nature (PPSN X). Springer, Berlin, Germany, pp 123–132

Emmerich M, Giotis A, Oezdemir M, Bäck T, Giannakoglou K (2002) Metamodel-assisted evolution strategies. In: Merelo J et al. (eds) Proceedings of the 7th international conference on parallel problem solving from nature (PPSN VII). Springer, Berlin, Germany, pp 361–370

Fang KT, Kotz S, Ng KW (1990) Symmetric multivariate and related distributions. Chapman and Hall, London, UK

Hansen N (2009) The CMA evolution strategy: A tutorial. Continuously updated technical report. Available via http://www.lri.fr/~hansen/cmatutorial.pdf. Accessed April 2009

Hansen N, Ostermeier A (1996) Adapting arbitrary normal mutation distributions in evolution strategies: the covariance matrix adaptation. In: Proceedings of the 1996 IEEE international conference on evolutionary computation (ICEC '96). IEEE Press, Piscataway, NJ, pp 312–317

Hansen N, Ostermeier A (2001) Completely derandomized self-adaptation in evolution strategies. Evolut Comput 9(2):159–195

Hansen N, Ostermeier A, Gawelczyk A (1995) On the adaptation of arbitrary normal mutation distributions in evolution strategies: the generating set adaptation. In: Eshelman L (ed) Proceedings of the 6th international conference on genetic algorithms (ICGA 6). Morgan Kaufmann, San Fransisco, CA, pp 57–64

Hoffmann F, Hölemann S (2006) Controlled model assisted evolution strategy with adaptive preselection. In: Proceedings of the 2006 international symposium on evolving fuzzy systems. IEEE Press, Piscataway, NJ, pp 182–187

Igel C, Hansen C, Roth S (2006) Covariance matrix adaptation for multi-objective optimization. Evolut Comput 15(1):1–28

Jastrebski G, Arnold D (2006) Improving evolution strategies through active covariance matrix adaptation. In: Proceedings of the 2006 IEEE congress on evolutionary computation (CEC 2006). IEEE Press, Piscataway, NJ

Kramer O, Schwefel HP (2006) On three new approaches to handle constraints within evolution strategies. Nat Comput 5:363–385

McKay M, Beckman R, Conover W (1979) A comparison of three methods for selecting values of input variables in the analysis of output from a computer code. Technometrics 21(2):239–245

Ostermeier A (1992) An evolution strategy with momentum adaptation of the random number distribution.

In: Manderick B, Männer R (eds) Proceedings of the 2nd international conference on parallel problem solving from nature (PPSN II). Elsevier, Amsterdam, The Netherlands, pp 197–206

Ostermeier A, Gawelczyk A, Hansen N (1994a) A derandomized approach to self adaptation of evolution strategies. Evolut Comput 2(4):369–380

Ostermeier A, Gawelczyk A, Hansen N (1994b) Step-size adaptation based on non-local use of selection information. In: Davidor Y et al. (eds) Proceedings of the 3rd international conference on parallel problem solving from nature (PPSN III). Springer, Berlin, Germany, pp 189–198

Poland J, Zell A (2001) Main vector adaptation: a CMA variant with linear time and space complexity. In: Proceedings of the genetic and evolutionary computation conference (GECCO 2001). Morgan Kaufmann, San Fransisco, CA, pp 1050–1055

Preuss M (2006) Niching prospects. In: Filipic B, Silc J (eds) Bioinspired optimization methods and their applications (BIOMA 2006). Jozef Stefan Institute, Ljubljana, Slovenia, pp 25–34

Preuss M, Schönemann L, Emmerich M (2005) Counteracting genetic drift and disruptive recombination in $(\mu + /, \lambda)$-EA on multimodal fitness landscapes. In: Beyer HG et al. (eds) Proceedings of the 2005 genetic and evolutionary computation conference (GECCO 2005), vol 1. ACM Press, New York, pp 865–872

Rao C (1973) Linear statistical inference and its applications, 2nd edn. Wiley, New York

Rechenberg I (1965) Cybernetic solution path of an experimental problem. Royal Aircraft Establishment, Library Translation 1122, Farnborough

Rechenberg I (1973) Evolutionsstrategie: optimierung technischer Systeme nach Prinzipien der biologischen Evolution. Frommann-Holzboog, Stuttgart

Rechenberg I (1978) Evolutionsstrategien. In: Schneider B, Ranft U (eds) Simulationsmethoden in der Medizin und Biologie. Springer, Berlin, Germany, pp 83–114

Rudolph G (1992) On correlated mutations in evolution strategies. In: Männer R, Manderick B (eds) In: Proceedings of the 2nd international conference on parallel problem solving from nature (PPSN II). Elsevier, Amsterdam, The Netherlands, pp 105–114

Rudolph G (1994) An evolutionary algorithm for integer programming. In: Davidor Y et al. (eds) Proceedings of the 3rd conference on parallel problem solving from nature (PPSN III). Springer, Berlin, Germany, pp 63–66

Rudolph G (1997) Convergence properties of evolutionary algorithms. Kovač, Hamburg

Rudolph G (2005) Parallel evolution strategies. In: Alba E (ed) Parallel metaheuristics: a new class of algorithms. Wiley, Hoboken, NJ, pp 155–170

Rudolph G, Schwefel HP (2008) Simulated evolution under multiple criteria revisited. In: Zurada J (ed) WCCI 2008 Plenary/Invited Lectures. Springer, Berlin, Germany, pp 248–260

Salomon R (1998) Evolutionary algorithms and gradient search: similarities and differences. IEEE Trans Evolut Comput 2(2):45–55

Schwefel HP (1964) Kybernetische Evolution als Strategie der experimentellen Forschung in der Strömungstechnik. Diplomarbeit, Technische Universität Berlin, Hermann Föttinger–Institut für Strömungstechnik

Schwefel HP (1977) Numerische Optimierung von Computer-Modellen mittels der Evolutionsstrategie. Birkhäuser, Basel

Schwefel HP (1981) Numerical optimization of computer models. Wiley, Chichester, UK

Schwefel HP (1987) Collective phenomena in evolutionary systems. In: Checkland P, Kiss I (eds) Problems of constancy and change – the complementarity of systems approaches to complexity, Papers presented at the 31st annual meeting of the International Society for General System Research, International Society for General System Research, vol 2. Budapest, pp 1025–1033

Schwefel HP, Rudolph G (1995) Contemporary evolution strategies. In: Morán F et al. (eds) Advances in artificial life – proceedings of the third European conference on artificial life (ECAL'95). Springer, Berlin, Germany, pp 893–907

Shir O, Emmerich M, Bäck T (2007) Self-adaptive niching CMA-ES with Mahalanobis metric. In: Proceedings of the 2007 IEEE congress on evolutionary computation (CEC 2007). IEEE Press, Piscataway, NJ

Smit S, Eiben A (2009) Comparing parameter tuning methods for evolutionary algorithms. In: Proceedings of the 2009 IEEE congress on evolutionary computation (CEC 2009). IEEE Press, Piscataway, NJ

Tong Y (1990) The multivariate normal distribution. Springer, New York

Ulmer H, Streichert F, Zell A (2004) Model assisted evolution strategies. In: Jin Y (ed) Knowledge incorporation in evolutionary computation. Springer, Berlin, Germany, pp 333–358

Wolpert D, Macready W (1997) No free lunch theorems for optimization. IEEE Trans Evolut Comput 1(1):67–82

23 Evolutionary Programming

Gary B. Fogel
Natural Selection, Inc., San Diego, CA, USA
gfogel@natural-selection.com

G. Rozenberg et al. (eds.), *Handbook of Natural Computing*, DOI 10.1007/978-3-540-92910-9_23,
© Springer-Verlag Berlin Heidelberg 2012

Abstract

Evolutionary programming (EP) has a long history of development and application. This chapter provides a basic background in EP in terms of its procedure, history, extensions, and application. As one of the founding approaches in the field of evolutionary computation (EC), EP shares important similarities and differences with other EC approaches.

1   Background

Evolutionary programming (EP) is an approach to simulated evolution that iteratively generates increasingly appropriate solutions in the light of a stationary or nonstationary environment and desired fitness function. Standard EP makes use of the same four components that are standard to all evolutionary algorithms (EAs): initialization, variation, evaluation, and selection. This approach to population-based search has its roots in the early work of Lawrence J. Fogel, as described later in ❷ Sect. 2. Over generations of simulated evolution, the population of solutions learns about its environment through stochastic trial and error, with a process of selection continually driving the population in more useful directions in the search space.

The basic procedure for canonical EP evolved finite state machines (FSMs) (also called "finite-state automata") (Mealy 1955; Moore 1957), which were a standard transducer representation in the literature at the time. Finite-state machines operate on a finite set of input symbols, possess a finite number of internal states, and produce symbols from a finite set. The corresponding input-output symbol pairs and set of state transitions specify the behavior of the FSM (❷ Fig. 1). A population of solutions (i.e., FSM representations) was initialized, generally at random, with memory limitations on the possible number of states in the FSM due to the computers that were being used at the time of its invention in the early 1960s. A known sequence of symbols from the environment was defined as an observed sequence

❏ **Fig. 1**

A finite-state machine (FSM) representation. Starting in state "A," input symbols are shown to the left of each slash and response symbols are shown to the right of the slash. Arrows provide connections between states. The path of connections through the FSM produces a series of output symbols from a given input symbol pattern that can be used as a predictor of that symbol environment.

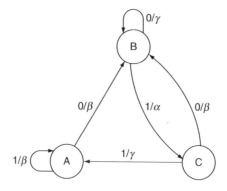

or time series. Existing FSMs in the population were mutated as "parent" solutions to generate "offspring" FSMs. Both parents and offspring were then scored in light of how well they predicted the next symbol in a known sequence of symbols. The best half of the FSMs in the population were retained as parents for the next iteration, with selection removing the remaining half of poor-quality solutions. As required, the best available machine in the population was used to predict the next symbol in the symbol sequence, and the actual symbol was then added to the known available data for model training and optimization. The process was iterated for as long as time permitted (❷ *Fig. 2*). This initial approach to EP was first offered in the literature in Fogel (1962a, b) and was studied through a variety of experiments in Fogel et al. (1964, 1965a, b, c) and summarized in Fogel et al. (1966).

1.1 The Canonical EP

For EP, the size of the evolving population, μ, ranges in size depending on the problem and computational processing speed available. A population of solutions was used in all of the early EP experiments, although small populations were typically used in initial experiments. The original process was written in Fortran II and processed on an IBM 7094, which had roughly half the processing speed of an Apple II or Commodore 64 (Fogel 1999). Each solution is evaluated with respect to a fitness function and offspring solutions are generated through a process of variation. For the initial FSM representations, this variation was typically via mutation as there were clear ways in which to vary these representations. The first was to add a state and then randomly assign all of the input-output and input-transition pairs for this

◻ **Fig. 2**
A typical output series for the evolution of finite-state machines using evolutionary programming (EP) to predict primeness of numbers. The evolving population rapidly converged on a useful finite state machine (FSM) after about 100 symbols experienced during the learning process.

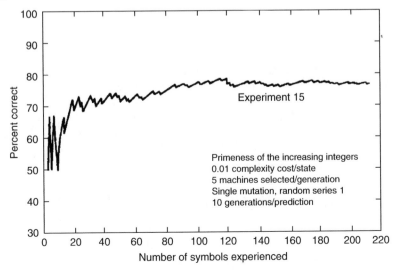

state, provided the FSM maximum size does not exceed memory constraints. A second mutation type was to delete a state if there is more than one state in the FSM. A third was to change an output symbol coupled with a specific input symbol in a single state. The fourth was to alter a state transition associated with an input symbol in a single state. Finally, a fifth was to change the starting state itself, if the FSM had more than one state. By adjusting the probability for each of these five mutation strategies and/or the number of mutations per offspring, a diverse set of possible FSMs for offspring solutions was possible. In the canonical version of EP, each parent member i generated λ_i offspring with the above mutation strategies. It should be noted that Fogel et al. (1966) suggested the possibility of other mutation strategies such as recombining representations to generate new solutions; however, these FSM-specific operators were sufficient for early experimentation to demonstrate the utility of the process. In addition, Fogel considered the FSM representations to be equivalent to individual phenotypes being evaluated in an environmental context by selection. As such, the actual methods of modification were less important than simply offering phenotypic change over time in the population of solutions. Thus, the EP process was designed to simulate evolution at the individual level and given that nature only scores phenotypic behavior in a given environment, so long as FSMs were able to generate useful predictions their increasing utility would be guaranteed by the process of evolutionary optimization.

For convenience and with respect to limiting computational requirements, early experiments in EP maintained a constant small population size of three to five FSMs. Each of the m parents was mutated to generate a single offspring, with the resulting $2m$ solutions all competing together for survival to the next generation. The best m solutions were retained as parents. It should be noted that Fogel (1964) and Fogel et al. (1966) noted the possibility of retaining less-fit solutions in the population, but computational hardware requirements were already difficult. Later experiments incorporated tournament selection or proportional-based selection to retain a sampling of m best solutions. The evolutionary process was terminated once a predetermined number of generations had expired, or other criterion had been satisfied.

2 History

Evolutionary programming (EP) was conceived by Lawrence J. Fogel in the early 1960s as an alternative approach to classic artificial intelligence (AI) in computers. While on leave from General Dynamics in 1960 as an assistant to the director of research for the National Science Foundation, Fogel was given the challenge of determining how much America should invest in basic research (Fogel 1999). A portion of this survey focused on methods of artificial intelligence. At the time, two different schools of thought prevailed in AI. The first, a "bionics" approach, attempted to build an artificial brain, neuron by neuron. These approaches were considered by Fogel to show promise for pattern recognition, but not for the generation of intelligence in computers. The second approach, "heuristic programming," attempted to achieve AI through a series of if-then rules. Modern expert system methods continue to pursue this approach, but there are many issues involved with heuristics such as the quality and qualifications of the experts themselves, how well rules that are written for current situations translate when the environment is nonstationary, and how to make decisions when experts disagree. In light of these shortcomings, Fogel conceived the idea of simulating the evolutionary process on computers – the same process that generated human intelligence over millions of years of evolution. Such a process does not rely on heuristics, but generates solutions of increasing intellect over time. This approach to AI was presented to his colleagues at the National Science Foundation in June of 1961 but the response was far from positive.

Upon returning to General Dynamics at San Diego, California in June of 1961, the first computer experiments were conducted, together with Al Owens, who was in charge of the computer laboratory. Jack Walsh, a mathematician at General Dynamics became interested in mathematical representations to determine the capabilities of the evolutionary process. The first experiments using FSMs resulted in the publication of *Industrial Research* in 1962 (Fogel 1962a). Several publications resulted from the team of Fogel, Owens, and Walsh between 1962 and 1966 (Fogel 1962b, 1963; Fogel et al. 1964, 1965a, b, c, 1966). Fogel's dissertation in 1964 at the University of California, Los Angeles (the first Ph.D. dissertation in the field of evolutionary computation) (Fogel 1964) described experiments with a population of FSMs where each FSM was exposed to a sequence of observed symbols (representing an environment). A predefined payoff function was used to score the worth of each FSM after input symbols were provided to each FSM and an output prediction was compared to the actual next symbol in the sequence. After all symbols are predicted, a function of the payoff values was used to calculate the overall worth of each FSM. From this set of parent FSMs, offspring FSMs were generated by mutating each parent machine using five possible methods: (1) change an output symbol, (2) change a state transition, (3) add a state, (4) delete a state, and (5) change the initial state. The mutation operators and the number of mutation operations were chosen using a uniform and Poisson probability distribution, respectively, or other specified probability distributions. Like the parent solutions, each offspring solution was then evaluated over the sequence of symbols and scored. FSMs providing the highest payoff were retained as parent solutions for a subsequent generation of FSMs. Standard EP approaches kept the best half of the population to serve as parents for the subsequent generation of solutions and one offspring was generated from each parent solution to maintain a constant population size. The process is continued until all symbols were trained and a prediction of a symbol yet to be encountered is required (equivalent to a testing example after a series of training examples has been learnt). Many of Fogel's early experiments applied evolved FSMs to successively more difficult tasks ranging from simple two-symbol cyclic sequences to eight-symbol cyclic sequences, with noise, to sequences of symbols generated by other FSMs, and even to nonstationary sequences and other number series taken from an article by Flood (1962).

In 1965, Fogel, Owens, and Walsh left General Dynamics to form a company, Decision Science, Inc., in San Diego, California to explore applications of evolutionary programming. This was the first company dedicated to the application of evolutionary computation. The company continued for 17 years applying EP for time series prediction and modeling aircraft combat. Many of these applications were aided by the expertise of George Burgin, another employee of the company. The company was eventually acquired by Titan Systems in 1982.

In 1966, Fogel, Owens, and Walsh published the first book in the field of evolution computation entitled *Artificial Intelligence through Simulated Evolution*, which described the evolution of FSMs in detail and introduced evolutionary computation broadly to the AI community. Surprisingly, the reaction by the AI community was largely negative (see Fogel 1999 for additional details).

Fogel and Burgin (1969) applied EP to discover globally optimal strategies in two-player, zero-sum games. This early application of evolutionary computation to game theory was later extended to gaming situations that outperformed human subjects in nonzero-sum games such as pursuit evasion for aerial combat. Such models evolved the path for an interceptor toward a moving target in three dimensions and were capable of predicting the position of the target quite successfully, even at this early stage in processing speed. This process, resulting in an intelligently interactive opponent, was called an adaptive maneuvering logic (AML) (Burgin et al. 1975). AML achieved a level of proficiency that even impressed pilots at the level of accomplished U.S. Navy Blue Angels (Burgin 2008). It was at about this same time that Fogel became the second president of the American Cybernetics Society, succeeding Warren S. McCulloch.

Walsh et al. (1970) presented an early application of EP for automatic control of a system with direct comparison to human control. FSMs were used to generate predictions of a future symbol from a sequence of input symbols. In this project, the overall output of the evolved system was itself a combination of optimized submachines. When compared to humans, the evolved models consistently outperformed their human counterparts in terms of prediction accuracy.

Throughout the 1970s, Don Dearholt of New Mexico State University supervised several masters theses on the subject of EP (Cornett 1972; Lyle 1972; Montez 1974; Root 1970; Trellue 1973; Vincent 1976; Williams 1977) in addition to Wirt Atmar's (1976) Ph.D. dissertation, also on EP and its relation to machine intelligence. Dearholt's interest in EP was reviewed in Fogel (2004), and during the 1970s, more evolutionary computation research was published in EP than any other form of simulated evolution. Atmar (1976) provided an early example of simulating evolution in an artificial life setting, and suggested how EP could be designed for what is now termed "evolvable hardware."

2.1 Philosophical Differences

As mentioned previously, EP was constructed as a simulation of phenotypic evolution in accordance with the neo-Darwinian paradigm that selection operates only on phenotypic behavior rather than directly at the level of the genotype. The underlying code of the phenotype (i.e., the genotype) is only affected indirectly by natural selection. This philosophy represents a "top-down" rather than "bottom-up" approach to optimization. Later approaches, such as Holland's genetic algorithm (GA), focused on a decidedly "bottom-up" view of evolution, relying on the identification, combination, and survival of "good" building blocks (i.e., schemata) to iteratively form better and larger building blocks. Thus, in GA, the importance was directed specifically at the level of the genotype and methods of recombination of representations, rather than the phenotype. The resulting building block hypothesis carries an implicit assumption that fitness is a separable part of the simulated genome. In EP, solutions are only judged on their phenotypic behavior in light of an environment without credit apportionment to the representation. EP and evolution strategies (ES) share great similarity in this regard.

3 Extensions

The canonical EP was extended in many ways over time. For example, in continuous EP, new individuals in the population are added without iterative generations on a continual basis (Fogel and Fogel 1995). This approach is similar to $(\pi+1)$ selection in ES. Self-adaptive strategies have been added to EP to adjust the probabilities of mutation through their own process of evolutionary optimization (Fogel et al. 1991, 1992). Commonly, in standard evolutionary algorithms, the probabilities associated with variation operators are fixed by the user in advance of the evolutionary process. With self-adaptation, these probabilities are modified concurrent with the evolutionary process, retaining a record of which probabilities and parameters worked best in previous generations and adjusting future probabilities in light of those successes. This approach avoids the necessity for user-tuned probabilities associated with each variation operator and adapts the variances via

$$x_i(k+1) := x_i(k) + v_i(k) \times N_i(0, 1)$$

$$v_i(k+1) := v_i(k) + [\sigma v_i(k)] 1/2 \times N_i(0, 1)$$

where x_i and v_i are the ith components of x and v, respectively, and $N_i(0,1)$ represents a standard normal distribution. To ensure that the variance v_i remains nonnegative, σ was introduced. Further information on this process is offered in Fogel (2006) and Porto (1997). Another extension of EP focused on altering the probability distribution associated with EP applied to continuous parameter optimization where the mutation operator is Gaussian with zero mean and variance. By altering this distribution to Cauchy or other distributions (even combinations of distributions), the efficiency of the search can be improved for some functions (Yao and Liu 1996, 1999; Fogel 1999). Other extensions include the addition of random extinction events to the evolving population (Greenwood et al. 1999). EP has also been used to optimize a wide range of modeling representations from simple FSMs to neural networks (Fogel et al. 1991), to fuzzy if-then rule sets (Fogel and Cheung 2005; Porto et al. 2005). The first real-world application of evolved neural networks with EP was offered by Porto (1989), who evolved neural network classifiers for active sonar returns to classify underwater metal objects from background. This approach was further developed in Porto et al. (1995), and also by Yao (1991), Yao and Liu (1997a, b), and others. The reader is directed to the award-winning review by Yao (1999) for an extensive survey of this approach. Yao and Liu (1997a, b) developed EPNet, an evolutionary system for evolving neural networks that makes use of EP. This approach made use of many concepts of EP described above including continuous EP and was reviewed in Yao and Islam (2008) for evolved neural network ensembles. Further, the approach has been extended by Yao and others to include the evolution of neural network ensembles. Martinez-Estudillo et al. (2006) generated an extension of neural networks optimized by EP via of nodes whose basis function units are products of the inputs raised to a real number power (termed "product units").

4 Recent Applications

Like many evolutionary algorithms, EP has a wide breadth of applications. For example, within robotic control, Chen et al. (2009) used EP coupled with fuzzy logic for the control of wheeled robots. EP was also used to optimize the control gains for the kinematic control system. In the area of mobile network strategic planning, Portilla-Figueras et al. (2008) applied EP to discover optimal cell size determination. This approach was compared to several standard approaches, and obtained significantly better results in many scenarios. Hoorfar (2007) reviewed applications of EP continuous, discrete, and mixed parameter electromagnetic optimization problems including RF circuit design. Guo et al. (2008) introduced a method to optimize wavelet filters using a combination of EP and a chaos (a so-called chaos evolutionary programming) algorithm. Jamnejad and Hoorfar (2004) optimized multifrequency feed horns for reflector antennas of the Jet Propulsion Laboratory/NASA Deep Space Network. The resulting designs were optimized in the light of constraints on return loss, antenna beam width, pattern circularity, and low cross-polarization.

EP has also received broad application in power plant control and optimization in the last decade. Sahoo (2008) evaluated the ability of EP to optimize the performance of a cogeneration system. The method used a canonical EP to search exergoeconomic parameters for the system, and relate these to the monetary cost of running the facility at a particular efficiency of electricity generation. Repeated results on several cogeneration systems demonstrated that EP was able to identify useful parameter settings. Similarly, Contreras-Hernández and Cedeño-Maldonado (2008) used EP to calculate voltage magnitudes for power systems following a power outage. This approach was evaluated on a modified IEEE 30-bus system and found to provide accurate solutions with minimal computational resources. Fong et al. (2007) optimized the installation of solar panels for centralized solar water-heating systems in high-rise buildings to maximize energy savings using EP.

Other applications include optimized methods for impedance-based structural health monitoring approaches (Xu et al. 2004), flow-shop scheduling (Wang and Zheng 2003), and economic load dispatch (Sinha et al. 2003). EP has a long history of application in biological, medical, and chemical applications including molecular docking (Gehlhaar et al. 1995; Kissinger et al. 1999) and optimization of neural networks for quantitative structure–activity prediction (Hecht and Fogel 2007) and prediction of viral tropism from sequence information (Lamers et al. 2008).

Given the no free lunch theorem (Wolpert and Macready 1997) and the realization of the basic similarities of all evolutionary algorithms (population-based random variation and selection), the importance of one particular EA approach versus another has become blurred as the field of evolutionary computation has grown. EP remains a historically important and currently valuable form within this larger community.

References

Atmar JW (1976) Speculation on the evolution of intelligence and its possible realization in machine form. Doctoral dissertation, New Mexico State University, Las Cruces

Burgin GH (2008) In memoriam – reflections on Larry Fogel at Decision Science, Inc. (1965–1982). IEEE Computational Intelligence Magazine 69–72

Burgin GH, Fogel LJ, Phelps JP (1975) An adaptive maneuvering logic computer program for the simulation of one-on-one air-to-air combat, NASA Contractor Report NASA CR-2582, Washington, DC

Chen C-Y, Li T-HS, Yeh Y-C (2009) EP-based kinematic control and adaptive fuzzy sliding-mode dynamic control for wheeled mobile robots. Info Sci 179:180–195

Contreras-Hernández EJ, Cedeño-Maldonado JR (2008) Evolutionary programming applied to branch outage simulation for contingency studies. Int J Power Energy Syst 28:48–53

Cornett FN (1972) An application of evolutionary programming to pattern recognition. Master's thesis, New Mexico State University, Las Cruces

Flood MM (1962) Stochastic learning theory applied to choice experiments with cats, dogs and men. Behav Sci 7:289–314

Fogel DB (2004) In memoriam Don Dearholt. IEEE Trans Evol Comput 8:97–98

Fogel DB (2006) Evolutionary computation: toward a new philosophy of machine intelligence, 3rd edn. Wiley-IEEE Press, Piscataway, NJ

Fogel DB, Fogel LJ, Atmar JW (1991) Meta-evolutionary programming. In: Chen RR (ed) Proceedings of 25th Asilomar conference on signals, systems, and computers. IEEE Computer Society, Los Alamitos, CA, pp 540–545

Fogel DB, Fogel LJ, Atmar JW, Fogel GB (1992) Hierarchic methods of evolutionary programming. In: Fogel DB,

Atmar W (eds) Proceedings of 1st annual conference on evolutionary programming. Evolutionary Programming Society, La Jolla, CA, pp 175–182

Fogel GB, Cheung M (2005) Derivation of quantitative structure-toxicity relationships for exotoxicological effects of organic chemicals: evolving neural networks and evolving rules. In: 2005 IEEE congress on evolutionary computation. IEEE, Edinburgh, pp 274–281

Fogel GB, Fogel DB (1995) Continuous evolutionary programming: analysis and experiments. Cybernet Syst 26:79–90

Fogel LJ (1962a) Autonomous automata. Indus Res 4:14–19

Fogel LJ (August, 1962b) Toward inductive inference automata. In: Proceedings of the congress, International Federation for Information Processing, Munich

Fogel LJ (1963) Biotechnology: concepts and applications. Prentice-Hall, Englewood, NJ

Fogel LJ (1964) On the organization of intellect. Doctoral dissertation, University of California, Los Angeles, CA

Fogel LJ (1990) The future of evolutionary programming. In: Chen RR (ed) Proceedings of 24th Asilomar conference on signals, systems, and computers. Maple Press, Pacific Grove, CA, pp 1036–1038

Fogel LJ (1999) Intelligence through simulated evolution: forty years of evolutionary programming. Wiley, New York

Fogel LJ, Burgin GH (1969) Competitive goal-seeking through evolutionary programming. Air Force Cambridge Research Labs, Final Report, Contract AF 19(628)–5927

Fogel LJ, Owens AJ, Walsh MJ (1964) On the evolution of artificial intelligence. In: Proceedings of 5th National Symp. on Human Factors in Engineering, IEEE, San Diego, CA, pp 63–67

Fogel LJ, Owens AJ, Walsh MJ (1965a) Intelligent decision-making through a simulation of evolution. Behav Sci 11:253–272

Fogel LJ, Owens AJ, Walsh MJ (1965b) Intelligent decision-making through a simulation of evolution. IEEE Trans Hum Factors Electron 6:13–23

Fogel LJ, Owens AJ, Walsh MJ (1965c) Artificial intelligence through a simulation of evolution. In: Maxfield M, Callahan A, Fogel LJ (eds) Biophysics and cybernetic systems: Proceedings of 2nd cybernetic science symposium. Spartan Books, Washington, DC, pp 131–155

Fogel LJ, Owens AJ, Walsh MJ (1966) Artificial intelligence through simulated evolution. Wiley, New York

Fong KF, Chow TT, Hanby VI (2007) Development of optimal design of solar water hearing system by using evolutionary algorithm. J Solar Energy Eng 129:499–501

Gehlhaar DK, Verkhivker GM, Rejto PA, Sherman CJ, Fogel DB, Fogel LJ, Freer ST (1995) Molecular recognition of the inhibitor AG-1343 by HIV-1 protease: Conformationally flexible docking by evolutionary programming. Chem Biol 2:317–324

Greenwood GW, Fogel GB, Ciobanu M (1999) Emphasizing extinction in evolutionary programming. In: Proceedings of 1999 congress on evolutionary computation. IEEE, Washington DC, pp 666–671

Guo S-M, Chang W-H, Tsai JSH, Zhuang B-L, Chen L-C (2008) JPEG 2000 wavelet filter design framework with chaos evolutionary programming. Signal Process 88:2542–2553

Hecht D, Fogel GB (2007) High-throughput ligand screening via preclustering and evolved neural networks. IEEE/ACM Trans Comput Biol Bioinform 4:476–484

Hoorfar A (2007) Evolutionary programming in electromagnetic optimization: a review. IEEE Trans Antennas Propagation 55:523–537

Jamnejad V, Hoorfar A (2004) Design of corrugated horn antennas by evolutionary optimization techniques. IEEE, Antennas Wireless Propagation Lett 3:276–279

Kissinger CR, Gehlhaar DK, Fogel DB (1999) Rapid automated molecular replacement by evolutionary search. Acta Crystallogr D Biol Crystallogr 55:484–491

Lamers SL, Salemi M, McGrath MS, Fogel GB (2008) Prediction of R5, X4, and R5X4 HIV-1 coreceptor usage with evolved neural networks. IEEE/ACM Trans Comput Biol Bioinform 5:291–299

Lyle MR (1972) An investigation into scoring techniques in evolutionary programming. Master's thesis, New Mexico State University, Las Cruces

Mealy GH (1955) A method of synthesizing sequential circuits. Bell Syst Tech J 34:1054–1079

Martinez-Estudillo A, Martinez-Estudillo F, Hervas-Martinez C, Garcia-Pedrajas N (2006) Evolutionary product unit based neural networks for regression. Neural Netw 19:477–486

Montez J (1974) Evolving automata for classifying electrocardiograms. Master's thesis, New Mexico State University, Las Cruces

Moore EF (1957) Gedanken-experiments on sequential machines: automata studies. Ann Math Stud 34: 129–153

Portilla-Figueras JA, Salcedo-Sanz S, Oropesa-García A, Bousoño-Calzón C (2008) Cell size determination in WCDMA systems using an evolutionary programming approach. Comput Oper Res 35:3758–3768

Porto VM (1989) Evolutionary methods for training neural networks for underwater pattern classification. 24th Ann. Asilomar conf. on signals, systems, and computers, vol 2, pp 1015–1019

Porto VW (1997) Evolutionary programming. In: Handbook on evolutionary computation. IOP, Bristol. Oxford University Press, New York

Porto VW, Fogel DB, Fogel LJ (1995) Alternative neural network training methods. IEEE Expert 10(3):16–22

Porto VW, Fogel DB, Fogel LJ, Fogel GB, Johnson N, Cheung M (2005) Classifying sonar returns for the presence of mines: evolving neural networks and evolving rules. In: Fogel DB Piuri V (eds) 2005 IEEE symposium on computational intelligence for homeland security and personal safety, IEEE Press, Piscataway, NJ, pp 123–130

Root R (1970) An investigation of evolutionary programming. Master's thesis, New Mexico State University, Las Cruces

Sahoo PK (2008) Exergoeconomic analysis and optimization of a cogeneration system using evolutionary programming. Appl Thermal Eng 28: 1580–1588

Sinha N, Chakrabarti R, Chattopadhyay PK (2003) Evolutionary programming techniques for economic load dispatch. IEEE Trans Evol Comput 7:83–94

Trellue RE (1973) The recognition of handprinted characters through evolutionary programming. Master's thesis, New Mexico State University, Las Cruces

Vincent RW (1976) Evolving automata used for recognition of digitized strings. Master's thesis, New Mexico State University, Las Cruces

Walsh MJ, Burgin GH, Fogel LJ (1970) Prediction and control through the use of automata and their evolution. U.S. Navy Final Report Contract N00014-66-C–0284

Wang L, Zheng D-Z (2003) A modified evolutionary programming for flow shop scheduling. Int J Adv Manuf Technol 22:522–527

Williams GL (1977) Recognition of hand-printed numerals using evolving automata. Master's thesis, New Mexico State University, Las Cruces

Wolpert DH, Macready WG (1997) No free lunch theorems for optimization. IEEE Trans Evol Comput 1:67–82

Xu J, Yang Y, Soh CK (2004) Electromechanical impedance-based structural health monitoring with evolutionary programming. J Aerospace Eng 17:182–193

Yao X (1991) Evolution of connectionist networks. In: Proceedings of the international symposium on AI, reasoning and creativity. Griffith University, Queensland, Australia, pp 49–52

Yao X (1999) Evolving artificial neural networks. Proc IEEE 87:1423–1447

Yao X, Islam M (February, 2008) Evolving artificial neural network ensembles. IEEE Comput Mag 3(1):31–42

Yao X, Liu Y (1996) Fast evolutionary programming. In: Fogel LJ, Angeline PJ, Bäck T (eds) Evolutionary programming V: Proceedings of 5th annual conference on evolutionary programming. MIT Press, Cambridge, MA, pp 451–460

Yao X, Liu Y (1997a) EPNet for chaotic time-series prediction. In: Yao X, Kim J-H Furuhashi T (eds) Simulated evolution and learning. Springer, Berlin, pp 146–156

Yao X, Liu Y (1997b) A new evolutionary system for evolving artificial neural networks. IEEE Trans Neural Netw 8:694–713

Yao X, Liu Y (1999) Evolutionary programming made faster. IEEE Trans on Evol Prog 3:82–102

24 Genetic Programming — Introduction, Applications, Theory and Open Issues

Leonardo Vanneschi[1] · *Riccardo Poli*[2]
[1]Department of Informatics, Systems and Communication, University of
Milano-Bicocca, Italy
vanneschi@disco.unimib.it
[2]Department of Computing and Electronic Systems, University of Essex,
Colchester, UK
rpoli@essex.ac.uk

G. Rozenberg et al. (eds.), *Handbook of Natural Computing*, DOI 10.1007/978-3-540-92910-9_24,
© Springer-Verlag Berlin Heidelberg 2012

Abstract

Genetic programming (GP) is an evolutionary approach that extends genetic algorithms to allow the exploration of the space of computer programs. Like other evolutionary algorithms, GP works by defining a goal in the form of a quality criterion (or fitness) and then using this criterion to evolve a set (or population) of candidate solutions (individuals) by mimicking the basic principles of Darwinian evolution. GP breeds the solutions to problems using an iterative process involving the probabilistic selection of the fittest solutions and their variation by means of a set of genetic operators, usually crossover and mutation. GP has been successfully applied to a number of challenging real-world problem domains. Its operations and behavior are now reasonably well understood thanks to a variety of powerful theoretical results. In this chapter, we introduce the main definitions and features of GP and describe its typical operations. We then survey some of its applications. We also review some important theoretical results in this field, including some very recent ones, and discuss some of the most challenging open issues and directions for future research.

1   Introduction

Genetic algorithms (GAs) are capable of solving many problems competently. Furthermore, in their simplest realizations, they have undergone a variety of theoretical studies, so that a solid understanding of their properties is now available. Nevertheless, the fixed-length string-type representation of individuals that characterizes GAs is unnatural and overly constraining for a wide set of applications. These include the evolution of computer programs, where, rather obviously, the most natural representation for a solution is a hierarchical, variable size structure rather than a fixed-length character string. In particular, fixed-length strings do not readily support the hierarchical organization of tasks into subtasks typical of computer programs; they do not provide a convenient way of incorporating iteration and recursion; and so on. Above all, traditional GA representations cannot be dynamically varied at run time. The initial choice of string-length limits in advance the number of internal states of the system, thereby setting an upper bound on what the system can learn.

This lack of representation power (already recognized two decades ago, for example, in De Jong (1988)) is overcome by genetic programming (GP), an extension of GAs that largely owes its success to Koza (1992a), who defines GP as:

▶   a systematic method for getting computers to automatically solve a problem starting from a high-level statement of what needs to be done (Koza and Poli 2003).

In some senses, GP represents an attempt to accomplish one of the most ambitious goals of computer science: being able to specify what one wants a program to do, but not how, and have the computer figure out an implementation.

At a slightly lower level of abstraction, GP basically works like a GA. As GAs do for fixed length strings of characters, GP genetically breeds a population of computer programs to solve a problem. The major difference between GAs and GP is that in the former the individuals in the population are fixed-length strings of characters, in the latter they are hierarchical variable-size structures that represent *computer programs.*

Every programming language (e.g., Pascal or C) is capable of expressing and executing general computer programs. Koza, however, chose the *LISP* (*LISt* Processing) language to code his GP implementation. Besides the fact that many sophisticated programming tools were available for LISP at the time, the language presented a variety of advantages. Both programs and data have the same form in LISP, so that it is possible and convenient to treat a computer program as data in the genetic population. This common form of programs and data is the list. Lists can be nested and, therefore, can easily represent hierarchical structures such as *syntax trees.* Syntax trees are particularly suitable as representations of computer programs. Their size, shape, and primitives can be changed dynamically by genetic operators, thereby imposing no *a priori* limit on the complexity of what's evolved by GP. LISP facilitates the programming of such manipulations.

While modern GP implementations are based on C, C++, Java, or Python, rather than LISP, and syntax trees are nowadays often represented using a flattened array-based representation rather than linked lists (typical of LISP), from a logical point of view programs are still treated and manipulated as trees as Koza did. This form of GP will be called *tree-based GP* hereafter.

While the tree-based representation of individuals is the oldest and most common one, it is not the only form of GP that has been employed to evolve computer programs. In particular, in the last few years, a growing attention has been dedicated by researchers to linear genomes (see for instance Brameier and Banzhaf 2001) and graph-based genomes (see for instance Miller 2001). Many other forms of GP, for example, based on grammars or based on the estimation of probability distributions, have also been proposed. However, for space limitations these forms are not covered in this chapter. The interested reader is referred to Poli et al. (2008) for a more comprehensive review.

This chapter is structured as follows. ❷ Sect. 2 introduces the main definitions and features of GP and its operators. ❷ Sect. 3 offers an outline of some of the most significant real-world applications of GP. ❷ Sect. 4 reviews the most significant theoretical results to date on GP, including some recent and exciting new advances. ❷ Sect. 5 lists some important open issues in GP. ❷ Sect. 6 provides some conclusions.

2 The Mechanics of Tree-Based GP

In synthesis, the GP paradigm works by executing the following steps:

1. Generate an initial population of computer programs (or individuals).
2. Perform the following steps until a termination criterion is satisfied:
   (a) Execute each program in the population and assign it a fitness value according to how well it solves the problem.
   (b) Create a new population by applying the following operations:
      i. Probabilistically select a set of computer programs for mating, on the basis of their fitness (selection).
      ii. Copy some of the selected individuals, without modifying them, into the new population (reproduction).

iii. Create new computer programs by genetically recombining randomly chosen parts of two selected individuals (crossover).

iv. Create new computer programs by substituting randomly chosen parts of some selected individuals with new randomly generated ones (mutation).

3. The best computer program which appeared in any generation is designated as the result of the GP process. This result may be a solution (or an approximate solution) to the problem.

In the following sections, each step of this process is analyzed in detail.

2.1 Representation of GP Individuals

The set of all the possible structures that GP can generate includes all the trees that can be built recursively from a set of function symbols $\mathscr{F} = \{f_1, f_2, \ldots, f_n\}$ (used to label internal tree nodes) and a set of terminal symbols $\mathscr{T} = \{t_1, t_2, \ldots, t_m\}$ (used to label the leaves of GP trees). Each function in the function set $\mathscr{F}$ takes a fixed number of arguments, known as its *arity*. Functions may include arithmetic operations $(+, -, \times, \text{etc.})$, mathematical functions (such as sin, cos, log, exp), Boolean operations (such as *AND, OR, NOT*), conditional operations (such as If-Then-Else), iterative operations (such as While-Do), and other domain-specific functions that may be defined to suit the problem at hand. Each terminal is typically either a variable or a constant.

For example, given the set of functions $\mathscr{F} = \{+, -\}$ and the set of terminals $\mathscr{T} = \{x, 1\}$, a legal GP individual is represented in ❷ *Fig. 1*. This tree can also be represented by the LISP-like S-expression $(+ \ x \ (- \ x \ 1))$ (for a definition of LISP S-expressions see, for instance, Koza (1992)).

It is good practice to choose function and terminal sets that satisfy two requirements: *closure* and *sufficiency*.

The *closure property* requires that each of the functions in the function set be able to accept, as its arguments, any value and data type that may possibly be returned by any function in the function set and any value and data type that may possibly result from the evaluation of any terminal in the terminal set. In other words, each function should be defined (and behave properly) for any combination of arguments that it may encounter. This property is essential since, clearly, programs must be executed in order to assign them a fitness. A failure in the execution of one of the programs composing the population would lead either to a failure of the whole GP system or to unpredictable results.

The function and terminal sets used in the previous example clearly satisfy the closure property. The sets $\mathscr{F} = \{*, /\}$ and $\mathscr{T} = \{x, 0\}$, however, do not satisfy this property. In fact, each evaluation of an expression containing an operation of division by zero would lead to unpredictable behavior. In practical applications, the division operator is often modified in order to make sure the closure property holds.

■ **Fig. 1**
A tree that can be built with the sets $\mathscr{F} = \{+, -\}$ and $\mathscr{T} = \{x, 1\}$.

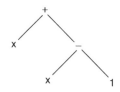

Respecting the closure property in real-world applications is not always straightforward, particularly if the use of different data types is required by the problem domain. A common example is the mixing of Boolean and numeric functions. For example, if one used a function set composed by Boolean functions (*AND, OR, . . .*), arithmetic functions ($+, -, *, /, . . .$), comparison functions ($>, <, =, . . .$), and conditionals (*IF THEN ELSE*), expressions such as:

$$IF \ ((x > 10*y) \ AND \ (y > 0)) \ THEN \ z + y \ ELSE \ z*x$$

might easily be evolved. In such cases, introducing typed functions and type-respecting genetic operations in the GP system can help enforce closure. This is achieved, for example, by *strongly typed GP* (Banzhaf et al. 1998), a GP system in which each primitive carries information about its type as well as the types it can call, thus forcing functions calling it to cast arguments into the appropriate types. Using types make even more sense in GP than for human programmers, since human programmers have a mental model of what they are doing, whereas the GP system is completely random in its initialization and variation phases. Furthermore, type consistency reduces the search space, which is likely to improve chances of success for the search.

The *sufficiency property* requires that the set of terminals and the set of functions be capable of expressing a solution to the problem. For instance, a function set including the four arithmetic operations combined with a terminal set including variables and constants respects the sufficiency property if the problem at hand requires the evolution of an analytic function. (Any function can be approximated to the desired degree by a polynomial via Taylor's expansion.) The primitive set $\mathscr{F} = \{+, -\}$, on the other hand, would not be sufficient, since it can only represent linear functions.

For many domains, there is no way of being sure whether sufficiency holds for the chosen set. In these cases, the definition of appropriate sets depends very much on prior knowledge about the problem and on the experience of the GP designer.

2.2 Initialization of a GP Population

The initialization of the population is the first step of the evolution process. It involves the creation of the program structures that will later be evolved. The most common initialization methods in tree-based GP are the *grow* method, the *full* method and the *ramped half-and-half* method (Koza 1992a). These methods will be explained in the following paragraphs. All are controlled by a parameter: the maximum depth allowed for the trees, d.

When the *grow method* is employed, each tree of the initial population is built using the following algorithm:

- A random symbol is selected with uniform probability from the function set $\mathscr{F}$ to be the root of the tree; let n be the arity of the selected function symbol.
- n nodes are selected with uniform probability from the union of the function set and the terminal set, $\mathscr{F} \cup \mathscr{T}$, to be its children;
- For each function symbol within these n nodes, the grow method is recursively invoked, that is, its children are selected from the set $\mathscr{F} \cup \mathscr{T}$, unless the node has a depth equal to $d-1$. In the latter case, its children are selected from $\mathscr{T}$.

While this method ensures that no parts of the tree are deeper than d, the drawing of primitives from $\mathscr{F} \cup \mathscr{T}$ allows branches to be fully leafed before they reach this maximum depth. So, initial trees of varied shape and size typically result from the use of the grow method. One should note, however, that the exact distribution of tree shapes depends in a nontrivial way on the ratio between the number of primitives of each arity and the number of terminals. If the primitive set contains a large number of terminals, the trees produced by the grow method tend to be small. On the contrary, if there are a few terminals and many functions with an arity of 2 or more, branches will tend to reach the maximum depth d, resulting in almost full trees.

Instead of selecting nodes from $\mathcal{F} \cup \mathcal{T}$, the *full method* chooses only function symbols until the maximum depth is reached. Then it chooses only terminals. The result is that every branch of the tree goes to the full maximum depth.

As first noted by Koza (1992a), population initialized with the above two methods may lack diversity. To fix this, Koza suggested to use a technique he called *ramped half-and-half*. This works as follows. With the ramped half-and-half method, a fraction $\frac{1}{d}$ of the population is initialized with trees having a maximum depth of 1, a fraction $\frac{1}{d}$ with trees of maximum depth 2, and so on. For each depth group, half of the trees are initialized with the full technique and half with the grow technique. In this way, the initial population is composed by a mix of large, small, full, and unbalanced trees, thus ensuring a certain amount of diversity.

2.3   Fitness Evaluation

Each program in the population is assigned a fitness value, representing the degree to which it solves the problem of interest. This value is calculated by means of some well-defined procedure. Two fitness measures are most commonly used in GP: the *raw fitness* and the *standardized fitness*. They are described below.

Raw fitness is the measurement of fitness that is stated in the natural terminology of the problem itself. In other words, raw fitness is the simplest and most natural way to calculate the degree to which a program solves a problem. For example, if the problem consists of controlling a robot so that it picks up as many of the objects contained in a room as possible, then the raw fitness of a program could simply be the number of objects actually picked up at the end of its execution.

Often, but not always, raw fitness is calculated over a set of *fitness cases*. A fitness case corresponds to a representative situation in which the ability of a program to solve a problem can be evaluated. For example, consider the problem of generating an arithmetic expression that approximates the polynomial $x^4 + x^3 + x^2 + x$ over the set of natural numbers smaller than 10. Then, a fitness case is one of those natural numbers. Suppose $x^2 + 1$ is the functionality of a program one needs to evaluate. Then, $2^2 + 1 = 5$ is the value taken by this expression over the fitness case 2. Raw fitness is then defined as the sum over all fitness cases of the distances between the target values expected from a perfect solution and the values actually returned by the individual being evaluated. Formally, the raw fitness, f_R, of an individual i, calculated over a set of N fitness cases, can be defined as

$$f_R(i) = \sum_{j=1}^{N} |S(i,j) - C(j)|^k \tag{1}$$

where $S(i,j)$ is the value returned by the evaluation of individual i over fitness case j, $C(j)$ is the correct output value associated to fitness case j and k is some natural number (often either $k=1$ or $k=2$).

Fitness cases are typically a small sample of the entire domain space. The choice of how many fitness cases (and which ones) to use is often a crucial one since whether or not an evolved solution will generalize over the entire domain depends on this choice. In practice, the decision is made on the basis of knowledge about the problem and practical performance considerations (the bigger the training set, the longer the time required to evaluate fitnesses). A first theoretical study on the suitable number of fitness cases to be used has been presented in Giacobini et al. (2002).

Standardized fitness is a reformulation of the raw fitness so that lower numerical values are better. When the smaller the raw fitness the better (as in the case in ❯ Eq. 1), then the standardized fitness, f_S, can simply be equal to the raw fitness, that is $f_S = f_R$. It is convenient and desirable to ensure the best value of standardized fitness is equal to zero. When this is not the case, this can be obtained by subtracting or adding a constant to the raw fitness. If, for a particular problem, the bigger the raw fitness the better, the standardized fitness f_S of an individual can be defined as $f_S = f_R^{max} - f_R$, where f_R is the raw fitness of the individual and f_R^{max} is the maximum possible value of raw fitness (which is assumed to be known).

2.4 Selection

Each individual belonging to a GP population can undergo three different operations: genetic operators can be applied to that individual, it can be copied into the next generation as it is, or it can be discarded. *Selection* is the operator that decides whether a chance of reproduction is given to an individual. It is a crucial step for GP, as is for all EAs, since the success and pace of evolution often depends on it.

Many selection algorithms have been developed. The most commonly used are *fitness proportionate selection, ranking selection,* and *tournament selection.* All of them are based on fitness: individuals with better fitness have higher probability of being chosen for mating. Selection mechanisms used for GP are typically identical to the ones used for GAs and for other EAs. They will not be discussed in detail in this chapter, since they are introduced elsewhere in this book.

2.5 Crossover

The *crossover* or sexual *recombination* operator creates variation in the population by producing offspring that consist of genetic material taken from the parent. The parents, T_1 and T_2, are chosen by means of a selection method.

Standard GP crossover (often called *subtree crossover*) (Koza 1992a) begins by independently selecting a random node in each parent – we will call it the *crossover point* for that parent. Usually crossover points are chosen in such a way that internal nodes are picked with a probability of 0.9 and any node (internal node or terminal) with a probability 0.1. The *crossover fragment* for a particular parent is the subtree rooted at the crossover point. An offspring is produced by deleting the crossover fragment of T_1 from T_1 and inserting the crossover fragment of T_2 at the crossover point of T_1. Some implementations also return a second offspring, which is produced in a symmetric manner. Figure ❯ 2 shows an example of standard GP crossover. Because entire subtrees are swapped and because of the closure property of the primitive set, crossover always produces syntactically legal programs.

☑ **Fig. 2**

An example of standard GP crossover. Crossover fragments are indicated by the *shaded areas*.

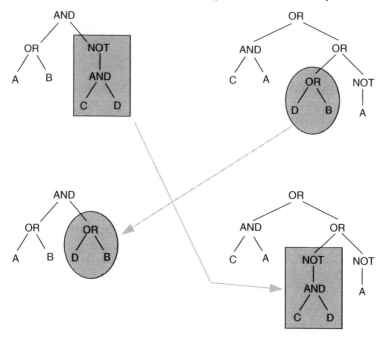

It is important to remark that when one of the crossover points is a leaf while the other is the root of a tree the offspring can be much bigger and deeper than the parents. While this may be desirable at early stages of a run, in the presence of the phenomenon of *bloat* – the progressive growth of the code size of individuals in the population generation after generation (more on this in ❷ Sect. 4.4) – some limit may have to be imposed on offspring size. Typically, if the offspring's size or depth is beyond a limit, the offspring is either rejected (and crossover attempted again) or accepted in the population but is given a very low fitness.

Many variants of the standard GP crossover have been proposed in the literature. The most common ones consist in assigning probabilities of being chosen as crossover points to nodes, based on node depth. In particular, it is very common to assign low probabilities to crossover points near the root and the leaves, so as to reduce the occurrence of the phenomenon described above. Another kind of GP crossover is *one-point crossover*, introduced in Poli and Langdon (1997), where the crossover points in the two parents are forced to be at identical positions. This operator was important it the development of a solid GP theory (see ❷ Sect. 4), since it provided a stepping stone that allowed the extension to GP of corresponding GA theory.

2.6 Mutation

Mutation is asexual, that is, it operates on only one parental program.

Standard GP mutation, often called *subtree mutation*, begins by choosing a *mutation point* at random, with uniform probability, within a selected individual. Then, the subtree rooted at the mutation point is removed and a new randomly generated subtree is inserted at that point.

■ **Fig. 3**
An example of standard GP mutation.

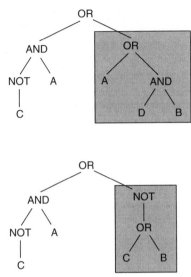

● *Figure 3* shows an example of standard GP mutation. As it is the case for standard crossover this operation is controlled by a parameter that specifies the maximum depth allowed, thereby limiting the size of the newly created subtree that is to be inserted. Nevertheless, the depth of the generated offspring can be considerably larger than that of the parent.

Many variants of standard GP mutation have been developed. The most commonly used are the ones aimed at limiting the probability of selecting as mutation points the root and/or the leaves of the parent. Another common form of mutation is *point mutation* (Poli and Langdon 1997) that randomly replaces nodes with random primitives of the same arity. Other commonly used variants of GP mutation are *permutation* (or *swap* mutation) that exchanges two arguments of a node and *shrink mutation* that generates a new individual from a parent's subtree. Another form of mutation is *structural mutation* (Vanneschi 2003, 2004). Being consistent with the edit distance measure, this form of mutation is useful in the study of GP problem difficulty (see ● Sect. 4.6).

2.7 Other Variants

In GP systems using a *steady state* scheme, the individuals produced by variation operators are immediately inserted into the population without waiting for a full generation to be complete. To keep the population size constant, a corresponding number of individuals need to be removed from the population. The removed individuals are typically the ones with the worst fitness, but versions of GP systems also exist where the removed individuals are randomly selected or where offspring replace their parents. After the new individuals have been inserted into the population, the process is iterated. A GP system using the traditional (non-steady-state) strategy is called *generational*.

Automatically defined functions (ADFs) are perhaps the best-known mechanisms by which GP can evolve and use subroutines inside individuals. In GP with ADFs, each program is represented by a set of trees, one for each ADF, plus one for the main program. Each ADF may have zero, one, two or more variables that play the role of formal parameters. ADFs allow parametrized reuse of code and hierarchical invocation of evolved code. The usefulness of ADFs has been shown by Koza (1994).

In the same vein, Koza defined automatically defined iterations (ADIs), automatically defined loops (ADLs), automatically defined recursions (ADRs), and automatically defined stores (ADSs) (Koza et al. 1999). All these mechanisms may allow a high degree of reusability of code inside GP individuals. However, their use is still not very widespread.

2.8 GP Parameters

The user of a GP has to decide the set of functions $\mathscr{F}$ and the set of terminals $\mathscr{T}$ to be used to represent potential solutions of a given problem and the exact implementation of the genetic operators to be used to evolve programs. This is not all, though. The user must also set some parameters that control evolution.

The most important parameters are population size, stopping criterion, technique used to create the initial population, selection algorithm, crossover type and rate, mutation type and rate, maximum tree depth, steady state vs. generational update, presence or absence of ADFs or other modularity-promoting structures, and presence or absence of elitism (i.e., the guaranteed survival of the best individual(s) found in each generation). The setting of these parameters represents, in general, an important choice for the performance of the GP system, although, based on our experience, we would suggest that population size and anything to do with selection (e.g., elitism, steady state, etc.) are probably the most important ones.

2.9 GP Benchmarks

Koza (1992a) defines a set of problems that can be considered as "typical GP problems," given that they mimic some important real-life applications and they are simple to define and to apply to GP. For this reason, they have been adopted by the GP research community as a, more or less, agreed-upon set of benchmarks, to be used in experimental studies. They include the following:

- *The Even k Parity Problem*, whose goal is to find a Boolean function of k arguments that returns *true* if an even number of its arguments evaluates to true, and *false* otherwise.
- *The h Multiplexer Problem*, where the goal is to design a Boolean function with h inputs that approximates a multiplexer function.
- *The Symbolic Regression Problem*, that aims at finding a program that matches a given target mathematical function.
- *The Intertwined Spirals problem*, the goal of which is to find a program to classify points in the $x-y$ plane as belonging to one of two spirals.
- *The Artificial Ant on the Santa Fe trail*, whose goal is to find a target navigation strategy for an agent on a limited toroidal grid.

3 Examples of Real-World Applications of GP

Since its early beginnings, GP has produced a cornucopia of results. Based on the experience of numerous researchers over many years, it appears that GP and other evolutionary computation methods have been especially productive in areas having some or all of the following properties:

- The relationships among the relevant variables is unknown or poorly understood.
- Finding the size and shape of the ultimate solution is a major part of the problem.
- Significant amounts of test data are available in computer-readable form.
- There are good simulators to test the performance of tentative solutions to a problem, but poor methods to directly obtain good solutions.
- Conventional mathematical analysis does not, or cannot, provide analytic solutions.
- An approximate solution is acceptable (or is the only result that is ever likely to be obtained).
- Small improvements in performance are routinely measured (or easily measurable) and highly prized.

The literature, which covers more than 5,000 recorded uses of GP, reports an enormous number of applications where GP has been successfully used as an automatic programming tool, a machine learning tool, or an automatic problem-solving engine. It is impossible to list all such applications here. Thus, to give readers an idea of the variety of successful GP applications, below a representative sample is listed (see Poli et al. (2008) for a more detailed analysis):

- *Curve Fitting, Data Modelling, and Symbolic Regression.* A large number of GP applications are in these three fields (see, e.g., Lew et al. 2006 and Keijzer 2004). GP can be used as a stand-alone tool or coupled with one of the numerous existing feature selection methods, or even GP itself can be used to do feature selection (Langdon and Buxton 2004). Furthermore, GP can also be used in cases where more than one output (prediction) is required. In that case, linear GP with multiple output registers, graph-based GP with multiple output nodes, or a GP operating on single trees with primitives on vectors can be used.
- *Image and Signal Processing.* The use of GP to process surveillance data for civilian purposes, such as predicting motorway traffic jams from subsurface traffic speed measurements was suggested in Howard and Roberts (2004); GP found recurrent filters in Esparcia-Alcazar and Sharman (1996); the use of GP to preprocess images, particularly of human faces, to find regions of interest for subsequent analysis was presented in Trujillo and Olague (2006). Classifications of objects and human speech with GP was achieved in Zhang and Smart (2006) and Xie et al. (2006), respectively. A GP system in which the image processing task is split across a swarm of evolving agents and some of its applications have been described in Louchet (2001). Successful GP applications in medical imaging can be found, for instance, in Poli (1996).
- *Financial Trading, Time Series, and Economic Modelling.* Recent papers have looked at the modeling of agents in stock markets (Chen and Liao 2005), evolving trading rules for the S&P 500 (Yu and Chen 2004), and forecasting the Hang Seng index (Chen et al. 1999). Other examples of financial applications of GP include, for example, Jin and Tsang (2006) and Tsang and Jin (2006).

- *Industrial Process Control.* GP is frequently used in industrial process control, although, of course, most industrialists have little time to spend on academic reporting. A notable exception is Dow Chemical, where a group has been very active in publishing results (Castillo et al. 2006; Jordaan et al. 2006). Kordon (2006) describes where industrial GP stands now and how it will progress. Other interesting contributions in related areas are, for instance, Dassau et al. (2006) and Lewin et al. (2006). GP has also been applied in the electrical power industry (Alves da Silva and Abrao 2002).

- *Medicine, Biology, and Bioinformatics.* Among other applications, GP is used in biomedical data mining. Of particular medical interest are very wide data sets, with typically few samples and many inputs per sample. Recent examples include single nuclear polymorphisms (Shah and Kusiak 2004), chest pain (Bojarczuk et al. 2000), and Affymetrix GeneChip microarray data (Langdon and Buxton 2004; Yu et al. 2007).

- *Computational Chemistry.* Computational chemistry is important in the drug industry. However, the interactions between chemicals that might be used as drugs are beyond exact calculation. Therefore, there is great interest in approximate models that attempt to predict either favorable or adverse interactions between proto-drugs and biochemical molecules. These models can be applied very cheaply in advance of the manufacturing of chemicals, to decide which of the myriad of chemicals is worth further study. Examples of GP approaches include Barrett and Langdon (2006), Hasan et al. (2006), and Archetti et al. (2007).

- *Entertainment, Computer Games, and Art.* Work on GP in games includes, among others, (Azaria and Sipper 2005; Langdon and Poli 2005). The use of GP in computer art is reported for instance in Jacob (2000, 2001). Many recent techniques are described in Machado and Romero (2008).

- *Compression.* GP has been used to perform lossy compression of images (Koza 1992b; Nordin and Banzhaf 1996) and to identify iterated functions system, which are used in the domain of fractal compression (Lutton et al. 1995). GP was also used to evolve wavelet compression algorithms (Klappenecker and May 1995) and nonlinear predictors for images (Fukunaga and Stechert 1998). Recently, a GP system called *GP-ZIP* has been proposed that can do for lossless data compression (Kattan and Poli 2008).

- *Human-Competitive Results.* Getting machines to produce competitive results is the very reason for the existence of the fields of AI and machine learning. Koza et al. (1999) proposed that an automatically created result should be considered "human-competitive" if it satisfies at least one of eight precise criteria. These include patentability, beating humans in regulated competitions, producing publishable results, etc. Over the years, dozens of GP results have passed the human-competitiveness test (see Trujillo and Olague 2006 and Hauptman and Sipper 2007).

4   GP Theory

As discussed in ❷ Sect. 3, GP has been remarkably successful as a problem-solving and engineering tool. One might wonder how this is possible, given that GP is a nondeterministic algorithm, and, as a result, its behavior varies from run to run. Why can GP solve problems and how? What goes wrong when it cannot? What are the reasons for certain undesirable behaviors, such as bloat? Below a summary of current research will be given.

4.1 Schema Theories

If we could visualize the search performed by GP, we would often find that initially the population looks like a cloud of randomly scattered points, but that, generation after generation, this cloud changes shape and moves in the search space. Because GP is a stochastic search technique, in different runs we would observe different trajectories. If we could see regularities, these might provide us with a deep understanding of how the algorithm is searching the program space for the solutions, and perhaps help us see why GP is successful in finding solutions in certain runs and unsuccessful in others. Unfortunately, it is normally impossible to exactly visualize the program search space due to its high dimensionality and complexity, making this approach nonviable.

An alternative approach to better understanding the dynamics of GP is to study mathematical models of evolutionary search. Exact mathematical models of GP are probabilistic descriptions of the operations of selection, reproduction, crossover, and mutation. They explicitly represent how these operations determine which areas of the program space will be sampled by GP, and with what probability. There are a number of cases where this approach has been very successful in illuminating some of the fundamental processes and biases in GP systems.

Schema theories are among the oldest and the best-known models of evolutionary algorithms (Holland 1975). These theories are based on the idea of partitioning the search space into subsets, called *schemata*. They are concerned with modelling and explaining the dynamics of the distribution of the population over the schemata. Modern GA schema theory (Stephens and Waelbroeck 1999) provides exact information about the distribution of the population at the next generation in terms of quantities measured at the current generation, without having to actually run the algorithm.

The theory of schemata in GP has had a difficult childhood. However, recently, the first exact GP schema theories have become available (Poli 2001), which give exact formulations for the expected number of individuals sampling a schema at the next generation. Initially, these exact theories were only applicable to GP with one-point crossover (see❯ Sect. 2.5). However, more recently, they have been extended to other types of crossover including most crossovers that swap subtrees (Poli and McPhee 2003a,b).

Other models of evolutionary algorithms include models based on Markov chain theory. These models are discussed in the next section.

4.2 Markov Chains

Markov chains are important in the theoretical analysis of evolutionary algorithms operating on discrete search spaces (e.g., Davis and Principe 1993). Since such algorithms are stochastic but Markovian (e.g., in GAs and GP the population at the next generation typically depends only on the population at the current generation), all one needs to do to study their behavior is to compute with which probability an algorithm in any particular state (e.g., with a particular population composition) will move any other state (e.g., another population composition) at the next generation. These probabilities are stored in a stochastic matrix M, which therefore describes with which probability the algorithm will exhibit all possible behaviors over one generation. By taking powers of the matrix M and multiplying it by the initial probability distribution over states determined by the initialization algorithm, it is then possible to study the behavior of the algorithm over any number of generations. It is, for example, possible to compute the probability of solving a problem within n generations, the first hitting time, the average and best fitness after n generations, etc.

Markov models have been applied to GP (Poli and McPhee 2003; Mitanskiy and Rowe 2006), but so far they have not been developed as fully as the schema theory model.

4.3   Search Spaces

The models presented in the previous two sections treat the fitness function as a black box. That is, there is no representation of the fact that in GP, unlike in other evolutionary techniques, the fitness function involves the execution of computer programs on a variety of inputs. In other words, schema theories and Markov chains do not tell how fitness is distributed in the search space.

The theoretical characterization of the space of computer programs explored by GP from the point of view of fitness distributions aims at answering this very question (Langdon and Poli 2002). The main result in this area is the discovery that the distribution of functionality of non Turing-complete programs approaches a limit as program length increases. That is, although the number of programs of a particular length grows exponentially with length, beyond a certain threshold, the fraction of programs implementing any particular functionality is effectively constant. This happens in a variety of systems and can be proven mathematically for LISP expressions (without side effects) and machine code programs without loops (e.g., see Langdon and Poli 2002). Recently, Poli and Langdon (2006) started extending these results to Turing complete machine code programs. A mathematical analysis of the halting process based on a Markov chain model of program execution and halting was performed, which derived a scaling law for the *halting probability* for programs as their length is varied.

4.4   Bloat

Very often, the average size (number of nodes) of the programs in a GP population, after a certain number of generations in which it is largely static, starts growing at a rapid pace. Typically, the increase in program size is not accompanied by any corresponding increase in fitness. This phenomenon is known as *bloat*. Its origin has been the focus of intense research for over a decade. Bloat is not only interesting from a scientific point of view: it also has significant practical deleterious effects, slowing down the evaluation of individuals, and often consuming large computational resources. There are several theories of bloat.

For instance, the *replication accuracy theory* (McPhee and Miller 1995) states that the success of a GP individual depends on its ability to have offspring that are functionally similar to the parent: as a consequence, GP evolves toward (bloated) representations that increase replication accuracy.

The nodes in a GP tree can be categorized into two classes: *inactive code* (code that is not executed or has no effect) and *active code* (all code that is not inactive). The *removal bias theory* (Soule and Foster 1998b) observes that inactive code in a GP tree tends to be low in the tree, residing in smaller-than-average-size subtrees. Crossover events excising inactive subtrees produce offspring with the same fitness as their parents. On average the inserted subtree is bigger than the excised one, so such offspring are bigger than average while retaining the fitness of their parent, leading ultimately to growth in the average program size.

The *nature of program search spaces theory* (Langdon et al. 1999) relies on the result mentioned above that above a certain size, the distribution of fitnesses does not vary with size. Since there are more long programs, the number of long programs of a given fitness is greater than the number of short programs of the same fitness. Over time GP samples longer and longer programs simply because there are more of them.

In Poli and Mcphee (2003), a *size evolution equation* for GP was developed, which provided an exact formalization of the dynamics of average program size. The original equation was derived from the exact schema theory for GP. This equation can be rewritten in terms of the expected change in average program size as

$$E[\Delta\mu(t)] = \sum_{\ell} \ell \times (p(\ell, t) - \Phi(\ell, t)) \tag{2}$$

where $\Phi(\ell, t)$ is the proportion of programs of size ℓ in the current generation and $p(\ell, t)$ is their selection probability. This equation does not directly explain bloat, but it constrains what can and cannot happen size-wise in GP populations. So, any explanation for bloat (including the theories above) has to agree with it.

The newest, and hopefully conclusive, explanation for bloat has been recently formalized in the so-called *crossover bias theory* (Dignum and Poli 2007; Poli et al. 2007), which is based on ❷ Eq. 2. On average, each application of subtree crossover removes as much genetic material as it inserts; consequently crossover on its own does not produce growth or shrinkage. While the *mean* program size is unaffected, however, *higher moments* of the distribution are. In particular, crossover pushes the population toward a particular distribution of program sizes, known as a *Lagrange distribution of the second kind*, where small programs have a much higher frequency than longer ones. For example, crossover generates a very high proportion of single-node individuals. In virtually all problems of practical interest, however, very small programs have no chance of solving the problem. As a result, programs of above average size have a selective advantage over programs of below average size, and the mean program size increases. Because crossover will continue to create small programs, the increase in average size will continue generation by generation.

Numerous empirical techniques have been proposed to control bloat (Soule and Foster 1998; Langdon et al. 1999). These include the use of size and depth limits, the use of genetic operators with an inbuilt anti-bloat bias, the use of multi-objective selection techniques to jointly optimize fitness and minimize program size. Among these are the famous parsimony pressure method (Koza 1992b; Zhang and Mühlenbein 1995) and a recent improvement of if, the covariant parsimony pressure method (Poli and McPhee 2008). The reader is referred to Poli et al. (2008), Chap. 11 for a survey.

4.5   Is GP Guaranteed to Find Solutions to Problems?

While often Markov transition matrices describing evolutionary algorithms are very large and difficult to manage numerically, it is sometimes possible to establish certain properties of such matrices theoretically. An important property in relation to the ability of an algorithm to reach all possible solutions to a problem (given enough time) is the property of *ergodicity*. What it means in practice is that there exist some $k \in \mathbb{N}$ such that all elements of M^k are nonzero. In this case, $\lim_{k \to \infty} M^k$ is a particular matrix where all elements are nonzero. This means that irrespective of the starting state, the probability of reaching a solution to a problem (e.g., the global optimum of a function, if one is using EAs for function optimization) is nonzero. As a result, given enough generations, the algorithm is guaranteed to find a solution.

In the context of EAs acting on traditional fixed-length representations, this has led to proving that EAs with nonzero mutation rates are global optimizers with guaranteed convergence (Rudolph 1994). While these results have been extended to more general search spaces (Rudolph 1996) and, as we discussed in ❷ Sect. 4.2, Markov chain models of some forms of GP have been recently derived (e.g., see Poli et al. (2004)), the calculations involved in setting up and studying these models are of considerable mathematical complexity. Furthermore, extending the work to infinite search spaces (which is required to model the traditional form of GP where operators can create trees of potentially any size over a number of generations) presents even bigger challenges. So, it is fair to say that at the present time there is no formal mathematical proof that a GP system can always find solutions.

Despite these formal obstacles, it is actually surprisingly easy to find good reasons in support of the conjecture that, with minor modifications, the traditional form of GP is guaranteed to visit a solution to a problem, given enough generations. In fact, if one uses mutation and the mutation operator is such that, albeit with a very low probability, any point in the search space can be generated by mutating any other point, then it is clear that sooner or later the algorithm will visit a solution to the problem. It is then clear that in order to ensure that GP has a guaranteed asymptotic ability to solve all problems, all we need to do is to add to the mix of operators already present in a GP system some form of mutation that has a nonzero probability of generating any tree in the search space. One such mutation operator is the subtree mutation operator described in ❷ Sect. 2.6.

4.6 Problem Difficulty in GP

What problems are solvable via GP? Is it possible to measure the difficulty of a problem for GP? In classical algorithms, we have a well-developed complexity theory that allows us to classify problems into complexity classes such that problems in the same class have roughly the same complexity, that is, they consume (asymptotically) the same amount of computational resources, usually time. Although, properly speaking, GP is a randomized heuristic and not an algorithm, it would be nice if we were able to somehow classify problems according to some measure of *difficulty*.

Difficulty studies have been pioneered in the related but simpler field of GAs, by Goldberg and coworkers (e.g., see Goldberg 1989 and Horn and Goldberg 1995). Their approach consisted in constructing functions that should *a priori* be easy or hard for GAs to solve. These approach has been followed by many others (e.g., Mitchell et al. 1992 and Forrest and Mitchell 1993) and have been at least partly successful in the sense that they have been the source of many ideas as to what makes a problem easy or difficult for GAs. One concept that underlies many measures of difficulty is the notion of *fitness landscapes*.

The metaphor of a fitness landscape (Stadler 2002), although not without faults, has been a fruitful one in several fields and is attractive due to its intuitiveness and simplicity. Probably, the simplest definition of fitness landscape in EAs is a plot where the points in the abscissas represent the different individual genotypes in a search space and the ordinates represent the fitness of each of these individuals (Langdon and Poli 2002). If genotypes can be visualized in two dimensions, the plot can be seen as a three-dimensional "map," which may contain peaks and valleys. The task of finding the best solution to the problem is equivalent to finding the highest peak (for maximization problems).

One early example of the application of the concept of fitness landscapes to GP is represented by the work of Kinnear (1994), where GP difficulty was related to the shape of the fitness landscape and analyzed through the use of the fitness *auto-correlation function*. Langdon and Poli (2002) took an experimental view of GP fitness landscapes in several works summarized in their book. After selecting important and typical classes of GP problems, they study fitness landscapes either exhaustively, whenever possible, or by randomly sampling the program space when enumeration becomes unfeasible. Their work highlights several important characteristics of GP spaces, such as density and size of solutions and their distribution. This is useful work. Even if the authors' goals were not establishing problem difficulty, it certainly has a bearing on it. More recently, Langdon (2003) has extended his studies for convergence rates in GP for simple machine models (which are amenable to quantitative analysis by Markov chain techniques) to convergence of program fitness landscapes for the same machine models using genetic operators and search strategies to traverse the space. This approach is rigorous because the models are simple enough to be mathematically treatable. The ideas are thus welcome, although their extension to standard GP might prove difficult.

The work of Nikolaev and Slavov (1998) represents the first effort to apply a difficulty measure called *fitness distance correlation* (FDC) (first studied by Jones (1995) for GAs) to GP, with the aim of determining which mutation operator, among a few that they propose, "sees" a smoother landscape on a given problem. Pros and cons of FDC as a general difficulty measure in GP were investigated in Vanneschi (2004), and Tomassini et al. (2005). Those contributions claimed the reliability of FDC and pointed out, as one of its major drawbacks, its lack of predictiveness of FDC (the genotype of the global optima must be known beforehand to calculate it).

In order to overcome this limitation, a new hardness measure, called *negative slope coefficient* (NSC) has been recently presented and its usefulness for GP investigated in Vanneschi et al. (2004, 2006). Results indicated that, although NSC may still need improving and is not without faults, it is a suitable hardness indicator for many well-known GP benchmarks, for some hand-tailored theoretical GP problems and also for some real-life applications (Vanneschi 2007).

Even though interesting, these results still fail to take into account many typical characteristics of GP. For instance, the NSC definition only includes (various kinds of) mutation, but it does not take crossover into account. A similarity/dissimilarity measure related to standard subtree crossover has been presented in Gustafson and Vanneschi (2005, 2008) and Vanneschi et al. (2006). This could be used in the future to build new measures of problem difficulty.

4.7   Does No-Free-Lunch Apply to GP?

Informally speaking, the no-free-lunch (NFL) theory originally proposed by Wolpert and Macready (1997) states that, when evaluated over all possible problems, all algorithms are equally good or bad irrespective of our evaluation criteria.

In the last decade-there have been a variety of results which have refined and specialized NFL (see Whitley and Watson 2005). One such result states that if one selects a set of fitness functions that are *closed under permutation*, the expected performance of any search algorithm over that set of problems is constant, that is, it does not depend on the algorithm we choose (Schumacher et al. 2001). What does it mean for a set of functions to be closed under permutation? A fitness function is an assignment of fitness values to the elements of the search space. A permutation of a fitness function is simply a rearrangement or reshuffling of the fitness values originally allocated to the objects in the search space. A set of problems/fitness functions is closed under permutation, if, for every function in the set, all possible shuffles of that function are also in the set.

The result is valid for any performance measure. Furthermore, Schumacher et al. (2001) showed that it is also the case that two arbitrary algorithms will have identical performance over a set of functions only if that set of functions is closed under permutation.

Among the many extension of NFL to a variety of domains, Woodward and Neil (2003) have made some progress in assessing the applicability of NFL to the search spaces explored by GP. In particular, they argued that there is a free lunch in a search space whenever there is a nonuniform many-to-one genotype–phenotype mapping, and that the mapping from syntax to functionality in GP is one such mapping. The reason why NFL would not normally be applicable to search in program spaces is that there are many more programs than functionalities and that not all functionalities are equally likely. So, interpreting syntax trees as genotypes and functionalities as phenotypes, the GP genotype–phenotype mapping is many-to-one and nonuniform, which invalidates NFL.

Beyond this interesting counterexample, to show that in general not all functionalities are equally likely in program search spaces, Woodward and Neil (2003) referred to the results on the limiting distribution of functionality reviewed above and to the universal distribution (Kirchherr et al. 1997) (informally this states that there are many more programs with a simple functionality than programs with a complex one). The latter result, however, applies to Turing complete languages, that is, to programs with memory and loops. So, Woodward and Neil proof applies only to Turing-complete GP, which is essentially a rarity.

In very recent work (Poli et al. 2009) it has been possible to extend these results to the case of standard GP problems and systems and to ascertain that a many-to-one genotype–phenotype mapping is not the only condition under which NFL breaks down when searching program spaces. The theory is particularly easy to understand because it is based on geometry. It is briefly reviewed in the rest of this section.

As seen above, the fitness of a program p in GP is often the result of evaluating the behavior of p in a number of fitness cases and adding up the results (Poli et al. 2008). That is

$$f(p) = \sum_{i=1}^{n} g(p(x_i), t(x_i)) \tag{3}$$

where f is the fitness function, $\{x_i\}$ is a set of fitness cases of cardinality n, and g is a function that evaluates the degree of similarity of its arguments. Almost invariably, the function g respects the axioms of a metric, the most common form of g being $g(a, b) = |a - b|^k$ for $k=1$ or $k=2$.

One can consider a finite space of programs $\Omega = \{p_i\}_{i=1}^{r}$, such as, for example, the space of all possible programs with at most a certain size or depth. A fitness function f over Ω can be represented as a vector $\mathbf{f} = (f_1, \ldots, f_r)$ where $f_i = f(p_i)$. Using this vector representation for fitness functions, we can write ❷ Eq. 3 as

$$\mathbf{f} = \left(\sum_{i=1}^{n} g(p_1(x_i), t(x_i)), \ldots, \sum_{i=1}^{n} g(p_r(x_i), t(x_i)) \right) \tag{4}$$

As we can see from ❷ Eq. 4, if n and the set of fitness cases $\{x_i\}$ are fixed a priori, then a fitness function is fully determined by the value of the vector of target behaviors, $\mathbf{t} = (t_1, t_2, \ldots, t_n)$, where $t_i = t(x_i)$ is fixed. If we focus on the most frequent type of program induction application in GP, symbolic regression, we typically have that the components t_i are scalars representing the desired output for each fitness case.

The function g is a distance (on $\mathbb{R}$). Because the sum of distances is also a distance, we can define the following distance function:

$$d(\mathbf{u}, \mathbf{v}) = \sum_{i=1}^{n} g(\mathbf{u}_i, \mathbf{v}_i) \tag{5}$$

where $\mathbf{u}, \mathbf{v} \in \mathbb{R}^n$. With this in hand, we can see that the fitness associated to a program p can be interpreted as the distance between the vector $\mathbf{t}$ and the vector $\mathbf{p} = (p(x_1), p(x_2), \ldots, p(x_n))$. That is

$$f(\mathbf{p}) = d(\mathbf{p}, \mathbf{t}) \tag{6}$$

Note that whenever we represent programs using their behavior vector we are essentially focusing on the phenotype-to-fitness mapping, thereby complementing the analysis of Woodward and Neil (2003) summarized above.

Using distances, ❷ Eq. 4 can be rewritten more concisely as

$$\mathbf{f} = (d(\mathbf{p}_1, \mathbf{t}), d(\mathbf{p}_2, \mathbf{t}), \ldots, d(\mathbf{p}_r, \mathbf{t})) \tag{7}$$

Note that if we know the fitness f of a program p, we know that the target behavior $\mathbf{t}$ that generated that fitness must be on the surface of a sphere centered on $\mathbf{p}$ (the vector representation of p) and of radius f. So, for every valid symbolic regression fitness function, the target behavior is at the intersection of the spheres centered on the behavior of each program in the search space (see ❷ *Fig. 4*).

◘ Fig. 4
If g measures the squared difference between two numbers, a valid fitness function requires the spheres centered on each program behavior and with radius given by their corresponding fitness to meet in one point: the target vector t.

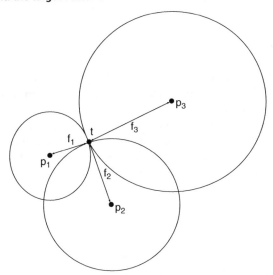

With this geometric representation of symbolic regression problems in hand, it is not difficult to find out under which conditions NFL does or does not hold for GP. In particular, Poli et al. (2009) showed that:

Theorem 1 (NFL for GP). *Let $\mathscr{F} = \{\mathbf{f}_1, \mathbf{f}_2, \ldots, \mathbf{f}_m\}$ be a set of fitness functions of the form in* ❯ *Eq. 7 and let $\mathscr{T} = \{\mathbf{t}_1, \mathbf{t}_2, \ldots, \mathbf{t}_m\}$ be the set of target vectors associated to the functions in $\mathscr{F}$, with $\mathbf{t}_i$ being the vector generating $\mathbf{f}_i$ for all i. The set $\mathscr{F}$ is closed under permutation (and NFL applies to it) if and only if for all target vectors $\mathbf{t} \in \mathscr{T}$ and for all permutations σ of $(1, 2, \ldots, r)$ there exists a target vector $\tilde{\mathbf{t}} \in \mathscr{T}$ such that:*

$$\sum_{i=1}^{n} g(p_{\sigma(j)}(x_i), t_i) = \sum_{i=1}^{n} g(p_j(x_i), \tilde{t}_i) \tag{8}$$

or, equivalently,

$$d(\mathbf{p}_{\sigma(j)}, \mathbf{t}) = d(\mathbf{p}_j, \tilde{\mathbf{t}}) \tag{9}$$

for all $j = 1, 2, \ldots, r$.

❯ Equation 9 is a mathematical statement of the geometric requirements for $\tilde{\mathbf{t}}$ to exist. Namely, the target vector $\mathbf{t} \in \mathscr{T}$ associated to a function $\mathbf{f} \in \mathscr{F}$ must be at the intersection of the spheres centered on programs $\mathbf{p}_1, \mathbf{p}_2, \ldots, \mathbf{p}_r$ and having radii $f_1, f_2, \ldots, f_r$, respectively. Permuting the elements of $\mathbf{f}$ via a permutation σ to obtain a new fitness function corresponds to shuffling the radii of the spheres centered on $\mathbf{p}_1, \mathbf{p}_2, \ldots, \mathbf{p}_r$. Note these centers remain fixed since they represent the behavior of the programs in the search space. After the shuffling, some spheres may have had their radius decreased, some increased, and some unchanged. If any radius has changed, then $\mathbf{t}$ can no longer be the intersection of the spheres. However, we must be able to find a new intersection $\tilde{\mathbf{t}} \in \mathscr{T}$ or else the new fitness function we have generated is not a symbolic regression fitness function and therefore cannot be a member of $\mathscr{F}$, which would imply that the set is not closed under permutation.

We should also note that Theorem 1 focuses on the set of program behaviors. So, it tells us something about the nature of the phenotype-to-fitness function. It really states under which conditions there can be an NFL for a searcher exploring program behaviors with no resampling. If a search algorithm instead explores the space of syntactic structures representing programs, in the presence of symmetries (such as those highlighted by Woodward and Neil), then the searcher will produce resampling of program behaviors even if it never resampled the same syntactic structure. So, in the presence of a set of behaviors for which NFL holds, this would unavoidably give the "syntactic" searcher lower average performance than an algorithm that never resampled behaviors.

This theorem is useful since, by designing fitness functions that break its assumptions, we can find conditions where there is a free lunch for GP. It turns out that such conditions are really easy to satisfy. In particular, it is easy to see that if two programs have identical behavior, they must have identical fitness or there cannot be any intersection between the spheres centered around them. Formally:

Theorem 2 *Consider a search space which includes at least two programs p_1 and p_2 such that $p_1(x) = p_2(x)$ for all x in the training set. Let a set of symbolic regression problems $\mathscr{F}$ contain a fitness function f induced by a target vector $\mathbf{t}$ such that there exists a third program p_3 in the* search space with fitness $f(p_3) \neq f(p_1) = f(p_2)$. Then $\mathcal{F}$ cannot be closed under permutation and NFL does not hold.

Note that this result is valid for search in the space of syntactic structures representing programs. As one can see, a many-to-one mapping is not required for a free lunch to be available.

Continuing with our geometric investigation of NFL, it is clear that a necessary condition for the existence of a target vector $\mathbf{t}$, which induces a particular fitness function, is that the triangular inequality be verified. More precisely:

Lemma 1 *If a target vector $\mathbf{t}$ induces a symbolic regression fitness function $\mathbf{f}$, then for every pair of program behaviors $\mathbf{p}_1$ and $\mathbf{p}_2$ the distance $d(\mathbf{p}_1, \mathbf{p}_2)$ between $\mathbf{p}_1$ and $\mathbf{p}_2$ (based on the same metric used to measure fitnesses) must not be greater than $f_1 + f_2$.*

From this result, we can see that another common situation where there is incompatibility between an assignment of fitness to programs and the fitness representing a symbolic regression problem is the case in which two programs have different behaviors (and so are represented by two distinct points in the $\mathbb{R}^n$), but the sum of their fitnesses is smaller than their distance. This leads to the following result:

Theorem 3 *Given a set of symbolic regression functions $\mathcal{F}$, if there exists any function $\mathbf{f} \in \mathcal{F}$ and any four program behaviors $\mathbf{p}_1, \mathbf{p}_2, \mathbf{p}_3, \mathbf{p}_4$ in a search space such that $d(\mathbf{p}_1, \mathbf{p}_2) > f_3 + f_4$ then the set is not closed under permutation and NFL does not apply.*

So, it is extremely easy to find realistic situations in which a set of symbolic regression problems is provably not closed under permutation. This implies that, in general, we can expect that there is a free lunch for techniques that sample the space of computer programs for the purpose of fitting data sets. This is particularly important since it implies that, for GP, unlike other areas of natural computing, it is worthwhile to try to come up with new and more powerful algorithms.

5 Open Issues

Despite the successful application of GP to a number of challenging real-world problem domains and the recent theoretical advancements in explaining the behavior and dynamics of GP, there remains a number of significant open issues. These must be addressed for GP to become an even more effective and reliable problem-solving method.

The following are some of the important open issues:

- Despite some preliminary studies trying to remove this limit, at present, GP is fundamentally a "static" process, in the sense that many of its characteristics are fixed once and for all before executing it. For instance, the population size, the language used to code individuals ($\mathcal{F}$ and $\mathcal{T}$ sets), the genetic operators used to explore the search space and the fitness function(s) used are all static.
- Directly related to the previous point, until now GP has been mainly used for static optimization problems. No solid explorations, neither theoretical nor applicative, exist of dynamic optimization problems, where for instance the target function (and thus the fitness value of the candidate solutions) change over time.

- While a large body of literature and well-established results exist concerning the issue of generalization for many non-evolutionary machine-learning strategies, this issue in GP has not received the attention it deserves. Only recently, a few papers dealing with the problem of generalization have appeared. Nonetheless, generalization is one of the most important performance evaluation criteria for artificial learning systems and its foundations for GP still have to be deeply understood and applied.

- Although some constructs for handling modularity in GP have been defined (some of them, like ADFs, have been discussed in ❷ Sect. 2.7) and widely used, still some questions remain unanswered. Can these constructs help in solving substantially more complex problems? Can we achieve modularity, hierarchy, and reuse in a more controlled and principled manner? Can we provide insights and metrics on how to achieve this?

In the rest of this section, some of these grand challenges and open issues will be expanded upon. It is hoped that this chapter will help focus future research in order to deepen our understanding of the method to allow the development of more powerful problem-solving algorithms.

5.1 Dynamic and/or Self-Adapting GP Frameworks

Although the difficulty measures discussed in ❷ Sect. 4.6 are useful to statically calculate the hardness of a given problem for GP starting from its high-level specifications, they have not been applied to dynamically improve the search as yet. Indeed, a dynamic modification "on the fly" (i.e., during the evolutionary process) of the fitness function, the population size, the genetic operators, or representation used, driven by one or more hardness measures represents one of the most challenging open issues in GP, as in many other fields of evolutionary computation. One of the major difficulties in the realization of such a dynamic and self-adapting GP environment is probably represented by the fact that both FDC and NSC have always been calculated, until now, using a large sample of the search space and not the evolving population that is typically much smaller. Furthermore, the lack of diversity in populations after some generations may seriously compromise the reliability of these indicators. Nevertheless, this issue remains a challenge and a first attempt in this direction has recently been presented in Wedge and Kell (2008).

Models using dynamically sized populations, although not based on the values of difficulty indicators, have been presented, among others, in Rochat et al. (2005), Tomassini et al. (2004), and Vanneschi (2004), where the authors show that fixed-size populations may cause some difficulties to GP. They try to overcome these difficulties by removing or adding new individuals to the population in a dynamic way during evolution. In particular, individuals are removed as long as the best or average population's fitness keeps improving, while new individuals (either randomly generated ones or mutations of the best ones in the population) are added when fitness stagnates. The authors show a set of experimental results that confirm the benefits of the presented methods, at least for the studied problems. These results suggest that fixed-size population unnaturally limits the GP behavior and that, according to what happens in nature, the evolution has a benefit when populations are left free to shrink or increase in size dynamically, according to a set of events. Some of the most relevant and macroscopic of these events are so called *plagues* that usually cause a violent decrease in natural population sizes. Plagues have been successfully applied to GP in Fernandez et al. (2003a,b) and Fernandez and Martin (2004). These contributions go in the same direction as the previous ones: fixed-size populations limit and damage the behavior of GP while the use of dynamically sized populations is often beneficial. Furthermore, the models presented in the above quoted references have been successfully used by Silva (2008) to study and control bloat.

5.2   GP for Dynamic Optimization

Dynamic environments abound and offer particular challenges for all optimization and problem-solving methods. A well-known strategy for survival in dynamic environments is to adopt a population-based approach (Dempsey 2007). This allows a diversity of potential solutions to be maintained, which increases the likelihood that a sufficiently good solution exists at any point in time to ensure the survival of the population in the long term. Dynamic environments can exhibit changes in many different ways including the frequency and degree/size of change. The types of changes might range from relatively small smooth transitions to substantial perturbations in all aspects of the domain (Dempsey 2007).

Given the power of the biological process of evolution to adapt to ever, changing environments, it is surprising that the number of studies applying and explicitly studying their artificial counterpart of GP in dynamic environments have been minimal (Dempsey 2007). Despite the existence of a recent special issue in the GP journal on dynamic environments (Yang et al. 2006), none of the four articles actually dealt with GP directly (e.g., Wang and Wineberg (2006)). While some applications in dynamic environments have been undertaken in the past two years (e.g., Wagner et al. 2007, Hansen et al. 2007, Jakobović and Budin 2006, and Kibria and Li 2006), there has been little analysis of the behavior of GP in these environments. The two main examples have examined bloat (Langdon and Poli 1998) and constant generation (Dempsey 2007).

This seems to hint that we are missing the boat by focusing on static problems where there is a single fixed target. Recent experiments in evolutionary biology simulations suggest that EC/evolution could work efficiently "because" of dynamic environments as opposed to "despite" them (Kashtan et al. 2007).

5.3   Generalization in GP

Generalization is one of the most important performance-evaluation criteria for artificial learning systems. Many results exist concerning the issue of generalization in machine learning, like, for instance, support vector machines (see for instance Smola and Scholkopf 1999). However, this issue in GP has not received the attention it deserves and only recently a few papers dealing with the problem of generalization have appeared (Eiben and Jelasity 2002; Kushchu 2002). A detailed survey of the main contributions on generalization in GP has been done by Kushchu (2002). Another important contribution to the field of generalization in GP is due to the work of Banzhaf and coworkers. In particular, in Francone et al. (1996) they introduce a GP system called *Compiling GP* and they compared its generalization ability with that of other machine-learning paradigms. Furthermore, in Banzhaf et al. (1996) they show the positive effect of an extensive use of the mutation operator on generalization in GP using sparse data sets. In 2006, Da Costa and Landry (2006) have proposed a new GP model called *Relaxed GP*, showing its generalization ability. In 2006, Gagné et al. (2006) have investigated two methods to improve generalization in GP-based learning: (1) the selection of the best-of-run individuals using a three-data-sets methodology, and (2) the application of parsimony pressure to reduce the complexity of the solutions.

A common design principle among ML researchers is the so called *minimum description length principle* (see for instance Rissanen 1978), which states that the best model is the one that minimizes the amount of information needed to encode it. In this perspective, preference for simpler solutions and over-fitting avoidance seem to be closely related. It is more likely that a complex solution incorporates specific information from the training set, thus over-fitting it, compared to a simpler solution. But, as mentioned in Domingos (1999), this argumentation should be taken with care as too much emphasis on minimizing complexity can prevent the discovery of more complex yet more accurate solutions. Finally, in Vanneschi et al. (2007), authors hint that GP generalization ability (or more precisely, *extrapolation* ability) can be improved by performing a multi-optimization on the training set. A related idea has been used in Gagné et al. (2006) in the domain of binary classification, where a two-objective sort is performed in order to extract a set of nondominated individuals. However, different objectives in the realm of regression problems are used in Vanneschi et al. (2007).

Despite the above mentioned studies, the issue of generalization remains an open one in GP, since no theoretical approach has ever been presented to tackle this problem until now.

5.4   Modularity in GP

The use of modules may be very important to improve GP expressiveness, code reusability, and performance. The best-known of these methods is Koza's ADFs, which is introduced in ❯ Sect. 2.7. The first step toward a theoretically motivated study of ADFs is probably represented by Rosca (1995), where an algorithm for the automatic discovery of building blocks in GP called *Adaptive Representation Through Learning* was presented. In the same year, Spector (1995) introduced techniques to evolve collections of automatically defined macros and showed how they can be used to produce new control structures during the evolution of programs and Seront (1995) extended the concept of code reuse in GP to the concept of generalization, showing how programs (or "concepts," using Seront's terminology) synthesized by GP to solve a problem can be reused to solve other ones. Altenberg offers a critical analysis of modularity in evolution in Altenberg (2009), stating that the evolutionary advantages that have been attributed to modularity do not derive from modularity *per se*. Rather, they require that there be an "alignment" between the spaces of phenotypic variation, and the selection gradients that are available to the organism. Modularity in the genotype–phenotype map may make such an alignment more readily attained, but it is not sufficient; the appropriate phenotype-fitness map in conjunction with the genotype–phenotype map is also necessary for evolvability. This contribution is interesting and stimulating, but its applicability to GP remains an open question.

In Jonyer and Himes (2006) the concept of ADFs is extended by using graph-based data mining to identify common aspects of highly fit individuals and modularizing them by creating functions out of the subprograms identified. In Hemberg et al. (2007) the authors state that the ability of GP to scale to problems of increasing difficulty operates on the premise that it is possible to capture regularities that exist in a problem environment by decomposition of the problem into a hierarchy of modules.

Although the use of modularity in GP has helped solve some problems and provide new data/control abstractions, for instance in the form of ADFs, still some issues remain open.

For instance: Has it solved really substantially more complex problems and does it hold the promise to do so in the near future? Are ADFs necessary/sufficient as a formalism to help solve grand-challenge problems, that is to provide scalability? And even more ambitiously: Can we achieve modularity, hierarchy, and reuse in a more controlled and principled manner? Can we provide insights and metrics on how to achieve this? Is it possible to routinely evolve algorithms? How to achieve iteration and/or recursion in a controlled and principled manner? What is the best way to handle the generation and maintenance of constants?

A first attempt to answer this last question is represented by Vanneschi et al. (2006) and Cagnoni et al. (2005) where the authors present a new coevolutionary environment where a GA dynamically evolves constants for GP. Although interesting, this contribution, as many others aimed at improving GP modularity, is mainly empirical, that is, it shows the benefits of the presented method(s) by means of a set of experimental results on some test functions. In order to be able to get a deeper insight of the real usefulness of modularity and code-reuse in GP, more theoretical studies are needed.

6 Conclusions

GP is a systematic method for getting computers to automatically solve a problem starting from its high-level specifications. It extends the model of genetic algorithms to the space of computer programs. In its almost 30 years of existence, it has been successfully applied to a number of challenging real-world problem domains. Furthermore, solid theory exists explaining its behavior and dynamics.

After a general introduction to the mechanics of GP, this chapter has discussed some examples of real-world applications of GP. In general, GP is particularly suitable for applications of curve fitting, data modeling, or symbolic regression, typically characterized by a large amount of data and where little is known about the relationships among the relevant features. Over the years, dozens of GP results have passed the human-competitiveness test, defined by Koza by establishing eight precise criteria, including patentability, beating humans in regulated competitions, producing publishable results, etc. Fields where GP have been successfully applied in the last few years include: image and signal processing; financial trading, time series, and economic modeling; industrial process control; medicine, biology and bioinformatics; computational chemistry; entertainment and computer games; arts; and compression.

Subsequently, we presented some of the most relevant GP theoretical results, including schema theories and Markov chains-based models. We discussed recent studies of the distribution of fitness in GP search spaces and of the distribution of programs' length. We presented various theories explaining the phenomenon of bloat, including the recently introduced crossover bias theory, which is based on a precise equation of the programs' size evolution, and that, hopefully, will give a conclusive explanation of this phenomenon. We gave arguments in support of the conjecture that GP is a global optimizer, that is, it is guaranteed to find a globally optimal solution, given enough time. Furthermore, we presented the most important results obtained in the field of GP problem difficulty, including a discussion of fitness distance correlation and negative slope coefficient, two of the most effective measures of problem difficulty for GP, and the recently introduced subtree crossover similarity/dissimilarity measure, that should help in defining new and hopefully more powerful difficulty measures. Finally, we summarized results on the existence of a "free lunch" for GP under some weak (and, in practice, often satisfied) conditions.

In the last part of this chapter, hot topics and open issues of GP have been discussed, including the development of a self-adapting framework to dynamically modify the GP configuration (population size, representation, genetic operators, etc.) during the evolution; the study of the usefulness of GP for dynamic optimization problems; the study and better understanding of the concept of generalization and its implications for GP; and the definition of more powerful and self-adaptive methods of using modularity in GP.

Although our chapter could not exhaustively cover the immensity of the field, it is hoped that this chapter will attract newcomers to the field of GP as well as help focus future research, leading to a wider application and deeper understanding of GP as a problem-solving strategy.

References

Altenberg L (2009) Modularity in evolution: Some low-level questions. In: Rasskin-Gutman D, Callebaut W (eds), *Modularity: Understanding the Development and Evolution of Complex Natural Systems.* MIT Press, Cambridge, MA

Alves da Silva AP, Abrao PJ (2002) Applications of evolutionary computation in electric power systems. In: Fogel DB et al. (eds), *Proceedings of the 2002 Congress on Evolutionary Computation CEC2002*, pp. 1057–1062. IEEE Press

Archetti F, Messina E, Lanzeni S, Vanneschi L (2007) Genetic programming for computational pharmacokinetics in drug discovery and development. *Genet Programming Evol Mach* 8(4):17–26

Azaria Y, Sipper M (2005) GP-gammon: Genetically programming backgammon players. *Genet Programming Evol Mach* 6(3):283–300, Sept. Published online: 12 August 2005

Banzhaf W, Francone FD, Nordin P (1996) The effect of extensive use of the mutation operator on generalization in genetic programming using sparse data sets. In: Ebeling W et al., (ed) *4th International Conference on Parallel Problem Solving from Nature (PPSN96)*, Springer, Berlin, pp. 300–309

Banzhaf W, Nordin P, Keller RE, Francone FD (1998) *Genetic Programming, An Introduction.* Morgan Kaufmann, San Francisco, CA

Barrett SJ, Langdon WB (2006) Advances in the application of machine learning techniques in drug discovery, design and development. In: Tiwari A et al. (eds), *Applications of Soft Computing: Recent Trends*, Advances in Soft Computing, On the World Wide Web, 19 Sept.–7 Oct. 2005. Springer, Berlin, 99–110

Bojarczuk CC, Lopes HS, Freitas AA (July–Aug. 2008) Genetic programming for knowledge discovery in chest-pain diagnosis. *IEEE Eng Med Biol Mag* 19(4):38–44

Brameier M, Banzhaf W (2001) A comparison of linear genetic programming and neural networks in medical data mining. *IEEE Trans Evol Comput* 5(1):17–26

Cagnoni S, Rivero D, Vanneschi L (2005) A purely-evolutionary memetic algorithm as a first step towards symbiotic coevolution. In: *Proceedings of the 2005 IEEE Congress on Evolutionary Computation (CEC'05)*, Edinburgh, Scotland, 2005. IEEE Press, Piscataway, NJ. pp. 1156–1163

Castillo F, Kordon A, Smits G (2006) Robust pareto front genetic programming parameter selection based on design of experiments and industrial data. In: Riolo RL, et al. (ed) *Genetic Programming Theory and Practice IV*, vol 5 of *Genetic and Evolutionary Computation*, chapter 2. Springer, Ann Arbor, 11–13 May

Chen S-H, Liao C-C (2005) Agent-based computational modeling of the stock price-volume relation. *Inf Sci* 170(1):75–100, 18 Feb

Chen S-H, Wang H-S, Zhang B-T (1999) Forecasting high-frequency financial time series with evolutionary neural trees: The case of Hang Seng stock index. In: Arabnia HR, (ed), *Proceedings of the International Conference on Artificial Intelligence, IC-AI '99*, vol 2, Las Vegas, NV, 28 June-1 July. CSREA Press pp. 437–443

Da Costa LE, Landry JA (2006) Relaxed genetic programming. In: Keijzer M et al., editor, *GECCO 2006: Proceedings of the 8th Annual Conference on Genetic and Evolutionary Computation*, vol 1, Seattle, Washington, DC, 8–12 July. ACM Press pp. 937–938

Dassau E, Grosman B, Lewin DR (2006) Modeling and temperature control of rapid thermal processing. *Comput Chem Eng* 30(4):686–697, 15 Feb

Davis TE, Principe JC (1993) A Markov chain framework for the simple genetic algorithm. *Evol Comput* 1(3): 269–288

De Jong KA (1988) Learning with genetic algorithms: An overview. Mach Learn 3:121–138

Dempsey I (2007) Grammatical evolution in dynamic environments. Ph.D. thesis, University College Dublin, Ireland

Dignum S, Poli R (2007) Generalisation of the limiting distribution of program sizes in tree-based genetic programming and analysis of its effects on bloat. In: Thierens, D et al. (eds), *GECCO '07: Proceedings of the 9th Annual Conference on Genetic and Evolutionary Computation*, vol 2 London, 7–11 July 2007. ACM Press, pp. 1588–1595

Domingos P (1999) The role of Occam's razor in knowledge discovery. *Data Mining Knowl Discov* 3(4):409–425

Eiben AE, Jelasity M (2002) A critical note on experimental research methodology in EC. In: *Congress on Evolutionary Computation (CEC'02)*, Honolulu, HI, 2002. IEEE Press, Piscataway, NJ, pp. 582–587

Esparcia-Alcazar AI, Sharman KC (Sept. 1996) Genetic programming techniques that evolve recurrent neural networks architectures for signal processing. In: *IEEE Workshop on Neural Networks for Signal Processing*, Seiko, Kyoto, Japan

Fernandez F, Martin A (2004) Saving effort in parallel GP by means of plagues. In: Keijzer M, et al. (eds), *Genetic Programming 7th European Conference, EuroGP 2004, Proceedings*, vol 3003 of *LNCS*, Coimbra, Portugal, 5–7 Apr. Springer-Verlag, pp. 269–278

Fernandez F, Tomassini M, Vanneschi L (2003) Saving computational effort in genetic programming by means of plagues. In: Sarker, R et al. (eds), *Proceedings of the 2003 Congress on Evolutionary Computation CEC2003*, Camberra, 8–12 Dec. 2003. IEEE Press, pp. 2042–2049

Fernandez F, Vanneschi L, Tomassini M (2003) The effect of plagues in genetic programming: A study of variable-size populations. In: Ryan, C et al. (ed) *Genetic Programming, Proceedings of EuroGP'2003*, vol 2610 of *LNCS*, Essex, 14–16 Apr. Springer-Verlag, pp. 317–326

Forrest S, Mitchell M (1993) What makes a problem hard for a genetic algorithm? Some anomalous results and their explanation. *Mach Learn* 13:285–319

Francone FD, Nordin P, Banzhaf W (1996) Benchmarking the generalization capabilities of a compiling genetic programming system using sparse data sets. In: Koza JR et al. (ed), *Genetic Programming: Proceedings of the First Annual Conference*, MIT Press, Cambridge, pp. 72–80

Fukunaga A, Stechert A (1998) Evolving nonlinear predictive models for lossless image compression with genetic programming. In: Koza, JR et al. (eds), *Genetic Programming 1998: Proceedings of the Third Annual Conference*, University of Wisconsin, Madison, WI, 22–25 July, Morgan Kaufmann pp. 95–102

Gagné C, Schoenauer M, Parizeau M, Tomassini M (2006) Genetic programming, validation sets, and parsimony pressure. In: Collet P et al. (ed), *Genetic Programming, 9th European Conference, EuroGP2006*, Lecture Notes in Computer Science, LNCS 3905, pp. 109–120. Springer, Berlin, Heidelberg, New York

Giacobini M, Tomassini M, Vanneschi L (2002) Limiting the number of fitness cases in genetic programming using statistics. In: Merelo JJ, et al. (eds), *Parallel Problem Solving from Nature – PPSN VII*, vol 2439 of Lecture Notes in Computer Science, Springer-Verlag, Heidelberg, pp. 371–380

Goldberg DE (1989) *Genetic Algorithms in Search, Optimization and Machine Learning*. Addison-Wesley, Boston, MA

Gustafson S, Vanneschi L (2005) Operator-based distance for genetic programming: Subtree crossover distance. In: Keijzer, M., et al. (ed), *Genetic Programming, 8th European Conference, EuroGP2005*, Lecture Notes in Computer Science, LNCS 3447, pp. 178–189. Springer, Berlin, Heidelberg, New York

Gustafson S, Vanneschi L (2008) Operator-based tree distance in genetic programming. *IEEE Trans Evol Comput* 12:4

Hansen JV, Lowry PB, Meservy RD, McDonald DM (Aug. 2007) Genetic programming for prevention of cyberterrorism through dynamic and evolving intrusion detection. *Decis Support Syst* 43(4): 1362–1374, Special Issue Clusters

Hasan S, Daugelat S, Rao PSS, Schreiber M (June 2006) Prioritizing genomic drug targets in pathogens: Application to mycobacterium tuberculosis. *PLoS Comput Biol* 2(6):e61

Hauptman A, Sipper M (2007) Evolution of an efficient search algorithm for the mate-in-N problem in chess. In: Ebner, M et al. (eds), *Proceedings of the 10th European Conference on Genetic Programming*, vol 4445 of Lecture Notes in Computer Science Valencia, Spain, 11–13 Apr. Springer pp. 78–89

Hemberg E, Gilligan C, O'Neill M, Brabazon A (2007) A grammatical genetic programming approach to modularity in genetic algorithms. In: Ebner, M et al. (eds), *Proceedings of the 10th European Conference on Genetic Programming*, vol 4445 of Lecture Notes in Computer Science, Valencia, Spain, 11–13 Apr. Springer pp. 1–11

Holland JH (1975) *Adaptation in Natural and Artificial Systems*. The University of Michigan Press, Ann Arbor, MI

Horn J, Goldberg DE (1995) Genetic algorithm difficulty and the modality of the fitness landscapes. In: Whitley D, Vose M (eds), *Foundations of Genetic Algorithms*, vol. 3, Morgan Kaufmann, pp. 243–269

Howard D, Roberts SC (2004) Incident detection on highways. In: O'Reilly, U-M et al., (eds), *Genetic*

Programming Theory and Practice II, chapter 16, Springer, Ann Arbor, 13–15 May pp. 263–282

Jacob C (May–June 2000) The art of genetic programming. *IEEE Intell Syst* 15(3):83–84, May–June

Jacob C (2001) *Illustrating Evolutionary Computation with Mathematica*. Morgan Kaufmann, San Francisco, CA

Jakobović D, Budin L (2006) Dynamic scheduling with genetic programming. In: Collet, P et al. (eds), *Proceedings of the 9th European Conference on Genetic Programming*, vol 3905 of Lecture Notes in Computer Science, Budapest, Hungary, 10–12 Apr. Springer pp. 73–84

Jin N, Tsang E (2006) Co-adaptive strategies for sequential bargaining problems with discount factors and outside options. In: *Proceedings of the 2006 IEEE Congress on Evolutionary Computation*, Vancouver, 6–21 July. IEEE Press, pp. 7913–7920

Jones T (1995) Evolutionary algorithms, fitness landscapes and search. Ph.D. thesis, University of New Mexico, Albuquerque

Jonyer I, Himes A (2006) Improving modularity in genetic programming using graph-based data mining. In: Sutcliffe GCJ, Goebe RG (eds), *Proceedings of the Nineteenth International Florida Artificial Intelligence Research Society Conference*, pp. 556–561, Melbourne Beach, FL, May 11–13 2006. American Association for Artificial Intelligence

Jordaan E, den Doelder J, Smits G (2006) Novel approach to develop structure-property relationships using genetic programming. In: Runarsson TP, et al. (eds), *Parallel Problem Solving from Nature – PPSN IX*, vol 4193 of *LNCS*, Reykjavik, Iceland, 9–13 Sept. Springer-Verlag pp. 322–331

Kashtan N, Noor E, Alon U (2007) Varying environments can speed up evolution. *Proceedings of the National Academy of Sciences*, 104(34):13711–13716, August 21

Kattan A, Poli R (2008) Evolutionary lossless compression with GP-ZIP. In *Proceedings of the IEEE World Congress on Computational Intelligence*, Hong Kong, 1–6 June. IEEE

Keijzer M (Sept. 2004) Scaled symbolic regression. *Genetic Programming and Evolvable Machines*, 5(3):259–269

Kibria RH, Li Y (2006) Optimizing the initialization of dynamic decision heuristics in DPLL SAT solvers using genetic programming. In: Collet P, et al. (eds), *Proceedings of the 9th European Conference on Genetic Programming*, vol. 3905 of Lecture Notes in Computer Science, Budapest, Hungary, 10–12 Apr. Springer. pp. 331–340

Kinnear KE Jr (1994) Fitness landscapes and difficulty in genetic programming. In: *Proceedings of the First IEEE Congress on Evolutionary Computation*, IEEE Press, Piscataway, NY, pp. 142–147

Kirchherr W, Li M, Vitanyi P (1997) The miraculous universal distribution. *Math Intell* 19:7–15

Klappenecker A, May FU (1995) Evolving better wavelet compression schemes. In: Laine, AF et al. (ed), *Wavelet Applications in Signal and Image Processing III*, vol 2569, San Diego, CA 9–14 July. SPIE

Kordon A (Sept. 2006) Evolutionary computation at Dow Chemical. *SIGEVOlution*, 1(3):4–9

Koza J, Poli R (2003) A genetic programming tutorial. In: Burke E (ed) *Introductory Tutorials in Optimization, Search and Decision Support*, Chapter 8. http://www.genetic-programming.com/jkpdf/burke2003tutorial.pdf

Koza JR (1992a) A genetic approach to the truck backer upper problem and the inter-twined spiral problem. In *Proceedings of IJCNN International Joint Conference on Neural Networks*, vol IV, IEEE Press, pp. 310–318

Koza JR (1992b) *Genetic Programming: On the Programming of Computers by Means of Natural Selection*. MIT Press, Cambridge, MA

Koza JR (1994) *Genetic Programming II*. The MIT Press, Cambridge, MA

Koza JR, Bennett FH III, Stiffelman O (1999) Genetic programming as a Darwinian invention machine. In: Poli R, et al. (eds) *Genetic Programming, Proceedings of EuroGP'99*, vol 1598 of *LNCS*, Goteborg, Sweden, 26–27 May. Springer-Verlag pp. 93–108

Koza JR, Bennett FH III, Andre D, Keane MA (1999) *Genetic Programming III: Darwinian Invention and Problem Solving*. Morgan Kaufmann, San Francisco, CA

Kushchu I (2002) An evaluation of evolutionary generalization in genetic programming. *Artif Intell Rev* 18(1):3–14

Langdon WB (2003) Convergence of program fitness landscapes. In: Cantú-Paz, E., et al. (ed) *Genetic and Evolutionary Computation – GECCO-2003*, vol 2724 of *LNCS*, Springer-Verlag, Berlin, pp. 1702–1714

Langdon WB, Buxton BF (Sept. 2004) Genetic programming for mining DNA chip data from cancer patients. *Genet Programming Evol Mach*, 5(3):251–257

Langdon WB, Poli R (1998) Genetic programming bloat with dynamic fitness. In: Banzhaf W, et al. (eds), *Proceedings of the First European Workshop on Genetic Programming*, vol 1391 of *LNCS*, Paris, 14–15 Apr. Springer-Verlag. pp. 96–112

Langdon WB, Poli R (2002) *Foundations of Genetic Programming*. Springer-Verlag

Langdon WB, Poli R (2005) Evolutionary solo pong players. In: Corne, D et al. (eds), *Proceedings of the 2005 IEEE Congress on Evolutionary Computation*, vol 3, Edinburgh, U.K., 2–5 Sept. IEEE Press pp. 2621–2628

Langdon WB, Soule T, Poli R, Foster JA (June 1999) The evolution of size and shape. In Spector, L et al. (eds), *Advances in Genetic Programming 3*, chapter 8, pp. 163–190. MIT Press, Cambridge, MA

Lew TL, Spencer AB, Scarpa F, Worden K, Rutherford A, Hemez F (Nov. 2006) Identification of response surface models using genetic programming. *Mech Syst Signal Process* 20(8):1819–1831

Lewin DR, Lachman-Shalem S, Grosman B (July 2006) The role of process system engineering (PSE) in integrated circuit (IC) manufacturing. *Control Eng Pract* 15(7):793–802 Special Issue on Award Winning Applications, 2005 IFAC World Congress

Louchet J (June 2001) Using an individual evolution strategy for stereovision. *Genet Programming Evol Mach* 2(2):101–109

Lutton E, Levy-Vehel J, Cretin G, Glevarec P, Roll C (1995) Mixed IFS: Resolution of the inverse problem using genetic programming. Research Report No 2631, INRIA

Machado P, Romero J (eds). (2008) *The Art of Artificial Evolution*. Springer

McPhee NF, Miller JD (1995) Accurate replication in genetic programming. In: Eshelman L (ed), *Genetic Algorithms: Proceedings of the Sixth International Conference (ICGA95)*, Pittsburgh, PA 15–19 July Morgan Kaufmann pp. 303–309

Miller J (2001) What bloat? Cartesian genetic programming on Boolean problems. In: Goodman ED (ed), *2001 Genetic and Evolutionary Computation Conference Late Breaking Papers*, pp. 295–302, San Francisco, CA 9–11 July

Mitavskiy B, Rowe J (2006) Some results about the Markov chains associated to GPs and to general EAs. *Theor Comput Sci* 361(1):72–110 28 Aug

Mitchell M, Forrest S, Holland J (1992) The royal road for genetic algorithms: Fitness landscapes and GA performance. In: Varela F, Bourgine P (eds), *Toward a Practice of Autonomous Systems, Proceedings of the First European Conference on Artificial Life*, The MIT Press, pp. 245–254

Nikolaev NI, Slavov V (1998) Concepts of inductive genetic programming. In: Banzhaf, W., et al. (ed), *Genetic Programming, Proceedings of EuroGP'1998*, vol 1391 of *LNCS*, Springer-Verlag, pp. 49–59

Nordin P, Banzhaf W (1996) Programmatic compression of images and sound. In: Koza JR, et al. (eds), *Genetic Programming 1996: Proceedings of the First Annual Conference*, Stanford University, CA 28–31 July. MIT Press pp. 345–350

Poli R (1996) Genetic programming for image analysis. In: Koza JR et al. (eds), *Genetic Programming 1996: Proceedings of the First Annual Conference*, Stanford University, CA 28–31 July MIT Press pp. 363–368

Poli R (2001) Exact schema theory for genetic programming and variable-length genetic algorithms with one-point crossover. *Genet Programming Evol Mach* 2(2):123–163

Poli R, Langdon WB (1997) Genetic programming with one-point crossover and point mutation. Tech. Rep. CSRP-97-13, University of Birmingham, B15 2TT, U.K., 15

Poli R, Langdon WB (2006) Efficient Markov chain model of machine code program execution and halting. In: Riolo RL, et al. (eds), *Genetic Programming Theory and Practice IV*, vol 5 of *Genetic and Evolutionary Computation*, chapter 13. Springer, Ann Arbor, 11–13 May

Poli R, Langdon WB, Dignum S (2007) On the limiting distribution of program sizes in tree-based genetic programming. In: Ebner, M et al. (eds), *Proceedings of the 10th European Conference on Genetic Programming*, vol 4445 of Lecture Notes in Computer Science, Valencia, Spain, 11–13 Apr. Springer pp. 193–204

Poli R, McPhee NF (Mar. 2003a) General schema theory for genetic programming with subtree-swapping crossover: Part I. *Evol Comput* 11(1):53–66

Poli R, McPhee NF (June 2003b) General schema theory for genetic programming with subtree-swapping crossover: Part II. *Evol Comput* 11(2):169–206

Poli R, McPhee NF (2008) Parsimony pressure made easy. In: *GECCO '08: Proceedings of the 10th Annual Conference on Genetic and Evolutionary Computation*, pp. 1267–1274, New York, NY, ACM

Poli R, McPhee NF, Rowe JE (Mar. 2004) Exact schema theory and Markov chain models for genetic programming and variable-length genetic algorithms with homologous crossover. *Genet Programming Evol Mach* 5(1):31–70

Poli R, McPhee NF, Graff M (2009) Free lunches for symbolic regression. In: *Foundations of Genetic Algorithms (FOGA)*. ACM, forthcoming

Poli R, Langdon WB, McPhee NF (2008) A Field Guide to Genetic Programming. Published via http://lulu.com and freely available at http://www.gp-field-guide.org.uk, (With contributions by J. R. Koza)

Rissanen J (1978) Modeling by shortest data description. *Automatica* 14:465–471

Rochat D, Tomassini M, Vanneschi L (2005) Dynamic size populations in distributed genetic programming. In: Keijzer M, et al. (eds), *Proceedings of the 8th European Conference on Genetic Programming*, vol 3447 of Lecture Notes in Computer Science, Lausanne, Switzerland, 30 Mar.–1 Apr. Springer. pp. 50–61

Rosca JP (1995) Towards automatic discovery of building blocks in genetic programming. In: *Working Notes for the AAAI Symposium on Genetic Programming*, AAAI, pp. 78–85

Rudolph G (1994) Convergence analysis of canonical genetic algorithm. *IEEE Trans Neural Netw* 5(1): 96–101

Rudolph G (1996) Convergence of evolutionary algorithms in general search spaces. In: *International Conference on Evolutionary Computation*, pp. 50–54

Schumacher C, Vose MD, Whitley LD (2001) The no free lunch and problem description length. In: *Proceedings of the Genetic and Evolutionary Computation Conference (GECCO)*, Morgan Kaufmann, pp. 565–570

Seront G (1995) External concepts reuse in genetic programming. In: Siegel EV, Koza JR (eds), *Working Notes for the AAAI Symposium on Genetic Programming*, MIT, Cambridge, MA 10–12 Nov. AAAI pp. 94–98

Shah SC, Kusiak A (July 2004) Data mining and genetic algorithm based gene/SNP selection. *Artif Intell Med* 31(3):183–196

Silva S (2008) Controlling bloat: individual and population based approaches in genetic programming. Ph.D. thesis, Universidade de Coimbra, Faculdade de Ciences e Tecnologia, Departamento de Engenharia Informatica, Portugal

Smola AJ, Scholkopf B (1999) A tutorial on support vector regression. Technical Report Technical Report Series – NC2-TR-1998-030, NeuroCOLT2

Soule T, Foster JA (1998a) Effects of code growth and parsimony pressure on populations in genetic programming. *Evol Comput* 6(4):293–309, Winter

Soule T, Foster JA (1998b) Removal bias: A new cause of code growth in tree based evolutionary programming. In *1998 IEEE International Conference on Evolutionary Computation*, Anchorage, Alaska 5–9 May IEEE Press. pp. 781–186

Spector L (1995) Evolving control structures with automatically defined macros. In: Siegel EV, Koza JR (eds), *Working Notes for the AAAI Symposium on Genetic Programming*, MIT, Cambridge, MA 10–12 Nov. AAAI pp. 99–105

Stadler PF (2002) Fitness landscapes. In: Lässig M, Valleriani A (eds), *Biological Evolution and Statistical Physics*, vol 585 of Lecture Notes Physics, pp. 187–207, Heidelberg, Springer-Verlag

Stephens CR, Waelbroeck H (1999) Schemata evolution and building blocks. *Evol Comput* 7(2):109–124

Tomassini M, Vanneschi L, Cuendet J, Fernandez F (2004) A new technique for dynamic size populations in genetic programming. In: *Proceedings of the 2004 IEEE Congress on Evolutionary Computation*, Portland, Oregon, 20–23 June. IEEE Press pp. 486–493

Tomassini M, Vanneschi L, Collard P, Clergue M (2005) A study of fitness distance correlation as a difficulty measure in genetic programming. *Evol Comput* 13(2):213–239, Summer

Trujillo L, Olague G (2006) Using evolution to learn how to perform interest point detection. In: X Y T et al. (ed), *ICPR 2006 18th International Conference on Pattern Recognition*, vol 1, IEEE, pp. 211–214. 20–24 Aug

Tsang E, Jin N (2006) Incentive method to handle constraints in evolutionary. In: Collet P, et al. (eds), *Proceedings of the 9th European Conference on Genetic Programming*, vol 3905 of Lecture Notes in Computer Science, Budapest, Hungary, 10–12 Apr. Springer. pp. 133–144

Vanneschi L (2004) Theory and practice for efficient genetic programming Ph.D. thesis, Faculty of Sciences, University of Lausanne, Switzerland

Vanneschi L (2007) Investigating problem hardness of real life applications. In: R. R. et al., (ed), *Genetic Programming Theory and Practice V*, Springer, Computer Science Collection, pp. 107–124, Chapter 7

Vanneschi L, Clergue M, Collard P, Tomassini M, Vérel S (2004) Fitness clouds and problem hardness in genetic programming. In: Deb K, et al. (eds), *Genetic and Evolutionary Computation – GECCO-2004, Part II*, vol 3103 of Lecture Notes in Computer Science Seattle, WA 26–30 June, Springer-Verlag pp. 690–701

Vanneschi L, Gustafson S, Mauri G (2006) Using subtree crossover distance to investigate genetic programming dynamics. In: Collet, P., et al. (ed), *Genetic Programming, 9th European Conference, EuroGP2006*, Lecture Notes in Computer Science, LNCS 3905, pp. 238–249. Springer, Berlin, Heidelberg, New York

Vanneschi L, Mauri G, Valsecchi A, Cagnoni S (2006) Heterogeneous cooperative coevolution: strategies of integration between GP and GA. In: Keijzer M, et al. (eds), *GECCO 2006: Proceedings of the 8th Annual Conference on Genetic and Evolutionary Computation*, vol 1, Seattle, Washington, DC, 8–12 July. ACM Press. pp. 361–368

Vanneschi L, Rochat D, Tomassini M (2007) Multioptimization improves genetic programming generalization ability. In: Thierens D, et al. (eds), *GECCO '07: Proceedings of the 9th Annual Conference on Genetic and Evolutionary Computation*, vol 2, London, 7–11 July. ACM Press. pp. 1759–1759

Vanneschi L, Tomassini M, Collard P, Clergue M (2003) Fitness distance correlation in structural mutation genetic programming. In: Ryan, C., et al., (ed), *Genetic Programming, 6th European Conference, EuroGP2003*, Lecture Notes in Computer Science, Springer-Verlag, Heidelberg, pp 455–464

Vanneschi L, Tomassini M, Collard P, Vérel S (2006) Negative slope coefficient. A measure to characterize genetic programming. In: Collet P, et al. (eds), *Proceedings of the 9th European Conference on Genetic Programming*, vol 3905 of Lecture Notes in Computer Science, Budapest, Hungary, 10–12 Apr. Springer. pp. 178–189

Wagner N, Michalewicz Z, Khouja M, McGregor RR (Aug. 2007) Time series forecasting for dynamic environments: The DyFor genetic program model. *IEEE Trans Evol Comput* 11(4):433–452

Wang Y, Wineberg M (2006) Estimation of evolvability genetic algorithm and dynamic environments. *Genet Programming Evol Mach* 7(4):355–382

Wedge DC, Kell DB (2008) Rapid prediction of optimum population size in genetic programming using a novel genotype–fitness correlation. In: Keijzer M, et al. (eds), *GECCO '08: Proceedings of the 10th Annual Conference on Genetic and Evolutionary Computation*, Atlanta, GA, ACM pp. 1315–1322

Whitley D, Watson JP (2005) Complexity theory and the no free lunch theorem. In: Burke EK, Kendall G (eds), *Search Methodologies: Introductory Tutorials in Optimization and Decision Support Techniques*, Chapter 11, pp. 317–339. Springer

Wolpert D, Macready W (1997) No free lunch theorems for optimization. *IEEE Trans Evol Comput* 1(1):67–82

Woodward JR, Neil JR (2003) No free lunch, program induction and combinatorial problems. In: Ryan C, et al. (eds), *Genetic Programming, Proceedings of EuroGP'2003*, vol 2610 of *LNCS*, Essex, 14–16 Apr. Springer-Verlag pp. 475–484

Xie H, Zhang M, Andreae P (2006) Genetic programming for automatic stress detection in spoken English. In: Rothlauf F, et al. (eds), *Applications of Evolutionary Computing, EvoWorkshops 2006: Evo-BIO, EvoCOMNET, EvoHOT, EvoIASP, EvoInteraction, EvoMUSART, EvoSTOC*, vol 3907 of *LNCS*, pp. 460–471, Budapest, 10–12 Apr. Springer Verlag

Yang S, Ong Y-S, Jin Y (Dec. 2006) Editorial to special issue on evolutionary computation in dynamic and uncertain environments. *Genet Programming Evol Mach* 7(4):293–294, Editorial

Yu T, Chen S-H (2004) Using genetic programming with lambda abstraction to find technical trading rules. In: *Computing in Economics and Finance*, University of Amsterdam, 8–10 July

Yu J, Yu J, Almal AA, Dhanasekaran SM, Ghosh D, Worzel WP, Chinnaiyan AM (Apr. 2007) Feature selection and molecular classification of cancer using genetic programming. *Neoplasia* 9(4):292–303

Zhang B-T, Mühlenbein H (1995) Balancing accuracy and parsimony in genetic programming. *Evol Comput* 3(1):17–38

Zhang M, Smart W (Aug. 2006) Using gaussian distribution to construct fitness functions in genetic programming for multiclass object classification. *Pattern Recog Lett* 27(11):1266–1274. Evolutionary Computer Vision and Image Understanding

25 The Dynamical Systems Approach — Progress Measures and Convergence Properties

Silja Meyer-Nieberg[1] · *Hans-Georg Beyer[2]*
[1]Fakultät für Informatik, Universität der Bundeswehr München, Neubiberg, Germany
silja.meyer-nieberg@unibw.de
[2]Department of Computer Science, Fachhochschule Vorarlberg, Dornbirn, Austria
hans-georg.beyer@fhv.at

G. Rozenberg et al. (eds.), *Handbook of Natural Computing*, DOI 10.1007/978-3-540-92910-9_25,
© Springer-Verlag Berlin Heidelberg 2012

Abstract

This chapter considers local progress and the dynamical systems approach. The approach can be used for a quantitative analysis of the behavior of evolutionary algorithms with respect to the question of convergence and the working mechanism of these algorithms. Results obtained so far for evolution strategies on various fitness functions are described and discussed before presenting drawbacks and limitations of the approach. Furthermore, a comparison with other analysis methods is given.

1 Introduction: Evolutionary Algorithms as Dynamical Systems

Evolutionary algorithms can be studied in various ways – theoretically and experimentally. This section gives an overview of an analysis approach that models the algorithms as dynamical systems. As in physics and engineering sciences, the time evolution of the systems is due to the forces acting on the systems. Thus, the forces, also referred to as local progress measures, are the basis to understand how evolutionary algorithms function.

The text is divided as follows: first, a short introduction to evolution strategies, the evolutionary algorithm to which the approach was mainly applied, is given. Afterward, a description of how the dynamics of evolution strategies can be transformed into a deterministic dynamical system is given. In the next step, the local progress measures are defined before some results are discussed. The following overview starts with a description of several commonly used fitness functions. Afterward, the dynamical systems and local progress approach is explained in more detail using a simple example. The next section lists the results obtained for undisturbed and uncertain environments. The approach presented has certain drawbacks and limitations. The remaining section is devoted to these aspects, setting the approach used in relation to other theoretical approaches.

1.1 Evolution Strategies

An *evolution strategy* (ES) is a specific evolutionary algorithm (EA) invented in 1963 by Bienert, Rechenberg, and Schwefel. The population-based search heuristic moves through the search space by means of variation, that is, mutation and recombination, and selection. A population consists of several individuals. Each individual represents a candidate solution coded in the object parameters.

The performance of an ES strongly depends on the choice of a so-called strategy parameter, the *mutation strength* also referred to as the *step size*, which controls the spread of the population due to mutation. During an optimization run, the mutation strength must be adapted continuously to allow the ES to travel with sufficient speed. To this end, several methods have been developed – for example, Rechenberg's well-known $1/5$th-rule (Rechenberg 1973), self-adaptation (Rechenberg 1973; Schwefel 1977), or the cumulative step-size adaptation (CSA) and covariance matrix adaptation (CMA) of Ostermeier, Gawelczyk, and Hansen, for example Ostermeier et al. (1995) and Hansen and Ostermeier (2001).

■ **Fig. 1**

The $(\mu/\rho \dotplus \lambda)$-$\sigma$SA-ES (cf. Beyer 2001b, p. 8).

```
BEGIN
    g:=0;
    INITIALIZATION (𝒫μ(0):= {(ym(0),σm(0),F(ym(0)))});
    REPEAT
        FOR EACH OF THE λ OFFSPRING DO
            𝒫ρ:= REPRODUCTION(𝒫μ(g));
            σ'l:= RECOMBσ (𝒫ρ);
            σl:= MUTATEσ (σ'l);
            y'l:= RECOMBy (𝒫ρ);
            yl:= MUTATEy (y'l,σl);
            Fl:= F(yl);
        END
        𝒫λ(g):= {(yl,σl,Fl)};
        CASE ","-SELECTION: 𝒫μ(g+1):=SELECT(𝒫λ(g));
        CASE "+"-SELECTION: 𝒫μ(g+1):=SELECT(𝒫μ(g),𝒫λ(g));
        g:=g+1;
    UNTIL stop;
END
```

Following Beyer (2001b, p. 8), ❏ *Fig. 1* illustrates the basic algorithm of a multi-parent $(\mu/\rho \dotplus \lambda)$-ES with σ-self-adaptation (σSA). In a self-adaptive ES, the tuning of the mutation strength is left to the evolution strategy itself. Similar to the object parameters, the strategy parameters are subject to variation. If an offspring is selected, it also has a chance to bequest its strategy parameters to the offspring generation. That is, self-adaptation assumes a statistic or probabilistic connection between strategy parameters and "good" fitness values.

As ❏ *Fig. 1* shows, a $(\mu/\rho \dotplus \lambda)$-ES maintains a population $\mathscr{P}_\mu^{(g)}$ of μ candidate solutions at generation g – with the strategy parameters used in their creation. Based on that parent population, λ offspring are created via variation.

Offspring are created as follows: For each offspring, ρ of the μ parents are chosen for recombination leading to the set $\mathscr{P}_\rho$. The selection of the parents may be deterministic or probabilistic (see, e.g., Beyer and Schwefel 2002; Eiben and Smith 2003).

First, the strategy parameters are changed. The strategy parameters of the chosen ρ parents are recombined and the result is mutated afterward. The change of the object parameters occurs in the next step. Again, the parameters are first recombined and then mutated. The newly created strategy parameter σ_l is used in the mutation process. Afterward, the fitness of the offspring is determined.

After the offspring population of λ individuals has been created, the μ best individuals with respect to their fitness values are chosen as the next parental population $\mathscr{P}_\mu^{(g+1)}$. Two selection schemes are generally distinguished: "comma" and "plus"-selection. In the case of "comma"-selection, only the offspring population is considered for selection, that is, only individuals of the offspring population have a chance to be selected into the next parent population. The old parent population is completely discarded.

In the case of "plus"-selection, members of the old parent population and the offspring population may be selected into the succeeding parent population.

There are two types of recombination that are mainly used in practice: intermediate and dominant multi-parent recombination denoted by $(\mu/\mu_I, \lambda)$ and $(\mu/\mu_D, \lambda)$, respectively. Using intermediate recombination for both the object parameters and the mutation strengths, the offspring are generated in the following manner:

1. Compute the mean, $\langle\sigma\rangle = \frac{1}{\mu}\sum_{m=1}^{\mu}\sigma_m$, of the mutation strengths σ_m of the parent population.
2. Compute the centroid, $\langle\mathbf{y}\rangle = \frac{1}{\mu}\sum_{m=1}^{\mu}\mathbf{y}_m$, of the object vectors $\mathbf{y}_m$ of the μ parents.
3. For all offspring $l \in \{1, \ldots, \lambda\}$:
   a. Derive the new mutation strength by mutating the mean $\langle\sigma\rangle$ according to $\sigma_l = \langle\sigma\rangle\zeta$ where ζ is a random variable, which should fulfill $E[\zeta] \approx 1$ (see Beyer and Schwefel (2002) for a discussion of this and further requirements). Typical choices of ζ's distribution include the log-normal distribution, the normal distribution, or a two-point distribution (Bäck 1997).
   b. Generate the object vector $\mathbf{y}_l$ according to $y_i = \langle y_i \rangle + \sigma_l \mathcal{N}(0,1)$ where y_i is the vector's ith component and $\mathcal{N}(0,1)$ stands for a standard normally distributed random variable.

Dominant recombination works differently. For each offspring, the recombination vector is created component-wise by randomly choosing one parent for each position and copying the value.

A nonrecombinant multi-parent strategy is denoted as a (μ, λ)-ES or a $(\mu + \lambda)$-ES. Offspring are generated by choosing one of the parents at random and adding a mutation vector.

After the offspring or the candidate solutions have been created, the μ best solutions ("plus": offspring and parents, "comma": offspring) are chosen, according to their fitness, and become the succeeding parent population.

Self-adaptive strategies also retain the strategy parameters for each selected individual. The strategy parameters can be seen as an additional part of an individual's genome – neutral with respect to the fitness but influencing the evolutionary progress. The mutation is usually realized by using the log-normal operator with *learning parameter* τ. The learning parameter τ is a parameter of the log-normal distribution. Similar to the standard deviation of the normal distribution, it controls the spread of the distribution (see ◉ *Table 1*). Larger values result in a larger spread whereas smaller values lead to a shrink toward an area including one.

◼ Table 1

Common mutation operators for self-adaptive evolution strategies. The parameter τ appearing in the log-normal operator is usually called the *learning parameter*. The random variable U for the two-point distribution is uniformly distributed in the interval (0, 1]. The second row for the two-point mutation gives an alternative definition, producing the same random process when using $\alpha = \ln(1 + \beta)$

Operator	Mutation	Random variable
Log-normal	$\sigma' := \zeta\sigma$	$\zeta = e^{\tau\mathcal{N}(0,1)}$
Meta-EP	$\sigma' := \zeta\sigma$	$\zeta = 1 + \tau\mathcal{N}(0,1)$
Two-point	$\sigma' := \begin{cases} \sigma(1+\beta) & \text{if } U(0,1] \leq 1/2 \\ \sigma/(1+\beta) & \text{if } U(0,1] > 1/2 \end{cases}$	
Two-point	$\sigma' := \zeta\sigma$	$\zeta = e^{\alpha\,\text{sign}(\mathcal{N}(0,1))}$

Note, however, that the log-normal distribution is restricted to positive values and is not symmetric. Furthermore, the learning parameter controls the spread but *is not* a central moment of the distribution, that is, τ^2 is not the variance. Further alternatives are given in ❯ *Table 1*.

Self-adaptation uses only one kind of information from the evolutionary search: the ranking of the fitness values. In contrast, the cumulative step-size adaptation (CSA) and the standard version of covariance matrix adaptation (CMA) use a so-called *evolution path*, which sums up the actually realized search steps. Therefore, these algorithms do not only use information from the fitness ranking but also related search space information. In the following, the basic CSA-algorithm is described for $(\mu/\mu_I, \lambda)$-ES:

$$\forall l = 1 \ldots \lambda : \mathbf{w}_l^{(g)} := \overrightarrow{\mathcal{N}}_l(0, 1) \tag{1}$$

$$\mathbf{x}^{(g)} := \sigma^{(g)} \mathbf{z}^{(g)} := \frac{\sigma^{(g)}}{\mu} \sum_{m=1}^{\mu} \mathbf{w}_{m;\lambda} \tag{2}$$

$$\mathbf{y}^{(g+1)} := \mathbf{y}^{(g)} + \mathbf{x}^{(g)} \tag{3}$$

$$\mathbf{s}^{(g+1)} := (1 - c)\mathbf{s}^{(g)} + \sqrt{\mu c(2 - c)}\mathbf{z}^{(g)} \tag{4}$$

$$\sigma^{(g+1)} := \sigma^{(g)} \exp\left(\frac{\|\mathbf{s}^{(g+1)}\| - \overline{\chi_N}}{D\overline{\chi_N}}\right) \tag{5}$$

with $\overrightarrow{\mathcal{N}}_l(0, 1)$ as a random vector with standard normally distributed components and D as the dampening constant (Beyer and Arnold 2003b). The mechanism of cumulative step size adaptation is based on the assumption that the step size is adapted optimally when consecutive search steps are uncorrelated or perpendicular: If the mutation strength is too small, the ES takes several steps into the same or similar direction. These steps could be substituted by a single larger step. The control rule should therefore increase the mutation strength and with it the step size. If otherwise the mutation strength is too large, the ES has to retrace its movements – consecutive steps may be antiparallel. In this case, the mutation strength must be decreased.

The *evolutionary path* or *accumulated progress vector* (❯ 4) and its length can be used to ascertain which situation is occurring at present. Let one neglect for a moment the influence of selection and assume that selection is purely random. In this case, the consecutive steps are normally distributed. Their weighted sum is therefore also normally distributed, and its length follows a χ-distribution with N degrees of freedom. This explains the appearance of the mean $\overline{\chi_N}$ of the χ_N-distribution in ❯ Eq. 5.

Selection changes the situation: If the mutation strength is too large, the ES has to retrace, and antiparallel movements may occur resulting in a path length smaller than the ideal case $\overline{\chi_N}$. The CSA decreases the step size according to ❯ Eq. 5. If the mutation strength is too small, the algorithm takes several steps into roughly the same direction, which results in a path length longer than that in the uncorrelated case. The CSA then increases the mutation strength.

Theoretical analyses of ESs performed so far have been devoted mainly to the isotropic mutation case that needs to consider a single mutation parameter – the mutation strength σ – only. The adaptation of arbitrary Gaussian mutation distributions is state-of-the-art in algorithm design, known as the covariance matrix adaptation-ES. This ES uses the concept of the evolutionary path and adapts additionally the full covariance matrix. Detailed descriptions of the CMA and CSA can be found in Hansen and Ostermeier (1996, 1997, 2001) and Hansen (2006). Theoretical performance analysis of the CMA-ES is, however, still in its infancy.

◻ **Table 2**

Types of ESs considered. Note a $(\mu/\mu, \lambda)$-ES and a (μ, λ)-ES always include the $(1, \lambda)$-ES as a special case. The groups for the $(1, \lambda)$-ES only list publications solely devoted to this ES

Evolution strategy	Fitness	References
$(1 + 1)$-ES	Sphere	Arnold and Beyer (2002a); Beyer (1989, 1993)
$(1 + \lambda)$-ES	Sphere	Beyer (1993)
$(1, \lambda)$-ES	Sphere	Beyer and Meyer-Nieberg (2006a); Beyer (1993)
$(\mu/\mu_I, \lambda)$-ES	Sphere	Arnold and Beyer (2003b, 2004, 2006a, b); Meyer-Nieberg and Beyer (2005); Beyer et al. (2003, 2004)
		Arnold and Beyer (2000, 2001b, 2002b, c); Beyer (1995a, 1996a); Beyer and Arnold (2003b)
$(\mu/\mu_D, \lambda)$-ES	Sphere	Beyer (1995a, 1996a)
(μ, λ)-ES	Sphere	Arnold and Beyer (2001a, 2003a); Beyer (1995b)
(λ_{opt})-ES	Sphere	Arnold (2006b)
(μ, λ)-ES	Linear	Arnold and Beyer (2001a, 2003a)
$(1, \lambda)$-ES	Ridge	Beyer and Meyer-Nieberg (2006b); Beyer (2001a); Oyman et al. (1998, 2000)
$(1 + \lambda)$-ES	Ridge	Oyman et al. (1998, 2000)
$(\mu/\mu_I, \lambda)$-ES	Ridge	Meyer-Nieberg and Beyer (2007, 2008); Arnold and Beyer (2008); Arnold (2006a); Oyman and Beyer (2000)
$(\mu/\mu_D, \lambda)$-ES	Ridge	Oyman and Beyer (2000)
$(1, \lambda)$-ES	General quadratic functions	Beyer and Arnold (1999)
$(\mu/\mu_I, \lambda)$-ES	General quadratic functions	Beyer (2004); Beyer and Arnold (2003a)
$(\mu/\mu_I, \lambda)$-ES	Positive definite quadratic forms	Arnold (2007); Beyer and Finck (2009)

❯ *Table 2* gives an overview of the ES variants analyzed up to now using the progress rate and dynamical system approaches. The analysis of the CSA-ES used a slightly different version of the CSA than previously presented. Instead of ❯ Eq. 5, the mutation strength update rule reads

$$\sigma^{(g+1)} = \sigma^{(g)} e^{\frac{\|s^{(g)}\|^2 - N}{2DN}}$$

(6)

See, e.g., Arnold (2002a) for a discussion and usual settings of the parameters c and D.

1.2   A Short Introduction to Dynamical Systems

The dynamical systems approach uses concepts that stem from the theory of dynamical systems. This section gives a short introduction to this area. To obtain more information, the reader is referred, for example, to Wiggins (1990) and Braun (1998). The term *dynamical* already points to one of the main characteristics of these systems: they change in time. But how does the system change? Are there any points where the system comes to rest? These are some of the questions that the theory of dynamical systems addresses.

The section starts with some illustrative examples. Afterward, the concept of equilibrium solutions, the rest points of a system, is addressed. The section closes with a short introduction to the stability of equilibrium points.

1.2.1 What Are Dynamical Systems?

First of all, the term *dynamical system* has to be explained. Dynamical systems can be approached in two ways. On the one hand, on an abstract level, a dynamical system is just a difference or differential equation, the properties of which shall be studied. On the other hand, this formal mathematical description generally appears as the result of a modeling process in the analysis of a time-dependent process, for example, the course of a chemical reaction, the spread of a forest fire, the growth of a population, the spreading of an epidemic disease, or the run of an evolutionary algorithm.

A modeling process aims at formalizing and quantifying the important characteristics of the system, for instance, the interactions between entities. If the modeler is interested in how the system evolves in time, the result is a dynamic system. Thus, a dynamical system is a model of a time-dependent system or process expressed in a formal mathematical way.

In general, several classes of dynamical systems can be distinguished. Dynamical systems can be grouped by the nature of the change, that is, whether the changes are deterministic or stochastic. This section considers deterministic systems. For a rigorous introduction to the field of random dynamical systems, the reader is referred to Arnold (2002b).

Dynamical systems can further be classified into discrete time and continuous time dynamical systems. In the first case, the change of the variable, $\mathbf{y} \in \mathbb{R}$, occurs at discrete time steps, leading to an *iterated map* or *difference equation*

$$\mathbf{y}(t + 1) = f(\mathbf{y}(t)) \tag{7}$$

with $f : \mathbb{R}^N \to \mathbb{R}^N$ and *starting point* $\mathbf{y}(0) = \mathbf{y}_0$. In this and the following sections, it is assumed that the time always starts at $t_0 = 0$.

A dynamical system in continuous time leads to a *differential equation*

$$\frac{d}{dt}\mathbf{y}(t) = g(\mathbf{y}(t)) \tag{8}$$

with $g : \mathbb{R}^N \to \mathbb{R}^N$ and starting point $\mathbf{y}(0) = \mathbf{y}_0$. The behavior of the system (⊙ 7) or (⊙ 8), respectively, depends on (a) the properties of f and g and (b) to some lesser extent on the initial value $\mathbf{y}_0$.

In the following, some examples of dynamic systems are discussed. Let one consider the situation that a saver opens a bank account with a starting capital y_0. The bank offers an interest rate of $p\%$ per annum. How does the capital increase?

The problem can be described by

$$y(t + 1) = y(t)\left(1 + \frac{p}{100}\right) \tag{9}$$

with the time, t, in discrete years. It is easy to see by performing a few iterations of ⊙ Eq. 9 starting with $y(0) = y_0$ that the solution of ⊙ Eq. 9 can be given in closed form as

$$y(t) = y_0\left(1 + \frac{p}{100}\right)^t \tag{10}$$

◻ **Fig. 2**

Examples for dynamical systems: (a) shows ❯ Eq. 10, a time-discrete exponential growth, with initial value $y_0 = 100$ and two choices for the interest rate. (b) shows ❯ Eq. 13 with carrying capacity $C = 2$ and initial value $y_0 = 0.2$.

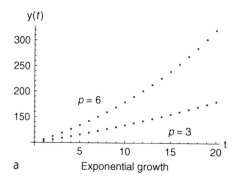

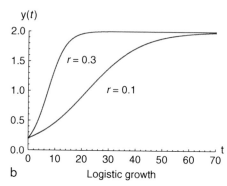

a   Exponential growth   b   Logistic growth

The capital grows exponentially, see ❯ Fig. 2a. The exact value at time t depends on the initial value, whereas the general behavior – the exponential growth – is a result of the form of the change. The form of ❯ Eq. 10 follows from the fact that ❯ Eq. 9 is a simple linear difference equation.

Let one now consider the growth of a population of bacteria in a petri dish. In experiments, it was found that a bacteria population first experiences an exponential growth which slows down in further progress. The decrease of the growth can be attributed among others to intra-species competition over increasingly scarce resources. Usually, an environment or a habitat (the petri dish) allows a maximal population density, the so-called carrying capacity C. (The population density is determined as (number of individuals)/(size of habitat).) The process can be described by a continuous model

$$\frac{\mathrm{d}}{\mathrm{d}t} y(t) = ry(t)\left(1 - \frac{y(t)}{C}\right) \tag{11}$$

with r standing for the exponential growth rate. Again, it is possible to obtain a solution in closed form. The calculations require partial fraction decomposition

$$\frac{\frac{\mathrm{d}}{\mathrm{d}t} y(t)}{ry(t)\left(1 - \frac{y(t)}{C}\right)} = \frac{\mathrm{d}}{\mathrm{d}t} y(t)\left(\frac{A}{ry(t)} + \frac{B}{\left(1 - \frac{y(t)}{C}\right)}\right) = 1 \tag{12}$$

with A and B to be determined and some knowledge in the area of differential equations. Finally,

$$y(t) = \frac{C y_0 e^{rt}}{C - y_0 + y_0 e^{rt}} \tag{13}$$

is obtained. ❯ Figure 2b shows two examples of a logistic growth which differ in the size of the parameter r. The larger r is, the faster the carrying capacity is reached. Let one consider a third example. This time, consider the interaction between a predator population and a prey population. Lotka and Volterra (see, e.g., Hofbauer and Sigmund 2002) used the following difference equation to describe the species' interactions

$$\frac{d}{dt}x(t) = ax(t) - bx(t)y(t)$$
$$\frac{d}{dt}y(t) = -cy(t) + dx(t)y(t)$$
(14)

with $x(t)$ denoting the density of the prey population and $y(t)$ the density of the predator. Lotka and Volterra assumed that in the absence of the predator the prey population would grow exponentially, whereas the predator would die out without any prey. In the absence of the prey, the predator decreases exponentially. The parameters a and c determine the growth rate of the prey and the loss rate of the predator. It is assumed that the population growth of the predator depends on its preying and therefore on its meeting the prey. The loss of the prey due to predation is also proportional to the meeting of predator and prey. The constants b and d regulate the respective loss and gain. The Lotka–Volterra predator–prey model is one of the earliest and simplest models in population biology. However, although it is extremely simple, there are no general analytical solutions of ❷ Eq. 14 which express x and y as time-dependent functions.

In the next case, the growth of a population, that is, the logistic growth, is reconsidered (Hofbauer and Sigmund 2002). For instance, in the case of insect populations with non-overlapping generations, the growth should be modeled with a difference equation instead of a differential equation leading to

$$y(t+1) = Ry(t)\left(1 - \frac{y(t)}{K}\right)$$
(15)

In contrast to the continuous model (❷ 13), a general analytical solution to ❷ Eq. 15 that depends on the time cannot be obtained.

There are many more examples where it is not possible to obtain a closed analytical solution for a differential or difference equation. The question that remains is what should be done in this case. One solution that comes to mind is to perform simulations or computer experiments, to let the system evolve inside the computer and to draw statistically sound conclusions from the observations. This established method may have some drawbacks, though. What if the behavior of the system depends strongly on the size of some parameters or the initial values, and the simulations did not consider these? Or what if the time horizon of the simulation is not sufficiently long for the system to show the interesting behavior? The study of equilibrium solutions provides another way to obtain more information about a system.

1.2.2 Rest Points: Equilibrium Solutions

Equilibrium solutions are solutions of ❷ Eqs. 8 or ❷ 7 that are the rest points of the system, that is, points where no more changes of the system occur. In other words, an equilibrium solution is a solution $\mathbf{y}_{eq}$ of the discrete system (❷ 7) that does not change in time, that is,

$$y(t+1) = y(t) = \mathbf{y}_{eq} \Rightarrow f(\mathbf{y}_q) = \mathbf{y}_q$$
(16)

or a solution $\mathbf{y}_{eq}$ of the continuous system (❷ 8)

$$\frac{d}{dt}\mathbf{y}(t) = g(\mathbf{y}(t)) = 0 \Rightarrow g(\mathbf{y}(t)) = g(\mathbf{y}_{eq}) = 0$$
(17)

There are many synonyms for equilibrium solutions – including the terms rest point, singularity, fix point, fixed point, stationary point (also solution or state), or steady state (Wiggins 1990, p. 6).

In the examples, the equilibrium solutions can be determined straightforwardly. In the case of the time-discrete exponential growth, ❷ Eq. 9

$$y(t + 1) = y(t)\left(1 + \frac{p}{100}\right) \tag{18}$$

requiring $y(t + 1) = y(t)$ leads only to one equilibrium solution, $y_{eq} = 0$: Only zero capital is not changed by interest yield.

In the case of the time-continuous logistic growth ❷ Eq. 11, one requires $d/(dt)y(t) = 0$ for no change, leading to

$$0 = ry(t)\left(1 - \frac{y(t)}{C}\right) \tag{19}$$

with two solutions $y_{eq1} = 0$ and $y_{eq1} = C$, the zero population and the carrying capacity. Again, this immediately makes sense. A population with zero individuals cannot grow and if the maximal allowed capacity is reached, a further increase is not possible.

In the case of the Lotka–Volterra model (❷ 14), the system only "rests" as a whole, if the predator equation and the prey equation have equilibrium solutions. Therefore, one has to set $d/(dt)x(t) = 0$ and $d/(dt)y(t) = 0$, leading to

$$0 = ax(t) - bx(t)y(t) = x(t)(a - by(t)) \tag{20}$$

$$0 = -cy(t) + dx(t)y(t) = y(t)(-c + dx(t)) \tag{21}$$

One solution, $x_{eq} = y_{eq} = 0$, follows immediately. If $x_{eq} \neq 0$, $a - by(t) = 0$ has to hold. Therefore, $y_{eq} = a/b$. For $a \neq 0$, $y_{eq} \neq 0$. In order to have $d/(dt)y(t) = 0$, $-c + dx = 0$ has to be fulfilled. This leads to the pair $x_{eq} = c/d$ and $y_{eq} = a/b$. Thus, the Lotka–Volterra model has two pairs of equilibrium solutions. The first one, $x_{eq} = y_{eq} = 0$ corresponds to an extinction of both populations, whereas the second corresponds to a coexistence of predator and prey.

Let one reconsider the time-discrete logistic growth model (❷ 15)

$$y(t + 1) = Ry(t)\left(1 - \frac{y(t)}{K}\right) \tag{22}$$

The demand $y(t + 1) = y(t)$ leads to the first solution $y_{eq1} = 0$ and

$$1 = R\left(1 - \frac{y}{K}\right) \Rightarrow \frac{K}{R} = K - y$$

$$\Rightarrow y_{eq} = K\left(\frac{R - 1}{R}\right) \tag{23}$$

if $R \geq 1$. The second solution starts in zero for $R = 1$ and increases afterward – approaching K for $R \to \infty$.

Note the difference of the equilibrium solutions between the time-discrete and the time-continuous logistic growth. As the examples showed, there is more than one equilibrium solution in some cases. This immediately raises the question of how the system behaves:

- Are these points ever attained by the system?
- What happens if the system is in one of these fix points and a disturbance occurs. Does it return?

In other words, one has to address the question of the stability of the equilibrium solutions.

1.2.3   On the Stability of Equilibrium Solutions

This section provides a very short introduction to the analysis of the stability of equilibrium solutions. The concept of a local stability, that is, the behavior of solutions that start close to one of the equilibrium solutions, is considered.

First of all, the term "stability" has to be defined. An equilibrium solution is *stable*, if solutions of the system which start close to it remain close for all times. This form of stability is also called *Liapunov stability*. An equilibrium solution is called *asymptotically stable*, if solutions of the dynamical system which start close to the equilibrium solution approach the equilibrium solution (Wiggins 1990, p. 6).

The concept is illustrated by the simple linear system $y(t + 1) = ay(t)$, $a \in \mathbb{R}$. The only equilibrium solution is $y_{eq} = 0$. The size of a decides the behavior of solutions starting close to the equilibrium. Consider a solution $y(t)$ of the linear system which starts close to $y_{eq} = 0$, for example, $y(t) = 0 + u(t)$ with $y(0) = u(0)$, $|u(0)|$ "small." Due to the form of the equation, the "new system" leads to $u(t) = a^t u(0)$.

If $|a| < 1$, the system is (asymptotically) stable, since $u(t)$ converges to zero. For $|a| > 1$ it is unstable; the disturbance explodes as the system moves toward infinity. The case of $|a| = 1$ is a special one: The system leaves the equilibrium solution, and does neither return nor does it go toward infinity. It remains at the distance which was given by $|u(0)|$ and may show oscillating behavior depending on the sign of a. This is not a case of asymptotical stability but only of (Liapunov) stability.

Let one consider now the discrete time logistic growth model (❯ 15)

$$y(t + 1) = Ry(t)\left(1 - \frac{y(t)}{K}\right) \tag{24}$$

with equilibrium solutions $y_{eq1} = 0$ and $y_{eq2} = K(1 - 1/R)$ for $R \geq 1$. In contrast to the previous example, the system is nonlinear. How can stability criteria be derived in this case? In general, one has a system with a nonlinear function $f: \mathbb{R} \to \mathbb{R}$

$$y(t + 1) = f(y(t)) \tag{25}$$

Let $y(t)$ be a solution which starts close to the equilibrium solution y_{eq}, $y(t) = y_{eq} + u(t)$ with $|u(0) - y_{eq}| \leq \delta$, $\delta > 0$. How can information be gained on how $y(t)$ behaves? Considering solutions close to the equilibrium solution allows us to apply a Taylor series expansion for the function f around y_{eq}

$$y(t + 1) = f(y_{eq}) + \frac{d}{dy}f(y)|_{y=y_{eq}} u + \mathcal{O}(|u|^2)$$

$$y(t + 1) - y_{eq} = \frac{d}{dy}f(y)|_{y=y_{eq}} u(t) + \mathcal{O}(|u|^2) \tag{26}$$

$$u(t + 1) = \frac{d}{dy}f(y)|_{y=y_{eq}} u(t) + \mathcal{O}(|u|^2)$$

Neglecting the quadratic and higher terms, it depends on the first derivative of f at the equilibrium solution whether the equilibrium solution is asymptotically stable or not. If

25

$|d/(dy)f(y_{eq})| < 1$, then y_{eq} is asymptotically stable. If $|d/(dy)f(y_{eq})| > 1$, then y_{eq} is unstable. (This can be proven formally, see, e.g., Wiggins (1990).) If $|d/(dy)f(y_{eq})| = 1$, the equilibrium solution is called *non-hyperbolic* and no information can be gained by using the approach presented. The opposite is a *hyperbolic* equilibrium solution. In this example of a nonlinear one-dimensional system, the equilibrium solution is called hyperbolic if the absolute value of the derivative of the function at the solution is not equal to one. In these cases, the linearization of the system via Taylor series expansion suffices completely for the analysis. If the point is non-hyperbolic, other methods have to be applied. A discussion of alternative ways to proceed can be found for instance in Wiggins (1990).

In the case of the discrete logistic growth $\bullet$ Eq. 15, the nonlinear function, f, reads

$$f(y) = Ry\left(1 - \frac{y}{K}\right) \text{ with} \tag{27}$$

$$\frac{d}{dy}f(y) = R\left(1 - 2\frac{y}{K}\right) \tag{28}$$

The equilibrium solutions are $y_{eq1} = 0$ and $y_{eq2} = K(R - 1)/R$ for $R \geq 1$. It follows immediately that the first solution, $y_{eq1} = 0$, is stable for $R < 1$ and unstable for $R > 1$. In the case of the second solution,

$$
\begin{aligned}
\frac{d}{dy}f(y_{eq2}) &= R\left(1 - \frac{2y_{eq2}}{K}\right) \\
&= R\left(1 - \frac{2K}{K}\left(\frac{R-1}{R}\right)\right) \\
&= R\left(1 - 2\left(\frac{R-1}{R}\right)\right) \\
&= R - 2R + 2 = 2 - R
\end{aligned}
\tag{29}
$$

follows. Since stability requires $|d/(dy)f(y_{eq2})| < 1$, one proceeds with determining the zero point of $2 - R$ as $R_0 = 2$. Considering first $R \leq 2, 2 - R < 1 \Rightarrow 1 < R$ has to hold in order that y_{eq2} is stable. The second equilibrium solution is stable for $R \in (1, 2]$. For $R > 2, R - 2 < 1 \Rightarrow R < 3$ must be fulfilled. In short, the second equilibrium point is stable for $R \in (1, 3)$. For all larger growth parameters, both equilibrium solutions are unstable. Note, though, that this system shows a very interesting behavior – exhibiting periodic solutions and moving toward a chaotic regime. Readers interested in a more thorough discussion of the behavior of the general logistic map are referred, for example, to Hofbauer and Sigmund (2002).

Let one now consider the N-dimensional linear system

$$\mathbf{y}(t + 1) = \mathbf{A}\mathbf{y}(t) \tag{30}$$

with $\mathbf{y}(t) \in \mathbb{R}^N$, $\mathbf{A} \in \mathbb{R}^{N \times N}$, and initial value $\mathbf{y}(0)$. The only equilibrium solution is $\mathbf{y}_{eq} = \mathbf{0}$. To analyze the stability, consider a solution, $\mathbf{y}(t)$, starting close to $\mathbf{y}_{eq}$ with $\mathbf{y}(t) = \mathbf{y}_{eq} + \mathbf{u}(t)$, $\|\mathbf{u}(0) - \mathbf{y}_{eq}\| \leq \delta, \delta > 0$ and system

$$\mathbf{u}(t + 1) = \mathbf{A}\mathbf{u}(t) \tag{31}$$

The solution is given by

$$\mathbf{u}(t) = \mathbf{A}^t\mathbf{y}(0) = \mathbf{A}^t\mathbf{u}(0) \tag{32}$$

Although the system is now in closed form, the behavior of the system is not easy to discern. One way to simplify the analysis is to switch to an alternative coordinate system. Let one consider the eigenvalues λ_i and eigenvectors $\mathbf{v}_i$ of the quadratic matrix $\mathbf{A}$ (to simplify the discussions, one assumes that $\mathbf{A}$ is diagonalizable) and recall that $\mathbf{A}\mathbf{v}_i = \lambda_i\mathbf{v}_i$. Setting $\mathbf{M} := (\mathbf{v}_1, \ldots, \mathbf{v}_N)$ and $\Lambda := (\lambda_1\mathbf{e}_1, \ldots, \lambda_N\mathbf{e}_N)$, with $\mathbf{e}_i$ denoting the ith unit vector, it follows that

$$\mathbf{A}\mathbf{M} = \mathbf{M}\Lambda \Rightarrow \mathbf{M}^{-1}\mathbf{A}\mathbf{M} = \Lambda \tag{33}$$

Setting $\mathbf{z}(t) := \mathbf{M}^{-1}\mathbf{u}(t)$, ❷ Eq. 31 changes to

$$\mathbf{z}(t+1) = \Lambda\mathbf{z}(t) \tag{34}$$

with the solution

$$\mathbf{z}(t) = \Lambda^t\mathbf{z}(0) \tag{35}$$

or

$$\mathbf{u}(t) = \mathbf{M}\Lambda^t\mathbf{M}^{-1}\mathbf{u}(0) \tag{36}$$

respectively. The eigenvalues of $\mathbf{A}$ determine the behavior. If $\max |\lambda_i| < 1$, then the system converges to zero: The equilibrium solution is stable. If $\max |\lambda_i| > 1$, the system diverges to infinity and the equilibrium solution is unstable. In both cases, equilibrium solution is hyperbolic. If $\max |\lambda_i| \leq 1$ and $|\lambda_i| = 1$ for at least one eigenvalue, the respective component of $\mathbf{z}$ remains at the initial value. In this case, the system is stable. Note, this only holds for diagonalizable matrices.

As in the one-dimensional case, the findings for the linear system can be transferred to the nonlinear system. Again, a Taylor series expansion around the equilibrium solution can be used. The equivalent to the first derivative in the one-dimensional case is the Jacobian matrix or Jacobian for functions $f: \mathbb{R}^N \rightarrow \mathbb{R}^M$: It gives the partial derivatives of the function components.

Concerning the question of stability, the Jacobian is decisive, or more correctly, its eigenvalues λ_i. If $\max |\lambda_i| < 1$ holds, then the equilibrium solution is asymptotically stable. If $|\lambda_i| > 1$ holds for at least one eigenvalue, then the solution is not stable.

Generally, in the case of time-discrete dynamic systems, an equilibrium solution is called *hyperbolic* if no eigenvalue λ_i of the Jacobian has an absolute value (modulus) equal to one, that is, $|\lambda_i| \neq 1$ for all eigenvalues. If $\max |\lambda_i| < 1$ and $|\lambda_i| = 1$ for at least one eigenvalue, then the equilibrium solution is non-hyperbolic and other methods than the one presented have to be applied.

Note, that the discussion has been limited, in this section, to the stability of the equilibrium solutions of maps or difference equations, respectively. The stability of equilibrium solutions of differential equations has not been considered. However, the discussions can be transferred with minor changes. This becomes clear when considering the simple linear system

$$\frac{\mathrm{d}}{\mathrm{d}t}y(t) = ay(t) \tag{37}$$

with $a \in \mathbb{R}$. The only equilibrium solution is $y_{\mathrm{eq}} = 0$. Consider a solution $y(t)$ of ❷ Eq. 37 that starts close to $y_{\mathrm{eq}} = 0$, that is, $y(t) = y_{\mathrm{eq}} + u(t)$ with $|u(0)| \leq \delta$, $\delta > 0$. Then

$$\frac{\mathrm{d}}{\mathrm{d}t}y(t) = \frac{\mathrm{d}}{\mathrm{d}t}u(t) = ay(t) = a(y_{\mathrm{eq}} + u(t)) = au(t)$$
$$\Rightarrow \frac{\frac{\mathrm{d}}{\mathrm{d}t}u(t)}{u(t)} = a \tag{38}$$

holds, leading to

$$\frac{\frac{\mathrm{d}}{\mathrm{d}t}u(t)}{u(t)} = a$$

$$\Rightarrow \frac{\mathrm{d}}{\mathrm{d}t}\ln(u(t)) = a \qquad (39)$$

$$\Rightarrow \ln\left(\frac{u(t)}{u(0)}\right) = at$$

$$\Rightarrow u(t) = u(0)e^{at}$$

Again, the constant a is decisive for the question of stability. For $a < 0$, the system (❯ 39) goes to zero: The equilibrium solution is asymptotically stable. For $a > 0$, the system diverges and the system is not stable. For $a = 0$, the solution remains at the starting point. Again, this is a case of Liapunov stability not of asymptotical stability. As in the case of maps, these findings can be transferred to the nonlinear case using Taylor series expansion and to the N-dimensional case via the eigenvalues of the Jacobian.

1.3 From Stochastic Processes to Dynamical Systems

The run of an evolutionary algorithm is a dynamical movement of the population in the search space. Due to random influences, for example mutation, the movement of an EA is stochastic. In other words, the algorithm induces a stochastic process which can be analyzed in several ways. The process itself takes place in a high-dimensional state space (dimensionality $\geq N$, N – search space dimension). However, it is often preferable to consider derived or aggregated quantities with respect to the task of the analysis: that is, whether the ES converges, how fast the optimizer is approached, how EA parameters influence the behavior, and whether the step-size adaptation mechanism works adequately.

The state of an evolution strategy can be characterized by so-called state variables: variables which characterize the important and interesting features and the system behavior completely. Common variables include the fitness values, the distance to the optimizer, $\hat{y}$, (depending on the fitness model), and, if the effects of step-size adaptation mechanism shall be considered, the mutation strength.

The task is to model and to analyze the evolution of these state variables over time. In the following, the sphere model (❯ 71) is used for further explanations. Since the sphere model consists of functions of the form $f(\mathbf{y}) = g(\|\mathbf{y} - \hat{\mathbf{y}}\|) = g(R)$, the state variables are chosen as the distance to the optimizer $R^{(g)} := \|\mathbf{y}^{(g)} - \hat{\mathbf{y}}\|$ and the mutation strength $\sigma^{(g)}$ at generation g. The dynamics of an ES generate the stochastic process

$$\begin{pmatrix} R^{(g)} \\ \sigma^{(g)} \end{pmatrix} \rightarrow \begin{pmatrix} R^{(g+1)} \\ \sigma^{(g+1)} \end{pmatrix} \qquad (40)$$

Till now, no closed solution for the transition kernels could be derived in general. The only exception is a $(1, 2)$-ES using the two-point rule for the mutation of the mutation strength (see Beyer 1996b or Beyer 2001b, p. 287).

Another way to proceed is to transform the stochastic process into a deterministic dynamical system in a first approximation (mean value dynamics) and into a Gaussian random process in a second approximation. This is done via a step-by-step approach introduced in Beyer (1996b).

The approach introduces the *evolution equations*: stochastic difference equations or iterated maps, respectively, used to describe the change of the state variables during one generation.

The change of the random variables can be divided into two parts: The first denotes the expected change. The second part covers the random fluctuations and is denoted by ε_R or ε_σ. In their most general form, the evolution equations read

$$R^{(g+1)} = R^{(g)} - E[R^{(g)} - R^{(g+1)}|R^{(g)}, \sigma^{(g)}] + \varepsilon_R(R^{(g)}, \sigma^{(g)}) \tag{41}$$

$$\sigma^{(g+1)} = \sigma^{(g)}\left(1 + E\left[\frac{\sigma^{(g+1)} - \sigma^{(g)}}{\sigma^{(g)}}|R^{(g)}, \sigma^{(g)}\right]\right) + \varepsilon_\sigma(R^{(g)}, \sigma^{(g)}) \tag{42}$$

in the case of self-adaptation. The case of CSA is described below. In ❷ Eq. 41, a well-known *progress measure* appears: the *progress rate* φ_R. The progress rate measures the expected change of the distance in one generation

$$\varphi_R(\sigma^{(g)}, R^{(g)}) := E[R^{(g)} - R^{(g+1)}|\sigma^{(g)}, R^{(g)}] \tag{43}$$

In the case of the evolution of the mutation strength, a different progress measure is used when self-adaptation is considered. Note, since the mutation of the mutation strength is generally realized by a multiplication with a random variable, ❷ Eq. 42 gives the relative change. The progress measure is called the (first-order) *self-adaptation response* (SAR) ψ. The SAR gives the expected relative change of the mutation strength in one generation

$$\psi(\sigma^{(g)}, R^{(g)}) := E\left[\frac{\sigma^{(g+1)} - \sigma^{(g)}}{\sigma^{(g)}}|\sigma^{(g)}, R^{(g)}\right]. \tag{44}$$

This progress measure is not used in analyses of CSA-evolution strategies. In CSA, the change of the mutation strength is derandomized. The stochastic influences stem from the mutation of the object parameters and enter the progress vector $\mathbf{z}^{(g)}$ ❷ Eq. 2. Analyses of the CSA usually consider the evolution of the length of the evolutionary path $\mathbf{s}^{(g)}$, ❷ Eq. 4, and the mutation strength ❷ Eq. 6

$$\|\mathbf{s}^{(g+1)}\|^2 = (1-c)^2\|\mathbf{s}^{(g)}\|^2 + 2(1-c)\sqrt{2\mu(2-c)}(\mathbf{s}^{(g)})^T\mathbf{z}^{(g)} + \mu c(2-c)\|\mathbf{z}^{(g)}\|^2 \tag{45}$$

$$\sigma^{(g+1)} = \sigma^{(g)}e^{\frac{\|\mathbf{s}^{(g+1)}\|^2 - N}{2DN}} \tag{46}$$

The fluctuation terms in ❷ Eqs. 41 and ❷ 42 present a problem: Their distribution is not known and must be approximated using a reference density. Common approaches comprise an expansion into a Gram–Charlier or Edgeworth series (see, e.g., Kolossa 2006). The reference distribution is usually (but not necessarily) chosen to be the normal distribution. In order to expand an unknown distribution at all, it must be possible to determine some of its moments or cumulants. If this is the case, the approach continues by standardizing the fluctuation terms using the mean and the standard deviations. Clearly, the conditional mean of ε_σ and ε_R is zero. Therefore, only the standard deviation remains to be determined.

The main points of the derivation are explained considering the case of ε_R. The case of the mutation strength is analogous. Let D_φ denote the standard deviation. Therefore, the standardized random part ε_R' is related to ε_R by $\varepsilon_R = D_\varphi\varepsilon_R'$. The standard deviation can be derived via ❯ Eq. 41 since its square equals the second conditional moment of ε_R (note $e[\varepsilon_R] = 0$)

$$
\begin{aligned}
D_\varphi^2(\sigma^{(g)}, R^{(g)}) = E\left[\varepsilon_R^2|\sigma^{(g)}, R^{(g)}\right] &= E\left[\left(R^{(g+1)} - R^{(g)} + \varphi_R(\sigma^{(g)}, R^{(g)})\right)^2|\sigma^{(g)}, R^{(g)}\right] \\
&= E\left[\left(R^{(g+1)} - R^{(g)}\right)^2|\sigma^{(g)}, R^{(g)}\right] - \varphi_R^2(\sigma^{(g)}, R^{(g)})
\end{aligned}
\tag{47}
$$

The distribution of ε_R' is expanded into an Edgeworth series. For the analysis, the expansion is cut off after the first term (cf. Beyer (2001b, p. 265)). That is to say, it is supposed that the deviations from the normal distribution are negligible in the analysis scenario the equations will be applied to. The random variable ε_R reads

$$
\varepsilon_R = D_\varphi(\sigma^{(g)}, R^{(g)})\mathcal{N}(0, 1) + \dots
\tag{48}
$$

The expectation $E[(R^{(g+1)} - R^{(g)})^2 \mid \sigma^{(g)}, R^{(g)}]$ appearing in ❯ Eq. 47 is called the *second-order progress rate*

$$
\varphi_R^{(2)}(\sigma^{(g)}, R^{(g)}) := E\left[\left(R^{(g+1)} - R^{(g)}\right)^2|\sigma^{(g)}, R^{(g)}\right]
\tag{49}
$$

The random variable ε_σ is obtained similarly. As in the case of the distance, a first-order approach (i.e., the first term of the series expansion) is used

$$
\varepsilon_\sigma = D_\psi(\sigma^{(g)}, R^{(g)})\mathcal{N}(0, 1) + \dots.
\tag{50}
$$

The derivation of the standard deviation is exactly the same. One has

$$
\begin{aligned}
D_\psi^2(\sigma^{(g)}, R^{(g)}) = E\left[\varepsilon_\sigma^2|\sigma^{(g)}, R^{(g)}\right] &= E\left[\left(\sigma^{(g+1)} - \sigma^{(g)} - \sigma^{(g)}\psi(\sigma^{(g)}, R^{(g)})\right)^2|\sigma^{(g)}, R^{(g)}\right] \\
&= (\sigma^{(g)})^2 E\left[\left(\frac{\sigma^{(g+1)} - \sigma^{(g)}}{\sigma^{(g)}}\right)^2|\sigma^{(g)}, R^{(g)}\right] - (\sigma^{(g)})^2\psi^2(\sigma^{(g)}, R^{(g)})
\end{aligned}
\tag{51}
$$

(cf. ❯ Eq. 42). Again, this introduces a new measure, the *second-order SAR*

$$
\psi^{(2)}(\sigma^{(g)}, R^{(g)}) := E\left[\left(\frac{\sigma^{(g+1)} - \sigma^{(g)}}{\sigma^{(g)}}\right)^2|\sigma^{(g)}, R^{(g)}\right]
\tag{52}
$$

Using the results obtained so far, the evolution equations can be rewritten as

$$
R^{(g+1)} = R^{(g)} - \varphi_R(\sigma^{(g)}, R^{(g)}) + D_\varphi(\sigma^{(g)}, R^{(g)})\mathcal{N}(0, 1) + \dots
\tag{53}
$$

$$
\begin{aligned}
\sigma^{(g+1)} &= \sigma^{(g)}\left(1 + \psi(\sigma^{(g)}, R^{(g)})\right) + D_\psi(\sigma^{(g)}, R^{(g)})\mathcal{N}(0, 1) + \dots \\
&= \sigma^{(g)}\left(1 + \psi(\sigma^{(g)}, R^{(g)}) + D_\psi'(\sigma^{(g)}, R^{(g)})\mathcal{N}(0, 1) + \dots\right)
\end{aligned}
\tag{54}
$$

with $D'_\psi = D_\psi/\sigma^{(g)}$.

1.3.1 The Deterministic Evolution Equations

In most analyses, the fluctuation parts are neglected with the exception of Beyer (1996b). The evolution equations without perturbation parts are generally termed *deterministic evolution equations* (Beyer 2001b). This approach serves well to extract the general characteristics of self-adaptive evolution strategies. The deterministic evolution equations read

$$R^{(g+1)} = R^{(g)} - \mathrm{E}[R^{(g)} - R^{(g+1)}|\sigma^{(g)}, R^{(g)}] \tag{55}$$

$$\sigma^{(g+1)} = \sigma^{(g)}\left(1 + \mathrm{E}\left[\frac{\sigma^{(g+1)} - \sigma^{(g)}}{\sigma^{(g)}}|\sigma^{(g)}, R^{(g)}\right]\right) \tag{56}$$

for σSA. In the case of the CSA, ❯ Eq. 56 is replaced by the system (❯ 45), (❯ 46). Analyses can usually be conducted only in the case that the evolution approaches a time-invariant distribution or a stationary (steady) state. In the deterministic approach, this is characterized by $R^{(g+1)} = R^{(g)}$, $\|\mathbf{s}^{(g+1)}\|^2 = \|\mathbf{s}^{(g)}\|^2$ and $\sigma^{(g+1)} = \sigma^{(g)}$. Note, demanding stationarity of the $R^{(g)}$-evolution equals a complete standstill of the ES. Often more interesting is the evolution equation of the normalized mutation strength $\sigma^{*(g)} := \sigma^{(g)} N / R^{(g)}$

$$\sigma^{*(g+1)} = \sigma^{*(g)}\left(\frac{1 + \psi(\sigma^{*(g)}, R^{(g)})}{1 - \dfrac{\varphi_R^*(\sigma^{*(g)}, R^{(g)})}{N}}\right) \quad (\sigma\text{SA}) \tag{57}$$

$$\sigma^{*(g+1)} = \sigma^{*(g)}\frac{R^{(g)}}{R^{(g+1)}}e^{\frac{\|\mathbf{s}^{(g+1)}\|^2 - N}{2DN}} \quad (\text{CSA}) \tag{58}$$

with $\varphi_R^* := \varphi_R N / R^{(g)}$ and $\sigma^{*(g+1)} = \sigma^{(g+1)} N / R^{(g+1)}$ since it admits a stationary state without requiring a stationary state of the $R^{(g)}$-evolution. In the case of the CSA, $\|\mathbf{s}^{(g+1)}\|^2 = \|\mathbf{s}^{(g)}\|^2$ is additionally demanded.

The assumption of the existence of a stationary state is motivated by the observation that it is optimal in many cases for the mutation strength to scale with the distance to the optimizer. Optimal in this case refers to a local progress measure, that is, to a maximal expected gain during one generation.

1.3.2 Including the Fluctuations

Considering the perturbation parts complicates the analysis. ❯ Equations 53 and ❯ 54 describe a Markov process or a Gaussian random field, the transition densities p_{tr} of which have to be determined. The variables $R^{(g+1)}$, $R^{(g)}$, $\sigma^{(g+1)}$, and $\sigma^{(g)}$ are all random variables. Assuming decomposability of the joint density of σ and R, the density of the distance R at generation g is denoted with $p(R^{(g)})$ and the density of the mutation strength with $p(\sigma^{(g)})$. As pointed out in Beyer (2001b, p. 313), it generally suffices to concentrate on some of the moments, generally the expected values, and not to determine the complete distribution

$$\overline{R^{(g+1)}} = \int_0^\infty R^{(g+1)} p(R^{(g+1)})\, \mathrm{d}R^{(g+1)}$$
$$= \int_0^\infty \int_0^\infty \left(R^{(g)} - \varphi_R(R^{(g)}, \sigma^{(g)})\right) p(\sigma^{(g)}) p(R^{(g)})\, \mathrm{d}R^{(g)}\, \mathrm{d}\sigma^{(g)} \tag{59}$$

$$\overline{\sigma^{(g+1)}} = \int_0^\infty \sigma^{(g+1)} p(\sigma^{(g+1)})\, \mathrm{d}\sigma^{(g+1)}$$
$$= \int_0^\infty \int_0^\infty \sigma^{(g)}\left((1 + \psi(\sigma^{(g)}, R^{(g)}))\right) p(\sigma^{(g)}) p(R^{(g)})\, \mathrm{d}\sigma^{(g)}\, \mathrm{d}R^{(g)} \tag{60}$$

As can be inferred from ❯ Eqs. 59 and ❯ 60, the transition densities are not needed if only the expected values are to be determined.

An equilibrium or a time-invariant limit distribution of a stochastic process is then characterized by a convergence to an equilibrium distribution, that is, $\lim_{g \to \infty} p(\sigma^{(g+1)}) = \lim_{g \to \infty} p(\sigma^{(g)}) = p_\infty(\sigma)$. Note, only the normalized mutation strength converges toward an equilibrium as long as the ES progresses still. If a stationary state is reached, the invariant density solves the eigenvalue equation

$$cp_\infty(\sigma) = \int_0^\infty p_{\mathrm{tr}}(\sigma|\sigma)p_\infty(\sigma)\,\mathrm{d}\sigma \tag{61}$$

with $c = 1$ and p_{tr} the transition density. In general, the equilibrium distribution p_∞ is unknown. As pointed out in Beyer (2001b, p. 318), it is possible to determine p_∞ numerically or even analytically. The results, however, tend to be quite complicated and do not allow a further analytical treatment. Instead of trying to obtain the distribution itself, the expected value is obtained by analyzing the mean value dynamics of the system. Unfortunately, the form of the evolution equations hinders a direct determination of the expectation since in general lower order moments depend on higher order moments leading to a nonending recursion.

1.4 Local Progress Measures

This section briefly describes some local progress measures. Local refers to the concept of time: They describe the (expected) changes for two consecutive generations.

1.4.1 Quality Gain

The *quality gain* is a local performance measure or local progress measure in fitness space. Instead of the expected change in distance to the objective, it gives the expected change of the fitness from one parent population to the next. It depends on the state $\mathscr{P}^{(g)}$ of the evolution strategy at generation g

$$\overline{\Delta Q(\mathbf{y}^{(g)})} := \mathrm{E}[F(\mathbf{y}^{(g+1)})_P - F(\mathbf{y}^{(g)})_P|\mathscr{P}^{(g)}] \tag{62}$$

When considering the $(\mu/\mu_I, \lambda)$-ES, the fitness of the parent populations is usually defined as the fitness of the centroid, that is, $F(\mathbf{y}^{(g)})_P := F(\langle \mathbf{y}^{(g)} \rangle)$ with $\langle \mathbf{y}^{(g)} \rangle := (1/\mu) \sum_{m=1}^{\mu} \mathbf{y}_m^{(g)}$. Another possible definition for an ES with μ parents is for instance the average fitness $F(\mathbf{y}^{(g)})_P := \langle F(\mathbf{y}^{(g)}) \rangle := (1/\mu) \sum_{m=1}^{\mu} F(\mathbf{y}_m^{(g)})$. The definitions are not equivalent except for $\mu = 1$.

The parental population at $g + 1$ consists of the μ best candidate solutions with respect to their fitness – or with respect to the fitness change to a constant reference point. In the case of "plus"-strategies, the new parental population can contain old parental individuals of generation g. In the case of "comma"-strategies, the populations are disjunct and the determination of ❯ Eq. 62 is usually simpler. The expectations can be calculated by obtaining the distribution of the offspring conditional to the state at g and using order statistics (Arnold et al. (1992)). An important quantity is the *local quality change* or *local fitness change* induced by a mutation

$$Q_y(\mathbf{z}) = F(\mathbf{y} + \mathbf{z}) - F(\mathbf{y}) \tag{63}$$

being usually the starting point for the determination of the quality gain and the progress rate.

1.4.2 Progress Rate

The *progress rate* is a local performance measure in parameter space. Progress rates measure the expected change of the distance to the optimizer or other reference points $\hat{\mathbf{y}}$

$$\varphi := \mathrm{E}[\|\mathbf{y}_P^{(g)} - \hat{\mathbf{y}}\| - \|\mathbf{y}_P^{(g+1)} - \hat{\mathbf{y}}\| |\mathscr{P}^{(g)}] \tag{64}$$

This is not mandatory, however. On functions with no finite optimizer, for instance linear or ridge functions, the progress rate measures the speed (i.e., the rate of change) in a predefined direction. When considering ridge functions, for example, a progress rate can be defined to measure the expected difference of the axial component for two consecutive generations.

Like the quality gain, the progress rate considers the parent population. Again, there are several possibilities for a definition. On the one hand, it is possible to define the progress rate as the expected change of the distance of two parental centroids

$$\varphi := \mathrm{E}[\| \langle \mathbf{y}^{(g)} \rangle - \hat{\mathbf{y}}\| - \| \langle \mathbf{y}^{(g+1)} \rangle - \hat{\mathbf{y}}\| |\mathscr{P}^{(g)}] \tag{65}$$

Alternatively,

$$\varphi := \mathrm{E}[\langle \|\mathbf{y}^{(g)} - \hat{\mathbf{y}}\| \rangle - \langle \|\mathbf{y}^{(g+1)} - \hat{\mathbf{y}}\| \rangle |\mathscr{P}^{(g)}] \tag{66}$$

with $\langle \|\mathbf{y}^{(g)} - \hat{\mathbf{y}}\| \rangle := (1/\mu)\sum_{m=1}^{\mu} \|\mathbf{y}_m^{(g)} - \hat{\mathbf{y}}\|$ is also possible. As in the case of the quality gain, both definitions are not interchangeable except for $\mu = 1$. The determination of the progress rate requires either the determination of the distribution of the distances of the *m*th best individuals, $m = 1, \ldots, \mu$, or the determination of the distribution of the distance of the centroid of the μ best individuals. For "plus"-strategies, the best individuals are taken from the offspring and parent population. Again, "comma"-strategies lead to disjunct populations and are easier to analyze. It should be noted that the *m*th best candidate solution means the solution with *m*th best fitness, not distance. It is therefore necessary to derive the distribution of the distance connected to the *m*th best fitness. Unless there is a straightforward relationship of distance and fitness, this task may not be solvable.

The progress rate is often used to derive the efficiency of an ES. The *serial efficiency* is defined as

$$\eta := \frac{\max_\sigma \varphi(\sigma)}{\lambda} \tag{67}$$

that is, the maximal progress per offspring or fitness evaluation. The measure can be used to compare different strategies assuming an optimally working step-size adaptation mechanism.

1.4.3 Self-Adaptation Response (SAR)

Self-Adaptation refers to a specific way to adapt control parameters of evolutionary algorithms. In short, if parameters are changed dynamically and according to feedback from the evolutionary process, the control rule is *adaptive*, and it is *self-adaptive* if the control is left to the EA itself. The *self-adaptation response (SAR)* is the expectation of the one-generational change of the mutation strength $\sigma_P^{(g)}$ of the parent population

$$\psi(\sigma_P^{(g)}, \mathscr{P}^{(g)}) := \mathrm{E}\left[\frac{\sigma_P^{(g+1)} - \sigma_P^{(g)}}{\sigma_P^{(g)}} |\mathscr{P}^{(g)}\right] \tag{68}$$

given the state of the algorithm at generation g. The relative change is considered because the mutation of the mutation strength is usually realized by a multiplication with a random number.

The SAR is not used in analyses of CSA-evolution strategies. However, the *logarithmic adaptation response*

$$\Delta_\sigma := \ln\left(\frac{\sigma^{(g+1)}}{\sigma^{(g)}}\right) \tag{69}$$

is often considered. While it is not a progress measure per se, the measure is similar to the integrand in ❯ Eq. 68 for $\sigma^{(g+1)} \approx \sigma^{(g)}$, since the logarithm has the series expansion $\ln(x) = \sum_{k=1}^{\infty}(1/k)(x-1)^k(-1)^{k-1}$ for $x \neq 0$. For $x \approx 1$, $\ln(x) \approx x - 1$ and $\ln(\sigma^{(g+1)}/\sigma^{(g)}) \approx (\sigma^{(g+1)} - \sigma^{(g)})/\sigma^{(g)}$.

2 Results from the Local Progress and Dynamical Systems Approach

This section presents results obtained using the local progress and dynamical systems approach. Firstly, some fitness functions that were used in the analyses are introduced. Secondly, the $(\mu/\mu_I, \lambda)$-ES is used on the sphere to explain how to derive a local progress measure, the quality gain, in detail. Afterward, the case of undisturbed fitness environments is considered. In practical optimization tasks, exact function values or exact information on the position in the search space often cannot be obtained. Therefore, noise is an important factor that has to be taken into account. The results for noisy and robust optimization are presented in the remaining part of this section.

2.1 Fitness Environments Considered

This section gives a short overview of test functions considered in the analysis.

2.1.1 Linear Functions

Linear functions are given by

$$F_{lf}(\mathbf{y}) = \mathbf{c}^T\mathbf{y} \tag{70}$$

with $\mathbf{c} \in \mathbb{R}^N$. The task is usually maximization. Since the "optimum" lies in infinity, the ES has to increase the fitness perpetually. Linear functions can be used to analyze, for instance, the effectiveness of the step-size adaptation mechanism.

2.1.2 Sphere Model

The sphere model is given by

$$F_{sphere}(\mathbf{y}) = g(\|\hat{\mathbf{y}} - \mathbf{y}\|) =: g(R) \tag{71}$$

with $\hat{\mathbf{y}}$ the optimal state and g either a strictly monotonically increasing or strictly monotonically decreasing function of $\|\hat{\mathbf{y}} - \mathbf{y}\|$. The simplest form of the sphere is

$$F_{\text{sphere}}(\mathbf{y}) = \beta\|\hat{\mathbf{y}} - \mathbf{y}\|^{\alpha} =: \beta R^{\alpha} \tag{72}$$

The sphere model is a simple test function, generally modeling more general functions in the vicinity of the optimizer. It is used to gain deeper insights into the working mechanisms of the evolutionary algorithm, its convergence speed and its scaling behavior with respect to the search space dimensionality N.

2.1.3 General Quadratic Functions

An extension of the sphere model is the function class of general quadratic functions

$$F_{\text{GQF}}(\mathbf{y}) = \mathbf{b}^{\mathrm{T}}\mathbf{y} - \mathbf{y}^{\mathrm{T}}\mathbf{Q}\mathbf{y} \tag{73}$$

with $\mathbf{Q}$ a symmetric, positive definite $\mathbb{R}^{N \times N}$ matrix and $\mathbf{b}$ an N-dimensional vector with $b_i \in \mathbb{R}$. The sphere (❷ Eq. 72) with $\alpha = 2$ and $\hat{\mathbf{y}} = \mathbf{0}$ is therefore a special case of general quadratic functions with $\mathbf{b} = \mathbf{0}$ and $\mathbf{Q} = \beta\mathbf{I}$.

2.1.4 Biquadratic Functions

Biquadratic functions are defined by

$$F_{\text{BQF}}(\mathbf{y}) = \sum_{i=1}^{n} a_i y_i - c_i y_i^4 \tag{74}$$

with $a_i \in \mathbb{R}$ and $c_i \in \mathbb{R}$.

2.1.5 Ridge Functions

The general ridge function, with axis direction $\mathbf{v}$ and parameters α and d determining the shape of the ridge, is given by

$$F_{\text{gR}}(\mathbf{y}) := \mathbf{v}^{\mathrm{T}}\mathbf{y} - d\left(\sqrt{(\mathbf{v}^{\mathrm{T}}\mathbf{y}\mathbf{v} - \mathbf{y})^{\mathrm{T}}(\mathbf{v}^{\mathrm{T}}\mathbf{y}\mathbf{v} - \mathbf{y})}\right)^{\alpha} \tag{75}$$

with $d > 0$ and $\alpha > 0$. The vector $\mathbf{v} \in \mathbb{R}^N$ with $\|\mathbf{v}\| = 1$ is called the ridge direction. Usually, a rotated version of ❷ Eq. 75 is considered

$$F_{\text{ridge}}(\mathbf{y}) = y_1 - d\left(\sum_{i=2}^{N} y_i^2\right)^{\frac{\alpha}{2}} \tag{76}$$

In this model, the ridge axis aligns with the coordinate axis $y_1\mathbf{e}_1$ (Beyer 2001a). Ridge functions consist of two parts: a linear gain part and a nonlinear loss part. The loss part in ❷ Eq. 76 resembles an $(N-1)$-dimensional sphere. The $(N-1)$-terms which make up the sphere component of the ridge can be interpreted as an $(N-1)$-dimensional distance to the axis $y_1\mathbf{e}_1$. Two parameters appear in ❷ Eq. 76: the weighting constant $d > 0$ and the parameter $\alpha > 0$. The latter determines the topology of the fitness landscape. Ridge functions with $\alpha = 1$ are called *sharp ridges*, with $\alpha = 2$ *parabolic ridges*, and with $\alpha = 3$ *cubic ridges*. The constant d weights the influence of the embedded sphere. In general, if $d \to 0$, the problem degenerates to the hyperplane $F(\mathbf{y}) = y_1$, whereas for increasing d, the isofitness lines appear more and more parallel to the axis and the problem approaches a sphere model with $F(\mathbf{y}) = -d\left(\sum_{i=2}^{N} y_i^2\right)^{\alpha/2}$.

Ridge functions do not have a finite optimum and therefore may be considered an "ill-posed" problem for ES (Arnold and Beyer 2008): Since the "optimum" lies in infinity, the fitness of the ES must be steadily increased. Improvement is possible in many ways. Generally, there are two viewpoints that may be taken (Beyer 2001a). The first viewpoint states (Oyman 1999, p. 32) the "object variable for the optimum [...] reads

$$\hat{y}_1 \to \infty, \forall i \neq 1: \; \hat{y}_i = 0."$$

This viewpoint derives its justification from seeing ridge functions as the limit of

$$F_c(\mathbf{y}) = y_1 - c y_1^2 - d\left(\sum_{i=2}^{N} y_i^2\right)^{\alpha/2} \tag{77}$$

for $c \to 0$ (cf. Oyman 1999 and Beyer 2001a). For every finite c, F_c has an optimal point at $(1/(2c), 0, \ldots, 0)^{\mathrm{T}}$. If c decreases, the position on the axis moves toward infinity.

Evolution strategies use local information. They sample the search space randomly and select the μ best offspring, that is, the candidate solutions with the μ highest fitness values. This is the foundation of the second viewpoint which takes a more process-oriented view: The ridge does not have a finite optimum. The algorithm is required to increase the fitness perpetually. This does not necessarily mean that it has to find the ridge. Although the highest fitness value is on the axis for every finite interval, the situation changes if an unbounded search space is considered. Actually, it is not even necessary to require a finite distance to the ridge. Since the search space is infinite, there are infinitely many points in arbitrary distance to the ridge with exactly the same fitness as a position on the axis. As a result, the ES may diverge from the axis – as long as it increases the linear component faster than the loss components, it still increases the overall fitness.

2.2 Example: The $(\mu/\mu_I, \lambda)$-Evolution Strategy on the Sphere

The determination of the quality gain of the $(\mu/\mu_I, \lambda)$-ES is used in the following as an illustrative example. The fitness environment considered is the sphere model with $g(\mathbf{y}) = -\|\mathbf{y} - \hat{\mathbf{y}}\|^2$. The quality gain for the $(\mu/\mu_I, \lambda)$-ES is defined as the expected change

$$\overline{\Delta Q} := \mathrm{E}\left[-\|\langle \mathbf{y}^{(g+1)} \rangle - \hat{\mathbf{y}}\|^2 - \left(-\|\langle \mathbf{y}^{(g)} \rangle - \hat{\mathbf{y}}\|^2\right)\right]$$
$$= \mathrm{E}[\|\langle \mathbf{y}^{(g)} \rangle - \hat{\mathbf{y}}\|^2 - \|\langle \mathbf{y}^{(g+1)} \rangle - \hat{\mathbf{y}}\|^2] =: \mathrm{E}[(R^{(g)})^2 - (R^{(g+1)})^2] =: \mathrm{E}[R^2 - r^2] \tag{78}$$

of the squared distances of two consecutive centroids $\langle \mathbf{y} \rangle = (1/\mu)\sum_{m=1}^{\mu} \mathbf{y}_m$ to the optimizer $\hat{\mathbf{y}}$. The determination of the quality gain requires the determination of the expected value of $\mathrm{E}[r^2]$, the squared distance of the centroid of the μ best offspring. An offspring l is created by adding a *mutation vector* $\mathbf{x}_l$ to the centroid of the parent population $\langle \mathbf{y} \rangle$: $\mathbf{y}_l = \langle \mathbf{y} \rangle + \mathbf{x}_l$. Due to its form, the sphere model allows for a very useful simplification of the representation of the original N-dimensional system by a change of the coordinate system: At generation g, the population's centroid is in a position $\langle \mathbf{y} \rangle$. The distance vector to the optimizer $\hat{\mathbf{y}}$ is well defined and given by $\mathbf{R}$. First of all, every vector can be expressed by another vector with origin in $\mathbf{R}$. The vector $\mathbf{R}$ also defines a perpendicular plane. Together, the vector and the base of the plane build an alternative base for the N-dimensional space. Changing the coordinate system, every other vector can be expressed by a vector with origin in $\mathbf{R}$ and with a component in $\mathbf{R}$-direction and a perpendicular component.

This holds also for $\mathbf{r}_l$, the distance vector of an offspring l, which can be written as $\mathbf{r}^l = \mathbf{R} - x_R^l \mathbf{e}_R + \mathbf{h}_R^l$ with $x_R^l \mathbf{e}_R$ the part in the plane of $\mathbf{R}$ and $\mathbf{h}_R^l$ in the perpendicular plane (see ❷ *Fig. 3*). The same decomposition can be applied to the distance vector $\mathbf{r}$ of the centroid which is given by $\mathbf{r} = \langle \mathbf{y} \rangle - \hat{\mathbf{y}} + \langle \mathbf{x} \rangle = \mathbf{R} + \langle \mathbf{x} \rangle$. The vector $\langle \mathbf{x} \rangle$ can be decomposed into a part $\langle x_R \rangle \mathbf{e}_R$ in the direction of $\mathbf{R}$ and into a perpendicular part $\langle \mathbf{h}_R \rangle$. The squared length is therefore

$$
\begin{aligned}
r^2 = \mathbf{r}^T \mathbf{r} &= (\mathbf{R} - \langle x_R \rangle \mathbf{e}_R + \langle \mathbf{h}_R \rangle)^T (\mathbf{R} - \langle x_R \rangle \mathbf{e}_R + \langle \mathbf{h}_R \rangle) \\
&= R^2 - 2R\langle x_R \rangle + \langle x_R \rangle^2 + \langle \mathbf{h}_R \rangle^2
\end{aligned}
\tag{79}
$$

The vector $\langle \mathbf{x} \rangle$ consists of the μ best mutations $\mathbf{x}_{m;\lambda}$, $m = 1, \dots, \mu$

$$
\langle \mathbf{x} \rangle = \frac{1}{\mu} \sum_{m=1}^{\mu} \mathbf{x}_{m;\lambda}
\tag{80}
$$

with $\mathbf{x}_{m;\lambda}$ standing for the mth best mutation with respect to the fitness change. The determination of the expected value of r^2 requires the determination of the expectation of the central component $\overline{\langle x_R \rangle}$, that is, of the mean of the central components of the μ best mutations and of the square of the radial component $\overline{\langle \mathbf{h}_R \rangle^2}$. Let one first consider $\overline{\langle \mathbf{h}_R \rangle^2}$. The expectation reads

$$
\overline{\langle \mathbf{h}_R \rangle^2} = E\left[\frac{1}{\mu^2} \sum_{m=1}^{\mu} \sum_{k=1}^{\mu} \mathbf{h}_{m;\lambda}^T \mathbf{h}_{k;\lambda} \right] = E\left[\frac{1}{\mu^2} \sum_{m=1}^{\mu} (\mathbf{h}_{m;\lambda})^2 \right] + E\left[\frac{1}{\mu^2} \sum_{m=1}^{\mu} \sum_{k=2, k\neq m}^{\mu} \mathbf{h}_{m;\lambda}^T \mathbf{h}_{k;\lambda} \right]
\tag{81}
$$

❑ **Fig. 3**
The decomposition of the mutation vector in the case of the sphere model.

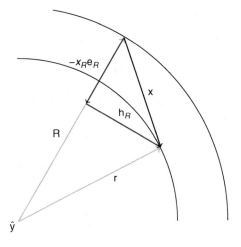

The expectation of the latter part is zero since $\mathrm{E}\left[\mathbf{h}_{m;\lambda}^{\mathrm{T}}\mathbf{h}_{k;\lambda}\right]$ for $m \neq k$. The perpendicular parts are selectively neutral and thus independent. Therefore,

$$\overline{\langle\mathbf{h}_R\rangle^2} = \mathrm{E}\left[\frac{1}{\mu^2}\sum_{m=1}^{\mu}(\mathbf{h}_{m;\lambda})^2\right] = \frac{1}{\mu}\overline{\langle\mathbf{h}_R^2\rangle} \tag{82}$$

holds. To obtain the expected value of the central and radial components of the μ best mutations, one considers the mutation-induced change of the fitness. The squared distance r_l^2 of an offspring is given by

$$r_l^2 = R^2 - 2Rx_R^l + (x_R^l)^2 + (\mathbf{h}_R^l)^2 \tag{83}$$

Without loss of generality, due to the isotropy of the mutations, one assumes that $\mathbf{e}_R$ aligns with the first component of the coordinate system. The first component is therefore normally distributed with standard deviation σ. The radial component consists of the remaining $(N-1)$-components – each also normally distributed

$$r_l^2 = R^2 - 2R\sigma z_x + \sigma^2 z_x^2 + \sigma^2\sum_{i=2}^{N}\mathcal{N}_i(0,1)^2 \tag{84}$$

with $z_x \sim \mathcal{N}(0,1)$. In the following, a normalization for σ is introduced, setting $\sigma^* := \sigma N/R$. Therefore, the equation changes to

$$\begin{aligned} r_l^2 &= R^2 - 2\frac{R^2}{N}\sigma^* z_x + \frac{R^2}{N^2}\sigma^{*2}z_x^2 + \frac{R^2}{N^2}\sigma^{*2}\sum_{i=2}^{N}\mathcal{N}_i(0,1)^2 \\ &= R^2 - 2\frac{R^2}{N}\sigma^* z_x + \frac{R^2}{N^2}\sigma^{*2}\sum_{i=1}^{N}\mathcal{N}_i(0,1)^2 \end{aligned} \tag{85}$$

The determination of the quality gain is restricted to $N \gg 1$. According to the law of large numbers, $\sum_{i=1}^{N}\mathcal{N}_i(0,1)^2/N$ goes to $\mathrm{E}[\mathcal{N}(0,1)^2] = 1$ for $N \rightarrow \infty$. Therefore, (R^2/N^2) $\sigma^{*2}\sum_{i=1}^{N}\mathcal{N}_i(0,1)^2$ approaches $(R^2/N)\sigma^{*2}$ for increasing N. Furthermore, according to the central limit theorem, $(1/N)\sum_{i=1}^{N}\mathcal{N}_i(0,1)^2$ approaches $\mathcal{N}(1,(2/N))$. Therefore, the contribution of the fluctuations of the squared random variables can be neglected and only the mean must be taken into account

$$r_l^2 = R^2 - 2\frac{R^2}{N}\sigma^* z_x + \frac{R^2}{N}\sigma^{*2} \tag{86}$$

The same holds for the contribution of $\langle\mathbf{h}_R^2\rangle$ which approaches

$$\frac{R^2}{N\mu}\sigma^{*2} \tag{87}$$

for increasing N. Only the contribution of the central component has to be considered. In the case of a single offspring, $z_x \sim \mathcal{N}(0,1)$ is the only remaining random variable. The mth best offspring or mutation is characterized by having the mth smallest r_l^2 or the mth largest z_x value, out of λ random samples from the standard normal distribution. The expected value of $\langle z_x\rangle$ is thus the mean of the expectation of the first μ order statistics $u_{1;\lambda}, \ldots, u_{\mu;\lambda}$ of the standard normal distribution

$$\overline{\langle z_x\rangle} = \frac{1}{\mu}\sum_{m=1}^{\mu}\mathrm{E}[u_{m;\lambda}] = \frac{1}{\mu}\sum_{m=1}^{\mu}\int_{-\infty}^{\infty} up_{m;\lambda}(u)\,\mathrm{d}u \tag{88}$$

The mth order statistic $u_{m;\lambda}$ is associated with the mth largest outcome. In other words, $(m-1)$ values must be larger than $u_{m;\lambda}$, and $(\lambda - m)$ values must be smaller. For this constellation, there are $\begin{pmatrix} \lambda - 1 \\ m - 1 \end{pmatrix}$ different possibilities. Since there are λ trials, it follows

$$p_{m;\lambda}(u) = \lambda \begin{pmatrix} \lambda - 1 \\ m - 1 \end{pmatrix} \frac{e^{-\frac{u^2}{2}}}{\sqrt{2\pi}} (1 - \Phi(u))^{m-1} \Phi(u)^{\lambda - m} \tag{89}$$

with $\Phi(u)$ the cumulative distribution function (cdf) of the standard normal distribution. The expectation of the mean reads

$$\begin{aligned}
\langle z_x \rangle &= \frac{1}{\mu} \sum_{m=1}^{\mu} \int_{-\infty}^{\infty} u \frac{e^{-\frac{u^2}{2}}}{\sqrt{2\pi}} \lambda \begin{pmatrix} \lambda - 1 \\ m - 1 \end{pmatrix} (1 - \Phi(u))^{m-1} \Phi(u)^{\lambda - m} \, du \\
&= \int_{-\infty}^{\infty} u \frac{e^{-\frac{u^2}{2}}}{\sqrt{2\pi}} \frac{\lambda}{\mu} \sum_{m=1}^{\mu} \begin{pmatrix} \lambda - 1 \\ m - 1 \end{pmatrix} (1 - \Phi(u))^{m-1} \Phi(u)^{\lambda - m} \, du
\end{aligned} \tag{90}$$

Let one first consider the sum in ❯ Eq. 90

$$\begin{aligned}
\sum_{m=1}^{\mu} \begin{pmatrix} \lambda - 1 \\ m - 1 \end{pmatrix} (1 - \Phi(u))^{m-1} \Phi(u)^{\lambda - m} &= \sum_{m=0}^{\mu-1} \begin{pmatrix} \lambda - 1 \\ m \end{pmatrix} (1 - \Phi(u))^m \Phi(u)^{\lambda - m - 1} \\
&= 1 - \sum_{m=\mu}^{\lambda - 1} \begin{pmatrix} \lambda - 1 \\ m \end{pmatrix} (1 - \Phi(u))^m \Phi(u)^{\lambda - m - 1}
\end{aligned} \tag{91}$$

This sum can be expressed by an integral. To this end, the regularized incomplete beta function $I_x(a, b)$ is considered

$$I_x(a, b) := \frac{1}{B(a, b)} \int_0^x t^{a-1}(1 - t)^{b-1} \, dt \tag{92}$$

with $B(a, b) = \Gamma(a)\Gamma(b)/\Gamma(a + b)$ as the complete beta function. The function $\Gamma(a)$ is the Gamma function with $\Gamma(a) = (a - 1)!$ for $a \in \mathbb{N}$. The series representation of ❯ Eq. 92 reads

$$I_x(a, b) := \sum_{j=a}^{a+b-1} \begin{pmatrix} a + b - 1 \\ j \end{pmatrix} x^j (1 - x)^{a+b-1-j} \tag{93}$$

(see, e.g., Abramowitz and Stegun 1984). The sum ❯ Eq. 91 is therefore

$$\begin{aligned}
\sum_{m=1}^{\mu} \begin{pmatrix} \lambda - 1 \\ m - 1 \end{pmatrix} (1 - \Phi(u))^{m-1} \Phi(u)^{\lambda - m} &= \sum_{m=0}^{\mu-1} \begin{pmatrix} \lambda - 1 \\ m \end{pmatrix} (1 - \Phi(u))^m \Phi(u)^{\lambda - m - 1} \\
&= 1 - \sum_{m=\mu}^{\lambda - 1} \begin{pmatrix} \lambda - 1 \\ m \end{pmatrix} (1 - \Phi(u))^m \Phi(u)^{\lambda - m - 1} \\
&= 1 - I_{1-\Phi(u)}(\mu, \lambda - \mu)
\end{aligned} \tag{94}$$

Since $I_x(a, b) = 1 - I_{1-x}(b, a)$, one has

$$\begin{aligned}
\sum_{m=1}^{\mu} \begin{pmatrix} \lambda - 1 \\ m - 1 \end{pmatrix} (1 - \Phi(u))^{m-1} \Phi(u)^{\lambda - m} &= \sum_{m=0}^{\mu-1} \begin{pmatrix} \lambda - 1 \\ m \end{pmatrix} (1 - \Phi(u))^m \Phi(u)^{\lambda - m - 1} \\
&= I_{\Phi(u)}(\lambda - \mu, \mu) \\
&= \frac{(\lambda - 1)! \int_0^{\Phi(u)} t^{\lambda - \mu - 1}(1 - t)^{\mu - 1} \, dt}{(\lambda - \mu - 1)!(\mu - 1)!}
\end{aligned} \tag{95}$$

The expectation (● Eq. 90) therefore reads

$$\overline{\langle z_x \rangle} = \int_{-\infty}^{\infty} \frac{u e^{-\frac{u^2}{2}}}{\sqrt{2\pi}} \binom{\lambda}{\mu} \frac{(\lambda - 1)!}{(\lambda - \mu - 1)!(\mu - 1)!} \int_0^{\Phi(u)} t^{\lambda - \mu - 1}(1 - t)^{\mu - 1}\, dt\, du$$

Changing the integration order leads to

$$\overline{\langle z_x \rangle} = \frac{\lambda!}{(\lambda - \mu)!\mu!} \int_0^1 \left(\int_{\Phi^{-1}(t)}^{\infty} \frac{u e^{-\frac{u^2}{2}}}{\sqrt{2\pi}}\, du \right) t^{\lambda - \mu - 1}(1 - t)^{\mu - 1}\, dt$$

Setting $t := \Phi(x)$, $dt = \exp(-x^2/2)/\sqrt{2\pi}\, dx$ leads to

$$\overline{\langle z_x \rangle} = \frac{\lambda!}{(\lambda - \mu - 1)!\mu!} \int_{-\infty}^{\infty} \int_x^{\infty} \frac{u e^{-\frac{u^2}{2}}}{\sqrt{2\pi}}\, du \Phi(x)^{\lambda - \mu - 1}(1 - \Phi(x))^{\mu - 1} \frac{e^{-\frac{x^2}{2}}}{\sqrt{2\pi}}\, dx$$

$$= \frac{\lambda!}{(\lambda - \mu - 1)!\mu!} \int_{-\infty}^{\infty} \Phi(x)^{\lambda - \mu - 1}(1 - \Phi(x))^{\mu - 1} \frac{e^{-x^2}}{2\pi}\, dx \tag{96}$$

$$= (\lambda - \mu)\binom{\lambda}{\mu} \int_{-\infty}^{\infty} \Phi(x)^{\lambda - \mu - 1}(1 - \Phi(x))^{\mu - 1} \frac{e^{-x^2}}{2\pi}\, dx =: c_{\mu/\mu,\lambda}$$

In general, the integral (● Eq. 96) cannot be solved analytically – only numerically. It is a special case of the so-called generalized progress coefficients introduced in Beyer (1995b). Plugging ● Eqs. 96 and ● 87 into ● Eq. 78, the expected value of the squared distance r^2 becomes

$$E[r^2] = R^2 - 2\frac{R^2}{N}\sigma^* c_{\mu/\mu,\lambda} + \frac{R^2}{N\mu}\sigma^{*2} \tag{97}$$

leading to the quality gain

$$\overline{\Delta Q} = R^2 - E[r^2] = R^2\left(\frac{2\sigma^*}{N}c_{\mu/\mu,\lambda} - \frac{\sigma^{*2}}{\mu N}\right) \tag{98}$$

Using the normalization $\overline{\Delta Q^*} := N\overline{\Delta Q}/(2R^2)$ gives the final expression

$$\overline{\Delta Q^*} = c_{\mu/\mu,\lambda}\sigma^* - \frac{\sigma^{*2}}{2\mu} \tag{99}$$

Note, due to the normalization, this quality gain is asymptotically equal to the normalized progress rate φ^*.

The result (● Eq. 99) can be used to discuss the influence of the normalized mutation strength on the performance, to determine optimal values, and to compare different evolution strategies. Since the derivation assumed a large search space dimensionality, ● Eq. 99 is asymptotically exact for $N \to \infty$. In the case of small N, ● Eq. 99 may be regarded as a simple approximation of the real quality gain. The quality gain ● Eq. 99 consists of two parts: a linear gain part and a nonlinear loss part. Therefore, ● Eq. 99 adheres to the *evolutionary progress principle* (EPP) which states that any evolutionary progress is a combination of progress gain and progress loss (see, e.g., Beyer 2001b, p. 20). Another effect that appears in ● Eq. 99 is the so-called *genetic repair effect* due to intermediate recombination. Recombination reduces the loss part, as indicated by the factor $1/\mu$. As a result, the ES can operate with larger mutation strengths. A more complete discussion of the genetic repair effect can be found in the following section.

2.3   Undisturbed Fitness Environments

This section describes the results obtained so far for undisturbed fitness environments. The overview starts with the sphere model and then proceeds to the discussion of other fitness functions. Afterward, analyses devoted to the step-size adaptation mechanisms are presented.

2.3.1   Sphere Model

The main focus of analyses on the undisturbed sphere concerns the derivation and analysis of the progress rate φ. This progress measure is of great importance because it can be used:

- To derive evolution criteria which have to be fulfilled for a convergence of the strategy
- To derive conditions for optimal performance
- To examine and compare the efficiency of different strategies
- To analyze the effects of recombination and
- To derive recommendations for population sizing

The sphere model was analyzed considering the following evolution strategies:

- The $(1 + 1)$-ES (Beyer 1989)
- The $(1 + \lambda)$-ES and the $(1, \lambda)$-ES (Beyer 1993)
- The (μ, λ)-ES (Beyer 1995b)
- The $(\mu/\mu, \lambda)$-ES (Beyer 1995a, 1996a) and
- The (λ_{opt})-ES (Arnold 2006b)

One of the first analyses concerned the $(1 + 1)$-ES (Beyer 1989). In contrast to other evolution strategies, this single point ES allows for an exact treatment with the caveat that the progress rate can only be obtained by numerical integration. The derivation relied on

(a) the decomposition of the mutation vector introduced in the previous section and
(b) on the *success probability*, that is, the probability that the mutation-created offspring substitutes the parent.

The analysis led to the following findings: Firstly, the progress rate rises steeply with the mutation strength to a clearly defined maximal value before it gradually declines. Secondly, the limit curve for $N \to \infty$ is already usable for search space dimensionalities greater than 30. This justifies the determination and analysis of an analytical expression of the progress rate for infinite-dimensional search spaces.

As first introduced in Beyer (1989), a step-by-step approach can be followed extracting the important characteristics of the process to develop a geometrically motivated model. This approach was applied in Beyer (1993) to derive an asymptotical progress rate for $(1, \lambda)$- and $(1 + \lambda)$-ESs.

First consider the $(1 + 1)$-ES. The approach followed relied again on the decomposition of the mutation vector into two parts: one into the direction of the objective and the other in the perpendicular plane. The length of the mutation vector or of its $(N - 1)$-dimensional perpendicular part can be replaced with its expected value when considering $N \to \infty$. This simplified the calculations. Next, normalized quantities were introduced, that is, normalized with respect to the distance to the objective and the search space dimensionality. Further derivations relied on expanding several expressions into their Taylor series. The higher order parts of these series scale with at least $1/N$. For $N \to \infty$, their contributions can be neglected. The approach gives the (normalized) asymptotical progress rate formula (Beyer 2001b, p. 67) which was already derived by Rechenberg (1973) following a different approach. The asymptotical progress rate (see ❯ *Table 3*) is decomposed into a gain and a loss part as indicated by the evolutionary progress principle (EPP). The loss part is weighted by the success probability. Interestingly, the gain part stems from the mutation component in the direction of the objective, whereas the loss part is the result of the perpendicular part. Both parts are zero for zero mutation strength and increase at first with the normalized mutation strength – the gain part linearly and the loss part quadratically. When the mutation strength increases further, both terms are dampened more and more so that the progress rate goes to zero for $\sigma^* \to \infty$, (see ❯ *Fig. 4a*). This can be explained as follows (Beyer 2001b): Large mutation strengths cause more unsuccessful mutations on average since it becomes harder to hit the region of success (cf. ❯ *Fig. 4b*). The parent survives for a longer time and the progress is close to zero (cf. ❯ *Fig. 4a*). The progress rate is significantly positive (>0) only for mutation strengths in the so-called *evolution window* (Rechenberg 1973). There exists an optimal mutation strength of around 1.224, which leads to a maximal progress rate of 0.202 (see ❯ *Fig. 4a*).

The approach described above can also be used to determine the progress rate for $(1, \lambda)$-ESs (see ❯ *Table 3*). And its application to $(\mu/\mu_I, \lambda)$-ESs has been discussed in the preceding subsection containing the progress rate of the $(1, \lambda)$-ES as a special case displayed in ❯ *Table 3*. In the progress rate formula, the so-called *progress coefficient* $c_{1,\lambda}$ appears. Mathematically, this is the expectation of the largest $x_{\lambda;\lambda}$ order statistic taken from λ standard, normally and independently distributed random trials (see ❯ Sect. 5).

◼ **Table 3**

Asymptotical progress rates and success probabilities for the sphere model. The normalizations are $\sigma^* = \sigma N/R$ and $\varphi_R^* = \varphi_R N/R$. The $c_{1,\lambda}$ is the progress coefficient (Beyer 2001b, p. 78) whereas $d_{1+\lambda}^{(1)}(x)$ denotes the first order progress function (Beyer 1993) or (Beyer 2001b, p. 78). The progress coefficient, $c_{\mu,\lambda}$, is a function of nine generalized progress coefficients (❯ Eq. 101) and μ. See ❯ Sect. 5 for a definition of the progress coefficients

ES	Progress rate	Success probability
$(1 + 1)$-ES	$\varphi_R^* = \dfrac{\sigma^*}{\sqrt{2\pi}} e^{-\frac{\sigma^{*2}}{8}} - \dfrac{\sigma^{*2}}{2}\left(1 - \Phi\left(\dfrac{\sigma^*}{2}\right)\right)$	$P_s = 1 - \Phi\left(\dfrac{\sigma^*}{2}\right)$
$(1 + \lambda)$-ES	$\varphi_R^* = \sigma^* d_{1+\lambda}^{(1)}\left(\dfrac{\sigma^*}{2}\right) - \dfrac{\sigma^{*2}}{2}\left(1 - \Phi\left(\dfrac{\sigma^*}{2}\right)\right)^\lambda$	$P_s = 1 - \Phi\left(\dfrac{\sigma^*}{2}\right)^\lambda$
$(1, \lambda)$-ES	$\varphi_R^* = c_{1,\lambda}\sigma^* - \dfrac{\sigma^{*2}}{2}$	–
(μ, λ)-ES	$\varphi_R^* = c_{\mu,\lambda}\sigma^* - \dfrac{\sigma^{*2}}{2}$	–
$(\mu/\mu_I, \lambda)$-ES	$\varphi_R^* = c_{\mu/\mu_I,\lambda}\sigma^* - \dfrac{\sigma^{*2}}{2\mu}$	–
$(\mu/\mu_D, \lambda)$-ES	$\varphi_R^* = c_{\mu/\mu,\lambda}\sqrt{\mu}\sigma^* - \dfrac{\sigma^{*2}}{2}$	–

◻ Fig. 4
The asymptotical progress rate φ^* of the $(1 + 1)$-ES on the sphere, **(a)** The progress rate is a nonlinear function of the normalized mutation strength, σ^*, with an optimal value of around 0.202 for $\sigma^* \approx 1.224$. **(b)** shows a two-dimensional sphere. The optimizer is located at $\hat{y}$. The ES is currently in position **y**. Operating with higher mutation strengths enlarges the region around **y** in which mutations most likely will fall.

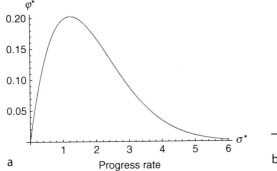
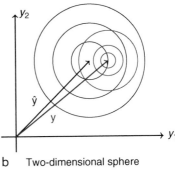

a   Progress rate   b   Two-dimensional sphere

◻ Fig. 5
(a) The asymptotical progress rate φ^* for the $(1, \lambda)$-ES on the sphere **(b)** and the efficiency as a function of λ.

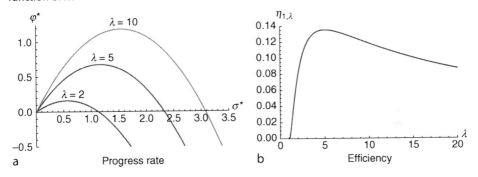

a   Progress rate   b   Efficiency

The progress rate of the $(1, \lambda)$-ES reveals a striking difference to that of the $(1 + 1)$-ES: The progress rate may become negative, which indicates divergence (see ❷ *Fig. 5*). This leads to the necessary *evolution criterion* which states that the mutation strength must be smaller than a strategy-specific limit (see ❷ *Table 4*). The limit increases with the population size (see ❷ *Fig. 5*). Again, the evolutionary progress principle is present in the progress rate expression. The progress rate consists of a (linear) gain part and a (quadratic) loss part. As long as the normalized mutation strength is small, the gain part outweighs the loss part. Increasing the mutation strength changes the relation: Finally, the progress rate becomes smaller than zero. The progress rate has a maximal value and an optimizer. Both depend on the strategy. The maximal value increases with the number of offspring and the location of the optimizer moves upward. While this would indicate that an increase in the number of offspring would be profitable, the increased costs of operating with larger populations have to be considered. These considerations introduce the concept of the progress rate per fitness

◻ **Table 4**

Several characteristic quantities for ES on the sphere model

ES	Evolution criterion	Maximal progress	Opt. mutation strength	Serial efficiency	Time complexity
$(1, \lambda)$	$\sigma^* < 2c_{1,\lambda}$	$\varphi^*_{max} = \dfrac{c^2_{1,\lambda}}{2}$	$\sigma^*_{max} = c_{1,\lambda}$	$\eta_{1,\lambda} = \dfrac{c^2_{1,\lambda}}{2\lambda}$	$G_{1,\lambda} = \mathcal{O}\left(\dfrac{N}{\ln(\lambda)}\right)$
$(\mu/\mu, \lambda)$	$\sigma^* < 2\mu c_{\mu/\mu,\lambda}$	$\varphi^*_{max} = \dfrac{\mu c^2_{\mu/\mu,\lambda}}{2}$	$\sigma^*_{max} = \mu c_{\mu/\mu,\lambda}$	$\eta_{\mu/\mu,\lambda} = \dfrac{\mu c^2_{\mu/\mu,\lambda}}{2\lambda}$	$G_{\mu/\mu,\lambda} = \mathcal{O}\left(\dfrac{N}{\lambda}\right)$
(μ, λ)	$\sigma^* < 2c_{\mu,\lambda}$	$\varphi^*_{max} = \dfrac{c^2_{\mu,\lambda}}{2}$	$\sigma^*_{max} = c_{\mu,\lambda}$	$\eta_{\mu/\mu,\lambda} = \dfrac{c^2_{\mu,\lambda}}{2\lambda}$	$G_{\mu,\lambda} = \mathcal{O}\left(\dfrac{N}{\ln(\lambda)}\right)$

evaluation φ^*/λ (Beyer 2001b, p. 73), which leads to the *serial efficiency* $\eta_{1,\lambda} := \varphi^*_{max}/\lambda$ with φ^*_{max} referring to the maximal progress rate for the specific $(1, \lambda)$-ES. Using numerical maximization, one finds that the efficiency has its maximum at $\lambda \approx 5$. This leads in turn to the conclusion that the $(1, 5)$-ES is the most efficient $(1, \lambda)$-ES (see ❯ *Fig. 5b*). The asymptotical behavior of the progress coefficient $c_{1,\lambda}$ can be used to derive expressions for the maximal progress rate and the efficiency. These are $c_{1,\lambda} = \mathcal{O}(\sqrt{\ln(\lambda)})$, $\varphi^*_{max} = \mathcal{O}(\ln(\lambda))$, and $\eta_{1,\lambda} = \mathcal{O}(\ln(\lambda)/\lambda)$. The progress rate can also be used to estimate the *EA time complexity*

$$G := \frac{N}{\varphi^*(\sigma^*)} \ln\left(\frac{R^{(g_0)}}{R^{(g)}}\right) \tag{100}$$

Since the progress rate scales with $\ln(\lambda)$, the EA time complexity of the $(1, \lambda)$-ES is of the order $G_{1,\lambda} = \mathcal{O}(N/\ln(\lambda))$.

The concepts developed for the $(1 + 1)$-ES and the $(1, \lambda)$-ES have also been applied to derive the progress rate for the $(1 + \lambda)$-ES (see ❯ *Table 3*). As for the $(1 + 1)$-ES, the progress rate approaches zero for increasing normalized mutation strengths. Considering the progress per fitness evaluation, it can be seen that operating with $\lambda > 1$ does not make any sense on the undisturbed sphere unless the algorithm is parallelized. It holds that the progress rate of the $(1 + \lambda)$-ES is greater than (or equal to) the progress rate of the $(1, \lambda)$-ES. Furthermore, concerning evolution strategies with a single parent on the undisturbed sphere, the $(1 + 1)$-ES is the most efficient ES.

In Beyer (1995b), the theory was extended to the (μ, λ)-ES. The progress rate in this case is defined as the expected change of the average distance to the optimizer. The derivation of the progress rate is inherently more difficult since the parent population must be modeled. The distribution of the parents is generation dependent and unknown. Leaving this difficulty aside, the determination of the progress rate requires obtaining the pdf (probability density function) of the distribution of the mth best offspring for which order statistics are to be used. To this end, the density of the offspring population $p(r)$ is needed, which changes from generation to generation. In general, there is no chance to get analytical exact solutions, approximations must be derived instead. Therefore, the random variables are standardized and expanded into a Gram–Charlier series with the normal distribution as reference. The expansions are cut off after the fourth term assuming that the unknown distributions are similar to a normal distribution; however, with a slight skewness.

The approach requires the determination of the mean, the standard deviation, and the third central moment. As a first step, the density of a single offspring $p(r \mid R_m)$, given a parent

R_m that is randomly drawn from the population, is considered. Any of the parents is drawn with the same probability. The offspring density, given the parent population, is therefore the mixture $p(r|R_1,...,R_\mu) = (1/\mu)\sum_{m=1}^{\mu} p(r|R_m)$ of the μ densities.

In this manner, the required moments can be obtained conditional to the distances of the μ parents. These, however, are random variables themselves produced by the random dynamics from the previous step and were selected as the μ best candidate solutions.

Determining the expectation requires a μ-fold integration and the determination of the joint density of the $R_1, \ldots, R_\mu$. In this expression, the pdf and cdf of the offspring of the previous generation appear. The moments of generation $g + 1$ depend, therefore, on the distribution of generation g. The next steps aim to express the moments as functions of the previous moments. To this end, the same series expansion for the pdf and cdf at time g is applied.

As a result, one obtains a mapping of the distribution parameters from generation g to $g + 1$. Assuming that the distributions approach a time-invariant limit distributions leads to nonlinear fix-point equations for the moments. The nonlinearity concerns primarily the parameter stemming from the third central moment. Assuming that the skewness of the distribution is not large, all higher order terms containing the expectation of this parameter can be dropped. The calculations still require the determination of the expectation of concomitants of order statistics and are quite lengthy. Finally, this leads to the progress rate for the (μ, λ)-ES and to the definition of the *generalized progress coefficients*

$$e_{\mu,\lambda}^{\alpha,\beta} = \frac{\lambda - \mu}{\sqrt{2\pi}^{\alpha+1}} \binom{\lambda}{\mu} \int_0^\infty t^\beta\, e^{-\frac{\alpha+1}{2}t^2} \Phi(t)^{\lambda-\mu-1}(1 - \Phi(t))^{\mu-\alpha}\, dt \qquad (101)$$

(Beyer 1995b) which reappear in the analysis of other strategies. The progress rate (see ❯ *Table 3*) itself depends on several generalized progress coefficients, which are finally aggregated in one general $c_{\mu,\lambda}$-coefficient. It behaves asymptotically as $c_{\mu,\lambda} \sim \sqrt{2\ln(\lambda/\mu)}$ for $\lambda \gg 1$ (Beyer 1996a). This scaling behavior shows that in the non-noisy case, retaining more than the best candidate solution does not lead to any advantage. As a consequence, see ❯ *Table 4*, the $(1, \lambda)$-ES and the (μ, λ)-ES share the same time complexity.

In (Beyer 1995a, 1996a), multi-recombinative evolution strategies were analyzed. Two types of recombinations were considered: dominant or discrete recombination and intermediate recombination. As for dominant recombination, the recombinant is created componentwise by choosing for each coordinate one of the parents randomly and copying the value. Evolution strategies with dominant/discrete recombination are denoted as $(\mu/\mu_D, \lambda)$-ESs. In the case where intermediate recombination is used, the recombination result is the centroid. These ESs are denoted as $(\mu/\mu_I, \lambda)$-ESs.

Let one first consider the progress rate of $(\mu/\mu_I, \lambda)$-ESs: Again, not only the distribution for the largest of λ trials has to be determined but also for all mth largest trials up to μ. For a $(\mu/\mu_I, \lambda)$-ES, the progress rate is defined as the expected change of the distance of the centroid for two consecutive generations. The new search point is generated by adding the centroid of the μ best mutations to the centroid of the previous generations. As the mutation vector in previous analyses, it can be decomposed into a component pointing toward the optimizer and into a perpendicular part. This decomposition followed by determining the pdf and cdf for the mth best offspring leads, finally, via several Taylor series expansions and neglection of higher order terms to the progress rate of the $(\mu/\mu_I, \lambda)$-ES (see ❯ *Table 3*). The derivation of the progress rate revealed an important performance-improving property regarding the perpendicular parts of the mutation vectors of the group of the μ best offspring: The sphere model is symmetric and the perpendicular parts are therefore selectively neutral when having the same length. The mutations are isotropic and all the directions are equally probable. The conclusion is that the perpendicular parts are statistically independent and the expected value of the product of two different vectors is zero. This results in

$$\overline{\langle \mathbf{h}_R \rangle}^2 = \frac{1}{\mu} \overline{\langle \mathbf{h}_R^2 \rangle} \qquad (102)$$

This will be important when discussing the effects of intermediate recombination.

The progress rate can be used to investigate whether recombinative $(\mu/\mu_I, \lambda)$-ESs have an advantage over nonrecombinative (μ, λ)-ESs. To ensure a fair comparison, several requirements (Beyer 1995a) should be met:

1. Problem equivalence: This states that the same fitness function and initial conditions should be used. Furthermore, the fitness function should not favor one of the algorithms per se.
2. Resource equivalence: Here, this may be satisfied by using the same number of offspring.
3. Maximal performance: Each algorithm should operate with its maximal performance (Beyer 1995a).

The comparison made for $\lambda = 50$ and $N = 100$ showed that the maximal progress rate of the recombinative strategies exceeds that of nonrecombinative strategies for all choices of μ, except $\mu = 1$. The $(1, \lambda)$-ES is the best performing nonrecombinative (μ, λ)-ES and is still outperformed by $(\mu/\mu_I, \lambda)$-ES for a wide range of μ. The maximal performance curve of ES with intermediate recombination indicates that an optimal value of μ exists with the largest maximal progress. Further investigations reveal that the optimal value depends on the search space dimensionality and the population size and increases with both.

The question remains, why recombinative ESs outperform the nonrecombinative comma strategies. This can be traced back to the *genetic repair principle* (GR). Intermediate recombination enables an ES to operate with larger mutation strengths. But why? Comparing the progress rates of the $(\mu/\mu_I, \lambda)$-ES with that of the (μ, λ)-ES reveals that the main factor to which ESs with intermediate recombination owe their better performance is the reduction of the loss part by $1/\mu$. The loss part of the progress rate stems from the perpendicular parts of the mutation vectors. They enter the derivation of the progress rate over the expected value of the squared length of their average. Intermediate recombination reduces this length to $1/\mu$th of the average of the squared lengths. This is beneficial: the perpendicular components deteriorate the progress. Let one compare two mutations which result in the same movement in the direction of the optimizer. If the perpendicular parts differ, the mutation with the larger perpendicular part results in a greater distance to the optimizer. Recombination reduces this harmful perpendicular part and offers a kind of *statistical error correction* (Beyer 1995a). ❯ *Figure 6* illustrates this mechanism graphically.

Evolution strategies with dominant/discrete recombination remain to be discussed. To enable a comparison with intermediate recombination, the progress rate is defined in the same manner. The calculation of the progress rate appears more difficult than in the previous case. The analysis requires the determination of the distribution of the parent and/or the offspring populations.

The analysis pursued in Beyer (1995a) considers the final effect of recombination and mutation as if it were a result of a *single* random variation. Thus, the concept of *surrogate mutations* was introduced. The surrogate mutations are applied to a virtual parental centroid.

◘ Fig. 6

A visualization of the genetic repair effect for intermediate recombination on the sphere. The ES is a $(4/4_I, \lambda)$-ES. The mutations, $z_{1;\lambda}$ to $z_{4;\lambda}$, which correspond to the four best offspring are displayed with their decompositions. As can be seen, intermediate recombination reduces the contribution of the harmful, perpendicular components (see Beyer and Schwefel 2002).

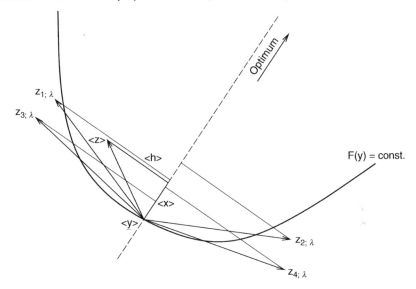

This enables us to decompose the mutation vector in the usual manner. However, the determination of the density of the surrogate mutations still remains. Its calculation appears very demanding and has not been solved for finite N, up until now. However, for $N \to \infty$ the progress rate can be determined: Within the limit, selection does not have a significant influence on the distribution of the offspring population. To determine the density of the population, the influence of selection can be neglected and only considered again for the progress rate.

The derivation of the distribution for the surrogate mutation vector leads to a very interesting finding: The distribution forgets any initialization effects and shrinks or expands to a limit distribution as $g \to \infty$. The standard deviation of the components approaches

$$\sigma_S := \sigma\sqrt{\mu} \tag{103}$$

being the "mutation strength" of the surrogate mutations.

The resulting progress rate differs from the progress rate for intermediate recombination in two ways (see **❷** *Table 3*): The loss part is not reduced by $1/\mu$. Instead, the gain part is increased by the factor $\sqrt{\mu}$. As a result, using intermediate recombination and using discrete recombination should lead to the same maximal performance. A closer look, though, reveals that this is not the case. The observed and better performance of $(\mu/\mu_I, \lambda)$-ESs is attributed to the explicit genetic repair (GR) present in intermediate recombination (Beyer 1995a).

However, genetic repair can be argued to be present implicitly in dominant recombination – leading to the *GR hypothesis for dominant recombination*. The surrogate mutations combine dominant recombination and mutation. Dominant recombination samples coordinate points from the μ parents. The mean of this sampling distribution is the respective coordinate point of the centroid. The drawn sample provides an estimate for the sampling distribution – mean included. Afterward mutation is applied. Mutation introduces no bias and therefore leaves the first moment of the distribution unchanged. The outcome of the mutation is again a sample providing an estimate for the distribution, the center of which is still the centroid of the parents – providing a kind of implicit center building and implicit genetic repair. Since it is a statistical estimate with statistical deviations, implicit genetic repair cannot work as well as explicit genetic repair.

Dominant recombination introduces a new hypothesis, the so-called *mutation-induced speciation by recombination* (MISR). In short this is a result of the limit value of the surrogate mutations (Beyer 1995a). Even without selection, recombination results in a stationary distribution with finite standard deviation (❷ Eq. 103): The offspring generated show cohesion. This can be interpreted as a kind of species formation.

The questions remain whether recommendations with regard to population sizing or truncation ratios, $\vartheta = \mu/\lambda$, can be given. Both recombination types achieve the same maximal progress for $N \to \infty$. The results have been used to determine the optimal truncation ratio $\vartheta_{\mathrm{opt}} = 0.27$ with a resulting serial efficiency of $\eta_{\mu/\mu,\lambda} = 0.202$. Comparing the result with the $(1 + 1)$-ES reveals that the recombinative ES achieves the same efficiency as the $(1 + 1)$-ES. Interestingly, the optimal success probability of the $(1 + 1)$-ES equals the optimal truncation ratio of the $(\mu/\mu, \lambda)$-ES. The $(\mu/\mu, \lambda)$-ES can, therefore, be claimed to be the equivalent of the $(1 + 1)$-ES for multi-parent strategies. As a consequence, using recombination does not improve the efficiency above that of the $(1 + 1)$-ES, unless parallel computation comes into play.

Arnold (2006b) investigated the performance of an ES with intermediate recombination, which uses weighted recombination taking *all* λ offspring into account. The progress vector $\langle \mathbf{z} \rangle$ is obtained by

$$\langle \mathbf{z} \rangle = \sum_{k=1}^{\lambda} w_{k;\lambda} \mathbf{z}_{k;\lambda} \tag{104}$$

The weights, $w_{k;\lambda}$, depend on the rank of the candidate solution. As performance measure, the quality gain was considered. The results were used to determine the optimal mutation strength and quality gain, which depends on the setting of the weights. The best performance is achieved by choosing the weights $w_{k;\lambda}$ proportionally to the expected values of the corresponding orders statistics $z_{\lambda+1-k;\lambda}$ of the standard normal distribution. This (λ_{opt})-ES is able to outperform the $(\mu/\mu, \lambda)$-ES and the $(1 + 1)$-ES.

2.3.2 General Quadratic and Non-quadratic Functions

The progress rate approach may be difficult for nonspherical fitness functions. Candidate solutions are selected based on the fitness values. Therefore, the induced order statistics appear in the determination of the progress rate: the distribution of μ distances connected with the μ best fitness values has to be obtained. On the sphere both quantities are monotonously related since the fitness is a monotonous function of the distance. But this is rather an exception than a rule. The quality gain, the expected change of the fitness for two consecutive generations, may behave differently and often the related optimizers of the strategy parameter do not agree with each other.

In Beyer (1994), the quality gain was derived for the $(1, \lambda)$-ES and the $(1 + \lambda)$-ES. The derivation of the quality gain requires the determination of the density of the fitness of the best offspring. This can be done by using order statistics. One of the tasks remaining is then, to find general expressions for the pdf and cdf of the fitness of a mutation-created offspring. The idea is to standardize the variable as a first step with its mean and standard deviation. Afterward, the distribution is expanded into a Gram–Charlier series with the standard normal distribution as a baseline. To this end, higher order moments or cumulants of the unknown distribution are to be determined.

This can be done for $(1 \dot{+} \lambda)$ strategies yielding general results which are not specific for a fitness function. An application, however, requires the computation of the mean, standard deviation and higher order moments. The convergence of the series expansion toward the unknown true distribution and the convergence speed depends on the fitness function considered. Therefore, the approximation quality and the point where the expansion can be cut off are function specific. Example applications (Beyer 1994) include the pseudo Boolean function OneMax $F(\mathbf{b}) = \sum_{i=1}^{l} b_i$ with $b_i \in \{0, 1\}$, biquadratic functions, and general quadratic fitness functions (see ❷ Table 5).

Arnold (2007) analyzed the performance of the $(\mu/\mu_I, \lambda)$-ES on a class of positive definite quadratic forms (PDQFs)

$$f(\mathbf{y}) = \xi \sum_{i=1}^{N\vartheta} y_i^2 + \sum_{i=N\vartheta+1}^{N} y_i^2 \tag{105}$$

with $\vartheta \leq 1$, $N\vartheta$ a natural number, and $\xi \geq 1$. The aim was to study the response to varying degrees of ill-conditioning. The condition number of a positive definite matrix is defined as the ratio of the largest eigenvalue to the smallest. In the case of ❷ Eq. 105, it is controlled by ξ.

❑ **Table 5**
The quality gain of the $(1, \lambda)$-ES for exemplary fitness functions. The symbols S_Q and M_Q denote the standard deviation and mean of the mutation-induced fitness change Q and κ_k the kth cumulant. The normalizations read $\sigma^* = \sigma \mathrm{Tr}[Q]/\|\mathbf{a}\|$ and $\overline{\Delta Q^*} = \overline{\Delta Q} \mathrm{Tr}[Q]/\|\mathbf{a}\|$ for general quadratic functions. In the case of biquadratic functions, they are given by $\sigma^* = \sigma \sqrt[3]{3\sum_i c_i}/\|\mathbf{a}\|$ and $\overline{\Delta Q^*} = \frac{\overline{\Delta Q}}{\|\mathbf{a}\|} \sqrt[3]{3\sum_i c_i}/\|\mathbf{a}\|$. For OneMax, they are defined as $\sigma^* = \sqrt{p_m} l/2 \left(2\frac{F_0}{l} - 1\right)$ and $\overline{\Delta Q^*} = 2\overline{\Delta Q}\left(2\frac{F_0}{l} - 1\right)$ with l the string length, p_m the bit-flipping probability, and F_0 the function value of the parent

ES	Fitness function	Quality gain
$(1, \lambda)$-ES	General functions	$\overline{\Delta Q_{1,\lambda}} = M_Q + S_Q \frac{\kappa_3}{6} +$ $c_{1,\lambda} S_Q \left(1 + \frac{5\kappa_3^2}{36} - \frac{\kappa_4}{8}\right) +$ $S_Q d_{1,\lambda}^{(2)} \frac{\kappa_3}{6} + d_{1,\lambda}^{(3)} S_Q \left(\frac{\kappa_4}{24} - \frac{\kappa_3^2}{18}\right) + \cdots$
$(1, \lambda)$-ES	General quadratic functions	$\overline{\Delta Q^*} = c_{1,\lambda} \sigma^* - \frac{\sigma^{*2}}{2}$
$(1, \lambda)$-ES	Biquadratic functions	$\overline{\Delta Q^*} = c_{1,\lambda} \sigma^* - \sigma^{*4}$
$(1, \lambda)$-ES	OneMax	$\overline{\Delta Q^*} = c_{1,\lambda} \sigma^* - \frac{\sigma^{*2}}{2}$

Larger values of ξ correspond with higher degrees of ill-conditioning. The performance measure used in Arnold (2007) was

$$\Delta = \mathrm{E}\left[-\ln\left(\frac{f(\mathbf{y}^{(g+1)})}{f(\mathbf{y}^{(g)})}\right)\right] \tag{106}$$

a logarithmic quality gain – being asymptotically equal to $\mathrm{E}[(f(\mathbf{y}^{(g)}) - f(\mathbf{y}^{(g)}))/f(\mathbf{y}^{(g)})]$ for $N \to \infty$. Reconsidering ❯ Eq. 105 shows that it is the sum of two spheres $f(R_1, R_2) = \xi R_1^2 + R_2^2$ – one $N\vartheta$-dimensional, the other $N(1 - \vartheta)$-dimensional and evolution takes place in two subspaces. The mutation vector of an offspring can be decomposed into two parts, one for each sphere. Then, the same decomposition as in the case of the sphere can be used with respect to the distance vectors to the centers of the spheres. This allows the determination of the quality change of an offspring as a function of two random components. The same decomposition can be applied to the progress vectors required for the determination of $\mathrm{E}\left[R_1^2\right]$ and $\mathrm{E}\left[R_2^2\right]$.

The resulting quality gain

$$\Delta^* = \frac{c_{\mu/\mu,\lambda}\sigma^*}{\sqrt{1 + \left(\frac{R_1}{R_2\xi}\right)^2}} - \frac{\sigma^{*2}}{2\mu} \tag{107}$$

with $\sigma^* := \sigma N\vartheta/R_1$ and $\Delta^* := \Delta N\vartheta/2$ can be used to determine the optimal mutation strength and quality gain as a function of the condition number. For most values of ξ, these values can only be obtained numerically. For large ξ-values, asymptotical expressions can be derived (see ❯ Table 9).

2.3.3 Ridge Functions

Ridge functions (❯ 76) can be considered using the progress rate approach. The concept of the progress rate, however, has to be widened since ridge functions do not have any finite optimum. The form of the ridge functions automatically leads to the definition of two performance measures. One component of the ridge is made up by the position on y_1 or parallel to the axis. The respective progress rate, the *axial progress rate* φ_{y_1} gives the expectation of the change of this component during one generation. The other component can be interpreted as a distance R of the remaining $N - 1$ dimensions to the axis. In this case, the *radial progress rate* φ_R measures the expected change of two consecutive distances.

The optimization of ridge functions consists of two subgoals: (a) decreasing the radial distance to the axis – optimizing the sphere component, (b) enlarging the axis position – optimizing the linear part.

Ridge functions without noise were considered in the following publications:

- Oyman et al. (1998, 2000) considered the performance of the $(1, \lambda)$-ES and $(1 + \lambda)$-ES on the parabolic ridge.
- Oyman and Beyer (2000) investigated the $(\mu/\mu, \lambda)$-ES on the parabolic ridge.
- Beyer (2001a) considered the $(1, \lambda)$-ES on general ridge functions.
- Arnold (2006a) analyzed the $(\mu/\mu_I, \lambda)$-ES on general ridge functions.

❯ *Table 6* shows the progress rates obtained. Oyman et al. (1998) analyzed the performance of the $(1, \lambda)$-ES and $(1 + \lambda)$-ES on the parabolic ridge. Not considering any step-size adaptation mechanism, the following results were obtained: The axial progress rate is always nonnegative. Provided that the step-size adaptation mechanism generates positive mutation strengths,

◻ **Table 6**

Radial and axial progress rates φ_R and φ_{y_1} for ridge functions. Explanation of variables and notation: $c_{\mu/\mu,\lambda}$ progress coefficient (see ❷ Sect. 5), σ^* normalized mutation strength (with respect to distance to optimizer and search space dimensionality N)

ES	Ridge		Progress rate
$(1, \lambda)$-ES	Parab.	Asymp.	$\varphi_{y_1}(\sigma) = \dfrac{c_{1,\lambda}\sigma}{\sqrt{1 + (2dR)^2}}$
$(1, \lambda)$-ES	Parab.		$\varphi_{y_1}(\sigma) = \dfrac{c_{1,\lambda}\sigma}{\sqrt{1 + (2dR)^2 + 2d^2(N-1)\sigma^2}}$
$(\mu/\mu_I, \lambda)$-ES	Parab.		$\varphi_{y_1}(\sigma) = \dfrac{c_{\mu/\mu,\lambda}\sigma}{\sqrt{1 + (2dR)^2 + 2d^2(N-1)\sigma^2}}$
$(\mu/\mu_D, \lambda)$-ES	Parab.		$\varphi_{y_1}(\sigma) = \dfrac{c_{\mu/\mu,\lambda}\sqrt{\mu}\sigma}{\sqrt{1 + (2dR)^2 + 2\mu d^2(N-1)\sigma^2}}$
$(\mu/\mu_I, \lambda)$-ES	Gen.	Asymp.	$\varphi_{y_1}^*(\sigma^*) = \dfrac{c_{\mu/\mu,\lambda}\sigma^*}{\sqrt{1 + \alpha^2 d^2 R^{2\alpha-2}}}$
$(\mu/\mu_I, \lambda)$-ES	Gen.	Asymp.	$\varphi_R^*(\sigma^*) = \dfrac{\alpha dR^{\alpha-1}c_{\mu/\mu,\lambda}\sigma^*}{\sqrt{1 + \alpha^2 d^2 R^{2\alpha-2}}} - \dfrac{\sigma^{*2}}{2\mu}$

the ES progresses in axis direction. Evolution strategies generally show a stationary state of the R-dynamics after a transient phase. That is, after a while, the ES fluctuates at a certain distance to the ridge and travels on an average, parallel to the axis. The progress that is achieved depends on two quantities: the residual distance and the mutation strength. The analysis considered large mutation strengths and a large ridge parameter d. Under these conditions, the linear y_1 component can be treated as noise and the ridge appears as a sphere. This assumption enables us to determine the residual distance to the axis in analogy to the sphere model. The axial progress rate for the $(1, \lambda)$-ES is then obtained by using the following argument: in the steady state of the R-dynamics, the isofitness landscape appears locally as a hyperplane. Progress in gradient direction can therefore be determined as the hyperplane progress rate φ_{hp} multiplied with the gradient direction as $\varphi_{hp}\nabla F/\|\nabla F\|$. The axial progress is then given as the scalar product of the first unit vector with the progress in gradient direction.

The results show an interesting response of the progress rate with respect to increasing mutation strengths: The progress rate reaches a finite strategy-dependent limit.

Comparing experimentally the performance of the $(1, \lambda)$-ES and the $(1 + \lambda)$-ES reveals an interesting behavior: The performance of the nonelitist $(1, \lambda)$-ES increases with the mutation strength to a limit value. In contrast, the progress of the elitist strategy is below the progress of the $(1, \lambda)$-ES, and decreases with the mutation strength. The $(1 + \lambda)$-ES emphasizes the short-term goal of minimizing the distance to the axis. This goal is better achieved by elitist strategies. Performing one generation experiments with a static mutation strength and measuring the quality gain starting from various distances show that the optimization behavior of $(1 + \lambda)$-ES achieves a larger quality gain. The progress in axis direction is a "byproduct of the selection process" (Oyman et al. 1998) and is smaller than that of the $(1, \lambda)$-ES.

The analysis was extended in Oyman et al. (2000) considering the $(1, \lambda)$-ES in more detail. The focus lay on obtaining and analyzing the progress rate, the quality gain, the success probability, and the success rate. In the case of the $(1, \lambda)$-ES, the success probability denotes the probability that the selected offspring has a better fitness than the parent, whereas the success rate gives the probability of progress by a single mutation (Oyman et al. 2000).

The progress rate was not obtained using a local approximation with a hyperplane. Instead the local quality function and order statistics were used. The so obtained progress rate depends on the distance to the axis and decreases with increasing distance. Therefore, an expression for the stationary distance was obtained by considering the evolution of the distance to the axis and applying a steady state criterion. The experimental comparison of the $(1, \lambda)$-ES with the $(1 + \lambda)$-ES again revealed progress deficiencies for the $(1 + \lambda)$-ES. This is also present in the success probability and success count. Both values deteriorate faster with the normalized mutation strength. This can be traced back to the optimization behavior of the $(1 + \lambda)$-ES. Further experiments revealed the existence of a stationary distance, which is smaller than for $(1, \lambda)$-ES. After this stationary state is reached, progress in axis direction is only small since the smaller distance lowers the success probability.

Oyman and Beyer (2000) presented an analysis of the $(\mu/\mu, \lambda)$-ES on the parabolic ridge. Two types of recombination: intermediate and dominant were considered.

The analysis aimed at providing insights in the progress behavior of these strategies using the axial progress rate and the steady state distance to the ridge axis. The results can be used to give recommendations with regard to the truncation ratio and for a comparison with the progress of the $(1, \lambda)$-ES on the one hand and for a comparison of the two recombination types on the other.

The analysis approach made use of the methods established for the sphere (Beyer 1995a). The analysis for intermediate recombination required the determination and approximation of the local quality function, the mutation-induced fitness change. First of all, the general function was expanded into its Taylor series and the resulting series was cut off after the quadratic term. In this approximation, the local quality function is a quadratic polynomial in the mutation vectors components. This was used to determine the required cdfs and pdfs in the derivation of the progress rate via a normal approximation of the true distribution.

The case of dominant recombination was treated using the result for intermediate recombination and applying the concept of surrogate mutations (Beyer 1995a), that is, the step size of the surrogate mutation $\sigma_S = \sqrt{\mu}\sigma$ is plugged into the progress rate for the intermediate case.

So far, the steady state dynamics of the distance to the ridge were not included in the analysis. Considering the square of the distance and following a similar approach as for the sphere model leads to an expression for the residual distance for the $(\mu/\mu_I, \lambda)$-ES. The distance depends on the population parameters μ and λ, on the search space dimensionality, on the mutation strength, and on the ridge parameter d. For $d \to \infty$, the residual distance approaches the mutation-induced residual distance of the sphere model $(N - 1)\sigma/(2\mu c_{\mu/\mu,\lambda})$. Recombination reduces the distance to the axis. This is similar to the genetic repair effect on the sphere model. Considering the distance to the axis, the mutation vector and the progress vector can be written again as the direct sum of two perpendicular components: one pointing to the axis and the perpendicular part. Averaging effectively reduces the perpendicular parts. The stationary distance for the $(\mu/\mu_D, \lambda)$-ES is obtained in a similar manner as before. Comparing the distances exemplarily reveals that the intermediate ES achieves smaller distances to the axis than the $(\mu/\mu_D, \lambda)$-ES.

In order to compare the stationary progress rates, the obtained residual distances are plugged into the respective formulas. In the case of intermediate recombination, the resulting axial progress rate shows a saturation behavior for increasing mutation strengths. Instead of increasing indefinitely, a strategy-dependent limit value is approached. A similar behavior was found for the $(1, \lambda)$-ES. Comparing these limits leads to the conjecture that the performance of the $(\mu/\mu_I, \lambda)$-ES exceeds that of the $(1, \lambda)$-ES if the mutation strength is large and $\mu c^2_{\mu/\mu,\lambda} > c^2_{1,\lambda}$, holds.

The result for the $(\mu/\mu_D, \lambda)$-ES is interesting since it indicates that using dominant recombination instead of intermediate recombination is beneficial on the parabolic ridge: The progress curve for dominant recombination is above the curve for intermediate recombination, a gap which closes for increasing mutation strengths.

In Beyer (2001a), the performance of the $(1, \lambda)$-ES on general ridge functions was analyzed for the case of finite N and constant mutation strength. The analysis provided a simple but more accurate approach that allowed a treatment of the radial dynamics. It was shown that the y_1-dynamics are controlled by the R-dynamics whereas the latter do not depend on the former. First of all, the progress rates were determined. The results were then used to discuss the general properties, and the mean value and steady state behavior of the ES.

The axial progress rate depends on the distance to the ridge. The asymptotical behavior for $R \to 0$ and $R \to \infty$ depends on the ridge topology parameter α. For $\alpha > 1$, for instance, the progress rate for the hyperplane is obtained for $R \to 0$, whereas for $\alpha < 1$, it appears as the limit for $R \to \infty$.

Any discussion of the mean value dynamics of the y_1-evolution requires a closer investigation of the R-dynamics. Experiments showed that the R-evolution approaches a steady state, that is, a time-invariant limit distribution, and the mean value converges to a constant. This results in a linear progress of the mean value of the y_1-dynamics after a transient phase.

The velocity by which the ES travels along the ridge axis depends on the value of the stationary distance R. Using the radial progress rate and setting it to zero gives a general equation to be solved for the stationary distance. However, the mutation strength appears as an unknown variable.

The steady state progress was also considered. For the sharp and the parabolic ridge, the influence of the distance can be eliminated, yielding analytically expressions for the steady state progress as functions of the mutation strength. This allows for an analysis of the influence of the mutation strength.

Other values of α have to be treated numerically. The size of α determines the qualitative behavior of the progress rate with respect to the mutation strength. While the progress rate increases continuously for $\alpha < 2$, it appears to reach a saturation behavior for $\alpha = 2$. Increasing α further leads to a unimodal progress rate with an optimal mutation strength.

Arnold (2006a) considered the general ridge function class. He derived the axial progress rate and a stationary condition for the distance to the ridge axis. The results were used to determine optimal values for the mutation strength, residual distance, and progress rate (see ❯ *Table 10*).

2.3.4 Step-Size Adaptation Mechanisms

In the following, the results of applying the dynamical systems approach in the analysis of step-size mechanisms is described. First, some results for self-adaptation are presented, before the cumulative step-size adaptation is considered.

⬛ **Table 7**

Self-adaptation response functions (SARs) for different fitness environments. Explanation of variables and notation: σ^* normalized mutation strength (with respect to distance to optimizer and search space dimensionality). For the coefficients $c_{\mu/\mu,\lambda}$, $e^{1,1}_{\mu,\lambda}$ and $d^{(2)}_{1,\lambda}$ see ❷ Sect. 5

Fitness	ES	SAR
Sphere	$(1, \lambda)$-ES	$\psi(\sigma^*) = \tau^2 \left(d^{(2)}_{1,\lambda} - \dfrac{1}{2} - c_{1,\lambda}\sigma^* \right)$
Sphere	$(\mu/\mu_I, \lambda)$-ES	$\psi(\sigma^*) = \tau^2 \left(\dfrac{1}{2} + e^{1,1}_{\mu,\lambda} - c_{\mu/\mu,\lambda}\sigma^* \right)$
Ridge	$(\mu/\mu_I, \lambda)$-ES	$\psi(\sigma^*) = \tau^2 \left(\dfrac{1}{2} + e^{1,1}_{\mu,\lambda} - c_{\mu/\mu,\lambda}\sigma^* \sqrt{\dfrac{\alpha^2 d^2 R^{2\alpha-2}}{(1 + \alpha^2 d^2 R^{2\alpha-2})}} \right)$
PDQF	$(\mu/\mu_I, \lambda)$-ES	$\psi(\sigma^*) = \tau^2 \left(\dfrac{1}{2} + e^{1,1}_{\mu,\lambda} - c_{\mu/\mu,\lambda}\sigma^* \dfrac{\vartheta(\xi - 1) + 1}{\vartheta\xi\sqrt{1 + \zeta^2}} \right)$

Self-Adaptation

Self-adaptation was considered in the following publications:

- Beyer (1996b) was the first analysis of the self-adaptation mechanism and introduced the analysis approach also followed in later analyses. The evolution strategy considered was the $(1, \lambda)$-ES.
- In Meyer-Nieberg and Beyer (2005), the analysis was extended to the $(\mu/\mu_I, \lambda)$-ES on the sphere.
- Meyer-Nieberg and Beyer (2007) and Beyer and Meyer-Nieberg (2006b) provided an analysis of the self-adaptive $(\mu/\mu_I, \lambda)$-ES on ridge functions.
- And Beyer and Finck (2009) extended the method to PDQFs.

An overview of the self-adaptation response functions obtained is given in ❷ *Table 7*. The analyses aimed to provide insights on the functioning of self-adaptation, to analyze the influence of the learning parameter τ, and the effects of recombination. Primarily, the analyses are devoted to the log-normal operator, cf. ❷ *Table 1*, which is a very common choice. The two-point rule was considered in Beyer (1996b, 2001b). Most of the results for the log-normal operator can be expected to be transferred for the two-point rule rather easily. In the case of the $(1, \lambda)$-ES, the correspondence between the mutation operator was shown in Beyer (1996b). In the following, the results for the sphere are described before considering ridge functions.

On the sphere, self-adaptation generally drives the mutation strength to scale with the distance to the optimizer (disregarding fluctuations of course). Using the distance as normalization, leads to the finding that the normalized mutation strength reaches a stationary state. In Beyer (1996b), the deterministic evolution equations were used to model the ES's time dynamics. To this end, the self-adaptation response function (❷ 68) had to be obtained which can be approximated as a linear function of the mutation strength for small learning parameters. The evolution equations were then used to determine conditions for the stationary state. As it has been shown, self-adaptation steers the normalized mutation strength toward a stationary point between the second zero point of the progress rate and the zero point of the self-adaptation response. On the sphere, it always holds that the zero point of the self-adaptation response is smaller than that of the progress rate. In this process, the learning parameter τ serves as control parameter: For large values of τ, the stationary normalized mutation strength is close to the zero of the SAR. Conversely, it is close to the zero of the progress rate if τ is small. Therefore, for small values of τ, there is nearly no progress at all. The maximum of the progress rate lies in between the two zeros. Therefore, a value of τ exists for which optimal expected progress can be obtained (Beyer 1996b). ❖ *Figure 7a* illustrates the behavior. It can be shown that choosing $\tau \propto 1/\sqrt{N}$ is roughly optimal which provides a theoretical basis for Schwefel's recommendation for choosing the learning parameter. If τ is increased above this value, the zero of the SAR is approached. This does not deteriorate the performance significantly: The zero of the SAR is close to the optimal point of the progress rate. This provides an explanation as to why $(1, \lambda)$-ESs are robust with respect to the choice of the learning parameter.

This is not the case anymore if $(\mu/\mu_I, \lambda)$-ESs are considered: As Grünz and Beyer found experimentally, $(\mu/\mu_I, \lambda)$-ESs are very sensitive with respect to the choice of τ (Grünz and Beyer 1999). Instead of still showing nearly optimal progress if τ is chosen too large, these evolution strategies slow down: Nearly optimal behavior of an ES is only attainable for a narrow range of learning parameters.

In Meyer-Nieberg and Beyer (2005), an explanation has been provided. The decisive point is that the zero of the SAR is not close to the optimizer of the progress rate anymore. Increasing the number of parents usually shifts the optimizer to larger values. That is, intermediate recombination enables the ES to work with higher mutation strengths which transfers to the optimal point. The zero of the SAR does not behave accordingly, though. It is less sensitive toward variations of the parent number and, even worse, it tends to decrease for the usual choices. There is a widening gap between the zero of the SAR and the optimizer if the number of parents is increased. The consequences of this are twofold: on the one hand, there is still an optimal choice of the learning rate, on the other hand deviating from this choice will soon result in severe progress deterioration. Self-adaptation

◻ **Fig. 7**
The influence of the learning rate, τ, on the stationary state of the mutation strength σ^* on the sphere. The case of a (1,10)-ES is depicted in (a), (b) shows a multi-parent ES. The stationary state is governed by the equation $-\psi(\sigma^*) = \varphi^*(\sigma^*)/N$, that is, by the intersection of the negative self-adaptation response with the progress rate φ^* (divided by the search space dimensionality N). The SAR is linear in the mutation strength. The slope and the y-axis intercept are scaled with τ^2. Shown are the progress rate and several SARs for different choices of τ. The search space dimensionality is $N = 100$.

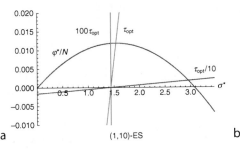

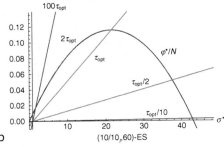

a   (1,10)-ES   b   (10/10$_I$,60)-ES

as a mechanism still works, however. The progress remains positive and the ES moves toward the optimizer. However, as ❯ Fig. 7b shows exemplarily, it is important to choose τ close to the optimal value. Provided that the parent number is neither too close to one nor too close to the number of offspring, $\tau \approx 1/\sqrt{2N}$ leads to nearly optimal progress for large search space dimensionalities.

One question that remains is, why is the zero of the SAR behaving in this manner? The answer lies in reconsidering the mechanism of self-adaptation. The mutation strength is variated *before* the recombination of the object parameters. No detailed information is passed to the self-adaptation of the mutation strength that – due to intermediate recombination – higher mutation strengths are possible. The feedback is only indirect over the selection of the individuals with higher fitness.

In Beyer (1996b), the dynamical behavior of the $(1, \lambda)$-ES with self-adaptation was considered. Two approaches have been pursued: One using the deterministic evolution equations, the other using the second-order approach – modeling the perturbation parts of the evolution equations with normally distributed random variables. The induced stochastic process was studied by considering some moments – the mean for most cases.

In the case of the evolution of the distance R, it was shown that the mean converges log-linearly as soon as the normalized mutation strength has reached its stationary state. Using the deterministic approximation, it can be shown that the time an ES needs to approach the steady state depends on $1/\tau^2$. As a result, the τ-scaling law $\tau \propto 1/\sqrt{N}$ causes the ES to have an adaptation time $g_A \propto N$. This might take too long for large search space dimensionalities. A short adaptation time and an optimal performance in the stationary state cannot be reached with a single learning parameter τ. This leads to the recommendation that in the initial phase, an N-independent learning parameter should be used and when approaching the steady state, one should switch to $\tau \propto 1\sqrt{N}$.

The second-order approach provides an explanation for the observation that the experimentally measured rate is somewhat smaller than the theoretically expected value. The random dynamics of the mutation strength introduce a detrimental variance term to the average progress rate

$$\overline{\varphi_R^*(\sigma^{*(g)})} = c_{1,\lambda}\overline{\sigma^{*(g)}} - \frac{(\overline{\sigma^{*(g)}})^2}{2} - \frac{D^2[\sigma^{*(g)}]}{2} \tag{108}$$

The R-dynamics are driven by the mutation strength. Obtaining expressions for the mean and higher moments proved difficult in the case of the mutation strength because lower order moments depend functionally on higher order moments. Therefore, an *ansatz* using the log-normal distribution can serve as an approximation of the true density. The results were used to show that the stationary mean value of the normalized mutation strength is always less than the deterministic prediction due to the fluctuations. Furthermore, the scaling law of the learning parameter also proved to be valid for this second-order model.

In Beyer and Meyer-Nieberg (2006b) and Meyer-Nieberg and Beyer (2007), ridge functions were considered: the sharp ridge and $(1, \lambda)$-ESs in Beyer and Meyer-Nieberg (2006b) and $(\mu/\mu_I, \lambda)$-ESs and additionally the parabolic ridge in Meyer-Nieberg and Beyer (2007). Both ridge functions must be described by two state variables, the distance to the ridge (radial) and the position parallel to the axis direction (axial). Three state variables are therefore decisive: the axial position, the radial position, and the mutation strength. The analyses (Beyer and Meyer-Nieberg 2006b; Meyer-Nieberg and Beyer 2007) revealed that the evolution of the axial position does not influence any other evolution. That is, there is no direct feedback from the present axial position. Only the mutation strength and the radial distance are the governing forces of the evolution. Therefore, in terms of a self-adaptive ES, a ridge appears as a distorted sphere.

Similar to the sphere, self-adaptation steers the mutation strength to scale on average with the distance to the ridge with the result that a stationary state of the normalized mutation strength is approached in the long run. And as before, self-adaptation drives the mutation strength toward a stationary point between the zero of the self-adaptation response (SAR) and the zero of the radial progress rate.

As for the sharp ridge ($\alpha = 1$) two quantitatively different behaviors can be observed: The ES either converges toward the axis or diverges from it. If it converges, the scaling of the mutation strength with the distance results in mutation strengths that are far too small for any significant axial progress: The search stagnates. Divergence, on the other hand, is coupled with an increasing mutation strength leading on an average to a positive quality or fitness change. The questions remain why the ES shows these different behaviors and what is the factor deciding which one occurs? These questions are answered by taking a closer look at the stationary state behavior.

In the case of the parabolic ridge, the ES attains a stationary state of the mutation strength and the distance (see $\circ$ *Table 8*). That is, there are (R, σ^*)-combinations which are stationary states of the R-evolution (zero progress $\varphi_R{}^* := \varphi_R/N$) and the σ^*-evolution ($\sigma^* = \sigma/N$, zero SAR ψ) (see $\circ$ *Fig. 8*). In the case of the sharp ridge, no such states exist. Other than $R = 0$ and $\sigma^* = 0$, no combination fulfills simultaneously both stationary state criteria. The ES either converges or diverges. The parameter d of the ridge function is the decisive parameter which determines the behavior. It can be shown that for a strategy-dependent critical value, the lines for $\varphi_R^* = 0$ and $\psi = 0$ overlap. If d is larger than this critical value, the line for $\varphi_R^* = 0$ is consistently above the line for $\psi = 0$. This is a similarity to the sphere model, where the second zero of the progress rate is always larger than the zero of the SAR. Depending on the choice of the learning rate τ, a self-adaptive ES adapts a normalized stationary mutation strength somewhere between these values. The attained mutation strength is connected with a positive radial progress, that is, the ES moves in expectation toward the axis. In $\circ$ *Fig. 9*, this indicates that the ES moves into the inner cone delimited by $\varphi_R^* = 0$ and $\psi = 0$. If d is smaller than the critical value, however, the situation is reversed. The line for $\psi = 0$ is above the line for $\varphi_R^* = 0$. Instead of operating with mutation strengths that are connected with positive progress, the ES adapts mutation strengths which lead toward a divergence from the axis.

◻ Table 8

Stationary state values of the self-adaptive $(\mu/\mu_I, \lambda)$-ES on ridge functions. The normalizations are $\sigma^* := \sigma R/N$ with R the distance to the axis, $\varphi_{y_1}^* := \varphi_{y_1} R/N$, and $\varphi_R^* := \varphi_R R/N$

Mutation strength	$\sigma^* = \sqrt{2\mu(1/2 + e_{\mu,\lambda}^{1,1})}\left(\dfrac{1/2 + e_{\mu,\lambda}^{1,1}}{\alpha^2 d^2(2\mu c_{\mu/\mu,\lambda}^2 - 1/2 - e_{\mu,\lambda}^{1,1})}\right)^{1/(2\alpha-2)}$
Distance	$R = \left(\dfrac{1/2 + e_{\mu,\lambda}^{1,1}}{\alpha^2 d^2(2\mu c_{\mu/\mu,\lambda}^2 - 1/2 - e_{\mu,\lambda}^{1,1})}\right)^{1/(2\alpha-2)}$
Progress	$\varphi_{y_1}^* = \sqrt{(1/2 + e_{\mu,\lambda}^{1,1})(2\mu c_{\mu/\mu,\lambda}^2 - 1/2 - e_{\mu,\lambda}^{1,1})}\left(\dfrac{1/2 + e_{\mu,\lambda}^{1,1}}{\alpha^2 d^2(2\mu c_{\mu/\mu,\lambda}^2 - 1/2 - e_{\mu,\lambda}^{1,1})}\right)^{1/(2\alpha-2)}$

☐ Fig. 8

(σ^*, R)-combinations with $\psi = 0$ and $\varphi_R^* = 0$ for (1, 10)-ES with $d = 1$. Shown are the cases for two ridge functions: $\alpha = 2$ in (a) and $\alpha = 3$ in (b). The normalizations are $\sigma^* := \sigma/N$ and $\varphi_R^* := \varphi_R/N$. The intersection indicates the stationary state. The lines divide the (σ^*, R)-phase space into several regions: Region I_1 is characterized by positive expected changes of R, $\Delta R > 0$, and positive changes of σ^*, $\Delta\sigma^* > 0$, I_2 by $\Delta R < 0$, $\Delta\sigma^* > 0$, I_3 by $\Delta R < 0$, $\Delta\sigma^* < 0$, and finally I_4 by $\Delta R > 0$, $\Delta\sigma^* < 0$. The system either leaves every region I_k again, that is, it oscillates, or it converges to the equilibrium point.

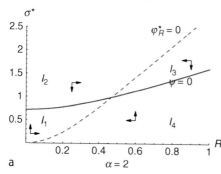
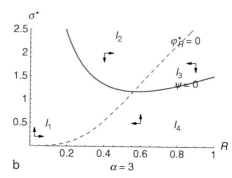

a $\alpha = 2$

b $\alpha = 3$

☐ Fig. 9

(σ^*, R)-combinations for zero progress and zero SAR for (1, 10)-ES on the sharp ridge. The normalizations are $\sigma^* := \sigma/N$ and $\varphi_R^* := \varphi_R/N$. The zero expected change combinations subdivide the (σ^*, R)-phase space. In (a) where $d > d_{\text{crit}}$ region I_1 is characterized by $\Delta R > 0$, $\Delta\sigma^* < 0$, I_2 by $\Delta R < 0$, $\Delta\sigma^* < 0$, and I_3 by $\Delta R < 0$, $\Delta\sigma^* > 0$. Possible movements between the regions are $I_3 \to I_2$ and $I_1 \to I_2$. It is easy to see that I_1 and I_3 will be left eventually. The region I_2 cannot be left and the system in σ^* and R approaches the origin. In (b) where $d < d_{\text{crit}}$ region I_1 is characterized by $\Delta R > 0$, $\Delta\sigma^* < 0$, I_2 by $\Delta R > 0$, $\Delta\sigma^* > 0$, and I_3 by $\Delta R < 0$, $\Delta\sigma^* > 0$. Possible movements are $I_1 \to I_2$ and $I_3 \to I_2$, but I_2 cannot be left. The system diverges to infinity.

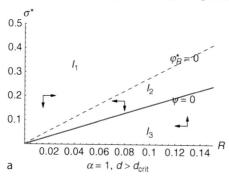
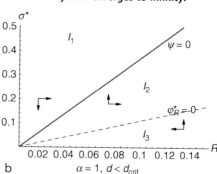

a $\alpha = 1, d > d_{\text{crit}}$

b $\alpha = 1, d < d_{\text{crit}}$

Cumulative Step-Size Adaptation

In Arnold and Beyer (2004), the performance of $(\mu/\mu_I, \lambda)$-ES on the sphere was investigated for the undisturbed and the noisy case. Concerning the noise-free case, the analysis is aimed at investigating whether the CSA succeeds in adapting a mutation strength that guarantees optimal progress and deriving recommendations for appropriate population sizes.

During an evolutionary run, the ES approaches a steady state: The squared length of the accumulated progress vector, its signed length, and the normalized mutation strength attain a limit distribution. The CSA should strive to increase too small mutation strengths and to decrease too large mutation strengths. In the analysis, the logarithmic adaptation response

$$\Delta_\sigma = \log\left(\frac{\sigma^{(g+1)}}{\sigma^{(g)}}\right) \tag{109}$$

which compares two succeeding mutation strengths is used. The target mutation strength of the CSA is the optimal mutation strength with respect to the progress rate. The ES does not succeed in realizing this objective: Due to the dynamically changing distance to the optimizer, the adaptation of the mutation strength lags behind and the target mutation strength is not reached. The realized normalized mutation strength exceeds the target by a factor of $\sqrt{2}$. In the undisturbed case, the ES with CSA achieves a progress rate of approximately 83% of the optimal progress rate (Arnold and Beyer 2004). Since the optimal mutation strength is not realized, the recommendations (Arnold and Beyer 2002b) for the population sizing with respect to maximal efficiency are slightly to be revised. However, basically the same conclusions as for the sphere can be drawn: The optimal number of offspring is well below the search space dimensionality and increases gradually with it. Furthermore, the recommendation of choosing the truncation ratio about $\mu/\lambda = 0.27$ applies as well.

Arnold (2007) considered the performance of the $(\mu/\mu_I, \lambda)$-ES on positive quadratic forms (❷ Eq. 73). Being based on the analysis of the expected quality change, the asymptotically optimal mutation strength and the quality gain have been derived as a function of large ξ-values (see ❷ Table 9). After a transient phase, the $(\mu/\mu_I, \lambda)$-ES with CSA yields a stationary mutation strength. This mutation strength and the quality gain depend on the condition number. For $\xi = 1$, the same findings as for the sphere are recovered: The generated mutation strengths are too large by a factor of $\sqrt{2}$. For the asymptotic $\xi \to \infty$, the ES generates too small mutation strengths (see ❷ Table 9). However, the adaptation mechanism still works in that, mutation strengths are realized with result in positive quality gain.

Arnold (2006a) investigated the performance of the $(\mu/\mu_I, \lambda)$-ES with CSA on the (nearly) complete ridge function class. First, the case of static mutation strengths was considered. After obtaining the axial progress rate and a stationarity condition for the standardized distance to the ridge, optimal values for the progress, distance, and mutation strength can be determined. Optimal refers here to optimal axial progress. The results are given in ❷ Table 10 and are applicable for general $\alpha > 2$. As it can be seen, they depend on the value of α. The CSA algorithm would work perfectly on the ridge function class if it succeeded in adopting the optimal mutation strength. This is, however, not the case. An ES using CSA tracks the ridge in unit standardized distance, regardless of α. This in turn results in an α-independent

❑ Table 9

Optimal settings and realized values of the CSA-ES for the quality gain on PDQFs with large condition numbers ξ. The values are normalized quantities $\sigma^* := \sigma N\vartheta/R_1$ and $\Delta^* := \Delta N\vartheta/2$

Opt. mutation strength	Optimal quality gain	Mut.strength (CSA)	Quality gain (CSA)
$\sigma_{\mathrm{opt}}^* = 2\mu c_{\mu/\mu,\lambda}$	$\Delta_{\sigma,\mathrm{opt}}^* = \dfrac{2\mu c_{\mu/\mu,\lambda}^2}{\xi}$	$\sigma_{\mathrm{CSA}}^* = \sqrt{2}\mu c_{\mu/\mu,\lambda}$	$\Delta_{\sigma,\mathrm{CSA}}^* = \dfrac{\mu c_{\mu/\mu,\lambda}^2}{\xi}$

◻ **Table 10**

The $(\mu/\mu_I, \lambda)$-ES with CSA on the ridge function class. Shown are the values for optimal axial progress and the values achieved by the CSA-mechanism. The following normalizations and abbreviations were used: $\rho = (\alpha d)^{1/(\alpha-1)}R$, $\sigma^* = \sigma N(\alpha d)^{1/(\alpha-1)}/(\mu c_{\mu/\mu,\lambda})$, $\varphi^* = \varphi N(\alpha d)^{1/(\alpha-1)}/(\mu c_{\mu/\mu,\lambda}^2)$ (Arnold 2006a)

	Distance	Mutation strength	Progress
Optimal	$\rho = \left(\dfrac{\alpha}{\alpha-2}\right)^{\frac{1}{2(\alpha-1)}}$	$\sigma^* = \sqrt{\dfrac{2\alpha^{\frac{\alpha}{\alpha-1}}}{(\alpha-1)(\alpha-2)^{\frac{1}{\alpha-1}}}}$	$\varphi^* = \dfrac{\alpha^{\frac{\alpha}{2(\alpha-1)}}(\alpha-2)^{\frac{\alpha-2}{2(\alpha-1)}}}{\alpha-1}$
CSA	$\rho = 1$	$\sigma^* = \sqrt{2}$	$\varphi^* = 1$

normalized mutation strength and progress. For small α, this causes a severe underestimation of the optimal values which gradually diminishes with increasing α.

2.4   Uncertain Environments

This section considers the optimization in uncertain environments, that is, noisy and robust optimization. If noise is involved, the optimizer is often required to show a robustness against perturbations, that is, its quality should decrease only gradually (Beyer and Sendhoff 2006, 2007a; Jin and Branke 2005). Noise is a common problem in practical optimization tasks. Noise may occur for various reasons, for instance, measurement errors or limitations, production tolerances, or incomplete sampling of large search spaces. Noise may also be encountered when optimizing parameters in computer simulations where either numerical errors appear or the simulation technique itself may lead to stochastic results (Arnold and Beyer 2006a; Beyer 2000).

2.4.1   Preliminaries: Accounting for Noise

In the following, several approaches to model noisy fitness measurements are described. Since the noise is modeled by random variable(s) its distribution must capture the main characteristics of real-world situations, for example, an unbiased measurement error. Furthermore, one has to decide how the noise influences the fitness evaluation. For instance, is the noise added afterward, does noise impair the exact localization in the search space or does it even change the functional structure of the fitness function?

Noise Distributions
In order to model practical situations where noise occurs, several noise distributions may be regarded. Each assumes different characteristics of the noise which range from a symmetric distribution without bias over the occurrence of outliers to skewed noise.

The most common noise model is normally distributed noise with zero mean and standard deviation σ_ε

$$Z \sim \mathcal{N}(0, \sigma_\varepsilon^2), \quad p(z) = \frac{1}{\sqrt{2\pi}\,\sigma_\varepsilon} e^{-\frac{z^2}{2\sigma_\varepsilon^2}} \tag{110}$$

The standard deviation σ_ε is usually referred to as the *noise strength*. However, as remarked in Arnold and Beyer (2006a), the Gaussian noise model may not be sufficient to model all practically relevant cases. Not only may the noise be highly non-normal, but also the assumption of a finite variance may be too restrictive. Recommendations obtained for Gaussian noise may not hold, for instance, for noise distributions with heavy tails and frequently occurring outliers. An alternative to using Gaussian noise is to assume Cauchy distributed noise. Its pdf and cdf read

$$p(z) = \frac{\sigma_\varepsilon}{\pi\left(z^2 + \sigma_\varepsilon^2\right)} \tag{111}$$

$$P(z) = \frac{1}{2} + \frac{1}{\pi}\arctan\left(\frac{z}{\sigma_\varepsilon}\right) \tag{112}$$

The Cauchy distribution has usually two parameters, the *location parameter*, set to zero in ❯ Eqs. 111 and ❯ 112, and the *scale parameter* σ_ε. The Cauchy distribution has a far heavier tail than the normal distribution and can therefore be used as a model for noise with frequently occurring outliers. All moments, including mean and variance, are undefined. The median, however, exists and is equal to the location parameter. Cauchy distributed noise is symmetric around the location parameter.

To model biased and skewed noise, the χ_1^2 distribution may be considered. The χ_1^2 distribution appears when taking the square of a normally distributed random variable. The pdf is

$$p(z) = \frac{e^{-\frac{z}{2\sigma_\varepsilon}}}{\sqrt{2\pi\sigma_\varepsilon z}} \text{ for } z \geq 0 \tag{113}$$

Since the distribution is not symmetric, it matters whether the random term is added or subtracted to the fitness function, or in other words, whether the noise term itself or its negative is χ_1^2-distributed.

Noise Models
The *standard noise model* assumes that noise is added to the result of the fitness evaluation

$$F_\varepsilon(\mathbf{x}) = F(\mathbf{x}) + z(\varepsilon, g) \tag{114}$$

where the noise term, $z(\varepsilon, g)$, may follow one of the distributions above. As the assumption of Gaussian noise, the additive noise model may be insufficient to capture all characteristics of real-word situations.

The additive noise model may be regarded as a special case of the *general noise model*

$$F_\varepsilon(\mathbf{x}) = F(\mathbf{x}, z(\varepsilon, g)) \tag{115}$$

In this case, the noise may directly influence the fitness function itself. As a result, the fitness function has to be interpreted as a random function. The approach provides more flexibility. For any computations, however, assumptions must be made.

A special case of ❯ Eq. 115 is *systematic noise* (or *actuator noise*) which refers to noise on the coordinates of the objective

$$F_\varepsilon(\mathbf{x}) = F(\mathbf{x} + \mathbf{z}(\varepsilon, g)) \tag{116}$$

(Beyer and Sendhoff 2007b). Actuator noise is often considered in the context of *robust optimization*. Robustness means that the obtained solution is relatively stable with respect to perturbations. Several measures can be introduced to evaluate the quality of a solution. In the context of robust optimization, several types of probabilistic criteria can be distinguished, for example: *threshold measures* and *statistical momentum measures*. The former require that the probability of F-realizations below a certain threshold is maximal

$$P(\{F \le q|\mathbf{y}\}) \to \max \tag{117}$$

if minimization is considered. The *statistical momentum measures*

$$E[F^k|\mathbf{y}] \to \min \tag{118}$$

require minimality with respect to the kth moment (see, e.g., Beyer and Sendhoff 2006, 2007a).

2.4.2 Noise and Progress

In the following, some results from analyses of the influence of noise on the local performance starting with the sphere model are presented. The first task is usually to obtain expressions for the progress rate or the quality gain, which can be used, for instance, to derive necessary or sufficient convergence criteria. Further points include examining the effects of recombination, the efficiency of resampling compared to operating with larger offspring populations, and determining appropriate population sizes.

Sphere Model and Linear Functions

The effects of noise on the progress of an ES were analyzed in the following publications

- Beyer (1993) considered the performance of the $(1, \lambda)$-ES and the $(1 + \lambda)$-ES.
- Arnold and Beyer (2002a) analyzed the effects of overvaluation in the $(1 + 1)$-ES.
- In Arnold and Beyer (2001b), the local performance of the $(\mu / \mu_I, \lambda)$-ES in infinite-dimensional search spaces was investigated.
- Arnold and Beyer (2002b) extended the analysis of Arnold and Beyer (2001b) to an analysis in N-dimensional search spaces.
- Arnold (2002a) and Arnold and Beyer (2001a, 2003a) considered the behavior of the (μ, λ)-ES.
- In Arnold and Beyer (2006a), several noise distributions were considered. Deriving the progress rates provides a means to investigate the implications of different distributions on the behavior of the ES.

In Beyer (1993), the influence of noise on the performance of evolution strategies with only one parent was analyzed. Two strategies were considered: the $(1, \lambda)$-ES and the $(1 + \lambda)$-ES. The first task was to obtain the progress rate (see ❷ *Table 11*). The analysis used again, the decomposition of the mutation vectors and applied order statistics to derive the final expressions. Due to the noise, induced order statistics, also referred to as noisy order statistics, were used (see Arnold 2002a). Instead of considering the fitness change induced by mutation, the perceived fitness change had to be taken into account. Therefore, two random processes: mutation and noise have to be considered.

The progress rate of the $(1, \lambda)$-ES consists again of an approximately linear gain part and a quadratic loss part (cf. evolutionary progress principle (EPP)). Only the gain part is

☐ **Table 11**

Asymptotical progress rates for the ES on the noisy sphere. Here, noise type *standard* refers to the additive noise model (❯ 114) whereas *systematic* is described by ❯ Eq. 116

ES	Noise type	Progress rate
$(1, \lambda)$-ES	Standard	$\varphi_R^* = \dfrac{c_{1,\lambda}\sigma^{*2}}{\sqrt{\sigma^{*2}+\sigma_\varepsilon^{*2}}} - \dfrac{\sigma^{*2}}{2}$
$(1 + 1)$-ES	Standard	$\varphi_R^* = \dfrac{\sigma^*}{\sqrt{2\pi}} \dfrac{\exp\left[-\dfrac{1}{2}\left(\dfrac{\sigma^*/2}{\sqrt{1+2\left(\frac{\sigma_\varepsilon^*}{\sigma^*}\right)^2}}\right)\right]}{\sqrt{1+2\left(\frac{\sigma_\varepsilon^*}{\sigma^*}\right)^2}}$ $-\dfrac{\sigma^{*2}}{2}\left(\left(1-\Phi\left(\dfrac{\frac{\sigma^*}{2}}{\sqrt{1+2\left(\frac{\sigma_\varepsilon^*}{\sigma^*}\right)^2}}\right)\right)\right)$
$(\mu/\mu_I, \lambda)$-ES	Standard	$\varphi_R^* = \dfrac{c_{\mu/\mu,\lambda}\sigma^{*2}}{\sqrt{\sigma^{*2}+\sigma_\varepsilon^{*2}}} - \dfrac{\sigma^{*2}}{2\mu}$
$(\mu/\mu_I, \lambda)$-ES	Standard	$\varphi_R^* = \dfrac{c_{\mu/\mu,\lambda}\sigma^{*2}\left(1+\frac{\sigma^{*2}}{2\mu N}\right)}{\sqrt{1+\frac{\sigma^{*2}}{\mu N}}\sqrt{\sigma^{*2}+\sigma_\varepsilon^{*2}+\frac{\sigma^{*4}}{2N}}} - N\left(\sqrt{1+\dfrac{\sigma^{*2}}{2\mu N}}-1\right)$
$(\mu/\mu_I, \lambda)$-ES	Systematic	$\varphi_R^* = \dfrac{c_{\mu/\mu,\lambda}\sigma^{*2}\left(1+\frac{\sigma^{*2}}{2\mu N}\right)}{\sqrt{1+\frac{\sigma^{*2}}{\mu N}}\sqrt{\sigma^{*2}+\sigma_\varepsilon^{*2}+\frac{(\sigma^{*2}+\sigma_\varepsilon^{*2})^2}{2N}}} - N\left(\sqrt{1+\dfrac{\sigma^{*2}}{2\mu N}}-1\right)$

influenced by the noise. Noise decreases the gain part which can be measured by the so-called selection sharpness

$$S := \frac{1}{\sqrt{1+(\sigma_\varepsilon^*/\sigma^*)^2}} \tag{119}$$

where σ_ε^* is the normalized noise strength which reads for the sphere model (❯ 71)

$$\sigma_\varepsilon^* = \frac{N\sigma_\varepsilon}{R\left|\frac{dg}{dR}\right|} \tag{120}$$

For "comma" strategies, there is a certain critical noise strength above which any positive mutation strength causes negative progress and expected divergence. The transition at which this appears is determined by the necessary evolution criterion. Below the critical noise strength, there are regions of $(\sigma^*, \sigma_\varepsilon^*)$-combinations which allow positive progress. Clearly, the border is defined by zero progress rate. The critical noise strength can be used to determine the *residual location error*, the stationary expected distance to the optimizer. The residual location error depends on σ and σ_ε. The mutation strength σ can generally be obtained only by analyzing the step-size adaptation mechanism.

Since the necessary evolution criterion depends on the size of the population, one way to increase the convergence region of the $(1, \lambda)$-ES would be to increase λ. It can be shown, however, that this is not very efficient. Another way to cope with noise is to reduce the noise strength itself which can be done by m-times resampling and averaging (provided that the statistical moments of the noise exist). The analysis revealed that except for small values of λ and m, a $(1, \lambda)$-ES with a lower population size but with repeated resampling and averaging might be preferable.

In the case of the $(1 + \lambda)$-ES, the perceived fitness of the apparent best offspring is compared with the perceived fitness of the parent. In Beyer (1993), it was assumed that the parent is evaluated anew when comparing it with the offspring. To determine the progress rate, the apparent or perceived success probability had to be obtained. The calculations led to an asymptotical progress rate, which revealed that in contrast to the non-noisy case, the expected progress can be negative – depending on the noise and mutation strength. Due to the nature of the integrals involved, the numerical determination of the progress rate is cumbersome. For the special case of $(1 + 1)$-ES, the calculations could be performed and analytical results obtained. Interestingly, the perceived success probability increases with the noise strength. That is, the offspring increasingly replaces the parent – without being necessarily better. This can be counteracted to some extent with an increase of the mutation strength – albeit only to a certain point. Depending on the normalized noise strength, there is an optimal success probability that guarantees optimal progress. However, this is only possible up to a certain noise limit. The success probabilities obtained lead to the conclusion that the 1/5th-rule with the standard settings should be applied. Instead of lowering the success probability for noise closer to the limit value, resampling and averaging should be applied to decrease the noise strength (Beyer 2001b, p. 99).

Beyer (1993) considered the case that the fitness of the parent is reevaluated every time it is compared to the offspring. This results in more fitness evaluations than usually necessary. Since fitness evaluations are costly, the question arises whether the $(1 + 1)$-ES benefits from reevaluations. Arnold and Beyer (2002a) investigated the local performance of the $(1 + 1)$-ES on the sphere without reevaluation. If the parent is not evaluated anew, the fitness of the parent is systematically overvalued. Accounting for this effect makes the analysis more difficult than before. The degree of overvaluation Ξ of the parent is defined as the difference between the ideal, true fitness, and the perceived fitness. Likewise, ξ denotes the degree of overvaluation of the offspring. For the offspring, ξ is normally distributed. In the case of the parent, selection influences the distribution. Obtaining this distribution proved to be quite difficult; therefore, a Gram–Charlier expansion was used. This also required the determination of several moments of the unknown series. Following Beyer (1995b), self-consistency conditions were imposed, that is, it was assumed that a time-invariant distribution is approached and that the moments do not change over time. Investigations revealed that cutting off the series after the second term, thus, assuming a normal distribution, yields already good results. The remaining task consisted in determining the mean and variance of the distribution of the parent's degree of overvaluation.

The analysis revealed that in the case of the $(1 + 1)$-ES with reevaluation, noise increases the probability of an offspring replacing the parent. Without reevaluation, noise has the opposite effect. As a result, an ES with reevaluation shows negative quality gains for a wide range of the mutation strength once the noise strength is above a limit. Without reevaluation, the quality gain remains positive.

Comparing the two strategies using their respective optimal mutation strength showed that the ES with reevaluation never outperforms the ES without reevaluation. Its performance is even inferior for noise larger than a limit value. This can be traced back to the reduced success probabilities for ESs without reevaluation, which are caused by the systematic overvaluation of the parent. Since the parent is overvaluated, an offspring is only accepted if the perceived fitness gain is relatively large. In this case, the perceived better fitness has a good chance to be actually better. The ES is less biased to accept offspring, which have a truly smaller fitness than the parent. However, overvaluation also means that many offspring with true but small fitness improvements may not be accepted.

The question remained what degree of overvaluation may be useful provided that it could be calibrated. The analysis in Arnold and Beyer (2002a) showed that an ES may profit from occasional reevaluations. The frequency of reevaluation, however, depends on the mutation strength and therefore on the step-size adaptation mechanism. Experiments with the 1/5th-rule revealed that this mechanism is not suited for a $(1 + 1)$-ES without reevaluation, if the noise strengths are larger than a limit: The 1/5th-rule tries to achieve a certain success probability. If the success probabilities are too small, the rule reduces the mutation strength. If the parent is not reevaluated, the target success probability is simply not reachable once the noise is too large. The ES reduces the mutation strength permanently and the search stagnates.

So far the discussion has been on strategies with only one parent. Arnold and Beyer (2001b) considered the performance of the $(\mu/\mu_I, \lambda)$-ES in infinite-dimensional search spaces. The analysis led to an evolution criterion, that is, a maximal noise strength for a nonnegative quality gain, which depends on the population parameters. Keeping the truncation ratio fixed, it can be seen that the effects of any noise strength can be counteracted by increasing the population size. A further result was that the $(\mu/\mu_I, \lambda)$-ES showed an improved performance in comparison with the $(1, \lambda)$-ES. This can be traced back to the genetic repair effect and the resulting larger mutation strengths. Larger mutation strengths decrease the noise-to-signal ratio, which improves the progress.

Arnold and Beyer (2002b) carried the analysis over to finite-dimensional search spaces and derived an asymptotical correct expression for the progress rate (see ❯ *Table 11*). This could be used to give more accurate estimates for the $(\sigma^*, \sigma_\varepsilon^*)$-combinations guaranteeing positive progress. It was found that increasing the number of parents widens the convergence region. The efficiency $\eta = \max_{\sigma^*} \varphi^*(\sigma^*)/\lambda$ is also influenced by the parent number with an optimal efficiency for a truncation ratio of 0.27 (Arnold and Beyer 2002b). In contrast to the finding in Arnold and Beyer (2001b), the efficiency does not increase indefinitely with the number of offspring. Using optimal values for the mutation strength and parent number, it was shown that an optimal number of offspring exists which depends on the noise strength and search space dimensionality.

In the case of the undisturbed sphere, the $(1 + 1)$-ES is the most efficient of the standard types of evolution strategies. Using common multi-parent strategies does not have any advantage unless the algorithm makes use of parallel computation. This changes in the presence of noise. The efficiency of multi-parent strategies exceeds the efficiency of point-based strategies – except for low noise levels. On the one hand, this can be traced back to the genetic repair effect, which is caused by intermediate recombination. On the other hand, the question remains whether evolution strategies without recombination, but with populations, may also achieve a better performance than the $(1 + 1)$-ES. Experiments had already shown that retaining more than the best candidate solution can lead to an improved performance.

In Arnold (2002a) and Arnold and Beyer (2001a, 2003a) the behavior of the (μ, λ)-ES was analyzed in more detail. As in the non-noisy case, the determination of the progress rate or quality gain is a demanding task. The distribution of the apparently μ best offspring has to be derived or approximated. This not only requires the use of noisy order statistics, but also the treatment of the parental states as random variables. The time-variant distributions of succeeding populations under the influence of mutation, noise, and selection have to be taken into account and modeled. Usually, however, as in the non-noisy case, the assumption can be made that a time-invariant limiting distribution is approached. The ES populations are samples drawn from this distribution. Therefore, the moments of the population also approach a limit distribution.

In Arnold (2002a) a first moment-based analysis was presented. In a first step, infinite noise strength was assumed. In this case, selection becomes purely random and can be neglected. This allows for an exact determination of the expectation of central moments. Extending the analysis to the case of finite noise, the determination of the joint distributions of noisy order statistics is necessary, which complicates the calculations. An approximation of the density of the offspring distribution from which the candidates are drawn can be obtained using the Gram–Charlier series. This yields the expectation of the central moments of the population at generation $g + 1$ as a function of this expansion and the moments at generation g. The latter are of course random variables. In Arnold (2002a), a simple approach was followed by ignoring all fluctuations of these terms and considering only their expected values. This leads to a quality gain expression for linear fitness functions which agrees well with the result of experiments.

Arnold and Beyer (2001a, 2003a) analyzed the (μ, λ)-ES on a noisy linear function and the sphere model with respect to whether the use of populations is beneficial and how this comes to pass. Therefore, the (μ, λ)-ES was compared to the $(\mu/\mu_I, \lambda)$-ES. The analysis followed the approach introduced in Beyer (1995b). The distribution of the offspring cannot be determined exactly. Therefore, it was expanded into derivatives of a normal distribution. To this end, the moments had to be obtained. The approach presented in Arnold (2002a) differs from Beyer (1995b). Instead of considering a skewed distribution, it was assumed that the derived random variables are normally distributed. Therefore, only the mean and variance of the population have been calculated using the respective values of the parent population and taking the effect of mutation into account.

The linear function, $f(\mathbf{y}) = \mathbf{a}^{\mathrm{T}}\mathbf{y}$, effectively projects candidate solutions onto a line defined by the direction of $\mathbf{a}$. The problem is, therefore, basically one-dimensional and the quality gain and progress rate can be defined straightforwardly. In analogy to the axial progress rate for ridge functions, the progress rate measures the expected change of the position. First, a general expression for the mean and variance of the offspring population must be derived. Due to the assumption that the distribution of the offspring is normal, a result obtained in Arnold (2002a) for noisy order statistics can be directly applied, leading to the quality gain and the progress rate. Both depend on the progress coefficient

$$c_{\mu/\lambda}(\theta) = \frac{1 + \kappa_2}{\sqrt{1 + \kappa_2 + \theta^2}} \, e_{\mu,\lambda}^{1,0} \tag{121}$$

with $\theta = \sigma_\varepsilon/\sigma$ the noise-to-signal ratio and $\kappa_2 = D^2/\sigma^2$. The parameter D^2 denotes the variance of the parent population. The result can be used to discuss the effects of operating with a population of $\mu > 1$. Without noise, the increase of the population variance with its size μ is counterbalanced by the decrease of the progress coefficient $e_{\mu,\lambda}^{1,0}$. Noise increases the population variance additionally. Furthermore, evolution strategies with populations operate with a different noise-to-signal ratio. While this value is $\theta = \sigma_\varepsilon/\sigma$ for the $(1, \lambda)$-ES, it is $\theta/\sqrt{1 + \kappa_2}$ for the (μ, λ)-ES. That is, (μ, λ)-ESs algorithmically decrease the noise-to-signal ratio.

As for the calculation of the parental population variance and mean, the approach of Beyer (1995b) was applied. This led to fix-point or self-consistent equations to be solved. The result showed that the population variance increases with the size of the population and with the noise. A comparison with experiments showed that the normality assumption for the offspring distribution is usually not sufficient for higher noise strengths and population sizes.

A similar approach can be applied to the sphere model using the decomposition technique for the mutation vectors. The analysis is based on the assumption that the distance to the objective exceeds the population variance. To this end, large search space dimensionalities and population sizes μ, not too large, must be assumed in conjunction with a normal approximation of the offspring distribution. Experiments showed that the resulting progress rate formula is an accurate predictor.

The progress rate can be used to compare several strategies using the efficiency as comparison measure. It was found that for each λ and noise strength σ_ε an optimal μ exists, which is usually between 0.1λ and 0.3λ. Using this optimal μ, the efficiency of the $(\mu_{\mathrm{opt}}, \lambda)$-ES and the $(1, \lambda)$-ES can be compared. The efficiency of the $(\mu_{\mathrm{opt}}, \lambda)$-ES is greater than that of the $(1, \lambda)$-ES except for small noise strengths. Retaining more than one offspring enables the ES to operate with smaller offspring populations in the presence of noise. Considering a $(1 + 1)$-ES showed that the single point strategy is only superior for very small normalized noise strengths and shows worse performance for larger noise strengths.

Regarding the efficiency of the (μ, λ) and the $(\mu/\mu_I, \lambda)$-ES, it can be stated that both strategies are able to operate with a reduced noise-to-signal ratio. The efficiency formulas read

$$\eta_{\mu,\lambda} = \frac{1}{\lambda}\left(\frac{1+\kappa_2}{\sqrt{1+\kappa_2+\theta^2}}\sigma^* e_{\mu,\lambda}^{1,0} - \frac{\sigma^{*2}}{2}\right) \tag{122}$$

$$\eta_{\mu/\mu,\lambda} = \frac{1}{\lambda}\left(\frac{\sigma^* e_{\mu,\lambda}^{1,0}}{\sqrt{1+\theta^2}} - \frac{\sigma^{*2}}{2\mu}\right) \tag{123}$$

The mechanisms by which the improved efficiency is obtained differ: The (μ, λ)-ES reduces the noise-to-signal ratio by an increase of the population variance which enhances the effect of the mutations. The $(\mu/\mu_I, \lambda)$-ES reduces the ratio by operating with higher mutation strengths made possible by means of the genetic repair effect.

Arnold (2006b) investigated the (λ_{opt})-ES on the noisy sphere. This ES-type uses a weighted multi-recombination taking the whole offspring population into account. Optimal weights $w_{k;\lambda}$ are given by the expected value of the $(\lambda + 1 - k)$th order statistic of the standard normal distribution. It was shown that the (λ_{opt})-ES does not benefit from the genetic repair effect. However, a rescaling of the weights $w_{k;\lambda} \to w_{k;\lambda}/\kappa$, with $\kappa > 1$ can be applied. This leaves the optimal quality gain unchanged in the noise-free case (for $N \to \infty$). The optimal mutation strength, however, increases with the scaling factor κ. An increase of the mutation strength lowers the noise-to-signal ratio. Obtaining the quality gain for the (λ_{opt})-ES on the noisy sphere, it could be shown that the ES is capable of maintaining a positive quality change up to a normalized noise strength of $\sigma_\varepsilon^* = 2\kappa$. The analysis indicated that large values of κ should be preferred.

Arnold and Beyer (2003b) analyzed the effects of outliers on the outcome of optimization. To simulate a noise distribution with frequently occurring outliers, the Cauchy distribution was chosen. The analysis was extended in Arnold and Beyer (2006a). Three noise distributions were compared for the $(\mu/\mu_I, \lambda)$-ES: normally distributed noise, Cauchy noise, and χ_1^2

distributed noise. In the latter case, two different cases were considered: adding the noise term to the fitness function, which results in a random variable with a mean greater than the actual value, and subtracting the noise term leading to a random variable with a smaller mean.

One aim was to investigate whether the assumption of a Gaussian noise model is too restrictive to allow a transfer of the obtained results to other situations. For each model, the first task was to derive the progress rate. The results can be used to investigate:

- The effects of varying noise levels
- The influence of population sizing, the truncation ratio, and the number of offspring
- The efficiency of reevaluation/resampling of search points compared to upgrading offspring population sizes

and to compare the results with respect to transferability from one noise model to another.

As shown in Arnold and Beyer (2006a), Cauchy noise leads to similar qualitative responses of the ES as Gaussian noise. It should be noted that in the case of Cauchy noise, the role of the noise strength is different. Instead of being the standard deviation of the noise, it should be regarded as a scale parameter. (Actually, this interpretation can also be applied to the Gaussian noise case, thus, allowing for a direct comparison of the performances.) Therefore, a comparison of the scaling behavior has been made (Arnold and Beyer 2006a). As for Gaussian noise, progress is only possible up to a maximal normalized noise strength. This quantity depends on the truncation ratio and on the size of the offspring population. Strategies with a ratio of 0.5 achieve the best results. Increasing offspring population sizes appears to increase the maximal admissible noise strength. To achieve optimal performance, the truncation ratio has to be variated with respect to the noise level starting with 0.27 for zero noise and going up to 0.5 for the upper noise limit. The difference between Gaussian and Cauchy noise lies only in the shape of the response curves.

The question remains for which ranges of the truncation ratios positive progress occurs. For both distributions, the range narrows with increasing noise levels to smaller and smaller intervals, which encompass 0.5 (see ❯ *Fig. 10a, b*). The similarities between the two noise models transfers to the question whether resampling is more efficient than using an ES with a larger offspring population. For both noise models, this can be negated. Comparing ESs with the same computational costs and varying resampling sizes showed that larger populations have to be preferred. The difference between the approaches is more pronounced for Cauchy noise, however (Arnold and Beyer 2006a). For Gaussian noise, the ES benefits from averaging over the reevaluations of search points since the variance of the noise is reduced. For Cauchy noise, averaging has no effect since the resulting distribution is exactly the same.

One major difference between Gaussian and Cauchy noise exists. This is caused by a different response toward vanishing noise-to-signal ratios. In the case of Gaussian noise, the effects of the noise may be nearly eliminated, if the population size and the mutation strength are sufficiently large. In the case of Cauchy noise, this does not hold anymore. Increasing the population sizes will not allow the strategy to operate as if noise were not present.

Noise with a χ_1^2-distribution leads to two possible cases: adding (positively biased noise) or subtracting the random term (negatively biased noise). The ES accordingly responds in different ways. First of all, increasing the offspring population size still increases the limit noise strength. However, the first difference appears in the optimal truncation ratio and its response to varying noise levels (see ❯ *Fig. 10c*). For neither case of χ_1^2-distributed noise, does the truncation ratio approach 0.5. As for Cauchy and Gaussian noise, increasing noise narrows the band of truncation ratios for which positive progress is possible. Unlike the previous cases, however,

▣ Fig. 10
The progress rate $\hat{\varphi}^*$ for optimal mutation strengths as a function of the truncation ratio α (Arnold and Beyer 2006a). The y-axis is scaled with the maximal progress rate $\hat{\varphi}_0^*$ obtained for the noise-free case and a truncation ratio of 0. 27. Shown are the curves for three different noise strength σ_ε^* to limit noise strength $\bar{\sigma}_\varepsilon^*$ ratios. Figure (a) depicts the case of Gaussian noise, (b) the case of Cauchy noise, and (c) shows the case of a χ_1^2 distributed noise.

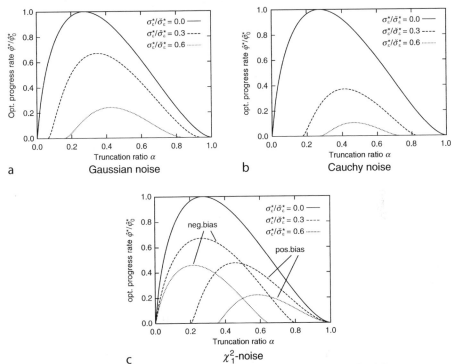

only one boundary of the interval moves: Positively biased noise extends upward to a ratio of 1.0, whereas negatively biased noise stretches downward to zero (Arnold and Beyer 2006a).

Comparing the efficiency of strategies with resampling and strategies with larger population sizes leads to the same finding as before: An increased population size is preferable. Noise with a χ_1^2-distribution has a similarity with Gaussian noise: The effects of noise can be eliminated with vanishing noise-to-signal ratio.

General Quadratic Functions and Other Functions
In Beyer and Meyer-Nieberg (2004), the quality gain of the $(1, \lambda)$-ES was derived under the assumption of additive, normally distributed noise. The approach used the concept of noisy order statistics and a Gram–Charlier expansion for the pdf of the fitness values. The result

$$\overline{\Delta Q_{1,\lambda}} = \frac{S_Q^2}{\sqrt{S_Q^2 + \sigma_\varepsilon^2}} c_{1,\lambda} + M_Q \tag{124}$$

with mean M_Q and standard deviation S_Q of the local quality change, that is, of the fitness change induced by a mutation – strongly resembles the progress rate equation for the noisy sphere. Instead of the mutation strength, the standard deviation of the local quality change appears. Noise only influences the first part in ❷ Eq. 124 and not the mean M_Q. A sufficient evolution criterion is obtained by demanding that $\overline{\Delta Q_{1,\lambda}} \geq 0$, leading to $S_Q^2 / \sqrt{S_Q^2 + \sigma_\varepsilon^2} c_{1,\lambda} \geq -M_Q$ (Beyer and Meyer-Nieberg 2004).

The remaining task for this kind of analysis is to determine the mean and standard deviation for the respective fitness function. This has been done for general quadratic, biquadratic, the L_1-norm, and – to test the ranges of the applicability – for the bit-counting function OneMax (see ❷ Table 12). For all these functions, the mean of the fitness corresponds with a negative, nonlinear loss term.

Ridge Functions

As in the undisturbed case, the ES is faced with two subgoals in the optimization process. Accordingly, there are two variables that have to be considered: The position on the y_1 axis and the distance R to the y_1 axis. In the long run, the latter (usually) approaches a stationary distribution, whereas the y_1-evolution continues to progress. The question remains how noise influences the stationary distribution of the distance and what the consequences are for the axial progress. The first question is similar to determining the localization error for the sphere model. The result, however, determines the axial progress, that is, the performance of the ES.

◩ **Table 12**

Quality change formulae for different fitness environments. Explanation of variables and notation: $c_{\mu/\mu,\lambda}$ progress coefficient ❷ Eq. 132, see Beyer (2001b, p. 172), σ^* normalized mutation strength, σ_ε^* standard deviation of noise distribution (normalized), higher order progress coefficients $d_{1,\lambda}^{(k)}$ (❷ Eq. 129) or see Beyer (2001b p. 119), S_Q and M_Q standard deviation and mean of the mutation-induced fitness change Q, κ_k kth cumulant. The normalizations are as follows: (a) sphere $\sigma^* = \sigma N/R$, $\sigma_\varepsilon^* = \sigma_\varepsilon N/R$, $\overline{\Delta Q^*} = \overline{\Delta Q} N/(R g'(R))$ (b) quadratic functions $\sigma^* = \sigma \mathrm{Tr}[\mathbf{Q}]/\|\mathbf{Q}(\mathbf{y} - \hat{\mathbf{y}})\|$, $\sigma_\varepsilon^* = \sigma_\varepsilon \mathrm{Tr}[\mathbf{Q}]/(2\|\mathbf{Q}(\mathbf{y} - \hat{\mathbf{y}})\|^2)$, $\overline{\Delta Q^*} = \overline{\Delta Q}/(2\|\mathbf{Q}(\mathbf{y} - \hat{\mathbf{y}})\|^2)$ (c) biquadratic functions $\sigma^* = \sigma \sqrt[3]{3 \sum_i c_i / \|\mathbf{a}\|^4}$, $\sigma_\varepsilon^* = \sigma_\varepsilon \sqrt[3]{3 \sum_i c_i / \|\mathbf{a}\|^4}$, $\overline{\Delta Q^*} = \overline{\Delta Q} \sqrt[3]{3 \sum_i c_i / \|\mathbf{a}\|^4}$, and (d) OneMax $\sigma^* = \sqrt{l p_m} 2(2F_0/l - 1)$, $\sigma_\varepsilon^* = \sigma_\varepsilon 2(2F_0/l - 1)$, $\overline{\Delta Q^*} = \overline{\Delta Q} 2(2F_0/l - 1)$ with F_0 the fitness of the parent, p_m the bit-flipping probability, and l the string length

Fitness	ES	Quality gain
General functions	$(1, \lambda)$-ES	$\overline{\Delta Q_{1,\lambda}} = \dfrac{c_{1,\lambda} S_Q}{\sqrt{S_Q^2 + \sigma_\varepsilon^2}} + M_Q$
Quadratic functions	$(1, \lambda)$-ES	$\overline{\Delta Q^*} = \dfrac{c_{1,\lambda} \sigma^*}{\sqrt{\sigma^{*2} + \sigma_\varepsilon^{*2}}} - \dfrac{\sigma^{*2}}{2}$
Biquadratic functions	$(1, \lambda)$-ES	$\overline{\Delta Q^*} = \dfrac{c_{1,\lambda} \sigma^*}{\sqrt{\sigma^{*2} + \sigma_\varepsilon^{*2}}} - \sigma^{*4}$
OneMax	$(1, \lambda)$-ES	$\overline{\Delta Q^*} = \dfrac{c_{1,\lambda} \sigma^{*2}}{\sqrt{\sigma^{*2} + \sigma_\varepsilon^{*2}}} - \dfrac{\sigma^{*2}}{2}$
Sphere	$(\mu/\mu, \lambda)$-ES	$\overline{\Delta Q^*} = \dfrac{c_{\mu/\mu,\lambda} \sigma^{*2}}{\sqrt{\sigma^{*2} + \sigma_\varepsilon^{*2}}} - \dfrac{\sigma^{*2}}{2\mu}$

Arnold and Beyer (2008) analyzed evolution strategies on the noisy parabolic ridge. The noise model considered was an additive, normally distributed noise model with different functional dependencies of the noise strength: constant noise strength, noise strength scaling quadratically with the distance, and noise scaling cubically with the distance.

Using the same decomposition of the progress vector as in the non-noisy case, the axial progress rate could be obtained (see ❷ *Table 13*). Instead of determining the radial progress rate, the expected change of the square of the distance was determined leading to an expression for the standardized distance to the ridge which is a polynomial of order four and influenced by the noise strength and the mutation strength.

Let one first consider the case of constant noise strength. In this case, increasing the mutation strength increases the performance as in the non-noisy case. Large mutation strengths achieve a better noise-to-signal ratio $\theta = \sigma_\varepsilon/\sigma$. Increasing the mutation strength eventually eliminates virtually the influence of noise on the progress. Furthermore, the ES achieves a stationary distance of $R_\infty = \sigma N/(2\mu c_{\mu/\mu,\lambda})$ and accordingly an axial progress rate of nearly $\varphi_{\max} = \mu c_{\mu/\mu,\lambda}^2/(dN)$. By increasing μ and λ, the location error to the axis can be decreased – if σ is constant. This is coupled with an increase of the performance. As on the sphere, it is optimal to choose a truncation ratio of $\mu/\lambda = 0.27$, since this maximizes $\mu c_{\mu/\mu,\lambda}^2$ (Arnold and Beyer 2008).

For quadratic noise strengths, an increase of the mutation strength is also beneficial and connected with an increased performance. While for the constant noise case the increase of the mutation strength eventually enabled the ES to reach nearly the same progress as in the undisturbed case, this is not the case for quadratic noise strengths. Increasing the mutation strength also increases the distance to the axis. Since the noise scales quadratically, the noise-to-signal ratio cannot be driven to zero by an increase of the mutation strength. Instead, a positive limit value is reached and the progress rate approaches a saturation value. The maximal achievable progress rate decreases with the noise. Furthermore, positive progress is only possible up to a certain noise level $\mu c_{\mu/\mu,\lambda}/(dN)$, with d the scaling parameter of the squared distance in ❷ Eq. 76. The noise level depends on the ES.

The case of cubic noise strengths is more involved. The stationary condition leads to two potential stationary solutions. For small mutation strengths, the smaller solution represents a stable stationary point, whereas the larger is unstable. Increasing the mutation strength increases the stable solution until it finally coincides with the unstable point. Thus, for

❑ Table 13

Asymptotical axial progress rates for the noisy parabolic ridge in the stationary case. The normalizations read $\varphi_{y_1}^* := \varphi_{y_1} dN/(\mu c_{\mu/\mu,\lambda}^2)$, $\sigma^* := \sigma dN/(\mu c_{\mu/\mu,\lambda})$, and $\sigma_\varepsilon^* := \sigma_\varepsilon dN/(\mu c_{\mu/\mu,\lambda})$ for constant noise. In the case of quadratic noise, $\sigma_\varepsilon^* = \zeta\rho^2$, $\zeta^* := \zeta dN/(\mu c_{\mu/\mu,\lambda})$ with ρ the standardized distance to the axis $\rho := 2Rd$

Noisy type	Progress rate
Constant noise	$\varphi_{y_1}^* = \dfrac{\sigma^{*2}}{\frac{\sigma^{*2}}{2} + \sqrt{\frac{\sigma^{*4}}{4} + \sigma^{*2} + \sigma_\varepsilon^{*2}}}$
Quadratic noise	$\varphi_{y_1}^* = \dfrac{\sigma^*(1 - \zeta^{*2})}{\frac{\sigma^{*2}}{2} + \sqrt{\frac{\sigma^{*4}}{4} + 1 - \zeta^{*2}}}$

mutation strengths above a noise-dependent limit value, no (stable) stationary solution exists. The ES diverges from the axis. The case of cubic noise is interesting, since it results in a new behavior of the noise-to-signal ratio. Here, increasing the mutation strength increases the ratio indefinitely. The influence of noise cannot be counteracted with an increase of the mutations strength as usual, but only with a decrease: the larger the noise, the smaller the mutation strength has to be. Depending on the noise, an increase of the mutation strength is only beneficial up to a certain limit. This limit decreases with the noise.

2.4.3 Residual Location Error and Fitness Error

Noise does not only influence the local performance of an ES but also the final solution quality. Provided that the noise strength is constant, the ES is unable to locate the optimizer. Instead, it typically converges to a time-invariant limit distribution. On an average, a residual location error, the expected distance to the optimizer and a residual fitness error, the expected difference between outcome and the optimum, are observed. In the following, how the local progress measures may be used to derive these values is described. The expressions can then be used to derive recommendations for population sizing with respect to the solution quality. As it will be shown, optimal settings with respect to the solution quality usually correspond to nonoptimal progress. Therefore, a compromise between convergence speed and solution quality must be made.

Sphere Model

The analyses for the sphere model aim to derive lower bounds for the residual location and fitness error and to investigate the influence of the truncation ratio. An important tool is the evolution criterion gained by the progress rate which gives the maximal noise strength (normalized). The limit noise strength corresponds with zero progress and therefore with a stationary state of the evolution of the distance. Undoing the normalization leads to an inequality in which the distance appears as the unknown. Therefore, solving the distance gives a lower bound for the residual location error.

In Arnold and Beyer (2002b), the simple sphere $f(R) = \beta R^\alpha$ was considered yielding lower bounds for the residual location error (see ❷ Table 14). The lower bound increases linearly with the noise strength and search space dimensionality. Further influences are the parent and offspring population sizes. Minimal values are obtained for truncation ratios of 0.5. The lower

❑ Table 14

Noise and residual location errors for the sphere. See ❷ Table 11 for the definition of the noise type. The progress coefficient is given by ❷ Eq. 132

Fitness function	Noise type	Residual location error
Sphere	Systematic	$R_\infty \geq \dfrac{N\sigma_\varepsilon}{\sqrt{8}\mu c_{\mu/\mu,\lambda}}\sqrt{1+\sqrt{1+\dfrac{8\mu^2 c_{\mu/\mu,\lambda}^2}{N}}}$
Sphere	Standard	$R_\infty = \sqrt{\dfrac{N\sigma_\varepsilon}{4\mu c_{\mu/\mu,\lambda}}}$

bound approximates the residual location error very closely. This is caused by the step-size adaptation mechanisms which operate with comparatively low mutation strengths as experiments showed.

General Quadratic Fitness Model

The analysis of general quadratic fitness models (❷ 73) is more difficult than the simple sphere model. While the residual location error can be obtained rather easily if the function is of the form $f(R) = \beta R^{\alpha}$, the general model requires more effort. Usually, the residual fitness error is derived instead of the location error. One approach, followed in Beyer and Arnold (1999, 2003a), is analogous to the sphere model case. The same evolution criterion is used substituting the radius with the mean radius to be obtained using differential geometry methods. This leads finally to a stationarity condition for general quadratic fitness functions. Switching to the eigen space of the **Q**-matrix simplifies the calculations.

In Beyer and Arnold (2003a) steady state conditions for the $(\mu/\mu_I, \lambda)$-ES were derived (see ❷ *Table 15*). Using the sphere model evolution criterion for this ES (Arnold and Beyer 2002b), an inequality for the square of the transformed location error $\|\mathbf{Q}(\mathbf{y} - \hat{\mathbf{y}})\|$ can be derived. Taking the expectation of the transformed location error leads to an expression that can be interpreted as the variance $\mathrm{E}[(y_i - \hat{y}_i)^2]$ of the object vector component y_i in the eigen space of **Q**, centered about the optimizer $\hat{y}_i$. Using the *equipartition assumption* which states that in expectation, every weighted component $q_i(y_i - \hat{y}_i)^2$ contributes the same amount to the fitness error, a lower bound for the fitness error could be derived. The equipartition assumption is similar to the equipartition assumption in statistical thermodynamics. Up to now, it has not been proven formally, but the following plausibility argument can be made (Beyer and Arnold 2003a): The ES is in the steady state. The mutations which generate the object vector components are not directed. Selection is not influenced by a single component since it only experiences the fitness as a whole and thus does not prefer a single weighted component. If one component dominated the rest, the corresponding individual would go extinct since this results in lower fitness values. The components fluctuate independently around the optimizer.

◻ Table 15

Noise and residual fitness errors for some fitness functions. The trace of the matrix **Q** is defined as $\mathrm{Tr}[\mathbf{Q}] = \sum_{i=1}^{N}(\mathbf{Q})_{ii}$, where $(\mathbf{Q})_{ii}$ are the diagonal elements of **Q**

Fitness function	Noise type	Residual fitness error
Biquadratic functions	Standard	$\mathrm{E}[\Delta F] \geq \dfrac{3\sigma_{\varepsilon}N}{40\mu c_{\mu/\mu,\lambda}}$
L_1-norm	Standard	$\mathrm{E}[\Delta F] \geq \dfrac{\sigma_{\varepsilon}N}{\pi\mu c_{\mu/\mu,\lambda}}$
General quadratic functions	Standard	$\mathrm{E}[\Delta F] \geq \dfrac{\sigma_{\varepsilon}N}{4\mu c_{\mu/\mu,\lambda}}$
General quadratic functions	Actuator	$\mathrm{E}[\Delta F] \geq \sigma_{\varepsilon}^2 \mathrm{Tr}[\mathbf{Q}] + \dfrac{N\sigma_{\varepsilon}^2 \mathrm{Tr}[\mathbf{Q}]}{8\mu^2 c_{\mu/\mu,\lambda}^2} \times \left(1 + \sqrt{1 + \dfrac{8\mu^2 c_{\mu/\mu,\lambda}^2 \mathrm{Tr}[\mathbf{Q}^2]}{\mathrm{Tr}[\mathbf{Q}]^2}}\right)$

For this reason, selection does not prefer any particular search direction. If there were such a preference, the ES would still exhibit a directed movement and would not have arrived at the steady state.

The lower bound obtained by the equipartition assumption depends linearly on the noise strength, on the search space dimensionality, and on the population parameters (see ❯ *Table 15*). Interestingly, there is no influence of the matrix **Q**. That is, the result holds for all ellipsoidal quadratic success domains. The agreement of the lower bound with experimentally determined residual fitness errors is quite good (Beyer and Arnold 2003a).

Other Fitness Models

Beyer and Meyer-Nieberg (2005) considered the residual fitness error of the biquadratic functions, and the L_1-norm. The expressions were obtained for the $(1, \lambda)$-ES. With a simple analogy argument, the results were transferred to the $(\mu/\mu_I, \lambda)$-ES which enabled a determination of the optimal truncation ratio.

The quality gain obtained in Beyer and Meyer-Nieberg (2004) can be used to derive a necessary evolution criterion $S_F^2 \geq \sigma_\varepsilon |M_F|/c_{1,\lambda}$. Under certain conditions, this lower bound can be assumed to be sharp. To obtain the fitness error, the mean and the standard deviation of the fitness value must be determined. For the fitness functions considered, the stationary state criterion is a nonlinear expression of the object vector components and the mutation strength. The determination of the fitness error requires some simplifications and assumptions, especially the equipartition assumption. All the function classes considered have a final fitness error of the form $E[\Delta F] = \sigma_\varepsilon N/(Cc_{1,\lambda})$, with C depending on the fitness function. Substituting $c_{1,\lambda}$ with $\mu c_{\mu/\mu,\lambda}$ gives the result for the intermediate ES. Owing to the form of the fitness error, the results obtained for the sphere can be transferred to biquadratic functions and the L_1-norm: A truncation ratio of $1/2$ is optimal with respect to the fitness error. Since working with $\mu/\lambda = 0.27$ does not significantly degrade the solution quality, this truncation ratio can still be used.

2.4.4 General and Systematic Noise

In the case of general noise, the fitness function itself is a random function F_ε which makes a general treatment difficult. In Beyer et al. (2003), however, the authors present a decomposition technique. Instead of using the exact random function, an approximate model is utilized. The random function F_ε is decomposed into two parts: a deterministic part, that is, the expectation $E[F_\varepsilon|\mathbf{y}]$, and a stochastic part. The stochastic part is expanded into an Edgeworth or Gram–Charlier series. In a first approximation, this leads to a normal distribution

$$F_a(\mathbf{y}) = E[F_\varepsilon|\mathbf{y}] + \mathcal{N}(0, \text{Var}[F_\varepsilon|\mathbf{y}]) \tag{125}$$

Thus, on the one hand, the conditional expectation and variance of F_ε given the state $\mathbf{y}$ have to be obtained. On the other hand, it is possible to apply the analysis techniques developed for the additive noise model.

Systematic noise was considered under the following aspects:

1. Beyer et al. (2003) introduced the decomposition approach.
2. Sendhoff et al. (2002) analyzed functions with noise-induced multi-modality (FNIMs).
3. Beyer et al. (2004) considered systematic noise on the quadratic sphere.

4. Beyer (2004) addressed actuator noise on general quadratic functions.
5. Beyer and Sendhoff (2006) introduced and analyzed a class of test functions for evolutionary robust optimization.

Beyer et al. (2003) considered the behavior of the $(\mu/\mu_I, \lambda)$-ES in two uncertain environment types, the sphere with actuator noise and a mixed noise model

$$F(\mathbf{y}) = a - (z + 1)\|\mathbf{y}\|^\alpha + bz \quad \text{with} \quad z \sim \mathcal{N}(0, \sigma_\varepsilon^2) \tag{126}$$

The task in this case is to maximize the fitness values. Two classes of probabilistic performance criteria, threshold measures (❱ Eq. 117) and statistical momentum measures (❱ Eq. 118), were applied in the analysis. In the case of the sphere model, it followed that the robust optimizer with respect to both measures equals $\mathbf{y} = \mathbf{0}$. A lower bound for the residual location error was obtained with a quite accurate predictive power. Again, the minimal location error is obtained for a truncation ratio of $1/2$. The standard recommendation $\mu / \lambda = 0.27$ still applies as a reasonable setting since the solution quality is only slightly degraded and the convergence velocity to the steady state is larger.

A further point of the analysis concerned the online fitness behavior and the robustness with respect to the statistical momentum measures. In the stationary state, the fitness values fluctuate around a mean value. In Beyer et al. (2003), the mean and the standard deviation of the fitness values were obtained. Using an *ansatz* for the variance of a single component of the centroid in object parameter space, it was possible to derive estimates for the expected value and standard deviation. Again, an ES with a truncation ratio of $1/2$ achieves the smallest mean and standard deviation.

Concerning ❱ Eq. 126, the analysis showed that threshold measures are not suitable as conflicting information from a distance increase or decrease. Consequently, there is no unique value of an optimal distance. Therefore, only momentum measures should be taken into account. However, this approach also leads to ambiguities. While the first moment can be decreased by a decrease in the distance, this results in an increased variance. The variance has a minimal value for $R^\alpha = b$. The analysis revealed that the ES shows a bifurcation behavior depending on the size of

$$\frac{2\alpha\mu c_{\mu/\mu,\lambda}}{\sigma_\varepsilon N} \tag{127}$$

If ❱ Eq. 127 is larger than one, only one stationary point exists and the ES shows a stable behavior with a well-defined lower bound for the residual location error. For values of ❱ Eq. 127 smaller than one, the system switches from having only one steady state to two steady states. Whether the ES converges or diverges depends on the initial distance. If the ES does not diverge, it compromises, in a sense, between a fitness maximal distance and a variance minimal distance: The residual location error lying somewhere in between and the population parameters can be used to bias the ES toward either of the extremes.

In Beyer et al. (2004), the influence of systematic noise on the performance and steady state behavior of ESs on the quadratic sphere was considered. The analysis was focused on the following questions:

1. Efficiency.
2. What are sufficient evolution criteria considering the dynamical behavior of the ES? That is, under which conditions do the ES *not* diverge?

3. What does this imply for σ control rules? And are there differences between CSA and σSA control?

Using the decomposition technique from above, it is possible to derive the progress rate showing close resemblance to the case of additive noise.

The progress rate obtained, the question of efficiency defined by $\eta = \max_\sigma \varphi(\sigma)/\lambda$ could be discussed. However, due to the form of the progress rate, the maximal efficiency can only be determined for cases where the noise strength scales with the distance to the optimizer. This special case leads to the recommendation of a truncation ratio $\mu/\lambda = 0.27$.

Considering general noise again, the progress rate allows for a derivation of a sufficient evolution criterion which leads to two main findings:

1. To increase the convergence probability in cases with high noise levels, any σ control mechanism should produce small step sizes. This, however, comes at the cost of reduced convergence speed. Again, there is a trade-off between convergence reliability and convergence speed.
2. Problems with higher search space dimensionality are evolutionarily more stable (Beyer et al. 2004).

The analysis then considered the case of constant noise strength. A steady state, that is zero progress, is attained if the equal sign in the evolution criterion is exactly fulfilled. As investigated empirically, both CSA and σSA eventually steer the ES toward the zero progress region. The criterion of zero progress can then be used to derive lower bounds for the residual location error yielding two results: First, the residual location error scales linearly with the noise strength. Second, recommendations concerning μ and λ can be given. It is optimal to have a truncation ratio of $\mu/\lambda = 1/2$. The region around the optimal truncation ratio is rather flat, allowing the use of smaller ratios with better efficiency. Furthermore, keeping the truncation ratio fixed, increasing the population size decreases the location error.

In Beyer (2004), the effects of actuator noise on the stationary state behavior of intermediate evolution strategies were considered using the general quadratic fitness model ($\bullet$ 73). The analysis showed that the expected deviation from the actual optimum consists of two parts: one part which is independent of the ES and depends only on the function and on the noise strength, and one part that can be influenced by the ES. Using a truncation ratio of $1/2$ yields a minimal deviation from the actual optimum. This ratio is valid as long as the assumption of (at least asymptotical) normality is guaranteed.

In Beyer and Sendhoff (2006), a new test function class, functions with noise-induced multi-modality (FNIMs), was introduced and analyzed. The criteria for the analysis were moment-based measures and threshold measures. The authors succeeded in developing a test function that captures the important properties of general FNIMs while being amenable to a theoretical analysis. The steady state performance of the $(\mu/\mu_I, \lambda)$-ES was investigated yielding the following findings:

- The resampling size should be kept as small as possible.
- A truncation ratio of about 0.3 is a good compromise between convergence speed and solution quality.
- Comparing σSA and CSA showed that each mechanism has its advantages and disadvantages: While using CSA leads to a faster convergence, σSA provides a better exploration behavior in the stationary state, albeit, with the risk of oscillatory behavior.

2.4.5 Noise and Step-Size Adaptation

This section considers the behavior of step-size adaptation mechanisms in the case of noisy fitness evaluations. Two mechanisms were primarily considered, self-adaptation and cumulative step-size adaptation. First, the results for self-adaptation are described.

Self-Adaptation
In Beyer and Meyer-Nieberg (2006a), the self-adaptive $(1, \lambda)$-ES on the noisy sphere was considered. The first task was to determine the self-adaptation response (see ❯ *Table 18*). Afterwards, the deterministic evolution equations were used in the analysis. The evolution equations model the dynamical behavior of the normalized mutation strength, the distance to the optimizer, and the normalized noise strength. It can be shown that the evolution of the normalized mutation strength and noise strength do not depend on the distance R. Therefore, the analysis can be restricted to the two-dimensional system of σ^* and σ_ε^*.

Three phases of the evolution can be distinguished for constant noise strength σ_ε.

(a) A first phase, characterized by large distances to the optimizer in which the ES nearly behaves as if the fitness evaluations were exact.
(b) A transient phase, in which the ES is more and more influenced by the noise as it moves toward the optimizer.
(c) A stationary state in which no progress on average occurs. In this case, the dynamics of the mutation strength σ^* and σ_ε^* approach stationary distributions.

The progress rate and the self-adaptation response can be used to determine the $(\sigma^*, \sigma_\varepsilon^*)$-combinations, which equal zero progress and zero expected change of the mutation strength. This leads to expressions for the stationary mutation strength, noise strength, and the residual location error, displayed in ❯ *Table 16*. In contrast to the result obtained in Beyer (1993), the nonzero mutation strength introduces a correction factor which increases the residual location error. However, the deviation is not large, that is, self-adaptation does not lead to a significant degradation of the solution quality.

There is a peculiarity in the behavior of the $(1, \lambda)$-ES: if the noise strength becomes too large, the ES loses step-size control. The mutation strength exhibits a random walk-like behavior biased toward small mutation strengths. This behavior cannot be predicted by the deterministic approach used in Beyer and Meyer-Nieberg (2006a). In spite of that, an explanation for this behavior can be given by considering the probability of a decrease in the mutation strength under random selection. This probability equals $1/2$. Selection changes the situation. The $(1, \lambda)$-ES accepts a decrease of the mutation strength with high probability.

◻ **Table 16**
The stationary state values of the $(1, \lambda)$-ES with σ-self-adaptation on the sphere $g(R) := -cR^\alpha$. The normalizations read $\sigma^* := \sigma N/R$ and $\sigma_\varepsilon^* := \sigma_\varepsilon N/(c\alpha R^\alpha)$

Mutation strength	Noise strength	Location error
$\sigma^* = \dfrac{2c_{1,\lambda}}{\sqrt{2\left(2c_{1,\lambda}^2+1-d_{1,\lambda}^{(2)}\right)}}$	$\sigma_\varepsilon^* = 2c_{1,\lambda}\sqrt{1 - \dfrac{1}{2\left(2c_{1,\lambda}^2+1-d_{1,\lambda}^{(2)}\right)}}$	$R = \sqrt[\alpha]{\dfrac{\sigma_\varepsilon N}{c\alpha 2c_{1,\lambda}}\sqrt{\dfrac{2\left(2c_{1,\lambda}^2+1-d_{1,\lambda}^{(2)}\right)}{2\left(2c_{1,\lambda}^2+1-d_{1,\lambda}^{(2)}\right)-1}}}$

Large mutation strengths coincide more often with a selective disadvantage and are punished on average. This results in a tendency of the $(1, \lambda)$-ES toward smaller mutation strengths: The ES performs some kind of biased random walk.

The aforementioned deficiency of $(1, \lambda)$-ESs leads to the question whether intermediate recombination ESs exhibit similar erratic behaviors. Interestingly, this is not the case. These strategies are biased (in probability) toward an increase of the mutation strength resulting in a more stable σ-adaptation behavior.

In Meyer-Nieberg and Beyer (2008) the performance of the $(\mu/\mu_I, \lambda)$-ES on the noisy sharp ridge was investigated (see ❯ Tables 17 and ❯ 18). The analysis provided an unexpected result: on the sharp ridge, noise can increase the performance. In order to understand this interesting property, one has to recall that the optimization consists of two subgoals:

1. Reducing the distance to the axis, that is, optimizing the embedded sphere model
2. Enlarging the linear axis component

Given a fixed strategy, the ridge parameter d decides which subgoal dominates the other. If d is large, the ES tries to optimize the sphere model which ultimately results in stagnation, since the mutation strength is decreased accordingly. Additive noise prevents the ES from achieving this subgoal. It attains a positive stationary $\sigma > 0$ and a "residual location error" (fluctuating at a certain distance from the ridge axis). Due to the positive mutation strength, the ES can serve the second subgoal: the enlargement of the linear part. Putting it another way, larger noise strengths result in worse achievements of the sphere subgoal which correspond to greater axial progress. Axial progress, mutation strength, and distance scale linearly with the noise. A remark concerning recombination is to be added. Recombination reduces the residual location error. This goes along with reduced mutation strengths and axial progress rates.

■ Table 17
The stationary state values of the $(\mu/\mu_I, \lambda)$-ES with σ-self-adaptation on the sharp ridge. Shown are the approximate values obtained for large population sizes λ and appropriate choices of μ. The normalizations read $\sigma^* := \sigma N/R$ and $\varphi^* := \varphi N/R$

Mutation strength	Distance	Progress
$\sigma^* = \sqrt{\mu}$	$R = \dfrac{\sigma_\varepsilon N}{2d\mu c_{\mu/\mu,\lambda}}$	$\varphi^* = \dfrac{1}{2d}$

■ Table 18
Self-adaptation response functions (SARs) for different fitness environments. Explanation of variables and notation: $c_{\mu/\mu,\lambda}$ progress coefficient (see ❯ Sect. 5), σ^* normalized mutation strength (with respect to distance to optimizer and search space dimensionality), σ_ε^* standard deviation of noise distribution (normalized), d and α parameters of the ridge function

Fitness	SAR
Sphere	$\psi(\sigma^*) = \tau^2 \left(\dfrac{1}{2} + \dfrac{e_{\mu,\lambda}^{1,1}\sigma^{*2}}{\sigma^{*2} + \sigma_\varepsilon^{*2}} - \dfrac{c_{\mu/\mu,\lambda}\sigma^{*2}}{\sqrt{\sigma^{*2} + \sigma_\varepsilon^{*2}}} \right)$
Ridge	$\psi(\sigma^*) = \tau^2 \left(\dfrac{1}{2} + \dfrac{e_{\mu,\lambda}^{1,1}(1 + \alpha^2 d^2 R^{2\alpha-2})\sigma^{*2}}{(1 + \alpha^2 d^2 R^{2\alpha-2})\sigma^{*2} + \sigma_\varepsilon^{*2}} - \dfrac{\alpha dR^{\alpha-1}c_{\mu/\mu,\lambda}\sigma^{*2}}{\sqrt{(1 + \alpha^2 d^2 R^{2\alpha-2})\sigma^{*2} + \sigma_\varepsilon^{*2}}} \right)$

However, recombination is necessary, since the $(1, \lambda)$-ES experiences a loss of step-size control. Therefore, the use of small truncation ratios $> 1/\lambda$ is recommended.

Cumulative Step-Size Adaptation (CSA)

Arnold and Beyer (2000, 2004, 2008) and Beyer and Arnold (2003b) analyzed the performance of the $(\mu/\mu, \lambda)$-ES with cumulative step-size adaptation (CSA). As far as the noise-free case is concerned, the reader is referred to ❷ Sect. 2.3.4. The work done focused on a deeper discussion of the underlying philosophy of CSA, progress rate optimality, and population sizing. Several fitness environments were considered, starting with the sphere (Beyer and Arnold 2003b; Arnold and Beyer 2004) and continuing to ridge functions Arnold and Beyer (2008).

This paragraph starts with the sphere model and the additive Gaussian noise model. In Arnold and Beyer (2000) a first analysis of the performance of adaptive $(\mu/\mu, \lambda)$-ESs was provided. Two major step-size adaptation mechanisms were considered: self-adaptive ESs (with arithmetic and geometric recombination of the mutation strength) and ESs with cumulative step-size adaptation on the noisy sphere. The noise strength was set to scale with the distance to the optimizer, that is, the normalized noise strength is constant. Demanding positiveness of the progress rate yields the necessary evolution criterion, that is, a criterion that must be fulfilled to have convergence to the optimizer at all. As a consequence, there is a maximal normalized noise strength above which the strategy diverges. This maximal noise strength depends on the population parameters and the fitness function. Experiments showed that the adaptation mechanisms encounter problems when noise comes into play. In the absence of noise, the algorithms achieve linear convergence order. For lower levels of noise, nearly optimal mutation strengths are generated. Increasing the noise level degrades the performance. For high noise-levels, the CSA generates decreasing normalized mutation strengths which first results in slower convergence and finally stagnation, if the noise-levels become too high. In contrast to the CSA-ES, the self-adaptive $(\mu/\mu, \lambda)$-ES achieves a stationary normalized mutation strength which results in divergent behavior.

In Arnold and Beyer (2004), the authors performed an in-depth analysis of the $(\mu/\mu, \lambda)$-CSA-ES on the sphere using the Gaussian noise model. The ES approaches a steady state which means that the progress vector, its length, and the normalized mutation strength have a time-invariant limit distribution. Calculating the target mutation strength, that is, the normalized mutation strength the value of which the CSA mechanism does not change in expectation, leads to a value smaller than the optimal mutation strength needed to maximize the progress rate. Furthermore, the logarithmic adaptation response indicates that for mutation strengths, which are significantly below the optimal value, there is only a small tendency toward an increase. This can be explained by the way the cumulative step-size adaptation works: It tries to adapt σ as if consecutive progress vectors were uncorrelated. Since noise overshadows the information gained from smaller step sizes, steps for small mutation strengths become nearly random and correlations disappear. This is in accordance with the CSA design philosophy, but the resulting search behavior is not always that what users desire when optimizing in noisy environments.

As in the non-noisy case, the ES cannot realize the target mutation strength in the steady state and it is usually not optimal. The mutation strength is too small for large noise strengths and too large for small noise strengths. The resulting convergence velocity is about 20–30% smaller than the optimum. Concerning the question of the population size, keeping the best 25–30% of the offspring population is a good choice.

In Beyer and Arnold (2003b), the fundamental working premises of the CSA algorithm were examined in detail. The analysis concerned the $(\mu/\mu_I, \lambda)$-ES on the *static* noisy sphere and tried to shed light on the question whether the perpendicularity or uncorrelation condition of the CSA guarantees optimal progress and if not to identify the reasons. As a first step in the analysis, the perpendicularity condition (Hansen 1998), stating that the expectation of the scalar product of two consecutive step vectors $z^{(g)}$ (❯ Eq. 2) is zero, was revisited. Considering two consecutive generations and decomposing the second search step, the angle between the two vectors can be calculated. Perpendicularity is achieved for a certain mutation strength which depends on the population parameters and the noise strength: In the absence of noise, the perpendicularity condition leads to an asymptotically optimal adaptation of the mutation strength. In the presence of noise, however, the optimal mutation strength is not reached. The perpendicularity condition leads to a CSA evolution criterion (see ❯ *Table 19*) resulting in a maximally admissible noise strength, which is only half the limit value necessitated by the *ES evolution criterion*. If the noise strength is in between, the CSA realizes angles below 90°. To "cope" with the apparently too large mutation strength, the CSA responds by reducing the mutation strength continuously to zero. This explains the observations made in experiments.

In Arnold (2006b), the performance of the (λ_{opt})-ES with CSA and weight rescaling was investigated on the noisy sphere. The realized mutation strength of the CSA differs from the optimal mutation strength. Again, there is a certain performance loss. However, it can be reduced by working with a large rescaling factor κ.

In Arnold and Beyer (2008), the performance of the $(\mu/\mu_I, \lambda)$-ES with CSA was analyzed for the noisy parabolic ridge. Three model functions describing different scalings of the standard deviation σ_ε (also referred to as noise strength) of the normally distributed noise have been investigated: (a) constant, (b) quadratically scaled, and (c) cubically scaled noise strengths. In the case of (b) and (c), the noise strength increases with the distance to the ridge axis. The steady state distance to the ridge axis, the mutation strength, and the axial progress rate were determined. Interestingly, it was found that the ES with CSA tracks the axis with a distance that is independent of the noise type. In the noise-free case, the $(\mu/\mu_I, \lambda)$-ES with CSA achieves a progress rate that is half the optimum. In the presence of noise, only for low levels of noise, a stationary distance and positive progress can be achieved.

For constant noise strength, see ❯ *Table 20*, this means that above a certain strategy-dependent noise strength, progress is not possible. Note that the analysis part, without considering step-size adaptation mechanisms, indicates that an ES could cope with any noise strength by increasing the mutation strength. However, this property cannot be

◻ **Table 19**

The perpendicularity condition of the CSA mechanism. According to the CSA philosophy, the expected angle, $\bar{\beta}$, between the parental change vector, and the local vector to the optimizer should be 90°. That is, $\cos(\bar{\beta}) = 0$ leads to the second column and in turn to the third

Angle	Mutation strength	CSA evolution criterion	ES evolution criterion
$\cos(\bar{\beta}) = \sigma^* \sqrt{\dfrac{\mu}{N}} \left(\dfrac{1}{\mu} - \dfrac{c_{\mu/\mu,\lambda}}{\sqrt{\sigma^{*2} + \sigma_\varepsilon^{*2}}} \right)$	$\sigma^* = \sqrt{\mu^2 c_{\mu/\mu,\lambda}^2 - \sigma_\varepsilon^{*2}}$	$\sigma_\varepsilon^* < \mu c_{\mu/\mu,\lambda}$	$\sigma_\varepsilon^* < 2\mu c_{\mu/\mu,\lambda}$

□ **Table 20**
The $(\mu/\mu_I, \lambda)$-ES with CSA on the noisy parabolic ridge in the case of constant noise $\sigma_\varepsilon =$ const

Distance	Mutation strength	Progress
$R = \dfrac{1}{2d}$	$\sigma = \sqrt{\dfrac{\mu^2 c_{\mu/\mu,\lambda}^2}{2d^2 N^2} - \dfrac{\sigma_\varepsilon^2}{2}}$	$\varphi = \dfrac{\mu c_{\mu/\mu,\lambda}^2}{2dN} - \dfrac{\sigma_\varepsilon^2 dN}{2\mu}$

conserved when switching the CSA part on. Increasing the strength of the noise results in a decrease of the mutation strength. Therefore, the CSA-ES fails to track the axis for higher noise strengths. Similar findings hold for the quadratically and cubically scaled noise cases. Above a certain limit of the scaling parameter, the CSA-ES fails to generate useful mutation strengths. This limit can be shifted to larger values by increasing the population sizes accordingly.

2.5 Dynamical Optimization: Optimum Tracking

This section considers *dynamical optimization* problems which arise in many practical optimization situations. In contrast to conventional, static optimization where the problem does not change with time, dynamical optimization considers problems where the position of the optimizer changes – either randomly or deterministically. While in static optimization, the goal is to locate the optimizer fast and reliably, the task in dynamical optimization is to track the moving optimizer closely. That is, the task is not to converge to a fixed state, but to adapt fast to dynamically changing environments. Examples of dynamical optimization problems include online job scheduling with new jobs arriving in the course of optimization or dynamical routing problems.

In Arnold and Beyer (2002c, 2006b), the dynamical systems approach was applied to two examples of dynamical optimization problems: a sphere model with a randomly moving target (Arnold and Beyer 2002c) and a sphere model with linearly moving target (Arnold and Beyer 2006b). The analyses aimed at answering the following questions:

1. Are ES with cumulative step length adaptation (CSA) capable of dealing with dynamical problems – although the CSA rule was designed for static optimization?
2. What are the effects of using populations and recombination?

In Arnold and Beyer (2002c), the performance of the $(\mu/\mu_I, \lambda)$-ES with CSA was investigated considering a sphere with random movements of the optimizer. Assuming that there exists a steady state of the ES once initialization effects have faded, an expression for the stationary tracking distance can be derived. The tracking distance depends on the mutation strength. Computing the derivative of the distance (with respect to the mutation strength) and setting it to zero, an optimal mutation strength and the minimal tracking error can be calculated (see ❍ *Table 21*). The question arises whether the CSA mechanism is able to tune the mutation strength to its optimal value. This is indeed the case (Arnold and Beyer 2002c). Therefore, it can be concluded that the CSA mechanism works perfectly for the $(\mu/\mu_I, \lambda)$-ES for tracking a randomly moving target.

In Arnold and Beyer (2006b), the analysis was extended to the tracking of a linearly moving target. The tracking distance of the stationary state of the ES as a function of the

◘ **Table 21**

Tracking behavior of the $(\mu/\mu_I, \lambda)$-CSA-ES on a randomly moving target (Arnold and Beyer 2002c) and a linearly moving target (Arnold and Beyer 2006b). The parameters μ and λ denote the sizes of the parent and offspring populations, N the search space dimensionality, $c_{\mu/\mu,\lambda}$ is the progress coefficient (see ❷ Sect. 5), and δ denotes the speed of the dislocation of the objective. As for the random case, δ is the standard deviation of the random change of a single component of the optimizer, that is, $\hat{y}_i^{(g+1)} = \hat{y}_i^{(g)} + \mathcal{N}(0, \delta^2)$. In the linear case, δ is simply the velocity of change, that is, $\hat{y}^{(g+1)} = \hat{y} + \delta v$ with v the unit vector

Dynamics	Tracking distance	Optimal track. distance	Optimal mutation strength	Mutation strength realized
Random	$R = \dfrac{N}{2c_{\mu/\mu,\lambda}}\left(\dfrac{\sigma}{\mu} + \dfrac{\delta^2}{\sigma}\right)$	$R = \dfrac{N\delta}{\sqrt{\mu}c_{\mu/\mu,\lambda}}$	$\sigma = \sqrt{\mu}\delta$	$\sigma = \sqrt{\mu}\delta$
Linear	$R = \dfrac{N\sigma^3 c_{\mu/\mu,\lambda}}{2\mu(\sigma^2 c_{\mu/\mu,\lambda}^2 - \delta^2)}$	$R = \dfrac{3\sqrt{3}N\delta}{4\mu c_{\mu/\mu,\lambda}^2}$	$\sigma = \dfrac{\sqrt{3}\delta}{c_{\mu/\mu,\lambda}}$	$\sigma = \dfrac{\sqrt{2}\delta}{c_{\mu/\mu,\lambda}}$

mutation strength, the optimal mutation strength, and distance were determined. Considering the CSA-rule, it was found that an ES with CSA is able to track the linearly moving target. However, the mutation strength adapted is not optimal. The ES realizes the same mutation strength as in the static case.

Summarizing, it was found that ESs with CSA are capable of tracking dynamically moving targets. In the case of randomly moving targets, the CSA even succeeds in adapting the optimal mutation strength. In both cases, recombination improves the behavior of the ES: Evolution strategies that make use of parent populations with size greater than one and of recombination track the target more closely than $(1, \lambda)$-ESs. These theoretical results are also of certain interest for other EAs, such as GAs, where it has been reported that dynamical problems often degrade the performance severely. As to the cases investigated so far, the ES does not experience such problems.

3 The Dynamical Systems Approach in Comparison

The approach presented so far used local progress measures to obtain analytical solutions, which can be used to discuss the dynamical behavior and convergence properties of the EAs. On the one hand, this enables us to consider the process in more detail: It is possible to analyze the relationship between parameter settings and the performance of the evolution strategy. On the other hand, due to the simplifications applied and assumptions made, ultimately, a "model" of the actual ES behavior is analyzed. The derivation of the model aims to take all important characteristics of the true ES into account and to transfer them to the final result. The most important assumption is the consideration of large search space dimensionalities N. This allows for the derivation of asymptotical progress measures, that is, measures obtained for infinite-dimensional search spaces. The so derived conclusions are valid as long as N is large. Considering small N will lead to deviations and usually the error made cannot be quantified analytically. That is why, the theoretical results must be supplemented with experiments.

Most analyses of self-adaptation consider only the expected changes. The ES dynamics are treated as if they were stemming from a deterministic model which may not be sufficient to capture all important characteristics of the process as the $(1, \lambda)$-ES on the noisy sphere has revealed.

Auger and Hansen (2006) presented a discussion on the limitations of the progress rate theory on the sphere model for the $(1, \lambda)$-ES. As they pointed out, for finite search space dimensionalities the usual definition of the progress rate theory corresponds to a *convergence in mean*. Connected with *almost sure convergence* is the *logarithmic progress rate* $\varphi_{\ln} = \mathrm{E}[\ln(R^{(g)}/R^{(g+1)})]$, instead. However, the discrepancy between both measures decreases with increasing search space dimensionality, and in the asymptotical limit, both measures are the same. While, using the logarithmic progress rate has a certain appeal (and it is implicitly used when plotting fitness dynamics in logarithmic scale), up until now, there is no method to calculate it analytically for finite N sphere. Monte Carlo simulations must be used for a quantification. This diminishes the usefulness of that progress measure.

While using local performance measures and treating the ES as a dynamical system has proven as a very fruitful approach yielding results of practical relevance, it is not the only one. Analysis methods from the field of stochastic systems have been tried as well. These either used the theory of Markov processes (Bienvenüe and François 2003; Auger 2005) or studied the induced supermartingales (Semenov 2002; Semenov and Terkel 2003; Hart et al. 2003). However, nearly all attempts had to resort to numerical Monte Carlo methods at some point. Therefore, also those results are not exact in that, the correctness of their conclusions can only be guaranteed probabilistically, that is, with a probability not equal to one.

Bienvenüe and François (2003) examined the global convergence of adaptive and self-adaptive $(1, \lambda)$-evolution strategies on the sphere. They showed that $(z_g)_{g \geq 1} = (\|x_g\|/\sigma_g)$ is a homogeneous Markov chain, that is, z_g only depends on z_{g-1}. This also confirms an early result obtained in Beyer (1996b) that the evolution of the mutation strength can be decoupled from the evolution of $\|x_g\|$. Furthermore, they showed that (x_g) converges or diverges log-linearly if the process has certain characteristics.

Auger (2005) followed this line of research. She analyzed a general model of an $(1, \lambda)$-ES on the sphere. Auger proved that the induced Markov process fulfills the required conditions for log-linear convergence or divergence if the offspring number λ is chosen appropriately. However, her final results for the $(1, \lambda)$-ES again depend on a Monte Carlo simulation, thus providing only a "probabilistic proof".

Semenov (2002) and Semenov and Terkel (2003) examined the convergence and the convergence velocity of evolution strategies using the theory of supermartingales. Because of the complicated nature of the underlying stochastic process, the authors did not succeed in a rigorous mathematical treatment of the stochastic process. Similar to the Markov chain approach, the authors had to resort to Monte Carlo simulations in order to show that the necessary conditions are fulfilled. In Semenov (2002) and Semenov and Terkel (2003), the $(1, \lambda)$-ES was considered. Offspring are generated according to

$$\begin{aligned} \sigma_{g,l} &= \sigma_g e^{\vartheta_{g,l}} \\ x_{g,l} &= x_g + \sigma_{g,l} \zeta_{g,l} \end{aligned} \tag{128}$$

The fitness function considered is given by $f(x) = -|x|$. The random variables $\vartheta_{g,l}$ and $\zeta_{g,l}$ are uniformly distributed with $\vartheta_{g,l}$ assuming values in $[-2, 2]$ whereas $\zeta_{g,l}$ is defined on $[-1, 1]$. For this problem, it could be shown that the object variable and the mutation strength converge almost surely to zero – provided that there are at least three offspring. Additionally, the convergence velocity of the mutation strength and the distance to the optimizer is bounded from above by a function of the form $\exp(-ag)$, which holds asymptotically almost surely.

Another viewpoint is to consider the evolutionary algorithm as a randomized algorithm and to conduct runtime analyses. The goal is to obtain upper and lower bounds for the runtime, bypassing the dynamical processes.

Concerning continuous search spaces, so far results for the $(1 + \lambda)$-ES, the $(\mu + 1)$-ES, and the $(1, \lambda)$-ES have been reported (see, e.g., Jägersküpper 2003, 2006b, 2007; Jägersküpper and Witt 2005). Some of the Jägersküpper (2006a) results are reproduced here:

- The $(1 + 1)$-ES performs with overwhelming probability $\mathcal{O}(N)$ steps to halve the approximation error in the search space.
- The $(1 + \lambda)$-ES as well as the $(1, \lambda)$-ES get along with $\mathcal{O}(N/\sqrt{\ln(1 + \lambda)})$ steps with overwhelming probability—when the $1/5$-rule is based on the number of successful mutations.
- The $(1 + \lambda)$-ES using a modified $1/5$-rule, which is based on the number of successful steps, is proved to be indeed capable of getting along with $\mathcal{O}(N/\sqrt{\ln(1 + \lambda)})$ steps with overwhelming probability, which is asymptotically optimal.
- The $(\mu + 1)$-ES using Gaussian mutations adapted by the $1/5$-rule needs $\mathcal{O}(\mu N)$ steps with overwhelming probability to halve the approximation error in the search space, which is also asymptotically optimal. (An *overwhelming probability* is defined as follows: An event occurs with overwhelming probability with respect to N if the probability of nonoccurrence is exponentially small in N (see Jägersküpper 2006a, p. 15).)

The analyzed fitness functions are the sphere and positive definite quadratic forms. The analyses also considered the adaptation mechanism of the mutation strength. So far, the $1/5$th-rule could be analyzed. Up to now, no results have been obtained considering self-adaptation or the cumulative step-size adaptation.

One general drawback of the approach is that the constants hidden in the $\mathcal{O}$-notation usually cannot be quantified.

So far, it can be stated that exact provable results can be obtained using runtime analysis or the theory of stochastic processes. However, this goes along with a coarser image of the dynamics. As a result, questions regarding the optimal population sizing or the influence of the mutation strength cannot be answered. In contrast, the progress rate theory and the dynamical systems approach can be used to obtain analytical results and to derive answers for these questions. These, however, are usually based on models of the real evolutionary processes and are valid as long as the assumptions made in the derivation hold. Each approach aims to obtain a specific type of result. Therefore, the different methodologies should be seen as complementary and not as conflictive approaches allowing for a gradually improving understanding of the convergence and optimization behavior of evolutionary algorithms.

4 Conclusions

This chapter presented an overview over the progress rate and dynamical systems approach mainly applied to the analysis of evolution strategies in real-valued search spaces. The approach aims to provide quantitative analyses of evolutionary algorithms. To this end, local progress measures, that is, expected changes of important quantities from one generation to the next, are introduced and analyzed. The results can be used to investigate the performance of different algorithms and its dependencies on the strategy parameter settings.

While the local progress measures may be regarded as the "microscopic forces" driving the EA as a dynamical system, the "macroscopic behavior" appears as the consequence of these quantities: The (self-)adaptive behavior of the EA is modeled as a dynamical system. The step-size adaptation mechanisms considered include mutative self-adaptation and the cumulative step-size adaptation which lies at the core of the covariance matrix adaptation. As a result, one gains insights into the functioning of the adaptation methods.

So far, the approach was mainly applied to evolution strategies. Further applications include real-coded genetic algorithms (Deb and Beyer 1999). However, the analysis techniques developed bear the potential to also gain insights into other natural computing paradigms, such as estimation of distribution algorithms (EDAs) and particle swarm optimization (PSO) algorithms.

5 The Progress Coefficients

For the readers' convenience, the definition of the progress coefficients are listed below:

$$d_{1,\lambda}^{(k)} = \frac{\lambda}{\sqrt{2\pi}} \int_{-\infty}^{\infty} t^k e^{-\frac{t^2}{2}} \Phi(t)^{\lambda-1} \, dt \tag{129}$$

$$d_{1+\lambda}^{(k)}(x) = \frac{\lambda}{\sqrt{2\pi}} \int_{x}^{\infty} t^k e^{-\frac{t^2}{2}} \Phi(t)^{\lambda-1} \, dt \tag{130}$$

$$e_{\mu,\lambda}^{\alpha,\beta} = \frac{\lambda-\mu}{\sqrt{2\pi}^{\alpha+1}} \binom{\lambda}{\mu} \int_0^{\infty} t^\beta e^{-\frac{\alpha+1}{2}t^2} \Phi(t)^{\lambda-\mu-1} \left(1 - \Phi(t)\right)^{\mu-\alpha} \, dt \tag{131}$$

$$c_{\mu,\lambda} = \frac{e_{\mu,\lambda}^{1,0} - \gamma_{\mu,\lambda}(1)e_{\mu,\lambda}^{1,1}}{\sqrt{1 - \frac{\mu-1}{\mu}\left[1 + e_{\mu,\lambda}^{1,1} - e_{\mu,\lambda}^{2,0} - 2\gamma_{\mu,\lambda}(1)(e_{\mu,\lambda}^{1,0} + e_{\mu,\lambda}^{1,2} - e_{\mu,\lambda}^{2,1})\right]}} \tag{132}$$

$$\gamma_{\mu,\lambda}(1) = \frac{e_{\mu,\lambda}^{1,0} - e_{\mu,\lambda}^{1,2} + 3e_{\mu,\lambda}^{2,1} - 2e_{\mu,\lambda}^{3,0}}{\frac{6\mu^2}{(\mu-1)(\mu-2)} - (6 + 3e_{\mu,\lambda}^{1,1} + 3e_{\mu,\lambda}^{1,3} - 6e_{\mu,\lambda}^{2,0} - 9e_{\mu,\lambda}^{2,2} + 6e_{\mu,\lambda}^{3,1})}. \tag{133}$$

The progress coefficient $c_{1,\lambda}$ is a special case of the progress coefficients $d_{1,\lambda}^{(k)}$ with $c_{1,\lambda} := d_{1,\lambda}^{(1)}$. The progress coefficient $c_{\mu/\mu,\lambda}$ is given as $c_{\mu/\mu,\lambda} = e_{\mu,\lambda}^{1,0}$.

References

Abramowitz M, Stegun IA (1984) Pocketbook of mathematical functions. Harri Deutsch, Thun

Arnold BC, Balakrishnan N, Nagaraja HN (1992) A first course in order statistics. Wiley, New York

Arnold DV (2002a) Noisy optimization with evolution strategies. Kluwer, Dordrecht

Arnold DV (2006a) Cumulative step length adaptation on ridge functions. In: Runarsson TP et al. (eds)

Parallel problem solving from nature PPSN IX. Springer, Heidelberg, pp 11–20

Arnold DV (2006b) Weighted multirecombination evolution strategies. Theor Comput Sci 361(1):18–37. Foundations of Genetic Algorithms

Arnold DV (2007) On the use of evolution strategies for optimising certain positive definite quadratic forms. In: Proceedings of the 9th annual conference on genetic and evolutionary computation: GECCO'07: London, July 7–11, 2007. ACM New York, pp 634–641

Arnold DV, Beyer H-G (2000) Efficiency and mutation strength adaptation of the $(\mu/\mu_I, \lambda)$-ES in a noisy environment. In: Schoenauer M (ed) Parallel problem solving from nature, vol 6. Springer, Heidelberg, pp 39–48

Arnold DV, Beyer H-G (2001a) Investigation of the (μ, λ)-ES in the presence of noise. In: Proceedings of the CEC'01 conference, Seoul, May 27–30, 2001. IEEE Piscataway, NJ, pp 332–339

Arnold DV, Beyer H-G (2001b) Local performance of the $(\mu/\mu_I, \lambda)$-ES in a noisy environment. In: Martin W, Spears W (eds) Foundations of genetic algorithms, vol 6. Morgan Kaufmann, San Francisco, CA, pp 127–141

Arnold DV, Beyer HG (2002a) Local performance of the $(1 + 1)$-ES in a noisy environment. IEEE Trans Evol Comput 6(1):30–41

Arnold DV, Beyer H-G (2002b) Performance analysis of evolution strategies with multi-recombination in high-dimensional $\mathbb{R}^n$-search spaces disturbed by noise. Theor Comput Sci 289:629–647

Arnold DV, Beyer H-G (2002c) Random dynamics optimum tracking with evolution strategies. In: Merelo Guervós JJ et al. (eds) Parallel problem solving from nature, vol 7. Springer, Heidelberg, pp 3–12

Arnold DV, Beyer H-G (2003a) On the benefits of populations for noisy optimization. Evolutionary Computation, 11(2):111–127

Arnold DV, Beyer H-G (2003b) On the effects of outliers on evolutionary optimization. In: Liu J, Cheung Y-M, Yin H (eds) IDEAL 2003: Fourth international conference on intelligent data engineering and automated learning, Hong Kong, March 21–23, 2003. Springer, Heidelberg, pp 151–160

Arnold DV, Beyer H-G (2004) Performance analysis of evolutionary optimization with cumulative step length adaptation. IEEE Trans Automatic Control, 49(4):617–622

Arnold DV, Beyer H-G (2006a) A general noise model and its effect on evolution strategy performance. IEEE Trans Evol Comput 10(4):380–391

Arnold DV, Beyer H-G (2006b) Optimum tracking with evolution strategies. Evol Comput 14:291–308

Arnold DV, Beyer H-G (2008) Evolution strategies with cumulative step length adaptation on the noisy parabolic ridge. Nat Comput 4(7):555–587

Arnold L (2002b) Random dynamical systems, 2nd printing. Springer, New York

Auger A (2005) Convergence results for the $(1, \lambda)$-SA-ES using the theory of ϕ-irreducible Markov chains. Theor Comput Sci 334:35–69

Auger A, Hansen N (2006) Reconsidering the progress rate theory for evolution strategies in finite dimensions. Seattle, WA, July 2006. Proceedings of the 8th annual conference on genetic and evolutionary computation: GECCO'06: ACM, New York, pp 445–452

Bäck T (1997) Self-adaptation. In: Bäck T, Fogel D, Michalewicz Z (eds) Handbook of evolutionary computation. Oxford University Press, New York, pp C7.1:1–C7.1:15

Beyer H-G (1989) Ein Evolutionsverfahren zur mathematischen Modellierung stationärer Zustände in dynamischen Systemen. Dissertation, Hochschule für Architektur und Bauwesen, Weimar, Reihe: HAB-Dissertationen, Nr. 16

Beyer H-G (1993) Toward a theory of evolution strategies: some asymptotical results from the $(1 \overset{+}{,} \lambda)$-theory. Evol Comput 1(2):165–188

Beyer H-G (1994) Towards a theory of 'evolution strategies': progress rates and quality gain for $(1 \overset{+}{,} \lambda)$-strategies on (nearly) arbitrary fitness functions. In: Davidor Y, Männer R, Schwefel H-P (eds) Parallel problem solving from nature, vol 3. Springer, Heidelberg, pp 58–67

Beyer H-G (1995a) Toward a theory of evolution strategies: on the benefit of sex – the $(\mu/\mu, \lambda)$-theory. Evol Comput 3(1):81–111

Beyer H-G (1995b) Toward a theory of evolution strategies: the (μ, λ)-theory. Evol Comput 2(4):381–407

Beyer H-G (1996a) On the asymptotic behavior of multi-recombinant evolution strategies. In: Voigt H-M, Ebeling W, Rechenberg I, Schwefel H-P (eds) Parallel problem solving from nature, vol 4. Springer, Heidelberg, pp 122–133

Beyer H-G (1996b) Toward a theory of evolution strategies: self-adaptation. Evol Comput 3(3):311–347

Beyer H-G (2000) Evolutionary algorithms in noisy environments: theoretical issues and guidelines for practice. Comput Methods Appl Mech Eng 186(2–4):239–267

Beyer H-G (2001a) On the performance of $(1, \lambda)$-evolution strategies for the ridge function class. IEEE Trans Evol Comput 5(3):218–235

Beyer H-G (2001b) The theory of evolution strategies. Natural computing series. Springer, Heidelberg

Beyer H-G (2004) Actuator noise in recombinant evolution strategies on general quadratic fitness models. In: Deb K et al. (eds) GECCO 2004: Proceedings of the genetic and evolutionary computation conference, Seattle, WA, June 26–30, 2004. Springer, Heidelberg, pp 654–665

Beyer H-G, Arnold DV (1999) Fitness noise and localization errors of the optimum in general quadratic fitness models. In: Banzhaf W, Daida J, Eiben AE, Garzon MH, Honavar V, Jakiela M, Smith RE (eds), GECCO-99: Proceedings of the genetic and evolutionary computation conference, Orlando, FL, July 1999. Morgan Kaufmann, San Francisco, CA, pp 817–824

Beyer H-G, Arnold DV (2003a) The steady state behavior of $(\mu/\mu_I, \lambda)$-ES on ellipsoidal fitness models disturbed by noise. In: Cantú-Paz E et al. (eds) Proceedings of the genetic and evolutionary computation conference, Chicago, IL, July 2003. Springer, Berlin, pp 525–536

Beyer H-G, Arnold DV (2003b) Qualms regarding the optimality of cumulative path length control in CSA/CMA-evolution strategies. Evol Comput 11 (1):19–28

Beyer H-G, Finck S (2009) Performance of the $(\mu/\mu_I, \lambda)$-σSA-ES on PDQFs. IEEE Trans Evol Comput (accepted) http://dx.doi.org/10.1109/TEVC.2009.2033581

Beyer H-G, Meyer-Nieberg S (2004) On the quality gain of $(1, \lambda)$-ES under fitness noise. In: Yao X, Schwefel H-P et al. (eds) PPSN VIII: Proceedings of the 8th international conference on parallel problem solving from nature, Birmingham. Springer, Berlin, pp 1–10

Beyer H-G, Meyer-Nieberg S (2005) On the prediction of the solution quality in noisy optimization. In: Wright AH, Vose MD, De Jong KA, Schmitt LM (eds), FOGA 2005: Foundations of genetic algorithms 8. Lecture notes in computer science, vol 3469. Springer, Springer, Berlin, pp 238–259

Beyer H-G, Meyer-Nieberg S (2006a) Self-adaptation of evolution strategies under noisy fitness evaluations. Genetic Programming Evolvable Mach 7(4):295–328

Beyer H-G, Meyer-Nieberg S (2006b) Self-adaptation on the ridge function class: first results for the sharp ridge. In: Runarsson TP et al. (eds) Parallel problem solving from nature, PPSN IX. Springer, Heidelberg, pp 71–80

Beyer H-G, Schwefel H-P (2002) Evolution strategies: a comprehensive introduction. Nat Comput 1(1):3–52

Beyer H-G, Sendhoff B (2006) Functions with noise-induced multimodality: a test for evolutionary robust optimization – properties and performance analysis. IEEE Trans Evol Comput 10(5):507–526

Beyer H-G, Sendhoff B (2007a) Evolutionary algorithms in the presence of noise: to sample or not to sample. In: McKay B et al. (eds) Proceedings of the 2007 IEEE symposium on foundations of computational intelligence, Honolulu, HI, April 2007. IEEE Press, Piscataway, NJ, pp 17–23

Beyer H-G, Sendhoff B (2007b) Robust optimization – a comprehensive survey. Comput Methods Appl Mech Eng 197:3190–3218

Beyer H-G, Olhofer M, Sendhoff B (2003) On the behavior of $(\mu/\mu_I, \lambda)$-ES optimizing functions disturbed by generalized noise. In: De Jong K, Poli R, Rowe J, (eds) Foundations of genetic algorithms, vol 7. Morgan Kaufmann, San Francisco, CA, pp 307–328

Beyer H-G, Olhofer M, Sendhoff B (2004) On the impact of systematic noise on the evolutionary optimization performance – a sphere model analysis. Genetic Programming Evolvable Mach 5:327–360

Bienvenüe A, François O (2003) Global convergence for evolution strategies in spherical problems: some simple proofs and difficulties. Theor Comput Sci 308:269–289

Braun M (1998) Differential equations and their applications. Springer, Berlin

Deb K, Beyer H-G (1999) Self-adaptation in real-parameter genetic algorithms with simulated binary crossover. In: Banzhaf W, Daida J, Eiben AE, Garzon MH, Honavar V, Jakiela M, Smith RE (eds) Proceedings of the genetic and evolutionary computation conference (GECCO-99), Orlando, FL, July 1999. Morgan Kaufmann, San Francisco, CA, pp 172–179

Eiben AE, Smith JE (2003) Introduction to evolutionary computing. Natural computing series. Springer, Berlin

Grünz L, Beyer H-G (1999) Some observations on the interaction of recombination and self-adaptation in evolution strategies. In: Angeline PJ (ed) Proceedings of the CEC'99 conference, Washington, DC, July 1999. IEEE, Piscataway, NJ, pp 639–645

Hansen N (1998) Verallgemeinerte individuelle Schrittweitenregelung in der Evolutionsstrategie. Doctoral thesis, Technical University of Berlin

Hansen N (2006) The CMA evolution strategy: a comparing review. In: Lozano JA, Larrañaga P, Inza I, Bengoetxea E (eds) Towards a new evolutionary computation. Advances in estimation of distribution algorithms. Springer, Berlin, Heidelberg, pp 75–102

Hansen N, Ostermeier A (1996) Adapting arbitrary normal mutation distributions in evolution strategies: the covariance matrix adaptation. In: ICEC'96: Proceedings of 1996 IEEE international conference on evolutionary computation, Japan, May 1996. IEEE Press, New York, pp 312–317

Hansen N, Ostermeier A (1997) Convergence properties of evolution strategies with the derandomized covariance matrix adaptation: the $(\mu/\mu_I, \lambda)$-CMA-ES. In: Zimmermann H-J (ed) EUFIT'97: 5th European congress on intelligent techniques and soft computing. Aachen, Germany, September, Mainz, pp 650–654

Hansen N, Ostermeier A (2001) Completely derandomized self-adaptation in evolution strategies. Evol Comput 9(2):159–195

Hart WE, DeLaurentis JM, Ferguson LA (2003) On the convergence of an implicitly self-adaptive evolutionary algorithm on one-dimensional unimodal problems. IEEE Trans Evol Comput

Hofbauer J, Sigmund K (2002) Evolutionary games and population dynamics. Cambridge University Press, Cambridge

Jägersküpper J (2003) Analysis of a simple evolutionary algorithm for minimization in euclidean spaces. In: Baeten J et al. (eds) ICALP 2003: Proceedings of the 30th international colloquium on automata, languages, and programming, Eindhoven, 2003. Lecture notes in computer science, vol 2719. Springer, Berlin, pp 1068–1079

Jägersküpper J (2006a) Probabilistic analysis of evolution strategies using isotropic mutations. Ph.D. thesis, Dortmund University

Jägersküpper J (2006b) Probabilistic runtime analysis of $(1 + \lambda)$-ES using isotropic mutations. In: GECCO'06: Proceedings of the 8th annual conference on genetic and evolutionary computation Seattle, WA, July 2006. ACM, New York, pp 461–468

Jägersküpper J (2007) Algorithmic analysis of a basic evolutionary algorithm for continuous optimization. Theor Comput Sci 379(3):329–347

Jägersküpper J, Witt C (2005) Rigorous runtime analysis of a $(\mu + 1)$ ES for the sphere function. In: Beyer H-G et al. (eds) GECCO 2005: Proceedings of the genetic and evolutionary computation conference Washington, DC, June 2005. ACM Press, New York, pp 849–856

Jin Y, Branke J (June 2005) Evolutionary optimization in uncertain environments—a survey. IEEE Trans Evol Comput 9(3):303–317

Kolossa JK (2006) Series approximation methods in statistics. Springer, New York

Meyer-Nieberg S, Beyer H-G (2005) On the analysis of self-adaptive recombination strategies: first results. In: McKay B et al. (eds) Proceedings of 2005 congress on evolutionary computation (CEC'05), Edinburgh, UK. IEEE Press, Piscataway, NJ, pp 2341–2348

Meyer-Nieberg S, Beyer H-G (2007) Mutative self-adaptation on the sharp and parabolic ridge. In: Stephens Ch et al. (eds) FOGA 2007: Foundations of genetic algorithms IX, Mexico City, January 2007. Springer Heidelberg, pp 70–96

Meyer-Nieberg S, Beyer H-G (2008) Why noise may be good: additive noise on the sharp ridge. In: Keijzer M et al. (eds) GECCO 2008: Proceedings of the 10th annual conference on genetic and evolutionary computation, Atlanta, GA, July 2008. ACM, New York, pp 511–518

Ostermeier A, Gawelczyk A, Hansen N (1995) A derandomized approach to self-adaptation of evolution strategies. Evol Comput 2(4):369–380

Oyman AI (1999) Convergence behavior of evolution strategies on ridge functions. Ph.D. Thesis, University of Dortmund

Oyman AI, Beyer H-G (2000) Analysis of the $(\mu/\mu, \lambda)$-ES on the parabolic ridge. Evol Comput 8(3):267–289

Oyman AI, Beyer H-G, Schwefel H-P (1998) Where elitists start limping: evolution strategies at ridge functions. In: Eiben AE, Bäck T, Schoenauer M, Schwefel H-P (eds) Parallel problem solving from nature, vol 5. Springer, Heidelberg, pp 34–43

Oyman AI, Beyer H-G, Schwefel H-P (2000) Analysis of a simple ES on the "parabolic ridge." Evol Comput, 8(3):249–265

Rechenberg I (1973) Evolutionsstrategie: Optimierung technischer Systeme nach Prinzipien der biologischen Evolution. Frommann-Holzboog, Stuttgart

Schwefel H-P (1977) Numerische Optimierung von Computer-Modellen mittels der Evolutionsstrategie. Interdisciplinary systems research; 26. Birkhäuser, Basel

Semenov MA (2002) Convergence velocity of evolutionary algorithms with self-adaptation. In: Langdon WB et al. (eds), GECCO 2002: Proceeding of the genetic and evolutionary computation conference. Morgan Kaufmann, New York City, July 2002, pp 210–213

Semenov MA, Terkel DA (2003) Analysis of convergence of an evolutionary algorithm with self-adaptation using a stochastic Lyapunov function. Evol Comput 11(4):363–379

Sendhoff B, Beyer H-G, Olhofer M (2002) On noise induced multi-modality in evolutionary algorithms. In: Wang L, Tan KC, Furuhashi T, Kim J-H, Sattar F (eds), Proceedings of the 4th Asia-Pacific conference on simulated evolution and learning – SEAL, vol 1, Singapore, November 18–22, pp 219–224

Wiggins S (1990) Introduction to applied nonlinear dynamical systems and chaos. Springer, New York

26 Computational Complexity of Evolutionary Algorithms

Thomas Jansen
Department of Computer Science, University College Cork, Ireland
t.jansen@cs.ucc.ie

G. Rozenberg et al. (eds.), *Handbook of Natural Computing*, DOI 10.1007/978-3-540-92910-9_26,
© Springer-Verlag Berlin Heidelberg 2012

Abstract

When applying evolutionary algorithms to the task of optimization it is important to have a clear understanding of their capabilities and limitations. By analyzing the optimization time of various variants of evolutionary algorithms for classes of concrete optimization problems, important insights can be gained about what makes problems easy or hard for these heuristics. Still more important than the derivation of such specific results is the development of methods that facilitate rigorous analysis and enable researchers to derive such results for new variants of evolutionary algorithms and more complex problems. The development of such methods and analytical tools is a significant and very active area of research. An overview of important methods and their foundations is presented together with exemplary applications. This enables one to apply these methods to concrete problems and participate in the theoretical foundation of evolutionary computing.

1 Introduction and Motivation

Evolutionary algorithms form a large and loosely defined class of algorithms, they can be and are used for many different purposes. Even though it can be argued that they are not really suited for optimization in the strong sense (De Jong 1992), optimization remains one of the main fields of application. There they compete with other optimization methods and can be classified as randomized optimization heuristics, while in the more general sense the label randomized search heuristic is more appropriate. (It can be argued that it would be more appropriate to speak of probabilistic heuristics since heuristics is not dealt with where randomization is introduced later but with heuristics where randomization is a central concept. However, since the notion of randomized algorithms is well established, this notion is adopted here, too.) When the performance of evolutionary algorithms is to be assessed in the context of optimization, it makes most sense to consider evolutionary algorithms as randomized *algorithms*, adopting the perspective that is common in the community concerned with the design and analysis of (randomized) algorithms (Cormen et al. 2001; Mitzenmacher and Upfal 2005; Motwani and Raghavan 1995). This implies that we perform rigorous analysis and present strictly proven results based on no unproven assumptions or simplifications. While such a rigorous approach makes it difficult to obtain results for complex evolutionary algorithms and complex optimization problems, it has several advantages that in summation make it much more attractive than a purely experimental approach or heuristic analyses based on reasonable yet unproven assumptions.

The most prominent and important advantage of the rigorous analytical approach is that it provides foundations, a firm ground of knowledge that can serve as a trustworthy basis for design and analysis of evolutionary algorithms. In most cases, the proofs of analytical results provide a level of insight and give leverage to understand the working principles of evolutionary algorithms that is difficult if not impossible to obtain in other ways. It advances common beliefs about evolutionary algorithms to proven facts and knowledge, advancing the field to a well-founded science. The results derived with the methods presented here come with a degree of generality that is impossible to achieve with experiments alone. Moreover, the methods themselves turn out to be applicable to a wide range of evolutionary algorithms and many different fitness functions. Another important aspect that is dealt with briefly is oriented toward complexity theory, that is, the derivation of fundamental lower bounds that hold for all algorithms operating within the considered scenario. This reveals fundamental limitations and can in practice save a lot of time since no effort is wasted for provably impossible tasks. Finally, a completely different aspect of equal importance is the usefulness of rigorously proven results in teaching. By learning analytical tools and methods for the analysis of evolutionary algorithms, students can develop a deep understanding and the ability to design and analyze their own algorithms in their own contexts.

It can be remarked that computational complexity of evolutionary algorithms can be taken to mean something entirely different, namely, aspects of the efficient implementation of evolutionary algorithms. While it is true that for some modules of evolutionary algorithms there are implementations known that are considerably more efficient than the obvious ones this is not at all the focus of this chapter. (One example is implementing standard bit mutations where each of n bits flips with probability $1/n$ by randomly determining the position of the next bit to be flipped. This decreases the (expected) number of random experiments from n to just 1.) While in concrete cases efficient implementations can decrease the run time considerably, in general, evolutionary algorithms are rather easy and straightforward to implement. In practical applications, often most time is spent in the evaluation of a complex fitness function so that the computational effort spent on the evolutionary algorithm's modules becomes negligible. This motivates that the analysis can be simplified by considering the number of evaluations of the fitness functions as an appropriate measure for the actual run time. This is more accurate than the number of generations, another measure that is often used, where the effect of the population size on the run time cannot be assessed. One has to be cautious, however, when dealing with complex evolutionary algorithms or hybridizations of evolutionary algorithms where the computational effort of the algorithm itself, apart from evaluations of the fitness function, becomes considerable or even dominant.

In the next section, we give a brief outline of the kind of evolutionary algorithms we consider and we describe the perspective adopted for analysis. Fundamental limitations that are inherent to the specific scenario that usually governs the application of evolutionary algorithms are discussed in ❷ Sect. 3. There are two ways of finding and describing such limits that are fundamentally different, namely, "no free lunch" arguments and black-box complexity. A brief introduction to both concepts will be given. ❷ Section 4 is devoted to the description of methods for the analysis of evolutionary algorithms. Clearly, this section contains the most important and hopefully useful parts of this chapter. Finally, this chapter concludes with a summary in ❷ Sect. 5.

2 Evolutionary Algorithms

While there is a large amount of different variants of evolutionary algorithms with new and related paradigms such as particle swarm optimization or ant colony emerging, here we restrict ourselves to simplified and rather basic algorithms. This restriction facilitates the analysis and allows for a clearer identification of causes for the effects that are observed. Yet, it is not too restrictive: it turns out that the methods, which are developed with a specific and quite restricted evolutionary algorithm in mind, can often be generalized and extended to more complex evolutionary algorithms.

The most basic evolutionary algorithm considered is the so-called $(1 + 1)$ EA. Its extreme simplicity makes it attractive for formal analysis, hence it shares important typical properties with much more complex evolutionary algorithms. Here we present it without any stopping criterion as this subject is typically not dealt with in the context of computational complexity of evolutionary algorithms.

Algorithm 1   (the (1 + 1) EA)

1. **Initialization**
   $t := 0$
   Choose $x_t \in \{0, 1\}^n$ uniformly at random.
2. Repeat forever
3. **Mutation**
   Create $y := x_t$. Independently for each $i \in \{1, 2, \ldots, n\}$,
   flip the bit $y[i]$ with probability $1/n$.
4. **Selection**
   If $f(y) \geq f(x_t)$ then $x_{t+1} := y$ else $x_t := y$.
   $t := t + 1$

The mutation operator employed in line 3 of the (1 + 1) EA (❷ *Algorithm 1*) is called standard bit mutation with mutation probability $1/n$. While $1/n$ is the most recommended choice, it is known that it can be far from optimal for some fitness functions (Jansen and Wegener 2000). When this mutation operator is replaced by a local operator that flips exactly one bit, we obtain an algorithm that is known as *randomized local search* (RLS). While being close to the (1+1) EA we consider RLS not to be an evolutionary algorithm since evolutionary algorithms are supposed to be able to perform a global search in the search space and randomized local search, by definition, is restricted to a local search. Since the (1 + 1) EA with mutation probability $1/n$ flips on average one bit in each mutation one may

---

Algorithm 2   (the ($\mu + \lambda$) EA)

1. **Initialization**
   $t := 0$
   Choose $x_t^{(1)}, x_t^{(2)}, \ldots, x_t^{(\mu)} \in \{0, 1\}^n$ independently uniformly at random.
2. Repeat forever
3. For $i \in \{1, 2, \ldots, \lambda\}$ do
4. **Selection**
   Select $z \in \{x_t^{(1)}, x_t^{(2)}, \ldots, x_t^{(\mu)}\}$ uniformly at random.
5. **Crossover**
   With probability p_c
   $z' := z$
   Select $z'' \in \{x_t^{(1)}, x_t^{(2)}, \ldots, x_t^{(\mu)}\}$ uniformly at random.
   $z := crossover(z', z'')$
6. **Mutation**
   Create $y^{(i)} := z$. Independently for each $j \in \{1, 2, \ldots, n\}$,
   flip the bit $y^{(i)}[j]$ with probability $1/n$.
7. **Selection**
   Sort $x_t^{(1)}, x_t^{(2)}, \ldots, x_t^{(\mu)}, y^{(1)}, y^{(2)}, \ldots, y^{(\lambda)}$ with respect to fitness in
   decreasing order, in case of equal fitness sort offspring y before parents x.
   Let the first μ points be the sequence $x_{t+1}^{(1)}, x_{t+1}^{(2)}, \ldots, x_{t+1}^{(\mu)}$.
   $t := t + 1$

speculate that the two algorithms behave similarly. It is know that this is indeed the case for many problems but that they differ in extreme ways on other problems and that it is difficult to give a precise description of problems where the two algorithms provably perform similarly (Doerr et al. 2008).

In general, evolutionary algorithms make use of a larger population size and produce more than one offspring per generation. A still very simple evolutionary algorithm that has both properties is described. Since the population size is larger than one, it can also make use of crossover.

It can be seen that the $(1 + 1)$ EA is the special case of the $(\mu + \lambda)$ EA that we obtain by setting $\mu = \lambda = 1$. As crossover operator, any operator acting on binary strings may be used. Uniform crossover will be considered. It is also possible to consider the $(\mu + \lambda)$ EA without crossover, that is, set the probability for crossover to 0, $p_c = 0$.

Finally, a third kind of evolutionary algorithm called (μ, λ) EA is introduced. It is similar to the $(\mu + \lambda)$ EA but employs a different selection at the end of each generation. The new population is selected from the offspring alone, the old generation is discarded in any case. Clearly, this implies that

$$\max\left\{ x_t^{(i)} \mid i \in \{1, 2, \ldots, \mu\} \right\}$$

may decrease with increasing t. A formal description is given just as was done for the other algorithms. Note that $\lambda \geq \mu$ is needed here.

Algorithm 3   (the (μ, λ) EA)

1. **Initialization**
   $t := 0$
   Choose $x_t^{(1)}, x_t^{(2)}, \ldots, x_t^{(\mu)} \in \{0, 1\}^n$ independently uniformly at random.
2. **Repeat forever**
3.   For $i \in \{1, 2, \ldots, \lambda\}$ do
4.   **Selection**
          Select $z \in \{x_t^{(1)}, x_t^{(2)}, \ldots, x_t^{(\mu)}\}$ uniformly at random.
5.   **Crossover**
          With probability p_c
             $z' := z$
             Select $z'' \in \{x_t^{(1)}, x_t^{(2)}, \ldots, x_t^{(\mu)}\}$ uniformly at random.
             $z := crossover(z', z'')$
6.   **Mutation**
          Create $y^{(i)} := z$. Independently for each $j \in \{1, 2, \ldots, n\}$,
          flip the bit $y^{(i)}[j]$ with probability $1/n$.
7.   **Selection**
          Sort $y^{(1)}, y^{(2)}, \ldots, y^{(\lambda)}$ with respect to fitness in decreasing order.
          Let the first μ points be the sequence $x_{t+1}^{(1)}, x_{t+1}^{(2)}, \ldots, x_{t+1}^{(\mu)}$.
          $t := t + 1$

Since in our description of evolutionary algorithms they never stop it is not obvious how we want to measure their computational complexity. We decide to consider the first point of time (measured by the number of evaluations of the fitness function) when a global optimum is sampled. We call this the *optimization time*. Note that it is usually not difficult to change this into other notions of achieving a goal like approximation. For the $(1 + 1)$ EA, the optimization time on a fitness function $f: \{0, 1\}^n \to \mathbb{R}$ is given by the following equation.

$$T_{(1+1)\ \text{EA},f} = \min\{t \mid f(x_t) = \max\{f(x') \mid x' \in \{0, 1\}^n\}\}$$

For the $(\mu + \lambda)$ EA, we are slightly less accurate when defining the optimization time. We define it based on the number of generations (but still close to the actual number of f-evaluations) in the following way.

$$T_{(\mu+\lambda)\ \text{EA},f} = \min\left\{\mu + \lambda t \mid \max\left\{f\left(x_t^{(i)}\right) \mid i \in \{1, \ldots, \mu\}\right\} = \max\{f(x) \mid x \in \{0, 1\}^n\}\right\}$$

For the (μ, λ) EA we use the same definition, $T_{(\mu, \lambda)\ \text{EA},f} = T_{(\mu+\lambda)\ \text{EA},f}$.

Clearly, $T_{A,f}$ is a random variable and we are mostly interested in its expectation $E(T_{A,f})$. In cases where this expected value is misleading due to exceptionally large values that occur only with small probabilities other properties are of interest. In particular, bounds on Prob $(T_{A,f} \leq t_u(n))$ and Prob $(T_{A,f} \geq t_l(n))$ for some upper bound t_u and some lower bound t_l are often of interest. One can consider an algorithm to be efficient if either $E\ (T_{(1+1)\ \text{EA},f})$ is polynomially bounded from above or Prob $(T_{A,f} \leq t_u(n)) \geq 1/p(n)$ holds for some polynomials $t_u(n)$ and $p(n)$. The reason for this rather generous notion of efficiency can be found in restarts: If one knows that Prob $(T_{A,f} \leq t_u(n)) \geq 1/p(n)$ holds for some evolutionary algorithm A, one can turn it into an evolutionary algorithm with polynomial expected optimization time at most $t_u(n)p(n)$ by stopping A after $t_u(n)$ fitness evaluations and restarting it.

In addition to considering simple evolutionary algorithms, the analysis is further simplified by not analyzing the optimization time with arbitrary precision but by restricting ourselves to an asymptotic analysis taking into account the dominating factors. One can make use of the well-known notions for the description of the order of growth of functions (see, e.g., Cormen et al. 2001) and perform analysis for a growing dimension of the search space n. Thus, the dimension of the search space n plays the same role as the length of the input in the analysis of problem-specific algorithms.

3   Fundamental Limits

Evolutionary algorithms are considered in the context of optimizing some unknown fitness function $f: S \to R$ where S is some finite search space and R typically is a subset of $\mathbb{R}$. Since S is finite it usually does not hurt to assume that R is finite, too. When looking for fundamental limits on the performance of evolutionary algorithms (and all other algorithms working in the same scenario) one has to be careful about what exactly the algorithm can make use of. It is assumed that the algorithm designer knows that the algorithm is going to be applied to some unknown function $f \in \mathscr{F} \subseteq \{g: S \to R\}$. So, any problem-specific knowledge the algorithm designer may have is in the framework expressed in this knowledge about the set of potential fitness functions $\mathscr{F}$. An algorithm is required to find some optimal point $s \in S$, that is, some $s \in S$ with $f(s) = \max\ \{f(x) \mid x \in S\}$. One can assume that the algorithm can make any use of the fact that the concrete fitness function f belongs to $\mathscr{F}$, $f \in \mathscr{F}$, but has no other way of obtaining knowledge about the concrete fitness function f but to query the function value $f(x)$ for arbitrary $x \in S$. The number of such queries (we call f-evaluations) are counted until an optimum is found for the first time. Note that we do not expect the algorithm to know or to notice that it found an optimum. This is the scenario that we consider and call *black-box optimization*. We call an algorithm A a *black-box algorithm* for $\mathscr{F}$ if for each function $f \in \mathscr{F}$ it finds an optimum of f within at most t f-evaluations with probability converging to 1 with t going to infinity, $\forall f \in \mathscr{F}: \lim_{t \to \infty} \text{Prob}(T_{A,f} \leq t) = 1$.

Adopting the usual worst-case perspective of computer science, one is interested in the best performance any algorithm can achieve for such a set of potential fitness functions $\mathscr{F}$ where performance is taken to be the worst-case performance. Formally, one can define for a class of functions $\mathscr{F}$ and a black-box algorithm for $\mathscr{F}$ the *worst case expected optimization time*

$$T_{A,\mathscr{F}} = \max\{E(T_{A,f}) \mid f \in \mathscr{F}\}$$

Finally, one can define for $\mathscr{F} \subseteq \{f : S \to R\}$ the *black-box complexity* $B_{\mathscr{F}}$ of $\mathscr{F}$:

$$B_{\mathscr{F}} = \min\{T_{A,\mathscr{F}} \mid A \text{ is a black-box algorithm for } \mathscr{F}\}$$

It is not difficult to see that restricting the attention to f-evaluations alone can lead to cases where NP-hard function classes $\mathscr{F}$ have only polynomial black-box complexity (Droste et al. 2006): if a polynomial number of f-evaluations is sufficient to reconstruct the concrete problem instance at hand an optimal solution can be computed (in exponential time) without any further f-evaluations. This seeming weakness is more than compensated by the fact that in contrast to classical complexity theory one is able to derive absolute lower bounds that are not based on any complexity theoretical assumption like $P \neq NP$.

Another noteworthy consequence from the definition is that for any class of functions $\mathscr{F}$ we have $B_{\mathscr{F}} \leq |\mathscr{F}|$: a black-box algorithm knows about all $f \in \mathscr{F}$, this includes knowledge about the global optima. Thus, it can sample for each function one of its global optima one by one sampling an optimum of the unknown objective function $f \in \mathscr{F}$ after at most $|\mathscr{F}|$ steps. Moreover, we have

$$\mathscr{F} \subseteq \mathscr{F}' \Rightarrow B_{\mathscr{F}} \leq B_{\mathscr{F}'}$$

since for the subset $\mathscr{F}$ the maximum in the definition of $T_{A,\mathscr{F}}$ is taken over a subset of functions.

Attention is restricted to deterministic black-box algorithms for a moment. Such algorithms can be described as a tree: The first point in the search space $x \in S$ they sample is the root of the tree. The second point the algorithm decides to sample is in general based on its f-value $f(x)$ so that there are up to $|R|$ possible second points. We represent this in the tree by attaching each second point as child to the root. We mark the corresponding edge with the f-value that led to this choice. Clearly, this way the root can become a node of degree at most $|R|$. Continuing this way by adding the third points the algorithm samples (depending on the first two points together with their function values) to the appropriate node we see that any deterministic black-box algorithm can be described as an $|R|$-ary tree. If we are interested in optimal black-box algorithms, it is safe to assume that on each path from the root to a leaf each point from S is encountered at most once. Thus, the trees have finite size. If necessary, one can enlarge the trees such that each path from the root to a leaf contains exactly $|S|$ nodes. If the criterion is not changed for assessing the performance of an algorithm, this formal change does not change its performance and may make a formal treatment simpler.

Considering deterministic black-box algorithms as trees we recognize that due to their finite size and finite possible markings at the nodes and edges there is only a finite number of such algorithms. In such cases, we describe any randomized black-box algorithm as a probability distribution over the deterministic black-box algorithms. From an algorithmic perspective, the idea is to modify randomized black-box algorithms in a way that they begin with making a sufficiently large number of random coin tosses, storing the outcomes in memory. After that, they work deterministically and every time the original randomized algorithm needed to make a random decision the modified version acts deterministically using the stored outcome of the random experiment.

3.1   "No Free Lunch"

For a fixed set of functions $\mathscr{F}$, in general $T_{A,\mathscr{F}}$ depends on the black-box algorithm A under consideration. There are, however, classes of functions, where this value is equal regardless of the choice of A – provided that we consider only distinct f-evaluations or, equivalently, black-box algorithms that do not revisit any point in the search space. Thinking of constant fitness functions this is clearly not surprising at all. But it is quite remarkable that this also holds for the class of all functions $\mathscr{F} = \{f : S \rightarrow R\}$. When this was first published (Wolpert and Macready 1997), using the colorful label "no free lunch theorem," it caused enormous discussions about its scope and meaning. If on average over all problems all algorithms perform equal, what sense does it make to design any specific randomized search heuristic? The result can even be strengthened. It is not necessary that one averages only over constant functions or over all functions. The same result can be proven if the set of functions $\mathscr{F}$ has the property to be closed under permutations of the search space. For a function $f: S \rightarrow R$ we can define the function $\sigma f: S \rightarrow R$ by $\sigma f(x) = f(\sigma^{-1}(x))$ for a permutation σ of S. A class of functions $\mathscr{F}$ is called *closed under permutations* of the search space if $f \in \mathscr{F}$ implies $\sigma f \in \mathscr{F}$ for all functions $f \in \mathscr{F}$ and all permutations σ. Moreover, it can be shown that being closed under permutations is not only a sufficient but also a necessary condition for a set $\mathscr{F}$ to allow for a "no free lunch" result. Note that it is not necessary to consider the average over the class of functions $\mathscr{F}$. One can make precise statements over the existence of "no free lunch" results for probability distributions over $\mathscr{F}$ different from the uniform distribution leading to the most general "no free lunch theorem" known (Igel and Toussaint 2004). Since the generalization is not too surprising, one can stick to the simpler result (Schumacher et al. 2001). In order to formulate the result precisely, the notion of a *performance measure* is needed.

Remember that any randomized black-box algorithm can be described as a probability distribution over the set of all deterministic black-box algorithms. Moreover, remember that any deterministic black-box algorithm can be described as an $|R|$-ary tree where each path from the root to a leaf contains exactly $|S|$ nodes. When one such deterministic black-box algorithm A and some function $f \in \mathscr{F}$ are fixed, this defines a path in the tree that one can call a *trace* $T(A, f) = ((x_1, f(x_1)), (x_2, f(x_2)), \ldots, (x_{|S|}, f(x_{|S|})))$. This trace can be projected onto a vector of function values $V(A, f) = (f(x_1), f(x_2), \ldots, f(x_{|S|}))$. If it is agreed that reasonable performance measures should only depend on the function values and not on the search points themselves then one can give a formal definition of a performance measure as a function mapping such vectors of function values to $\mathbb{R}$. Clearly, important performance measures, in particular the performance measure one can consider (the first point of time when a global maximum is found), can be defined in this way. Using this formal definition of a performance measure one can obtain the following result that is cited without a proof.

Theorem 1 (NFL Theorem (Schumacher et al. 2001)) *Let S, R be two finite sets. On average over all functions $f \in \mathcal{F} \subseteq \{g : S \to R\}$ all black-box algorithms perform equal for all performance measures $M : \{V | V = V(A, h)\} \to \mathbb{R}$ if and only if the set of functions $\mathcal{F}$ is closed under permutations of the search space.*

While this result has the obvious consequence that there is no way one can develop meaningful algorithms for function classes $\mathcal{F}$ that are closed under permutations, it is not so clear what the relevance of this result is. We mention that it is unusual for a class of functions to be closed under permutations. The fraction of classes where this is the case is exponentially small (Igel and Toussaint 2003). Moreover, classes of functions that are defined via some nontrivial notion of a neighborhood cannot be closed under permutations since neighborhoods are not preserved under permutations of the search space. On the other hand, it is not difficult to see that each black-box algorithm that performs well on a specific function performs very poorly on an exponentially large number of similar functions with probability very close to one (Droste et al. 2002b).

3.2 Black-Box Complexity

Much more interesting than the fundamental but in their relevance severely limited "no free lunch" results are concrete results on the black-box complexity $B_{\mathcal{F}}$ for some relevant class of functions $\mathcal{F}$. One can obtain upper bounds on $B_{\mathcal{F}}$ by proving that some specific black-box algorithms optimizes on average each function $f \in \mathcal{F}$ within some bounded number of f-evaluations. Lower bounds, however, are much more difficult to obtain. They require a formal proof that no algorithm at all is able to optimize all functions from $\mathcal{F}$ with less than a certain number of f-evaluations. If one sticks to this assumption that S and R are finite sets, though, Yao's minimax principle is a very useful tool to prove such lower bounds.

Theorem 2 (Yao's Minimax Principle (Motwani and Raghavan 1995; Yao 1977)) *If a problem consists of a finite set of instances of a fixed size and allows a finite set of deterministic algorithms, the minimal worst-case instance expected optimization time of a randomized algorithm is bounded below for each probability distribution on the instances by the expected optimization time of an optimal deterministic algorithm.*

Yao's minimax principle simplifies the proofs of lower bounds in two ways: First, there is no need to analyze a randomized algorithm but we can restrict our attention to deterministic algorithms. This, typically, is much simpler. Second, one is free to define a distribution over the class of functions $\mathcal{F}$ any way one likes. Two concrete examples will be considered to see how such results can be derived. To this end, two example functions and a general way of creating a class of functions based on a single fitness function are discussed.

Definition 1 Let $n \in \mathbb{N}$ be given. The function $\text{ONEMAX} : \{0, 1\}^n \to \mathbb{R}$ is defined by $\text{ONEMAX}(x) = \sum_{i=1}^{n} x[i]$. The function $\text{NEEDLE} : \{0, 1\}^n \to \mathbb{R}$ is defined by $\text{NEEDLE}(x) = \prod_{i=1}^{n} x[i]$.

For any function $f : \{0, 1\}^n \to \mathbb{R}$ and any $a \in \{0, 1\}^n$ let $f_a : \{0, 1\}^n \to \mathbb{R}$ be defined by $f_a(x) = f(x \oplus a)$, where $x \oplus a$ denotes the bit-wise exclusive or of x and a. For any function $f : \{0, 1\}^n \to \mathbb{R}$ let $f^* := \{f_a | a \in \{0, 1\}^n\}$.

The very well-known example function ONEMAX yields as function value the number of ones in the bit string. Its generalization ONEMAX_a can be described as maximizing the Hamming distance to the bit-wise complement of the global optimum $\bar{a}$, where the fitness value exactly equals this Hamming distance. Clearly, ONEMAX is rather easy to optimize since it gives clear hints into the direction of the unique global optimum. On the other, the example function NEEDLE is a flat plateau of fitness 0 with a single peak at the all ones bit string 1^n. Its generalization NEEDLE_a has this unique global optimum at $\bar{a}$. Since NEEDLE_a gives no hints at all with respect to the global optimum, it is rather difficult to optimize.

It is remarked that NEEDLE^* is closed under permutations of the search space. It follows from the NFL theorem (Theorem 1) that all algorithms that do not revisit any points make the same number of f-evaluations on NEEDLE^*. We consider a deterministic algorithm that enumerates the search space $\{0, 1\}^n$ in some fixed order. Clearly, for each $i \in \{1, 2, \ldots, 2^n\}$, there is exactly one function NEEDLE_a that is optimized by this algorithm with the ith f-evaluation. Thus, the average number of f-evaluations this algorithm (and due to $\bullet$ Theorem 1 any algorithm) makes on NEEDLE^* equals

$$\sum_{i=1}^{2^n} \frac{i}{2^n} = \frac{2^n(2^n+1)}{2 \cdot 2^n} = 2^{n-1} + \frac{1}{2}$$

and we obtain $B_{\mathscr{F}} = 2^{n-1} + \frac{1}{2}$. Nevertheless, the lower bound will be derived in the proof of the next theorem to see a very simple application of Yao's minimax principle ($\bullet$ Theorem 2).

Theorem 3 (Droste et al. 2006) $B_{\text{NEEDLE}^*} = 2^{n-1} + 1/2$. $B_{\text{ONEMAX}^*} = \Omega(n/\log n)$

Proof The upper bound for B_{NEEDLE^*} follows from the analysis of the deterministic algorithm enumerating the search space $\{0, 1\}^n$. For the lower bound, we use the uniform distribution over NEEDLE^* and consider a deterministic black-box algorithm A for NEEDLE^*. Clearly, such an algorithm is equivalent to a tree with at least 2^n nodes since otherwise there is some $x \in \{0, 1\}^n$ that is not a label of any node in this tree. Any function that has this x as its unique global optimum cannot be optimized by A. Since x is the unique global optimum of $\text{NEEDLE}_{\bar{x}}$, we see that we have at least 2^n nodes as claimed. Each node in the tree of an optimal black-box algorithm has at most one child. The only two possible function values are 0 and 1. After making an f-evaluation for some x with function value 1 the algorithm terminates since the global optimum is found. Thus, the tree is actually a list of 2^n nodes. Clearly, the average number of function evaluations made by this algorithm equals

$$\sum_{i=1}^{2^n} \frac{i}{2^n} = \frac{2^n(2^n+1)}{2 \cdot 2^n} = 2^{n-1} + \frac{1}{2}$$

and the lower bound follows.

For ONEMAX^* one can argue in a similar way, again using the uniform distribution over the class of functions under consideration. Any black-box algorithm for ONEMAX^* needs to contain at least 2^n nodes since otherwise there is some function ONEMAX_a that cannot be optimized by this algorithm. Now there are $n + 1$ different function values for ONEMAX_a. After sampling a point with function value n any optimal deterministic black-box algorithm for

OneMax* can terminate. Thus, the degree of each node in the tree of such an algorithm is bounded above by n. Since an n-ary tree of height h contains at most

$$\sum_{i=0}^{h} n^i = n^{h+1} - 1$$

nodes we have $n^{h+1} - 1$ and $h = \Omega(n/\log n)$ follows. Since the average height of this tree is still $\Omega(n/\log n)$, $\Omega(n/\log n)$ is obtained as lower bound on $B_{\text{OneMax*}}$.

Results on black-box complexity are not restricted to results for example functions and classes of functions obtained by generalizing such example functions. It is possible to derive nontrivial lower bounds on the black-box complexity of combinatorial optimization problems such as sorting and shortest paths. Moreover, for the class of unimodal problems an exponential lower bound of $2^{n^{\varepsilon}}$ for any positive constant $\varepsilon < 1$ is known (Droste et al. 2006).

4 Methods for the Analysis of Evolutionary Algorithms

The analysis of evolutionary algorithms is not so much concerned with the derivation of general lower bounds on the difficulty of problems as in black-box complexity but with proving concrete bounds on the (expected) optimization time of specific evolutionary algorithms for specific problems. While at the end of the day one is interested in results for one's own evolutionary algorithm on one's own specific problem or, from a broader perspective, in results on classes of evolutionary algorithms on relevant practical problems, it is worthwhile to start with very simple evolutionary algorithms on very simple example functions. The aim of this section is not to present the most advanced results obtained by clever applications of the most advanced methods. Here we aim at presenting the most important and useful methods in an accessible way following a hands-on kind of approach. The goal is to enable the reader of this section to apply these methods himself or herself and participate in the development of more advanced results and methods.

4.1 Fitness-Based Partitions

We begin with a very simple and basic method for the derivation of upper bounds on the expected optimization time. In spite of its simplicity, this method, called the method of fitness-based partitions, is very helpful and delivers strong bounds in many cases. Consider the $(1 + 1)$ EA (❷ *Algorithm 1*) on some arbitrary fitness function $f: \{0,1\}^n \to \mathbb{R}$. Due to the strict selection employed, the fitness value $f(x_t)$ is increasing with t, though not strictly, of course. Clearly, $\mathrm{E}(T_{(1+1)\ \mathrm{EA},f})$ equals the expected number of f-evaluations until the function value of the current search point x_t is increased to the maximum value for the first time. This together with the increasing nature of $f(x_t)$ leads one to the idea to structure the optimization of f by the $(1 + 1)$ EA in the following way. The search space is partitioned into disjoints sets of points with similar fitness values. We obtain an upper bound on $\mathrm{E}(T_{(1+1)\ \mathrm{EA},f})$ by summing up the average waiting times for leaving these sets. This idea is made more precise in the following definition and theorem.

Definition 2 Let $f: \{0,1\}^n \to \mathbb{R}$ be some fitness function. $L_0, L_1, \ldots, L_k$ (for $k \in \mathbb{N}$) is called an f-based partition if the following properties hold.

1. $L_0 \cup L_1 \cup \ldots \cup L_k = \{0, 1\}^n$
2. $\forall i < j \in \{0, 1, \ldots, k\}: \forall x \in L_i, y \in L_j: f(x) < f(y)$
3. $L_k = \{x \in \{0, 1\}^n \mid f(x) = \max\{f(y) \mid y \in \{0, 1\}^n\}\}$

Due to the strict selection, if the $(1 + 1)$ EA leaves some set L_i it can only reach a point in some set L_j with $j > i$. Moreover, with reaching L_k a global optimum is found. These observations lead directly to the following theorem.

Theorem 4 *For $f: \{0,1\}^n \to \mathbb{R}$ be some fitness function. Let $L_0, L_1, \ldots, L_k$ be some f-based partition. The probability of leaving L_i is defined as*

$$s_i := \min\left\{\sum_{j=i+1}^{k} \sum_{y \in L_j} \left(\frac{1}{n}\right)^{H(x,y)} \left(1 - \frac{1}{n}\right)^{n-H(x,y)} \mid x \in L_i\right\}$$

where $H(x, y)$ denotes the Hamming distance of $x, y \in \{0,1\}^n$

For the expected optimization time of the $(1 + 1)$ EA on f, the following holds:

$$\mathrm{E}\left(T_{(1+1)\ EA,f}\right) \le \sum_{i=0}^{k-1} \frac{1}{s_i}$$

Proof We observe that for any $x \in L_i$

$$\sum_{j=i+1}^{k} \sum_{y \in L_j} \left(\frac{1}{n}\right)^{H(x,y)} \left(1 - \frac{1}{n}\right)^{n-H(x,y)}$$

equals the probability to mutate $x \in L_i$ to some $y \in L_j$ with $j > i$. Thus, $f(y) > f(x)$ holds and such an event implies that the $(1 + 1)$ EA leaves L_i. Thus, s_i is a lower bound on the actual probability to leave the current set L_i by means of the next mutation. The expected waiting time for such an event is geometrically distributed with expectation $1/s_i$. Since each fitness layer needs to be left at most once

$$\mathrm{E}\left(T_{(1+1)\ EA,f}\right) \le \sum_{i=0}^{k-1} \frac{1}{s_i}$$

follows.

This method will be applied to the analysis of OneMax. Note that since the $(1 + 1)$ EA is completely symmetric with respect to 0- and 1-bits, the same result holds for any OneMax$_a$. We use the trivial f-based partition, namely, defining

$$L_i = \{x \in \{0, 1\}^n \mid \text{ONEMAX}(x) = i\}, i \in \{0, 1, \ldots, n\}$$

to have one fitness layer for each of the $n + 1$ fitness values. It is observed that in order to leave L_i, it suffices to flip exactly one of the $n - i$ 0-bits in one mutation. Thus,

$$s_i \ge \binom{n-i}{1} \frac{1}{n} \left(1 - \frac{1}{n}\right)^{n-1} \ge \frac{n-i}{en}$$

follows and we obtain

$$E\left(T_{(1+1)\text{ EA, OneMax}}\right) \leq \sum_{i=0}^{n-1} \frac{en}{n-i} = en \sum_{i=1}^{n} \frac{1}{i} = O(n \log n)$$

as an upper bound. In comparison with the lower bound obtained by means of black-box complexity, it can be seen that this upper bound is off by a factor of at most $O(\log^2 n)$. It will be seen in ❯ Sect. 4.2 that it is even tight.

We consider the function BinVal: $\{0, 1\}^n \to \mathbb{R}$ defined by

$$\text{BnVal}(x) = \sum_{i=1}^{n} x[i] 2^{n-i-1}$$

We observe that $x \in \{0, 1\}^n$ yields as function value the integer that has x as its binary representation. Using the trivial f-based partition, namely

$$L_i = \{x \in \{0, 1\}^n \mid \text{BnVal}(x) = i\}, i \in \{0, 1, \ldots, 2^n - 1\}$$

we obtain 2^n fitness layers. Since it is obvious that at least one generation is needed to leave each layer, the best one can hope for is $2^n - 1$ as upper bound. We learn that a large number of fitness layers cannot be used if small upper bounds are to be proved.

We try to be a bit less naive and define

$$L_i = \left\{ x \in \{0, 1\}^n \setminus \bigcup_{j=0}^{i-1} L_j \mid \text{BnVal}(x) < \sum_{j=0}^{i} 2^{n-1-j} \right\}, i \in \{0, 1, \ldots, n\}$$

as f-based partition for BinVal. We observe that L_i contains all $x \in \{0, 1\}^n$ that start with a sequence of i consecutive 1-bits followed by a 0-bit, so L_0 contains exactly all $x \in \{0, 1\}^n$ where the leftmost bit is 0 and $L_n = 1^n$. Thus, for each L_i with $i < n$, it suffices to flip the leftmost 0-bit in order to leave L_i. This implies

$$s_i \geq \frac{1}{n} \left(1 - \frac{1}{n}\right)^{n-1} \geq \frac{1}{en}$$

as lower bound on s_i and we obtain

$$E\left(T_{(1+1)\text{ EA,BnVal}}\right) \leq \sum_{i=0}^{n-1} en = en^2 = O(n^2)$$

as upper bound for the expected optimization time of the $(1 + 1)$ EA on BinVal. While being much improved over $2^n - 1$ it will be seen in ❯ Sect. 4.4 that this is still not tight.

Fitness-based partitions do also work for families of functions with similar structure. To see an example, we consider Jump_k: $\{0, 1\}^n \to \mathbb{R}$ (Droste et al. 2002a) with

$$\text{Jump}_k(x) = \begin{cases} n - \text{OneMax}(x) & \text{if } n - k < \text{OneMax}(x) < n \\ k + \text{OneMax}(x) & \text{otherwise} \end{cases}$$

where $k \in \{1, 2, \ldots, n\}$ is a parameter. These n different fitness functions all have the same main properties. The unique global optimum is the all one string 1^n, all points with exactly $n - k$ 1-bits are local optima (except for $k = 1$, since $\text{Jump}_1 = 1 + \text{OneMax}$ and there is no local optimum that is not also a global one). For $x \in \{0, 1\}^n$ with less than $n - k$ 1-bits the function equals $k + \text{OneMax}(x)$ so in that region, the local optima can be found in time $O(n\log n)$. There a direct mutation to the global optimum is necessary; such a mutation flips exactly all k remaining 0-bits. An upper bound on $E(T_{(1+1)\text{ EA,Jump}_k})$ is obtained using the trivial f-based partition where each fitness value gets its own fitness layer L_i by

$$L_i = \{x \in \{0, 1\}^n \mid \text{Jump}_k(x) = i\}, i \in \{1, 2, \ldots, n\}$$

and $L_{n+1} = \{1^n\}$. For all $i < k$ we have $n - k < \text{OneMax}(x) < n$ for all $x \in L_i$. Thus, for these i it suffices to flip one of the $n - i$ 1-bits in a mutation to leave L_i. For all i with $k \le i < n$ the situation is similar to OneMax, it suffices to flip one of the $k + n - i$ 0-bits to leave L_i. Finally, for L_n, it is necessary to flip exactly the k remaining 0-bits. These insights yield

$$s_i \ge \begin{cases} \binom{n-i}{1}\frac{1}{n}\left(1 - \frac{1}{n}\right)^{n-1} \ge \frac{n-i}{en} & \text{for } i < k \\ \binom{n-i+k}{1}\frac{1}{n}\left(1 - \frac{1}{n}\right)^{n-1} \ge \frac{n-i+k}{en} & \text{for } k \le i < n \\ \left(\frac{1}{n}\right)^k\left(1 - \frac{1}{n}\right)^{n-k} \ge \frac{1}{en^k} & \text{for } k = n \end{cases}$$

leading to

$$E\left(T_{(1+1)\text{ EA,Jump}_k}\right) \le \sum_{i=1}^{k-1}\frac{en}{n-i} + \sum_{i=k}^{n}\frac{en}{n-i+k} + en^k = O\left(n\log n + n^k\right)$$

as upper bound on the expected optimization time.

We have already seen for BinVal that different f-based partitions can lead to different upper bounds on $E(T_{(1+1)\text{ EA},f})$ for the same fitness function f. In some situations this may turn out to be useful. We consider the class of functions LongPath_k: $\{0, 1\}^n \to \mathbb{R}$ (Horn et al. 1994) with parameter $k \in \mathbb{N}, k > 1$ and $(n-1)/k \in \mathbb{N}$, given by

$$\text{LongPath}_k(x) = \begin{cases} n^2 + i & \text{for } x = x^{(i)} \\ n^2 - n\sum_{i=1}^{k}x[i] - \sum_{i=k+1}^{n}x[i] & \text{otherwise} \end{cases}$$

based on a path $P_k^n = \left(x^{(1)}, x^{(2)}, \ldots, x^{(l(k,n))}\right)$ with $x^{(i)} \in \{0, 1\}^n$ and $H\left(x^{(i)}, x^{(i+1)}\right) = 1$ for all i. The path P_k^n is recursively defined by $P_k^1 = (0, 1)$ for $n = 1$ and

$$P_k^n = \left(0^k x^{(1)}, 0^k x^{(2)}, \ldots, 0^k x^{(l(k,n-k))}, 0^{k-1}1 x^{(l(k,n-k))}, 0^{k-2}11 x^{(l(k,n-k))}, \ldots,\right.$$
$$\left. 1^k x^{(l(k,n-k))}, 1^k x^{(l(k,n-k)-1)}, \ldots, 1^k x^{(1)}\right)$$

for $n \in \mathbb{N}$ where $x^{(i)} \in \{0, 1\}^{n-k}$ is the ith point in the path P_k^{n-k}. The path P_k^n is called the long-k path of dimension n, it has length $l(k, n) = (k + 1)2^{(n-1)/k} - k + 1$ and the following interesting property. For each $x^{(i)} \in P_k^n$ with at least s successors on the path we have that $H\left(x^{(i)}, x^{(i+j)}\right) = j$ for all $j \in \{1, 2, \ldots, \min\{s, k - 1\}\}$ and no other successor of $x^{(i)}$ on the path has Hamming distance j to $x^{(i)}$ (Rudolph 1997).

Clearly, LongPath_k does not have any local optima. Either a point is the unique global optimum or it has a Hamming neighbor with strictly larger function value. Thus, it is possible to optimize LongPath_k by means of single-bit mutations. On the other hand, due to its recursive definition, the long k-path of dimension n consists of k segments where for each path point in any segment, it suffices to flip at most k specific bits to leave the current segment.

Thus, there are two different f-based partitions that align with the properties of LONGPATH$_k$. The trivial one has one fitness layer for each fitness value. Since the mutations of single bits are sufficient to leave a fitness layer, this yields $E(T_{(1+1) \text{ EA,LONGPATH}_k}) = O(n|P_k^n|)$ as upper bound. The other f-based partition uses the segments of the long k-path as fitness layers leading to $E(T_{(1+1) \text{ EA,LONGPATH}_k}) = O(n^{k+1}/k)$ as upper bound. Since both upper bounds are valid we have the following result.

Theorem 5 (Rudolph 1997) $E(T_{(1+1) \text{ EA,LONGPATH}_k}) = O(\min\{n|P_k^n|, n^{k+1}/k\})$

As a final example for the wide applicability of the method of fitness-based partitions the class of linear functions is considered. A fitness function $f: \{0,1\}^n \to \mathbb{R}$ is called *linear* if there exist weights $x_0, x_1, \ldots, x_n \in \mathbb{R}$ such that $f(x) = w_0 + \sum_{i=1}^{n} x[i]w_i$ holds. We observe that ONEMAX as well as BINVAL are linear functions.

Since the (1+1) EA is completely symmetric with respect to the roles of 0-bits and 1-bits one can change any linear function with some weights $w_i < 0$ to another linear function with no negative weights simply by replacing w_i by $-w_i$. This changes the roles of 0 and 1 at the ith position and does not influence the optimization time in any way. Moreover, one can assume $w_0 = 0$ since a constant summand has no influence on the selection employed in the (1+1) EA. Since the $(1 + 1)$ EA is also completely symmetric with respect to bit positions, one can reorder the bits such that $w_1 \geq w_2 \geq \ldots \geq w_n$ holds. Finally, if one is concerned with upper bounds on the expected optimization time, it can be assumed that even $w_1 \geq w_2 \geq \ldots \geq w_n > 0$ holds, since $w_i = 0$ implies that the ith bit has no influence on the function value. Clearly, this doubles the number of global optima and can only decrease the expected optimization time in comparison to the case where $w_i \neq 0$ holds.

Now, considering any linear function $f(x) = \sum_{i=1}^{n} w_i x[i]$ with $w_1 \geq w_2 \geq \ldots \geq w_n > 0$ one defines the following f-based partition:

$$L_i = \left\{ x \in \{0,1\}^n \mid \sum_{j=1}^{i-1} w_j x[j] \leq f(x) < \sum_{j=1}^{i} w_j x[j] \right\}, i \in \{1, 2, \ldots, n\}$$

and $L_{n+1} = \{1^n\}$. It is observed that now the situation is similar to the situation for BINVAL. For any $x \in L_i$ there is at least one bit among the leftmost i bits such that flipping this bits leads to an offspring $y \in L_j$ with $j > i$. This yields $s_i \geq (1/n)(1 - 1/n)^{n-1} \geq 1/(en)$ and we obtain

$$E(T_{(1+1) \text{ EA,}f}) \leq \sum_{i=1}^{n} en = O(n^2)$$

as upper bound. Note that this upper bound holds for any linear function f.

The method of fitness-based partitions is not necessarily restricted to the simple $(1 + 1)$ EA, but it is tightly linked to the strict plus-selection that does not accept any decrease in function values. It is not too difficult to generalize it to other evolutionary algorithms with such a selection. For example, it can be shown that for the $(1 + \lambda)$ EA, that is, the $(\mu + \lambda)$ EA with population size $\mu = 1$, the expected optimization time on ONEMAX is bounded above by $E(T_{1+\lambda \text{ EA,ONEMAX}}) = O(n \log n + n\lambda)$ using the method of fitness-based partitions (Jansen et al. 2005). But it does not work for the (μ, λ) EA due to the missing monotonicity in fitness of the current best member of the population.

4.2   General Lower Bound

In order to assess the strength of an upper bound on $E(T_{A,f})$, a lower bound is needed for comparison. One way of obtaining lower bounds is resorting to fundamental limits. This, however, will not lead to strong lower bounds since "no free lunch" results do not apply to function classes of practical interest and black-box complexity most often yields relatively weak lower bounds since it takes into account the performance of optimal algorithms, not the specific evolutionary algorithm one is dealing with. In order to obtain stronger lower bounds on $E(T_{A,f})$ one needs to take into account specific properties of evolutionary algorithm A.

We consider an evolutionary algorithm A that uses standard bit mutations as the only way of creating new offspring. This holds for the $(1 + 1)$ EA and the $(\mu + \lambda)$-EA without crossover. If A has a population of size $\mu > n \log n$ it makes $\Omega(n \log n)$ f-evaluations already in the initialization. It can be claimed that if the fitness function has a unique global optimum $x^* \in \{0, 1\}^n$, we have $E(T_{A,f}) = \Omega(n \log n)$ in any case, that is, for $\mu \leq n \log n$, too (Jansen et al. 2005).

We consider some $x \in \{0, 1\}^n$ from the initial population, that is, chosen uniformly at random. Each bit in x differs from the corresponding bit in the unique global optimum x^* with probability $1/2$. We can introduce random variables X_i with $i \in \{1, 2, \ldots, n\}$, one for each bit, that assumes the value 1 if $x[i] \neq x^*[i]$ and the value 0 otherwise. Clearly, the expected Hamming distance between x and x^* equals $E(H(x, x^*)) = \sum_{i=1}^{n} X_i = n/2$. Since the random variables X_i are independent, one is in a situation where Chernoff bounds can be applied (Motwani and Raghavan 1995). We have a random variable $X = \sum_{i=1}^{n} X_i$ with $0 < \text{Prob}\,(X_i = 1) < 1$ for all $i \in \{1, 2, \ldots, n\}$. Chernoff bounds yield strong bounds on the probability that X deviates considerably from its expected value. For example, for δ with $0 < \delta < 1$

$$\text{Prob}\,(X \geq (1 + \delta)E(X)) = e^{-E(X)\delta^2/3}$$
$$\text{Prob}\,(X \leq (1 - \delta)E(X)) = e^{-E(X)\delta^2/2}$$

hold. In this situation, we have that with probability $1 - e^{-\Omega(n)}$ the Hamming distance of x and x^* is bounded below by $n/3$. With a population of size $\mu \leq n \log n$ this holds for each member of the population. We get a bound on the probability that any member of the population has a Hamming distance of less than $n/3$ to the unique global optimum x^* by taking the union bound, that is, by summing up the probabilities for each member of the initial population. Thus, with probability $1 - \sum_{i=1}^{\mu} e^{-\Omega(n)} = 1 - e^{-\Omega(n)}$, each member of the initial population has Hamming distance at least $n/3$ to x^*. One can denote this event as B and have $\text{Prob}\,(B) = 1 - e^{-\Omega(n)}$.

If one wants to reach x^* in this situation within at most t f-evaluations, there has to be at least one member of the initial population where all those differing bits flip at least once within at most t mutations. Remember that the number of differing bits is bounded below by $n/3$. The probability for this event can be estimated in a very direct way. A single bit flips with probability $1/n$, thus it does not flip at all in t mutations with probability $(1 - 1/n)^t$. One can see that it flips at least once within these t mutations with probability $1 - (1 - 1/n)^t$. Since the bits are independent we have that with probability at most $\left(1 - (1 - 1/n)^t\right)^{n/3}$ the at least $n/3$ bits differing from x^* are all flipped at least once in these t mutations. Thus, finally, one can bound the probability that among those at least $n/3$ bits there is at least one that was never flipped in t mutations from below by $1 - \left(1 - (1 - 1/n)^t\right)^{n/3}$. We consider $t = (n-1)\ln n$ generations and see that $(1 - 1/n)^{(n-1)\ln n} \geq e^{-\ln n} = 1/n$ holds. Thus, in $t = (n-1)\ln n$ generations the probability that there is one bit that is needed to flip in order to reach x^* is not flipped at all is bounded below by $1 - (1 - 1/n)^{n/3} \geq 1 - e^{-1/3} = \Omega(1)$. We denote this event as C and have Prob $C = \Omega(1)$. For the optimization time we thus have

$$E\left(T_{A,f}\right) \geq E\left(T_{A,f}|B \wedge C\right) \text{Prob}(B \wedge C) = E\left(T_{A,f}|B \wedge C\right) \text{Prob}(B) \text{Prob}(C)$$
$$= (n-1)\ln(n)\left(1 - e^{-\Omega(n)}\right)\Omega(1) = \Omega(n \log n)$$

as claimed. It is not difficult to see that the proof can be generalized to fitness functions where the number of global optima is larger than 1 but polynomially bounded. It can be remarked that this proof is similar to the classical result on the coupon collector's problem (Mitzenmacher and Upfal 2005; Motwani and Raghavan 1995).

Clearly, this general lower bound $E(T_{A,f}) = \Omega(n \log n)$ that holds for any mutation-based evolutionary algorithm and any fitness function $f: \{0, 1\}^n \to \mathbb{R}$ is still rather general and as an immediate consequence rather weak. Nevertheless, it proves that the upper bound on ONEMAX is in fact asymptotically tight and we have $E\left(T_{(1+1) \text{ EA,ONEMAX}_a}\right) = \Theta(n \log n)$ for all $a \in \{0, 1\}^n$.

4.3 Expected Multiplicative Distance Decrease

We return to the often simpler task of proving upper bounds on the expected optimization time. The idea of fitness-based partitions is to measure the progress of $(1 + 1)$ EA by means of the increase in fitness values. To achieve this, the fitness values are grouped in appropriate layers. Considering BINVAL as the example, we have already seen that finding an appropriate grouping can make the difference between a good upper bound and a completely useless upper bound. While this method has the advantage that it is easy to apply and often yields very useful and sometimes even tight upper bounds, it has the intrinsic disadvantage that one needs to define in advance how the algorithm under consideration will most likely increase the function values. This ordering is in some sense implicit in the definition of the fitness layers. One can see this by considering BINVAL. The definition of L_i made it necessary to have the bits in the current bit string flip from left to right. Clearly, the real random process under consideration is less strict: any 0-bits can be flipped to increase the function value. It may well be the case that due to this greater degree of freedom, the real random process that governs the $(1 + 1)$ EA on BINVAL is faster in finding the optimum than the upper bound of $O(n^2)$ predicts.

We reconsider the main idea – measuring the progress the algorithm makes by means of an increase in function value – and capture it in a more general form. To be concrete, attention should be again restricted to the $(1 + 1)$ EA and it should be remarked that generalizations to other evolutionary algorithms like the $(1 + \lambda)$ EA are again not too difficult.

Consider the $(1 + 1)$ EA optimizing some fitness function $f: \{0, 1\}^n \to \mathbb{Z}$ with $f_{\text{opt}} = \max\{f(x)|x \in \{0, 1\}^n\}$. Note that the assumption that the function values are integers will be important in the following. When the current search point is x_t, we know that a global optimum is reached if the function value is increased by $f_{opt} - f(x_t)$. If, for any $x_t \in \{0, 1\}^n$, one is able to describe a sequence of at most l operations that can all be applied to x_t and that when applied simultaneously lead to the global optimum, then each of these operations causes an average increase of function value of at least $(f_{opt} - f(x_t))/l$. In different terms, each operation decreases the distance to the maximal fitness value $f_{opt} - f(x_t)$ on average by a factor of at least $1 - 1/l$. Since different potential points $x_t \in \{0, 1\}^n$ are dealt with, it does not really make sense to talk about $f(x_t)$; the average increase in function value $(f_{opt} - f(x_t))/l$ clearly depends on the x_t under consideration. Thus, we consider $f_{max} = \max\{f_{opt} - f(x) \mid x \in \{0, 1\}^n\}$. Now we have the distance to the maximal function value after t such operations is decreased in expectation by a factor of at least $(1 - 1/l)^t$. Since the function values are integers, we know that the global optimum is reached if the difference is decreased to less than 1. Thus, if $(1 - 1/l)^t f_{max} < 1$ holds, we know that in expectation the optimum is reached. So, a number of $l \ln f_{max}$ such local operations is on average sufficient to reach the optimum. If it is additionally assumed that the expected waiting time for each such local operation is bounded above by w we obtain

$$E\big(T_{(1+1)\,\text{EA},f}\big) = O(wl \ln f_{max})$$

as upper bound on the expected optimization time.

This method of the expected multiplicative distance decrease (Neumann and Wegener 2004) is applied to the class of linear functions with integer weights. Consider some $f: \{0, 1\}^n \to \mathbb{Z}$ with $f(x) = w_0 + \sum_{i=1}^{n} w_i x[i]$ with all $w_i \in \mathbb{Z}$. As we did when using the method of fitness-based partitions, we assume without loss of generality that $w_0 = 0$ and $w_1 \geq w_2 \geq w_2 \geq \cdots \geq w_n > 0$ holds since one wants to derive an upper bound on $E(T_{(1+1)\,\text{EA},f})$. Remember that we had $E\big(T_{(1+1)\,\text{EA},f}\big) = O(n^2)$ using fitness-based partitions.

Due to the assumptions on f, the unique global optimum is 1^* and we have

$$f_{max} = \max\{f_{opt} - f(x) \mid x \in \{0, 1\}^n\} = \sum_{i=1}^{n} w_i \leq n w_{max}$$

where $w_{max} = \max\{w_i \mid i \in \{1, 2, \dots, n\}\}$ denotes the maximal weight of a single bit. For each $x \in \{0, 1\}^n$ we consider the at most n mutations that flip single bits for positions i where $x[i] = 0$ holds. Clearly, applying these up to n mutations simultaneously leads to the unique global optimum 1^n. Moreover, the probability to have a mutation flipping a single bit equals

$$\binom{n}{1} \frac{1}{n} \left(1 - \frac{1}{n}\right)^{n-1} = \left(\frac{1}{n}\right)^{n-1} > e^{-1}$$

so we have $w < e$ in the notation. Since $l = n$ we have

$$E\big(T_{(1+1)\,\text{EA},f}\big) = O(wl \ln f_{max}) = O(n \ln(n w_{max}))$$

for any linear functions with integer weights. For BINVAL, we have $w_{max} = 2^{n-1}$ and $E\big(T_{(1+1)\,\text{EA},f}\big) = O(n^2)$ follows. This is no improvement over the bound of the same size obtained using f-based partitions. But this result is more powerful. If one restricts oneself to linear functions with polynomially bounded integer weights, one can obtain $E\big(T_{(1+1)\,\text{EA},f}\big) = O(n \log n)$, which is asymptotically tight since it matches the general lower bound $\Omega(n \log n)$.

While being more flexible than the method of fitness-based partitions, the method of the expected multiplicative distance decrease has the inherent disadvantage that the upper bounds one can obtain directly depend on the size of the function values. Since for example functions there are no bounds on the size of the function values, one may be tempted to believe that at least for some example functions, one cannot obtain any useful result at all using this method. This, however, is too pessimistic. It turns out that function values can be unnecessarily large when considering evolutionary algorithms or other randomized search heuristics where selection depends only on the ordering of the function values and not on the function values themselves. Another way to say this is the following. Given a fitness function $f: \{0,1\}^n \to \mathbb{Z}$ with arbitrary function values, it is possible to come up with another fitness function $f': \{0,1\}^n \to \mathbb{Z}$ such that the $(1 + 1)$ EA behaves the same on f and f' but the function values of f' are limited in size (Reichel and Skutella 2009). This way, much tighter bounds can be obtained, not only for example functions but also for combinatorial optimization problems.

4.4 Drift Analysis for Upper Bounds

The idea of both methods, fitness-based partitions as well as the method of the expected multiplicative distance decrease, is to measure the progress of the evolutionary algorithm A in some way. When using fitness-based partitions this progress has to be achieved in a pre-defined way that is implicit in the definition of the fitness layers. When using the method of the expected multiplicative distance decrease, the progress can be obtained by operations in any way but it is always measured in function values. We gain even more flexibility if we allow for different measures of progress. Consider some evolutionary algorithm A where Z denotes the set of all possible populations of A. We call a function $d: Z \to \mathbb{R}_0^+$ a *distance measure* if $d(P) = 0$ holds if and only if the population $P \in Z$ contains a point with maximal function value. Calling a unary operator a distance measure appears to be strange at first sight. However, what is really measured is the distance to the set of global optima. Thus, we can omit the obligatory second part in distance measuring because it is fixed for a fixed fitness function f. Note that this notion of distance is very weak, in particular, d need not be a metric. Again, we concentrate on the $(1 + 1)$ EA in order to simplify the considerations. It has to be remarked, though, that the method taken into consideration is not at all restricted to such simple evolutionary algorithms and that it can be easily applied to much more complex evolutionary algorithms.

For the $(1 + 1)$ EA we have $Z = \{0, 1\}^n$ since the current population consists of a single search point $x_t \in \{0, 1\}^n$. Clearly, we have $T_{(1+1)\,\text{EA},f} = \min\{t \mid d(x_t) = 0\}$ for any fitness function f. We consider the maximum distance M and want to bound the average time until initial distance that is at most M is decreased to 0. Clearly, $M = \max\{d(x)|x \in Z\}$ holds. We concentrate on a single generation and consider the decrease in distance in a single generation, that is, $D_t = d(x_{t-1}) - d(x_t)$. Note that D_t is a random variable that may take negative values with positive probability, actually indicating an increase in distance. One is interested in the expected decrease in distance in a single generation, $E(D_t | T_{(1+1)\,\text{EA},\,f} > t)$, and call this *drift*. Since one wants to derive an upper bound on the expected optimization time one can assume a pessimistic point of view and consider the worst-case drift Δ, that is, the smallest drift possible, $\Delta = \min\{E(D_t \mid T_{(1+1)\,\text{EA},f} > t) \mid t \in \mathbb{N}\}$. Even if this worst-case drift Δ is strictly positive, one can prove the drift theorem $E(T_{(1+1)\,\text{EA},f}) \le M/\Delta$ (He and Yao 2004), here for the $(1 + 1)$ EA. The result is intuitively plausible. If one wants to overcome a distance of at most M and can travel at a speed of at least Δ, the average travel time is bounded above by M/Δ. Since the proof will turn out to be helpful when looking for a similar method for lower bounds, the theorem is stated together with its proof here. Again, we restrict ourselves to the $(1 + 1)$ EA even though much more general results can be proved with essentially the same proof.

Theorem 6 (Drift Theorem for the $(1 + 1)$ EA (He and Yao 2004)) *Consider the $(1 + 1)$ EA on some function $f: \{0, 1\}^n \to \mathbb{R}$ and some distance measure $d: \{0, 1\}^n \to \mathbb{R}_0^+$ with $d(x) = 0 \Leftrightarrow f(x) = \max\{f(y) \mid y \in \{0, 1\}^n\}$. Let $M = \max\{d(x) \mid x \in \{0, 1\}^n\}$, $D_t = d(x_{t-1}) - d(x_t)$, and $\Delta = \min\{\mathrm{E}(D_t \mid T_{(1+1)\ \mathrm{EA},f} > t) \mid t \in \mathbb{N}\}$.*
$\Delta > 0 \Rightarrow \mathrm{E}(T_{(1+1)\ \mathrm{EA},f}) \leq M/\Delta$

Proof In order to simplify notation we define $T := T_{(1+1)\ \mathrm{EA},f}$. If we consider $\sum_{t=1}^{T} D_t$ we see that

$$\sum_{i=1}^{T} D_i = \sum_{t=1}^{T} d(x_{i-1}) - d(x_i) = d(x_0) - d(x_T) = d(x_0)$$

holds since most of the summands cancel out and we have $d(x_T) = 0$ by assumption on d and definition of T. Since M is an upper bound on $d(x)$ for all $x \in \{0, 1\}^n$, we have $M \geq d(x_0)$ and

$$M \geq \sum_{i=1}^{T} D_i$$

follows. Since M is not a random variable, we have $M = \mathrm{E}(M)$. By the definition of the expected value and the law of total probability, we get

$$M \geq \mathrm{E}\left(\sum_{i=1}^{T} D_i\right) = \sum_{t=1}^{\infty} \mathrm{Prob}(T = t)\mathrm{E}\left(\sum_{i=1}^{T} D_i \mid T = t\right)$$

Making use of the linearity of the expectation and changing the ordering in the sums, we get

$$\sum_{t=1}^{\infty} \mathrm{Prob}(T = t)\mathrm{E}\left(\sum_{i=1}^{T} D_i \mid T = t\right) = \sum_{t=1}^{\infty}\sum_{i=1}^{T} \mathrm{Prob}(T = t)\mathrm{E}(D_i \mid T = t)$$

Making use of the total law of probability a second time and reordering again, we get

$$\sum_{i=1}^{\infty}\sum_{t=i}^{\infty} \mathrm{Prob}\ (T = t)\mathrm{E}(D_i \mid T = t)$$

$$= \sum_{i=1}^{\infty}\sum_{t=i}^{\infty} \mathrm{Prob}(T \geq i)\mathrm{Prob}\ (T = t \mid T \geq i)\mathrm{E}(D_i \mid T = t \wedge T \geq i)$$

$$= \sum_{i=1}^{\infty} \mathrm{Prob}(T \geq i) \sum_{t=i}^{\infty} \mathrm{Prob}\ (T = t \mid T \geq i)\mathrm{E}(D_i \mid T = t \wedge T \geq i)$$

$$= \sum_{i=1}^{\infty} \mathrm{Prob}(T \geq i) \sum_{t=1}^{\infty} \mathrm{Prob}\ (T = t \mid T \geq i)\mathrm{E}(D_i \mid T = t \wedge T \geq i)$$

where the last equation makes use of the fact that $\mathrm{Prob}\ (T = t \mid T \geq i) = 0$ for $t < i$ holds.

Another application of the law of total probability (this time in the reverse direction) yields

$$\sum_{i=1}^{\infty} \text{Prob } (T \ge i) \sum_{t=1}^{\infty} \text{Prob } (T = t \mid T \ge i) \text{E}(D_i \mid T = t \wedge T \ge i)$$

$$= \sum_{i=1}^{\infty} \text{Prob } (T \ge i) \text{E}(D_i \mid T \ge i)$$

Clearly, $\Delta \le \text{E}(D_i | T \ge i)$ holds, so we have

$$\sum_{i=1}^{\infty} \text{Prob } (T \ge i) \text{E}(D_i \mid T \ge i) \ge \Delta \sum_{i=1}^{\infty} \text{Prob } (T \ge i) = \Delta \text{E}(T)$$

by the definition of the expectation. Putting things together we observe that we proved $M \ge \Delta \text{E}(T)$ and $\text{E}(T) \le M/\Delta$ follows for $\Delta > 0$.

One of the advantages of the drift theorem is that it allows for the definition of almost arbitrary distance measures. It was known that the expected optimization time of the (1+1) EA on linear functions is $O(n \log n)$ (Droste et al. 2002a) (different from the bound $O(n^2)$ that was proven here using f-based partitions) before the method of drift analysis was introduced to the field of evolutionary computation, but the proof is long, tedious, and difficult to follow. Applying drift analysis, the same result can be shown in a much simpler way making use of an appropriate distance measure.

As usual, we consider a linear function $f: \{0, 1\}^n \to \mathbb{R}$ with $f(x) = \sum_{i=1}^{n} w_i x[i]$ where $w_1 \ge w_2 \ge \cdots \ge w_n > 0$ holds. Clearly, the bits with smaller index have at least the potential to be more important than the other bits. In one extreme case, BINVAL is an example, any bit has a weight w_i that is so large that it dominates all bits to its right, that is, $w_i > \sum_{j=i+1}^{n} w_j$. In the other extreme case, ONEMAX is an example, all weights are equal. We try to prove an upper bound for all these cases simultaneously by defining a distance measure (similar to a potential function) that in some way expresses the greater importance of bits that are closer to the left of the bit string. We define (He and Yao 2004)

$$d(x) = \ln \left(1 + 2 \sum_{i=1}^{n/2} (1 - x[i]) + \sum_{i=(n/2)+1}^{n} (1 - x[i]) \right)$$

making bits in the left half twice as important as bits in the right half. It can be observed that

$$M = \max \left\{ \ln \left(1 + 2 \sum_{i=1}^{n/2} (1 - x[i]) + \sum_{i=(n/2)+1}^{n} (1 - x[i]) \right) \mid x \in \{0, 1\}^n \right\}$$

$$= \ln \left(1 + n + \frac{n}{2} \right) = \Theta(\log n)$$

holds. For an upper bound, one needs now a lower bound on

$$\text{E}\big(D_t \mid T_{(1+1) \text{ EA}, f} > t\big) = \text{E}\big(d(x_{t-1}) - d(x_t) \mid T_{(1+1) \text{ EA}, f} > t\big)$$

It is easy to observe that this expectation is minimal either for $x_{t-1} = 011 \cdots 1 = 01^{n-1}$ or $x_{t-1} = 11 \cdots 1011 \ldots 1 = 1^{n/2} 01^{(n/2)-1}$. One can skip the still somewhat tedious calculations for these two cases and remark that $\Omega(1/n)$ can be proved as the lower bound in both cases. So, we have $\Delta = \Omega(1/n)$ and $M = \Theta(\log n)$ and this leads to

$$E\left(T_{(1+1)\text{ EA},f}\right) = \frac{M}{\Delta} = \frac{\Theta(\log n)}{\Omega(1/n)} = O(n \log n)$$

as claimed.

Another advantage of this method is that it is not a problem if the distance is not monotonically decreased during the run. It completely suffices if the expected decrease in distance, that is, the drift, is positive. This can be illustrated with a small example that makes use of the (μ, λ) EA (❷ Algorithm 3).

Consider the $(1, n)$ EA, that is, the (μ, λ) EA with $\mu = 1$ and $\lambda = n$ on the function LEADINGONES : $\{0, 1\}^n \to \mathbb{R}$ that is defined by $\text{LEADINGONES}(x) = \sum_{i=1}^{n} \prod_{j=1}^{i} x[j]$. The function value LEADINGONES(x) equals the number of consecutive 1-bits in x counting from left to right. A simple application of fitness-based partitions yields $E(T_{(1+1)\text{EA,LEADINGONES}}) = O(n^2)$ (Droste et al. 2002a) and $E(T_{(1+\lambda)\text{ EA,LEADINGONES}}) = O(n^2 + n\lambda)$ (Jansen et al. 2005). For the $(1, \lambda)$ EA, however, this method cannot be applied. With drift analysis, however, an upper bound can be derived with ease.

We define $d(x) = n - \text{LEADINGONES}(x)$ as distance and observe that $M = n$ holds. Applying the drift theorem, as presented in Theorem 6, yields an upper bound on the number of expected *generations* the evolutionary algorithm needs to locate an optimum. For the $(1 + 1)$ EA this simply equals the number of f-evaluations. For the $(1, \lambda)$ EA, this number needs to be multiplied by λ in order to get the expected number of f-evaluations. For the estimation of $E(d(x_{t-1}) - d(x_t)|T > t)$ the following two simple observations are made. If among the $\lambda = n$ mutations generating offspring there is at least one where exactly the left most 0-bit is flipped, the distance is decreased by exactly 1. Since such a mutation has probability $(1/n)(1 - 1/n)^{n-1} \geq 1/(en)$ we have that this happens with probability at least $1 - (1 - (1/(en)))^n = \Omega(1)$. Second, the distance can only increase if in one generation there is no mutation that does not change a bit at all. This holds since such a mutation simply creates a copy of its parent. In this case, the parent belongs to the offspring and the fitness of the parent in the next generation cannot be worse. Such an event happens with probability at most $(1 - (1 - 1/n)^n)^n \leq (1 - 1/e)^n = e^{-\Theta(n)}$. Moreover, in this case, the distance cannot increase by more than n. Together this yields

$$E(d(x_{t-1}) - d(x_t)|T > t) \geq 1 \cdot \left(1 - \left(1 - \left(\frac{1}{en}\right)\right)\right) - n \cdot \left(\left(1 - \frac{1}{n}\right)^n\right)^n$$

$$\geq 1 \cdot \Omega(1) - n \cdot e^{-\Theta(n)} = \Omega(1)$$

and proves that $\Delta = \Omega(1)$ holds. Thus, we have $E(T_{(1, n)\text{ EA, LEADINGONES}}) = O(n)^2$.

4.5   Drift Analysis for Lower Bounds

It would be nice to have such a general and powerful method like drift analysis for lower bounds, too. We reconsider the proof of ❷ Theorem 6 and observe that this is not so hard to have. In the calculations in this proof, almost all steps are exact equalities, only at two places we made estimations and use of bounds. One is $\Delta \leq E(D_i | T \geq i)$ and this stems from the definition of Δ:

$$\Delta = \min\{E(D_t \mid T_{(1+1)\ \text{EA}, f} > t) \mid t \in \mathbb{N}\}$$

Clearly, for a lower bound on the expected optimization time we need a different definition that takes into account the maximal drift, not the minimal as for the upper bound. Thus, we replace Δ by

$$\Delta_l = \max\{E(D_t \mid T_{(1+1)\ \text{EA}, f} > t) \mid t \in \mathbb{N}\}$$

and have $\Delta_l \geq E(D_i | T \geq i)$ as needed.

The second place one can make use of a bound is $M \geq \sum_{i=1}^{T} D_i$. This, again is true due to the definition of M

$$M = \max\ \{d(x) \mid x \in \{0, 1\}^n\}$$

Clearly, one could again replace M by $M_l = \min\ \{d(x) | x \in \{0, 1\}^n\}$, but this, unfortunately, is pointless: $M_l = 0$ holds and only the trivial bound $E(T_{A, f}) \geq 0$ can be achieved. Reconsidering the proof of (❯ Theorem 6) a bit closer, we observe that actually the first step reads

$$E(M) = M \geq E\left(\sum_{i=1}^{T} D_i\right)$$

and we see that $E(M)$ could be replaced by $E(d(P_0))$, the expected distance of the initial population P_0. This yields $E(d(P_0)) = E\left(\sum_{i=1}^{T} D_i\right) \leq E(T_{A,f})\Delta_l$ and we obtain $E(T_{A, f}) \geq E(d(P_0))/\Delta_l$ as drift theorem for lower bounds.

We apply this lower bound technique at a simple example and derive a lower bound on $E(T_{(1+1)\ \text{EA, LEADINGONES}})$. Like for the upper bound for the $(1, n)$ EA on LEADINGONES we use the rather trivial distance measure

$$d(x) = n - \text{LEADINGONES}(x)$$

Now we need to bound the decrease in distance in one generation from above. To this end, it is observed that it is necessary that the leftmost 0-bit is flipped in a mutation to decrease d. This happens with probability $1/n$. In this case, clearly, the decrease can be bounded by n. This, however, does not yield a useful bound. It should be remembered that one only needs to bound the *expected* decrease and try to estimate this in a more careful way.

The decrease in distance stems from two sources. The leftmost 0-bits that is flipped decreases the distance by 1. Moreover, any consecutive bits right of this leftmost 0-bit that all happen to be 1-bits after the mutation decrease this distance further, it is decreased by the length of the sequence of consecutive 1-bits. One needs to estimate this random length.

The crucial observation for the estimation of the random length of this block of 1-bits is that all bits that are right of the leftmost 0-bit never had any influence on the selection. This is the case since the number of leading ones never decreases and the function value $\text{LEADINGONES}(x) = \sum_{i=1}^{n} \prod_{j=1}^{i} x[j]$ does not depend on any bit right of the leftmost 0-bit. It is claimed that this implies that these bits are distributed according to the uniform distribution. This can be proved by induction on the number of steps. Clearly, it holds in the beginning since this is the way the initial bit string x_0 is generated. For any $t > 0$ we have for an arbitrary $x \in \{0, 1\}^n$

$$\text{Prob}\,(x_t = x) = \sum_{y \in \{0,1\}^n} \text{Prob}\,(x_{t-1} = y)\,\text{Prob}\,(\text{mut}(y) = x)$$

$$= 2^{-n} \sum_{y \in \{0,1\}^n} \text{Prob}\,(\text{mut}(y) = x)$$

making use of the induction hypothesis. The crucial observation is that

$$\text{Prob}\,(\text{mut}(y) = x) = \text{Prob}\,(\text{mut}(x) = y)$$

holds for standard bit mutations. This is the only property of the mutation operator that is needed here. Note, however, that there are mutation operators known where this does not hold (Jansen and Sudholt 2010). Now we have

$$\text{Prob}\,(x_t = x) = 2^{-n} \sum_{y \in \{0,1\}^n} \text{Prob}\,(\text{mut}(y) = x)$$

$$= 2^{-n} \sum_{y \in \{0,1\}^n} \text{Prob}\,(\text{mut}(x) = y) = 2^{-n}$$

and see that these bits are indeed uniformly distributed.

Thus, the expected decrease in distance in a single mutation is

$$\sum_{i=1}^{n} i \text{Prob}\,(\text{decrease distance by } i) \leq \sum_{i=1}^{n} i \left(\frac{1}{n}\right)\left(\frac{1}{2^{i-1}}\right) < \frac{1}{n}\sum_{i=1}^{\infty} \frac{i}{2^{i-1}} = \frac{4}{n}$$

and we have $\Delta_l < 4/n$.

Since $E(d(x_0))/\Delta_l = \Theta(nE(d(x_0)))$, all one has to do is to estimate the expected initial distance. Clearly,

$$E(d(x_0)) < n - \sum_{i=1}^{n} \frac{i}{2^i} = n - 2$$

and we have $E(T_{(1+1)\,\text{EA, LeadingOnes}}) = \Theta(n^2)$. It can be seen that this implies that the upper bound on $E(T_{(1+1)\,\text{EA, LeadingOnes}})$ derived by means of f-based partitions is indeed tight.

4.6   Typical Events

Sometimes, a statement about the expected optimization time can be made and proved based on a good understanding of the evolutionary algorithm and the fitness function. Such proofs tend to be *ad hoc* and difficult to generalize. Sometimes, however, there is a common element that is worth pointing out. Here, we consider situations where the optimization time is dominated by something that typically happens in a run of the evolutionary algorithm, we call this a typical event.

To make things more concrete, an example is considered. As usual, we use the $(1 + 1)$ EA and reconsider a fitness function where we already proved an upper bound on the expected optimization time, Jump_k. Remember that using the method of fitness-based partitions, we were able to prove that

$$E(T_{(1+1)\,\text{EA, Jump}_k}) = O(n^k + n \log n)$$

holds. The dominating term n^k stems from the fact that in order to reach the unique global optimum starting in one of the local optima, a mutation of k specific bits is necessary. Such a mutation has probability $((1/n)^k (1 - 1/n)^{n-k})$ and leads to an expected waiting time of Θn^k. Remembering the structure of JUMP_k

$$\text{JUMP}_k(x) = \begin{cases} n - \text{ONEMAX}(x) & \text{if } n - k < \text{ONEMAX}(x) < n \\ k + \text{ONEMAX}(x) & \text{otherwise} \end{cases}$$

the following observations are made. If the current x_t contains at most $n - k$ 1-bits, in order to reach the global optimum by means of a direct mutation, such a mutation needs to flip exactly all 0-bits in x_t. In this case the expected optimization time is bounded below by $\Omega(n^k)$. Since for all $x \in \{0, 1\}^n$ with more than $n - k$ but less than n 1-bits the function value equals $n - \text{ONEMAX}(x)$ leading toward the local optima, we believe that it is quite likely that a local optimum will be reached.

We make this reasoning more concrete and define the typical event B that there exists some $t \in \mathbb{N}$, such that $\text{ONEMAX}(x_t) \leq n - k$ holds. It follows from the law of total probability that

$$E\left(T_{(1+1)\text{EA},\text{JUMP}_k}\right) \geq \text{Prob }(B)E\left(T_{(1+1) \text{ EA, JUMP}_k} \mid B\right)$$

holds. As already argued above, $E\left(T_{(1+1) \text{ EA, JUMP}_k} \mid B\right) = \Omega(n^k)$ holds. Since $E\left(T_{(1+1) \text{ EA, JUMP}_k}\right) = \Omega(n \log n)$ follows from the general lower bound (❯ Sect. 4.2), it suffices to prove that Prob $(B) = \Omega(1)$ holds.

We make a case distinction on the value of k. For $k < n/3$ we have

$$\text{Prob }(\text{ONEMAX}(x_0) \leq n - k) = 1 - 2^{-\Omega(n)}$$

as can easily be shown using Chernoff bounds. Clearly, Prob $(B) = 1 - 2^{-\Omega(n)}$ follows in this case. For the case $k > (n/3)$ one can first observe that still

$$\text{Prob }(\text{ONEMAX}(x_0) \leq n - k) = 1 - 2^{-\Omega(n)}$$

holds. It can be seen that Prob (B) is bounded below the probability that the unique global optimum is reached before a local optimum is reached. It can be observed that after an initialization with at most $(2/3)n$ 1-bits the global optimum can only be reached via a mutation of all 0-bits with at least $n/3$ 0-bits in each current x_t. Such a mutation has probability at most $1/n^{n/3}$. Since the function is similar to $-\text{ONEMAX}$ it can be concluded that some local optimum is reached in expectation within $O(n \log n)$ steps. More precisely, there is some constant c such that the expected time to reach a local optimum is bounded above by $cn \log n$. By Markov's inequality, it follows that with probability at least $1/2$, we reach a local optimum in at most $2cn \log n$ steps if the global optimum is not reached before. We consider $2cn^2 \log n$ steps as n independent trials to reach a local optimum, each consisting of $2cn \log n$ steps. It can be concluded that, given that the global optimum is not reached before, a local optimum is reached within $O(n^2 \log n)$ steps with probability $1 - 2^{-\Omega(n)}$. Since the probability to reach the optimum in a single mutation is bounded above by $1/n^{n/3}$, the probability to reach the optimum within $O(n^2 \log n)$ steps is bounded above by $O\left((n^2 \log n)/n^{n/3}\right) = 2^{-\Omega(n)}$. Together, we have Prob$(B) = 1 - 2^{-\Omega(n)}$ in this case, too. So we have

$$E\left(T_{(1+1) \text{ EA, JUMP}_k}\right) = \Theta\left(n^k + n \log n\right)$$

and observe that the upper bound derived by means of f-based partitions is indeed tight.

4.7 Typical Runs

The situation described in the previous section was in some sense exceptionally simple. The optimization time is governed by one specific event that needs to occur: Some point x with at most $n - k$ 1-bits needs to become x_t before the global optimum is found. Often the situation is less clear but still there is a rather clear idea how an evolutionary algorithm may optimize a specific fitness function. Often, it is not even necessary that the evolutionary algorithm does in fact optimize this fitness function in this way. Proving that one specific way is possible within some number of f-evaluations and with some probability is often sufficient to get at least an upper bound on the expected optimization time. This can be made clear by again considering JUMP_k but now a more complex evolutionary algorithm. Attention is restricted to JUMP_k for relatively small values of k, $k = O(\log n)$.

We consider the $(\mu + \lambda)$ EA with $\lambda = 1$, $p_c > 0$ and uniform crossover. Note that this is the first time that an evolutionary algorithm with crossover is dealt with. It will be seen that this makes the analysis considerably more difficult. It is assumed that the population size μ is polynomially bounded from above and not too small. It is assumed $\mu \geq k \log^2 n$ holds. Since $k = O(\log n)$, this allows for relatively small population sizes. Finally, it is assumed that the probability for crossover p_c is rather small. A proof that works for $p_c \leq 1/(ckn)$ where $c > 0$ is a sufficiently large positive constant is discussed.

We start by describing what we consider to be a typical run of the $(\mu + 1)$ EA with uniform crossover. We do so by describing different *phases* and what one expects in these phases to happen. For each phase, we define entry conditions that need to be fulfilled at the beginning of each phase and exit conditions that define the end of each phase. Here, three phases are described. For the first phase, the entry condition is empty. This has the advantage that we can assume any time to start with phase 1. In particular, if anything goes wrong in any phase, we can assume the algorithm to be restarted. This would be different if we had entry conditions that assume that the current population is actually randomly distributed according to the uniform distribution. The exit condition of the first phase is that each member of the current population contains exactly $n - k$ 1-bits. This implies that the population contains only local optima.

The entry condition for the second phase is equal to the exit condition of the first phase. This is usually the case since the exit of one phase should be the entry to the next phase. The exit condition of the second phase is that still all members of the current population contain exactly $n - k$ 1-bits each; additionally, we demand that

$$\forall i \in \{1, 2, \ldots, n\}: \sum_{j=1}^{\mu} 1 - x^{(j)}[i] \leq \mu/(4k) \qquad (1)$$

holds where $x^{(j)}$ denotes the jth member of the current population. It is observed that ❷ Eq. 1 ensures that at each of the n position, the number of 0-bits summed over the complete population is bounded above by $\mu/(4k)$. Note that this a relatively large upper bound. The complete population contains exactly μk 0-bits, since each member contains exactly k 0-bits in the second phase. Thus, if these bits were evenly distributed, one would expect to see $\mu k/n = O(\mu \log(n)/n)$ 0-bits at each position. The bound $\mu/(4k)$ is by a factor of $\Theta(n/k^2) = \Omega(n/\log^2 n)$ larger.

The entry condition for the third and final phase is the exit condition of the second case. Its exit condition is that an optimal search point is member of the current population.

In the first phase, the situation is similar to ONEMAX. It is not difficult to see that on average after $O(\mu n \log n)$ f-evaluations the exit condition of phase 1 is met (Jansen and Wegener 2002). Moreover, by considering longer runs as restarts like we did in the previous section, one can obtain an upper bound of $p_1 \leq \varepsilon$ for the probability not to complete the first phase within $O(\mu n \log n)$ steps for any positive constant ε.

The second phase is much more difficult to analyze. When the phase begins, each member of the population contains exactly k 0-bits. We have to prove that these 0-bits become somewhat distributed among the different positions. To this end, we consider some specific bit position, say the first. Let z denote the number of 0-bits at this position summed over the whole population. We want to show that $z \leq \mu/(4k)$ holds after some time.

Clearly, in one generation, this number of 0-bits z can change by at most by 1. Let A_z^+ resp. A_z^- be the event that (given z) the number of zeros (at position 1) increases resp. decreases in one round. One wants to estimate the probabilities of A_z^+ and A_z^-. Since one wants to prove an upper bound on the number of 0-bits, one needs an upper bound on Prob (A_z^+) and a lower bound on Prob (A_z^-). Since this seems to be rather complicated, one can consider six (classes of) events that have probabilities that one can estimate directly. From now on, the index z will be omitted to simplify the notation.

- B: crossover is chosen as operator, $\text{Prob}(B) = p_c$.
- C: an object with a one at position 1 is chosen for replacement, $\text{Prob}(C) = (\mu - z)/\mu$.
- D: an object with a zero at position 1 is chosen for mutation, $\text{Prob}(D) = z/\mu$.
- E: the bit at position 1 does not flip, $\text{Prob}(E) = 1 - 1/n$.
- F_i^+: there are exactly i positions among the $(k-1)$ 0-positions $j \neq 1$ that flip and exactly i positions among the $(n-k)$ 1-positions that flip, $\text{Prob}(F_i^+) = \binom{k-1}{i}\binom{n-k}{i}$

  $(1/n)^{2i}(1 - 1/n)^{n-2i-1}$.
- G_i^+: there are exactly i positions among the k 0-positions that flip and exactly $i - 1$ positions among the $(n-k-1)$ 1-positions $j \neq 1$ that flip, $\text{Prob}(G_i^+) = \binom{k}{i}\binom{n-k-1}{i-1}(1/n)^{2i-1}(1 - 1/n)^{n-2i}$.

Using this event, one can observe that

$$A^+ \subseteq B \cup \left(\overline{B} \cap C \cap \left[\left(D \cap E \cap \bigcup_{0 \leq i \leq k-1} F_i^+ \right) \cup \left(\overline{D} \cap \overline{E} \cap \bigcup_{1 \leq i \leq k} G_i^+ \right) \right] \right)$$

holds. Similarly, we get

$$A^- \supseteq \overline{B} \cap \overline{C} \cap \left[\left(D \cap \overline{E} \cap \bigcup_{1 \leq i \leq k} F_i^- \right) \cup \left(\overline{D} \cap E \cap \bigcup_{0 \leq i \leq k} G_i^- \right) \right]$$

where

- F_i^-: there are exactly $i - 1$ positions among the $(k-1)$ 0-positions $j \neq 1$ that flip and exactly i positions among the $(n-k)$ 1-positions that flip, $\text{Prob}(F_i^-) = \binom{k-1}{i-1}\binom{n-k}{i}(1/n)^{2i-1}(1 - 1/n)^{n-2i}$.

- G_i^-: there are exactly i positions among the k 0-positions that flip and exactly i positions among the $(n - k - 1)$ 1-positions $j \neq 1$ that flip, Prob $\left(G_i^-\right) = \binom{k}{i}\binom{n-k-1}{i}(1/n)^{2i}(1 - 1/n)^{n-2i-1}$.

Since we are interested in the situation where the number of 0-bits is too large, we assume $z \geq \mu/(8k)$ in the following. It is tedious but not difficult to see that Prob (A^-) − Prob $(A^+) = \Omega(1/(nk))$ holds (Jansen and Wegener 2002). We consider the random walk defined by the random changes of z and prove that after $\Theta(\mu n^2 k^3)$ steps, the number of 0-bits at the current position is sufficiently decreased with probability $1 - O(1/n)$. By increasing the constant factor hidden in the number of generations $\Theta(\mu n^2 k^3)$ we can decrease the probability not to decrease the number of 0-bits. We obtain a bound on the probability that this happens at at least one of the n positions by simply taking the union bound.

Unfortunately, this line of reasoning is valid only if $z \geq \mu/(8k)$ holds during the complete phase. If this is not the case, the probabilities Prob (A^+) and Prob(A^-) change. We cope with this additional difficulty in the following way. Consider the last time point of time when $z < \mu/(8k)$ holds. If at this point of time there are at most $\mu/(8k)$ generations left, the number of 0-bits can at most increase to $\mu/(8k) + \mu/(8k) = \mu/(4k)$, which is the bound that we wanted to prove. If there are more than $\mu/(8k)$ steps left, we can repeat the line of reasoning and now know that $z \geq \mu/(8k)$ holds all the time.

Finally, in the third phase, the probability that the global optimum is found is not too small. Note that this is the only phase where we actually take beneficial effects of uniform crossover into account. We can control the number of 0-bits at each position in a way that is similar to the second phase. We ensure that the number of 0-bits does not exceed $\mu/(2k)$ at each position. So we can concentrate on the creation of the global optimum. This is achieved if the following happens. We perform crossover (with probability p_c), we select two parents that do not share a 0-bit at any position (with some probability p), in uniform crossover the optimum 1^n is created (with probability $(1/2)^{2n}$), and this optimum is not destroyed by mutation (with probability $(1 - 1/n)^n$). We need to estimate the probability p of selecting two parents without a common 0-bit. After the first parent is selected, there are exactly k positions where the second parent may "collide" with this parent. Since at each position one has at most $\mu / (2k)$ 0-bits, the probability for this event is bounded below by $k \cdot (\mu/(2k))/\mu = 1/2$. Together, we have a probability of $\Omega\left(p_c 2^{-2k}\right)$ to create the global optimum for each generation of the third phase. Combining the results for the different phases, we obtain $O\left(\mu n^2 k^3 + 2^{2k}/p_c\right)$ as upper bound on the expected optimization time.

5   Summary

Evolutionary algorithms can be described as randomized search heuristics that are often applied as randomized optimization heuristics. This motivates the investigation of their efficiency as is done for other optimization algorithms, too. Placing evolutionary algorithms in the context of design and analysis of randomized algorithms in this way makes the question of their computational complexity one of immense importance.

The aim of this chapter is to demonstrate that the analysis of the expected optimization time of evolutionary algorithms is something that can be done. More importantly, it aims at demonstrating how it can be done. This encompasses rather general approaches like "no free lunch" theorems and results on the black-box complexity. For the latter, Yao's minimax principle is introduced and its application is demonstrated in practical examples. On the other hand, concrete tools and methods for the analysis of evolutionary algorithms' expected optimization times are discussed.

The method of fitness-based partitions is the conceptually simplest of the analytical tools introduced. It measures the progress of an evolutionary algorithm by means of fitness values in a predefined way. For a successful analysis, the definition of an appropriate grouping of fitness values into fitness layers is crucial. For many functions, tight upper bounds on the expected optimization time can be achieved this way. The method, however, is restricted to algorithms employing a kind of elitist selection, it cannot deal with decreases in function value.

The method of the expected multiplicative distance decrease also establishes upper bounds on the expected optimization time by measuring the progress of an evolutionary algorithm by means of fitness values. It does so, however, in a more flexible way. Instead of defining static fitness layers, one identifies ways of increasing the fitness in all situations. This yields an average multiplicative decrease of the difference between the current fitness and the fitness of an optimal solution that finally leads to an upper bound on the average time needed to decrease the difference to 0. This method is restricted to functions with integer function values. Since a finite search space is dealt with, this is not a fundamental limitation. Multiplication can turn rational function values into integer ones. More fundamental is that the proven bounds depend on the size of the function values. While this can be removed to some extent (Reichel and Skutella 2009), it often leads to bounds that are not asymptotically tight.

An even more flexible way of measuring the progress of an evolutionary algorithm is drift analysis. Here, progress can be measured in an almost arbitrary way. Moreover, it suffices to have bounds on the expected progress in a single generation for the proof of bounds on the expected optimization time. This allows for the analysis of algorithms that do not guarantee monotone progress as long as the expected progress is positive.

These three general methods for establishing upper bounds on the expected optimization time are complemented by lower bound techniques. There is a very general lower bound for all mutation-based algorithms of size $\Omega(n \, log \, n)$ that holds for almost all fitness functions of practical interest. While being useful, it does not clearly establish a technique. The most powerful technique for proving lower bounds on the expected optimization time is drift analysis. The differences between the drift techniques for establishing upper and lower bounds are rather small, and drift analysis is equally flexible and powerful for both upper and lower bounds. We concentrated on proving upper and lower bounds for expected optimization times that are polynomial. For proving exponential lower bounds, better-suited drift theorems are available that allow us to establish strong negative results at ease (Oliveto and Witt 2008).

Basing analysis on either typical events or typical runs are not upper or lower bound techniques per se. Here, a lower bound proven with the method of typical events and an upper bound using the method of typical runs are presented. Both methods require a good idea of how the evolutionary algorithms actually works on the concrete fitness function under consideration. It is noteworthy that using the method of typical runs one could derive proven results for an evolutionary algorithm with crossover. However, it is not the case that the other methods do not allow for such results.

All methods using example functions that are in some sense artificial are introduced. This has the advantage that these fitness functions have a very clear structure. They can easily be used to demonstrate certain effects. Here they allowed for a clear presentation of the analytical tools employed. Most methods described here most often have been developed using such example functions that facilitate the analysis. It is worth mentioning that the method of the expected multiplicative distance decrease is an exception. It was introduced not using any specific example function but for a combinatorial optimization problem, namely the minimum spanning tree problem (Neumann and Wegener 2004).

Clearly, these methods for the analysis of evolutionary algorithms are in no way restricted to the analysis of such example functions. They have been successfully applied to a large number of combinatorial optimization problems and a variety of different variants of evolutionary algorithms, see Oliveto et al. (2007) for an overview.

We concentrated on the presentation of analytical tools in a way that allows the reader to employ these methods for his or her own problems. This leads to a presentation using practical examples instead of presenting the methods in their most general form. Moreover, a complete overview of all methods ever employed in the analysis of evolutionary algorithms is not presented. Two methods not covered here worth mentioning are the method of delay sequences (Dietzfelbinger et al. 2003) that allows for very tight bounds and the method of family trees that facilitates the analysis of an evolutionary algorithm with a population size $\mu > 1$ (Witt 2006). Moreover, the various mathematical tools that are useful in the analysis of algorithms are not covered in detail, in particular the application of martingales is not discussed (Jansen and Sudholt 2010; Williams 1991).

While aiming at the analysis of evolutionary algorithms no method is actually restricted to this class of algorithms. It is possible to apply these methods to other randomized algorithms, successful examples include ant colony optimization (Doerr et al. 2007; Sudholt and Witt 2008a), memetic algorithms (Sudholt 2008), and particle swarm optimization (Sudholt and Witt 2008b; Witt 2009). We can be sure to see these methods presented here further applied to more algorithms and more problems and to see more powerful methods developed in the future.

Acknowledgment

This material is based upon works supported by Science Foundation Ireland under Grant No. 07/SK/I1205.

References

Cormen TH, Leiserson CE, Rivest RL, Stein C (2001) Introduction to algorithms, 2nd edn. MIT Press, Cambridge, MA

De Jong KA (1992) Genetic algorithms are NOT function optimizers. In: Whitley LD (ed) Proceedings of the second workshop on foundations of genetic algorithms (FOGA), Morgan Kaufmann, San Francisco, CA, pp 5–17

Dietzfelbinger M, Naudts B, Hoyweghen CV, Wegener I (2003) The analysis of a recombinative hill-climber on H-IFF. IEEE Trans Evolut Comput 7(5): 417–423

Doerr B, Neumann F, Sudholt D, Witt C (2007) On the runtime analysis of the 1-ANT ACO algorithm. In: Proceedings of the genetic and evolutionary computation conference (GECCO), ACM, New York, pp 33–40

Doerr B, Jansen T, Klein C (2008) Comparing global and local mutations on bit strings. In: Proceedings of the genetic and evolutionary computation conference (GECCO), ACM, New York, pp 929–936

Droste S, Jansen T, Wegener I (2002a) On the analysis of the (1+1) evolutionary algorithm. Theor Comput Sci 276:51–81

Droste S, Jansen T, Wegener I (2002b) Optimization with randomized search heuristics – the (A)NFL theorem, realistic scenarios, and difficult functions. Theor Comput Sci 287(1):131–144

Droste S, Jansen T, Wegener I (2006) Upper and lower bounds for randomized search heuristics in black-box optimization. Theory Comput Syst 39(4): 525–544

He J, Yao X (2004) A study of drift analysis for estimating computation time of evolutionary algorithms. Nat Comput 3(1):21–35

Horn J, Goldberg DE, Deb K (1994) Long path problems. In: Davidor Y, Schwefel HP, Männer R (eds) Proceedings of the 3rd international conference on parallel problem solving from nature (PPSN III), Springer, Berlin, Germany, LNCS 866, pp 149–158

Igel C, Toussaint M (2003) On classes of functions for which no free lunch results hold. Inf Process Lett 86:317–321

Igel C, Toussaint M (2004) A no-free-lunch theorem for non-uniform distributions of target functions. J Math Model Algorithms 3:313–322

Jansen T, Sudholt D (2010) Analysis of an asymmetric mutation operator. Evolut Comput 18(1):1–26

Jansen T, Wegener I (2000) On the choice of the mutation probability for the (1+1) EA. In: Schoenauer M, Deb K, Rudolph G, Yao X, Lutton E, Merelo-Guervos J, Schwefel HP (eds) Proceedings of the 6th international conference on parallel problem solving from nature (PPSN VI), Springer, New York, LNCS 1917, pp 89–98

Jansen T, Wegener I (2002) On the analysis of evolutionary algorithms – a proof that crossover really can help. Algorithmica 34(1):47–66

Jansen T, De Jong KA, Wegener I (2005) On the choice of the offspring population size in evolutionary algorithms. Evolut Comput 13(4):413–440

Mitzenmacher M, Upfal E (2005) Probability and computing. Cambridge University Press, Cambridge, MA

Motwani R, Raghavan P (1995) Randomized algorithms. Cambridge University Press, Cambridge, MA

Neumann F, Wegener I (2004) Randomized local search, evolutionary algorithms, and the minimum spanning tree problem. In: Proceedings of the genetic and evolutionary computation conference (GECCO), Springer, Berlin, Germany, LNCS 3102, pp 713–724

Oliveto PS, Witt C (2008) Simplified drift analysis for proving lower bounds in evolutionary computation. In: Proceedings of the 10th international conference on parallel problem solving from nature (PPSN X), Springer, Berlin, Germany, LNCS 5199, pp 82–91

Oliveto PS, He J, Yao X (2007) Time complexity of evolutionary algorithms for combinatorial optimization: a decade of results. Int J Automation Comput 4(3):281–293

Reichel J, Skutella M (2009) On the size of weights in randomized search heuristics. In: Garibay I, Jansen T, Wiegand RP, Wu A (eds) Proceedings of the tenth workshop on foundations of genetic algorithms (FOGA), ACM, New York, pp 21–28

Rudolph G (1997) How mutation and selection solve long path problems polynomial expected time. Evolut Comput 4(2):195–205

Schumacher C, Vose MD, Whitley LD (2001) The no free lunch and problem description length. In: Proceedings of the genetic and evolutionary computation conference (GECCO), Morgan Kaufmann, San Francisco, CA, pp 565–570

Sudholt D (2008) Memetic algorithms with variable-depth search to overcome local optima. In: Proceedings of the genetic and evolutionary computation conference (GECCO), ACM, New York, pp 787–794

Sudholt D, Witt C (2008a) Rigorous analyses for the combination of ant colony optimization and local search. In: Proceedings of ant colony and swarm intelligence (ANTS), Springer, Berlin, Germany, LNCS 5217, pp 132–143

Sudholt D, Witt C (2008b) Runtime analysis of binary PSO. In: Proceedings of the genetic and evolutionary computation conference (GECCO), ACM, New York, pp 135–142

Williams D (1991) Probability with martingales. Cambridge University Press, Cambridge

Witt C (2006) Runtime analysis of the (μ+1) EA on simple pseudo-boolean functions. Evolut Comput 14(1):65–86

Witt C (2009) Why standard particle swarm optimizers elude a theoretical runtime analysis. In: Garibay I, Jansen T, Wiegand RP, Wu A (eds) Proceedings of the tenth workshop on foundations of genetic algorithms (FOGA), ACM, New York, pp 13–20

Wolpert DH, Macready WG (1997) No free lunch theorems for optimization. IEEE Trans Evolut Comput 1(1):67–82

Yao AC (1977) Probabilistic computations: towards a unified measure of complexity. In: Proceedings of the 17th IEEE symposium on foundations of computer science (FOCS), New York, pp 222–227

27 Stochastic Convergence

Günter Rudolph
Department of Computer Science, TU Dortmund,
Dortmund, Germany
guenter.rudolph@tu-dortmund.de

G. Rozenberg et al. (eds.), *Handbook of Natural Computing*, DOI 10.1007/978-3-540-92910-9_27,
© Springer-Verlag Berlin Heidelberg 2012

Abstract

Since the state transitions of an evolutionary algorithm (EA) are of stochastic nature, the deterministic concept of the "convergence to the optimum" is not appropriate. In order to clarify the exact semantic of a phrase like "the EA converges to the global optimum" one has to, at first, establish the connection between EAs and stochastic processes before distinguishing between the various modes of stochastic convergence of stochastic processes. Subsequently, this powerful framework is applied to derive convergence results for EAs.

1 Introduction

Broadly speaking, the notion of "convergence" of an evolutionary algorithm (EA) simply means that the EA should approach some "limit" in the course of its evolutionary sequence of populations. Needless to say, this limit should represent something like the optimum of the optimization problem that one intends to solve. In this case, the "convergence" of an EA to a "limit" is a desirable property.

In order to endow the terms "convergence" and "limit" with a precise meaning in the context of EAs, it is necessary to embed EAs in the framework of "stochastic processes" (❯ Sect. 2) before the different modes of "stochastic convergence" and their basic results are introduced in ❯ Sect. 3. Subsequently, these basic results are applied, refined, and extended to EAs in ❯ Sect. 4, before ❯ Sect. 5 provides annotated pointers to the first results regarding the convergence theory of multi-objective EAs.

2 Stochastic Process Models of Evolutionary Algorithms

The analysis of the dynamic behavior of EAs benefits from the fact that it is always possible to establish a bijective mapping between an EA and a particular stochastic process. As a consequence, one can exploit the results and techniques of the well-founded theory of stochastic processes (Doob 1967).

Definition 1   Let $(X_t)_{t \in T}$ be a family of random variables (r.v.s) on a joint probability space $(\Omega, \mathscr{F}, \mathsf{P})$ with values in a set E of a measurable space $(E, \mathscr{B})$ and index set T. Then $(X_t)_{t \in T}$ is called a *stochastic process* with index set T.   □

In general, there is no mathematical reason for restricting index set T to be a set of numerical values. But here the index set T is identical with $\mathbb{N}_0$ and the indices $t \in T$ will be interpreted as points of time.

Definition 2   A stochastic process $(X_t)_{t \in T}$ with index set $T = \mathbb{N}_0$ is called a *stochastic process with discrete time*. The image space E of $(X_t)_{t \in T}$ is called the *state space* of the process. The *transition probability*

$$\mathsf{P}\{X_{t+1} \in A | X_t = x_t, X_{t-1} = x_{t-1}, \dots, X_1 = x_1, X_0 = x_0\} \tag{1}$$

describes the probability of the event of transitioning to some state in set $A \subseteq E$ in step $(t+1)$ subject to previous steps though the state space E.   □

Evidently, random variable X_t will represent the population at generation $t \geq 0$ and the function in ❷ Eq. 1 will indicate the likeliness that the population at generation $t + 1$ will be some $x \in A$ conditioned by all previous populations. Since the transition probability ❷ Eq. 1 must somehow capture the stochastic nature of the genetic operators and selection methods, it is clear that the modeling of the transition probability function is the main task when establishing a link between EA and stochastic process. Due to combinatorial complexity, the more difficult this task becomes the larger is the set of preceding populations on which the transition probabilities are conditioned. Fortunately, the vast majority of EAs are designed in a manner that does not require the most general formulation of the stochastic process model, although there are tools and results supporting an analysis (Iosifescu and Grigorescu 1990). But typically, the genetic operators and selection methods only act on the current population of individuals without taking into account the genomes of the antecedent populations of individuals. In other words: The phylogenetic trees of the current individuals do not affect the future. As a consequence, the transition probabilities only depend on the current population. Stochastic processes with this property are endowed with an extensive theory (see, e.g., Iosifescu (1980), Seneta (1981), Nummelin (1984), and Meyn and Tweedie (1993)) and they bear their own name:

Definition 3 A stochastic process $(X_t)_{t \in T}$ with discrete time whose transition probability satisfies

$$P\{X_{t+1} \in A | X_t = x_t, X_{t-1} = x_{t-1}, \ldots, X_1 = x_1, X_0 = x_0\} = P_t\{X_{t+1} \in A | X_t = x_t\}$$

for all $t \geq 0$ is called a *Markov process with discrete time*. As for notation, often $P_t(x, A) := P_t\{X_{t+1} \in A | X_t = x\}$. If the transition probability of a Markov process with discrete time does not explicitly depend on the time parameter $t \geq 0$, that is,

$$\forall t \geq 0 : P_t\{X_{t+1} = x | X_t = x_t\} = P\{X_{t+1} = x | X_t = x_t\},$$

then the Markov process is *time-homogeneous*, and *time-inhomogeneous* otherwise. A Markov process with countable state space is termed a *Markov chain*. □

Since EAs work with discrete time, their stochastic models also use discrete time. Therefore, the phrase "with discrete time" will be left out hereinafter when talking about Markov processes and Markov chains. Next, suppose that an EA is deployed for solving the optimization problem

$$f(x) \rightarrow \min ! \quad \text{subject to } x \in \mathcal{X} \tag{2}$$

where the objective function $f : \mathcal{X} \rightarrow \mathbb{R}$ is bounded from below. Now, solely, the type of the search space $\mathcal{X}$ determines which type of Markov model and which part of Markov theory can be used.

2.1 Finite Search Space

If the search space $\mathcal{X}$ is finite, so is the state space of the Markov chain. There are at least two approaches for modeling the state space. This will be exemplified by the binary search space $\mathcal{X} = \mathbb{B}^n$ with $\mathbb{B} = \{0, 1\}$.

2.1.1 Generic State Space Model

A straightforward generic model for the state space E of a Markov chain representing an EA with finite population size μ and search space $\mathcal{X} = \mathbb{B}^n$ is the μ-fold cartesian product $E = \mathcal{X}^\mu = (\mathbb{B}^n)^\mu = \mathbb{B}^{n \times \mu}$ that has been used in the early 1990s (see Eiben et al. (1991), Fogel (1994), Rudolph (1994b), and others). As a consequence, there are $|E| = 2^{n \times \mu}$ possible states and one needs $|E| \times |E|$ transition probabilities to characterize the stochastic effects caused by variation and selection operators. Evidently, these transition probabilities can be gathered in a square matrix $P = (p_{ij})$ of size $|E| \times |E|$, termed the *transition matrix*. Finally, one has to specify the *initial distribution* of the Markov chain to obtain a complete stochastic model of the EA. The initial distribution models the initialization of the EA.

The advantages of the generic state space model are that it also works for EAs with spatial structure and that it has a natural analogon for denumerable and innumerable search spaces.

2.1.2 Davis/Vose State Space Model

An alternative state space model was introduced independently by Davis (1991) and Vose (Nix and Vose 1992) also in the early 1990s. Its advantage is a smaller state space but its disadvantage is the fact that only panmictic EAs (every individual may mate with every other individual) without spatial structure can be modeled in this manner. Moreover, it has no analogon in case of innumerable search spaces.

Suppose that the position of the individuals in the population is of no importance and that the EA has a panmictic mating policy. Then, one only needs to know how many individuals of a certain type are present in the current population for deriving the transition probabilities. Since there are $|\mathcal{X}| = |\mathbb{B}^n| = 2^n$ possible individuals one needs a 2^n-tuple of nonnegative integers whose components count the number of occurrences of each type of individuals. Of course, the sum over all components must be exactly μ. As a consequence, the cardinality of the state space reduces to

$$|E| = \binom{2^n + \mu - 1}{\mu}$$

which is considerably smaller than the value $2^{n \times \mu}$ of the generic model. The transition matrix is of considerably smaller size as well.

Both approaches are thoroughly justified. But it is important to keep in mind that every EA that can be modeled by this approach can also be modeled in the generic setting whereas the converse is wrong in general. For example, see the efforts and insuperable difficulties in Muhammad et al. (1999) when trying to model a spatially structured EA with a Davis/Vose state space model.

2.2 Denumerable Search Space

An example of a denumerable but not finite search space is the integer search space $\mathcal{X} = \mathbb{Z}^n$. Apparently, this case has not been analyzed yet in the context of Markov chains although there exists a theory (see, e.g., part II in Seneta (1981)) using infinite-dimensional transition matrices. The reason might be due to the fact that in practice, the search space is almost always equipped with box constraints leading to a finitely sized subset. Exceptions without box constraints are rare (see, e.g., Rudolph 1994c).

The generic state space model for a population of μ individuals is $E = \mathcal{X}^\mu = \mathbb{Z}^{n \times \mu}$ whereas the analogon of the Davis/Vose state space model $E = \{e \in \mathbb{N}_0^\infty : \sum_{i=1}^\infty e_i = \mu\}$ requires an infinite-dimensional tuple of nonnegative integers whose components count the number of occurrences of each type of individual and must add up to μ. Since this model has the same limitations as in the finite case and both models are now of infinite cardinality, it is quite obvious that there is no advantage of the Davis/Vose model in the denumerable case.

2.3 Innumerable Search Space

The typical representative for an innumerable search space is $\mathcal{X} = \mathbb{R}^n$. Here, only the generic state space model is expedient: $E = \mathcal{X}^\mu = \mathbb{R}^{n \times \mu}$. Since the probability to sample a specific point in $\mathbb{R}^n$ from a continuous distribution is zero, the transition probabilities are specified by transition probability functions (in lieu of matrices) describing transitions from a state to a set of states with nonzero measure (see, e.g., Rudolph 1996).

3 Stochastic Convergence

The limit behavior of stochastic sequences requires a concept of convergence that captures the random nature of the sequence. This can be done in quite a different manner (Lukacs 1975). Here, only the most frequently used concepts are presented.

Definition 4 Let $Z, Z_0, Z_1, \ldots$ be random variables defined on a probability space $(\Omega, \mathcal{A}, \mathsf{P})$. The sequence $(Z_t : t \geq 0)$ is said

(a) to *converge completely* to random variable Z, denoted $Z_t \xrightarrow{c} Z$, if

$$\sum_{t=0}^\infty \mathsf{P}\{|Z_t - Z| > \varepsilon\} < \infty \text{ for every } \varepsilon > 0,$$

(b) to *converge with probability* 1, denoted $Z_t \xrightarrow{w.p.1} Z$, if

$$\mathsf{P}\left\{\lim_{t \to \infty} |Z_t - Z| = 0\right\} = 1,$$

(c) to *converge in probability* to Z, denoted $Z_t \xrightarrow{P} Z$, if

$$\lim_{t \to \infty} \mathsf{P}\{|Z_t - Z| > \varepsilon\} = 0 \text{ for every } \varepsilon > 0 \text{ and}$$

(d) to *converge in mean* to Z, denoted $Z_t \xrightarrow{m} Z$, if

$$\lim_{t \to \infty} \mathsf{E}[|Z_t - Z|] = 0. \qquad \square$$

Basic relationships between these concepts of stochastic convergence are summarized in Theorem 1 below.

Theorem 1 (see Lukacs (1975), pp. 33–36 and 51–52) $Z_t \xrightarrow{c} Z \Rightarrow Z_t \xrightarrow{w.p.1} Z \Rightarrow Z_t \xrightarrow{P} Z$ and $Z_t \xrightarrow{m} Z \Rightarrow Z_t \xrightarrow{P} Z$. *The converse is wrong in general.* $\qquad \square$

Under additional conditions, some implications may be reversed. Here, the last implication is of special interest:

Theorem 2 (see Williams 1991, pp. 127–130) *If* $|Z_t| \leq Y$ *with* $E[Y] < \infty$ *for all* $t \geq 0$ *and* $Z_t \overset{P}{\to} Z$ *then also* $Z_t \overset{m}{\to} Z$. □

Notice that the so-called *dominated convergence theorem* above also holds for the special case, where random variable Y is replaced by a finite constant $K \in (0, \infty)$. The example below is intended to provide a first impression about the differences between the modes of stochastic convergence introduced previously.

Example 1 (Modes of stochastic convergence) Let $(Z_t)_{t \geq 1}$ be a sequence of independent random variables. Depending on their probability distributions, the sequences converge to zero in different modes.

- $P\{Z_t = 0\} = 1 - \frac{1}{t}$ and $P\{Z_t = 1\} = \frac{1}{t}$
  At first, one can confirm convergence to zero in probability since $P\{Z_t > \varepsilon\} = P\{Z_t = 1\} = \frac{1}{t} \to 0$ for $t \to \infty$. But there is no complete convergence since the probability mass does not move quickly enough to zero: $\sum_{t=1}^{\infty} P\{Z_t > \varepsilon\} = \sum_{t=1}^{\infty} P\{Z_t = 1\} = \sum_{t=1}^{\infty} \frac{1}{t} = \infty$. Note that $0 \leq Z_t \leq 1$ for all $t \geq 0$. Evidently, the sequence is bounded from above by constant $K = 1$. Thanks to ❷ Theorem 1, one has convergence in mean.
- $P\{Z_t = 0\} = 1 - \frac{1}{t^2}$ and $P\{Z_t = 1\} = \frac{1}{t^2}$
  As can be seen from $P\{Z_t > \varepsilon\} = P\{Z_t = 1\} = \frac{1}{t^2} \to 0$ for $t \to \infty$ the sequence converges to zero in probability. Here, the convergence is quick enough to guarantee even complete convergence: $\sum_{t=1}^{\infty} P\{Z_t > \varepsilon\} = \sum_{t=1}^{\infty} P\{Z_t = 1\} = \sum_{t=1}^{\infty} \frac{1}{t^2} < \infty$. For the same reasons as above, the sequence also converges in mean.
- $P\{Z_t = 0\} = 1 - \frac{1}{t}$ and $P\{Z_t = t\} = \frac{1}{t}$
  This sequence also converges to zero in probability: $P\{Z_t > \varepsilon\} = P\{Z_t = t\} = \frac{1}{t} \to 0$ for $t \to \infty$. But the convergence is not quick enough to establish complete convergence: $\sum_{t=1}^{\infty} P\{Z_t > \varepsilon\} = \sum_{t=1}^{\infty} P\{Z_t = t\} = \sum_{t=1}^{\infty} \frac{1}{t} = \infty$. Note that the sequence is not bounded from above. Moreover, the fact that $E[Z_t] = 0 \cdot P\{Z_t = 0\} + t \cdot P\{Z_t = t\} = t \cdot \frac{1}{t} = 1$ for all $t \geq 1$ reveals that the sequence does not converge to zero in mean.
- $P\{Z_t = 0\} = 1 - \frac{1}{t^2}$ and $P\{Z_t = t\} = \frac{1}{t^2}$
  Thanks to $P\{Z_t > \varepsilon\} = P\{Z_t = t\} = \frac{1}{t^2} \to 0$ for $t \to \infty$ one may confirm not only convergence in probability but also complete convergence: $\sum_{t=1}^{\infty} P\{Z_t > \varepsilon\} = \sum_{t=1}^{\infty} P\{Z_t = t\} = \sum_{t=1}^{\infty} \frac{1}{t^2} < \infty$. Although the sequence is not bounded from above, one can ensure convergence in mean via $E[Z_t] = 0 \cdot P\{Z_t = 0\} + t \cdot P\{Z_t = t\} = t \cdot \frac{1}{t^2} = \frac{1}{t} \to 0$ for $t \to \infty$.
- $P\{Z_t = 0\} = 1 - \frac{1}{t}$ and $P\{Z_t = t^2\} = \frac{1}{t}$
  This sequence converges to zero in probability because of $P\{Z_t > \varepsilon\} = P\{Z_t = t^2\} = \frac{1}{t} \to 0$ for $t \to \infty$, but owing to $\sum_{t=1}^{\infty} P\{Z_t > \varepsilon\} = \sum_{t=1}^{\infty} P\{Z_t = t^2\} = \sum_{t=1}^{\infty} \frac{1}{t} = \infty$ and $E[Z_t] = 0 \cdot P\{Z_t = 0\} + t^2 \cdot P\{Z_t = t^2\} = t^2 \cdot \frac{1}{t} = t \to \infty$ for $t \to \infty$ there is neither complete convergence nor convergence in mean.
- $P\{Z_t = 0\} = 1 - \frac{1}{t^2}$ and $P\{Z_t = t^2\} = \frac{1}{t^2}$
  Finally, this sequence converges to zero in probability owing to $P\{Z_t > \varepsilon\} = P\{Z_t = t^2\} = \frac{1}{t^2} \to 0$ for $t \to \infty$ and it actually converges completely to zero on account

■ Table 1

The modes of stochastic convergence that are realized by the random sequences of Example 1 depending on their probability distributions

Probability distribution		$\xrightarrow{c}$	$\xrightarrow{p}$	$\xrightarrow{m}$
$P\{Z_t = 0\} = 1 - \frac{1}{t}$	$P\{Z_t = 1\} = \frac{1}{t}$	−	+	+
$P\{Z_t = 0\} = 1 - \frac{1}{t^2}$	$P\{Z_t = 1\} = \frac{1}{t^2}$	+	+	+
$P\{Z_t = 0\} = 1 - \frac{1}{t}$	$P\{Z_t = t\} = \frac{1}{t}$	−	+	−
$P\{Z_t = 0\} = 1 - \frac{1}{t^2}$	$P\{Z_t = t\} = \frac{1}{t^2}$	+	+	+
$P\{Z_t = 0\} = 1 - \frac{1}{t}$	$P\{Z_t = t^2\} = \frac{1}{t}$	−	+	−
$P\{Z_t = 0\} = 1 - \frac{1}{t^2}$	$P\{Z_t = t^2\} = \frac{1}{t^2}$	−	+	+

of $\sum_{t=1}^{\infty} P\{Z_t > \varepsilon\} = \sum_{t=1}^{\infty} P\{Z_t = t^2\} = \sum_{t=1}^{\infty} \frac{1}{t^2} < \infty$. Convergence to zero in mean, however, cannot be attested: $E[Z_t] = 0 \cdot P\{Z_t = 0\} + t^2 \cdot P\{Z_t = t^2\} = t^2 \cdot \frac{1}{t^2} = 1$ for all $t \geq 1$. □

❯ *Table 1* summarizes the results obtained from this example.

Stochastic processes with special properties can be useful devices in the analysis of the limit behavior of evolutionary algorithms.

Definition 5   Let $(\Omega, \mathscr{F}, P)$ be a probability space and $\mathscr{F}_0 \subseteq \mathscr{F}_1 \subseteq \ldots \mathscr{F}$ be an increasing family of sub–σ–algebras of $\mathscr{F}$ and $\mathscr{F}_\infty := \sigma\left(\bigcup_t \mathscr{F}_t\right) \subseteq \mathscr{F}$. A stochastic process $(Z_t)_{t \geq 0}$ that is $\mathscr{F}_t$–measurable for each t is termed a *supermartingale* if

$$E[\,|Z_t|\,] < \infty \text{ and } E[Z_{t+1} \,|\, \mathscr{F}_t] \leq Z_t \text{ w.p.1}$$

for all $t \in \mathbb{N}_0$. □

Nonnegative supermartingales, which means that additionally $Z_t \geq 0$ for all $t \geq 0$ have the following remarkable property:

Theorem 3   (see Neveu (1975, p. 26)) *If $(Z_t)_{t \geq 0}$ is a nonnegative supermartingale then* $Z_t \xrightarrow{w.p.1} Z < \infty$. □

Although nonnegative supermartingales do converge to a finite limit with probability 1, nothing can be said about the limit itself unless additional conditions are imposed. The example below reveals that even a strict inequality, that is, $E[Z_{t+1} \,|\, \mathscr{F}_t] < Z_t$, does not necessarily imply a zero limit for nonnegative supermartingales.

Example 2   (Monotone decrease does not imply zero limit) Let $Z_t = 1 + 2^{-t} \cdot X_t^2 \geq 0$ for $t \geq 1$ where the sequence $X_1, X_2, \ldots$ consists of independent and identically distributed (i.i.d.) random variables with standard normal distribution $N(0,1)$. Since $E[X_t^2] = 1$ for all $t \geq 1$ one has

$$E[E[Z_{t+1} \,|\, \mathscr{F}_t]] = 1 + 2^{-(t+1)} < E[Z_t] = 1 + 2^{-t}$$

for all $t \geq 1$ but $Z_t \xrightarrow{m} 1$ as $t \to \infty$. □

For later purposes, it is of interest under which conditions the limit is the constant zero.

Theorem 4 (see Rudolph (1997a, p. 52)) *If* $(Z_t)_{t \geq 0}$ *is a nonnegative supermartingale satisfying*

$$E[Z_{t+1} \mid \mathcal{F}_t] \leq c_t Z_t \quad \text{w.p.1}$$

for all $t \geq 0$ *with* $c_t \geq 0$ *and*

$$\sum_{t=1}^{\infty} \left(\prod_{k=0}^{t-1} c_k \right) < \infty \tag{3}$$

then $Z_t \xrightarrow{m} 0$ *and* $Z_t \xrightarrow{c} 0$.

It remains to provide some simple conditions that imply inequality (❯ Eq. 3).

Lemma 1 (see Rudolph (1997a, p. 53)) *The infinite series in* (❯ Eq. 3) *converges to a finite limit if*

(a) $\limsup_{t \geq 0} c_t < 1$ *or*
(b) $c_t \leq 1 - a/t$ *for some* $a > 1$ *and almost all* $t \geq 1$. ☐

The limit properties and stability of fixed points of deterministic dynamical systems can be analyzed via so-called Lyapunov functions. Bucy (1965) has shown that supermartingales play the role of Lyapunov functions in the stochastic setting. In fact, both concepts are closely related.

Theorem 5 (see Bucy (1965, pp. 153–154), and Bucy and Joseph (1968, pp. 83–84)) *Let* $h: \mathbb{R}^n \times \mathbb{R}^m \to \mathbb{R}^n$ *be a continuous function with* $h(0, \cdot) = 0$. *Consider the stochastic dynamical system*

$$X_{t+1} = h(X_t, Y_t) \tag{4}$$

for $t \geq 0$ *with* $X_0 \in \mathbb{R}^n$ *and where* $(Y_t)_{t \geq 0}$ *is a sequence of random vectors. If there exists a continuous nonnegative function* $V: \mathbb{R}^n \to \mathbb{R}$ *with the properties*

(a) $V(0) = 0$,
(b) $V(x) \to \infty$ *as* $\|x\| \to \infty$,
(c) $V(X_t)$ *is a supermartingale along the motion of* ❯ Eq. 4 *and*
(d) *there exists a continuous function* $\gamma: \mathbb{R}_+ \to \mathbb{R}_+$ *vanishing only at the origin and along the motion of* ❯ Eq. 4 *holds*

$$E[V(X_{t+1}) \mid \mathcal{F}_t] \leq V(X_t) - \gamma(\|X_t\|) \tag{5}$$

then $\|X_t\| \xrightarrow{w.p.1} 0$ *as* $t \to \infty$. ☐

In the theorem above, the function $V(\cdot)$ is a Lyapunov function. This result can be used to obtain a slightly weaker result than ❯ Theorem 4 without any effort: Set $Z_t = V(X_t)$ such that $(Z_t)_{t \geq 0}$ is a nonnegative supermartingale and choose $\gamma(z) = (1 - c)z$ with $c \in (0, 1)$. Then all conditions of ❯ Theorem 5 including ❯ Eq. 5 with

$$E[Z_{t+1} \mid \mathcal{F}_t] \leq Z_t - \gamma(Z_t) = Z_t - (1 - c) Z_t = c Z_t$$

are fulfilled and one may conclude that $Z_t \xrightarrow{w.p.1} 0$ as $t \to \infty$.

4   Convergence Results for Evolutionary Algorithms

With the definitions of the previous section, one can assign a rigorous meaning to the notion of the convergence of an evolutionary algorithm.

Definition 6   Let $(X_t : t \geq 0)$ be the sequence of populations generated by some evolutionary algorithm and let $F_t^* = \min\{f(X_{t,1}),\ldots,f(X_{t,\mu})\}$ denote the best objective function value of the population of size $\mu < \infty$ at generation $t \geq 0$. An evolutionary algorithm is said to *converge in mean (in probability, with probability 1, completely) to the global minimum* $f^* = \min\{f(x) : x \in \mathscr{X}\}$ of objective function $f: \mathscr{X} \to \mathbb{R}$ if the nonnegative random sequence $(Z_t : t \geq 0)$ with $Z_t = F_t^* - f^*$ converges in mean (in probability, with probability 1, completely) to zero. □

The convergence results for evolutionary algorithms depend on the type of the transitions and the search space.

4.1   Time-Homogeneous Transitions

4.1.1   Finite State Space

If the state space is finite and the transition matrix does not depend on the iteration counter, then the limit behavior of the Markov chain and its associated EA solely depend on the structure of the transitions matrix: Let $p^{(0)}$ denote the initial distribution and P the transition matrix. The *state distribution* $p^{(t)}$ of the Markov chain at step $t \geq 0$ is given by $p^{(t)} = p^{(t-1)} \cdot P$ for $t \geq 1$ and $p^{(0)}$ for $t = 0$. The Chapman–Kolmogorov equations (see, e.g., Iosifescu 1980, p. 65) reveal that actually $p^{(t)} = p^{(0)} \cdot P^t$ for $t \geq 0$, where P^t denotes the tth power of transition matrix P. Clearly, the structural properties of P determine the shape of the limit matrix P^∞, if any, for $t \to \infty$ and, therefore, also the limit behavior of the EA. In order to exploit these facts, some terminology is required first:

Definition 7   A square matrix $P: m \times m$ is called a *permutation matrix* if each row and each column contain exactly one 1 and $m - 1$ zeros. A matrix A is said to be *cogredient* to a matrix B if there exists a permutation matrix P, such that $A = P'BP$. A square matrix A is said to be *diagonal-positive* if $a_{ii} > 0$ for all diagonal elements and it is said to be *nonnegative (positive)*, denoted $A \geq 0 \ (>0)$, if $a_{ij} \geq 0 \ (>0)$ for each entry a_{ij} of A. A nonnegative matrix is called *reducible* if it is cogredient to a matrix of the form

$$\begin{pmatrix} C & 0 \\ R & T \end{pmatrix}$$

where C and T are square matrices. Otherwise, the matrix is called *irreducible*. An irreducible matrix is called *primitive* if there exists a finite constant $k \in \mathbb{N}$, such that its kth power is positive. A nonnegative matrix is said to be *stochastic* if all its row sums are 1. A stochastic matrix is termed *stable* if it has identical rows and it is called *column allowable* if each column contains at least one positive entry. □

Theorem 6 (Iosifescu 1980, p. 95) *Each transition matrix of a homogeneous finite Markov chain is cogredient to one of the following normal forms:*

$$P_1 = \begin{pmatrix} C_1 & & & \\ & C_2 & & \\ & & \ddots & \\ & & & C_r \end{pmatrix} \quad or \quad P_2 = \begin{pmatrix} C_1 & & & & \\ & C_2 & & & \\ & & \ddots & & \\ & & & C_r & \\ R_1 & R_2 & \cdots & R_r & T \end{pmatrix}$$

where submatrices $C_1,\ldots,C_r$ with $r \geq 1$ are irreducible and at least one of the submatrices R_i is nonzero. □

Submatrix T in matrix P_2 above is associated with the *transient set* of the state space, whereas the matrices C_i are associated with the non-transient or *recurrent* states. Each matrix C_i represents a *recurrent set* of states that cannot be left once it is entered – as is evident from the structure of the normal forms above. In contrast, transient states can be visited several times but as soon as there is a transition to a recurrent state, the transient state will never be visited again. Thus, a transient state will be left forever with probability 1 after a finite number of iterations. Not surprisingly, recurrent and transient sets will play a different role in the limit behavior of a homogeneous finite Markov chain. Before presenting basic limit theorems, some terms have to be introduced:

Definition 8 Let P be the transition matrix of a homogeneous finite Markov chain. A distribution p on the states of the Markov chain is called a *stationary distribution* if $pP = p$ and a *limit distribution* if the limit $p = p^{(0)} \lim_{t \to \infty} P^t$ does exists. □

Definition 9 Let $E = \mathcal{X}^\mu$ be the state space of a Markov chain representing some EA with population size $\mu \in \mathbb{N}$, finite search space $\mathcal{X}$ and objective function $f: \mathcal{X} \to \mathbb{R}$ to be minimized. A state $x^* \in E$ is said to be *optimal* if there exists an $i = 1,\ldots,\mu$ such that $f(x_i^*) = f^*$. The set E^* of all optimal states is termed the *optimal state set*. □

Now some limit theorems may be stated:

Theorem 7 (Iosifescu 1980, p. 126; Seneta 1981, p. 127) *Let P be a reducible stochastic matrix, where $C: m \times m$ is a primitive stochastic matrix and $R,T \neq 0$. Then*

$$P^\infty = \lim_{k \to \infty} P^k = \lim_{k \to \infty} \begin{pmatrix} C^k & 0 \\ \sum_{i=0}^{k-1} T^i R C^{k-i} & T^k \end{pmatrix} = \begin{pmatrix} C^\infty & 0 \\ R_\infty & 0 \end{pmatrix}$$

is a stable stochastic matrix with $P^\infty = 1' p^{(\infty)}$, where $p^{(\infty)} = p^{(0)} P^\infty$ is unique regardless of the initial distribution, and $p^{(\infty)}$ satisfies: $p_i^{(\infty)} > 0$ for $1 \leq i \leq m$ and $p_i^{(\infty)} = 0$ for $m < i \leq n$. The rate of approach to the limit is geometric. □

Thanks to this general result, it is a straightforward exercise to formulate general limit theorems for EAs: Suppose that the transition matrix for some EA is reducible and that all associated recurrent states are optimal states, that is, $E_r \subseteq E^*$. In this case

$$P\{Z_t \leq \varepsilon\} = P\{F_t^* - f^* \leq \varepsilon\} = P\{X_t \in E^*\} \geq P\{X_t \in E_r\} = \sum_{i \in E_r} p_i^{(t)}$$

holds for sufficiently small $\varepsilon > 0$. Since $P\{Z_t > \varepsilon\} = 1 - P\{Z_t \leq \varepsilon\}$ one obtains

$$P\{Z_t > \varepsilon\} \leq 1 - P\{X_t \in E_r\} = 1 - \sum_{i \in E_r} p_i^{(t)} \tag{6}$$

so that the limit of ❯ Eq. 6 is

$$\lim_{t \to \infty} P\{Z_t > \varepsilon\} \leq 1 - \sum_{i \in E_r} p_i^{(\infty)} = 1 - 1 = 0 \tag{7}$$

which is the defining condition for convergence to the optimum in probability. Thanks to ❯ Theorem 7, it is known that $p_i^{(\infty)} > 0$ if and only if $i \in E_r$. Therefore, the sum in ❯ Eq. 7 accumulates to 1. Moreover, ❯ Theorem 7 states that the approach to the limit is geometric. Since a geometric series (of probabilities < 1) converges to a finite limit, one obtains the stronger mode of complete convergence to the optimum. Convergence in mean follows from ❯ Theorem 2 since $Z_t \xrightarrow{P} 0$ and the value of Z_t must be bounded in a finite state space. As a result, we have proven:

Theorem 8 (Rudolph 1997a, p. 119) *If the transition matrix of an EA is reducible and the set of recurrent states is a subset of the set E^* of optimal states, then the EA converges completely and in mean to the global optimum regardless of the initial distribution.* □

The above theorem is formulated as a sufficient criterion. Actually, it is also a necessary criterion as shown in Agapie (1998a) and it also holds for stochastic processes more general than Markov chains ("random systems with complete connections").

Theorem 9 (Necessary and sufficient condition; see Agapie (1998b, 2007)) *A time-homogeneous EA with finite state space converges completely and in mean to the global optimum if and only if all recurrent states are optimal or, equivalently, if and only if all nonoptimal states are transient.* □

Another general result from Markov chain theory can be exploited for a different limit result regarding EAs.

Theorem 10 (Iosifescu 1980, p. 123; Seneta 1981, p. 119) *Let P be a primitive stochastic matrix. Then P^k converges as $k \to \infty$ to a positive stable stochastic matrix $P^\infty = 1' p^{(\infty)}$, where the limit distribution $p^{(\infty)} = p^{(0)} \cdot \lim_{k \to \infty} P^k = p^{(0)} P^\infty$ has nonzero entries and is unique regardless of the initial distribution. The rate of approach to the limit is geometric. Moreover, the limit distribution is identical with the unique stationary distribution and is given by the solution $p^{(\infty)}$ of the system of linear equations $p^{(\infty)} P = p^{(\infty)}$, $p^{(\infty)} 1' = 1$.* □

The weakest version of stochastic convergence considered here is convergence in probability. If this mode of convergence cannot be verified, then all stronger modes are precluded automatically. Now, suppose that the transition matrix for some EA is primitive. Owing to ❯ Theorem 10, one obtains

$$P\{Z_t > \varepsilon\} = 1 - P\{X_t \in E^*\} = 1 - \sum_{i \in E^*} p_i^{(t)} \tag{8}$$

for sufficiently small $\varepsilon > 0$. Note that $E^* \subset E$ unless the objective function is constant. Therefore

$$\sum_{i \in E^*} p_i^{(\infty)} < 1 \quad \text{and finally} \quad \lim_{t \to \infty} \mathsf{P}\{Z_t > \varepsilon\} = 1 - \sum_{i \in E^*} p_i^{(\infty)} > 0$$

falsifying convergence in probability. Thus, we have proven:

Theorem 11 (Rudolph 1997a, p. 120) *If the transition matrix of an EA is primitive, then the EA does not converge in probability to the global optimum regardless of the initial distribution.*

□

At this point, it should be noted that the property of visiting an optimal state with probability one is a precondition for convergence but that the additional property of convergence itself does not automatically indicate any advantage with respect to finding the global solution. Since all practical implementations of EAs store the best solution they have ever seen, it suffices to visit some optimal state only once. Therefore, EAs with primitive transition matrix also can be useful optimization algorithms.

Theorem 12 (Rudolph 1997a, p. 120) *Let $(X_t)_{t \geq 0}$ be the sequence of populations generated by an EA with primitive transition matrix. Then the stochastic sequence $(V_t)_{t \geq 0}$ of best objective function values ever found defined by $V_t = \min\{F_\tau^*: \tau = 0, 1, \ldots, t\}$ converges completely and in mean to the global optimum.*

□

As shown above, the convergence properties of EAs can be determined by a closer look at their transition matrices. Since the transition matrix can be decomposed into a product of intermediate transition matrices, each of them describing the probabilistic behavior of a single evolutionary operator (like crossover, mutation, and selection), it is sufficient to map each evolutionary operator on a transition matrix and to look at their products. For this purpose, the following result is useful.

Lemma 2 (Rudolph 1994b, p. 97; Agapie 1998a, p. 188) *Let I, D, C, P, A be stochastic matrices where I is irreducible, D is diagonal-positive, C column-allowable, P positive, and A arbitrary. Then the products*

(a) *AP and PC are positive,*
(b) *ID and DI are irreducible.*

□

After initialization, the typical cycle of an EA consists of selection for reproduction (transition matrix R), recombination / crossover (transition matrix C), mutation (transition matrix M), and selection for survival (transition matrix S). As a consequence, the transition matrix P of the EA can be decomposed via $P = R \cdot C \cdot M \cdot S$.

For example, the transition matrix M of standard bit-flipping mutation on binary strings has the entries $m_{ij} = p_m^{\|i-j\|}(1 - p_m^{n\mu - \|i-j\|})$, where $\| \cdot \|$ denotes the Hamming norm on binary strings. The entries m_{ij} are all positive if the mutation probability p_m is greater than 0 and less than 1. In this case, transition matrix M for mutation is positive. Now, ❯ Lemma 2(a) reveals, that regardless of the structure of the 'crossover matrix' C, the product $C \cdot M$ is positive. The same ❯ Lemma 2(a) implies that no structure of the transition matrix of selection for reproduction R can avoid that the product $R \cdot C \cdot M$ is positive. Finally, ❯ Lemma 2(a) also ensures that the entire transition matrix P is positive, if the transition matrix of selection for survival S is column allowable. Since every positive matrix is also primitive, one may invoke ❯ Theorem 11 to conclude that the EA will not converge to the optimum. Results of this type have been proven in Davis and Principe (1993) and Rudolph (1994b) for proportional survival selection: the corresponding transition matrix S is in fact column-allowable. The same result can be achieved for other popular selection operations.

Lemma 3 (Rudolph 1997a, p. 122) *If the selection operation chooses from offspring only then the associated transition matrix is column-allowable for proportional selection, q-ary tournament selection, q-fold binary tournament selection, and truncation selection.* □

The structure of matrix S is decisive for convergence if the preceding operations are represented by a positive matrix. If the selection operation is made *elitist* then the product of all transition matrices becomes a reducible matrix as required by ❯ Theorem 8. A proof of this result specialized to proportional selection was sketched in Eiben et al. (1991) and rigorously elaborated in Suzuki (1993, 1995). Elitism can be realized by at least two different mechanisms:

1. Select from parents and offspring and ensure that the best individual is selected with probability 1.
2. Select from offspring only; if the best selected individual is worse than the best parent, then replace the worst selected individual with the best parent.

Selection operations that can be used for elitism of the first kind are stochastic universal sampling, q-fold binary tournament selection, and truncation selection. Selection operations with elitism of the second kind can use almost any selection method in the first phase since the reinjection of the best parent, if necessary, in the second phase guarantees convergence once an optimum has been found for the first time.

Theorem 13 (Rudolph 1997a, p. 125) *If the transition matrix for mutation is positive and the selection operation for survival is elitist, then the EA converges completely and in mean to the optimum regardless of the initial distribution.* □

There are at least two reasons for nonconvergence: The optimum cannot be found with probability 1 or the optimum is as often lost as it is found. The second case can be eliminated by elitsm. Then it remains to make sure that the optimum is found with probability 1.

Without elitism, the mutation operator usually causes loss of an optimum previously found. Therefore, early research (prior to 1990) recommended to switch off mutation. But this maneuver does not help, since an EA without mutation may converge to an arbitrary (possibly nonoptimal) uniform population (Fogel 1994) that cannot be altered by crossover / recombination and selection. Nevertheless, for special cases (Rudolph 2005), it is possible to specify the probability (<1) of finding the optimum depending on the population and problem size.

But elitism is not a CONDITIO SINE QUA NON to achieve convergence to the optimum for Markovian optimization algorithms with time-homogeneous transition matrices and finite state space (Agapie 1998b) as demonstrated in the next example.

Example 3 (Convergence to optimum without elitism) The idea for the example is taken from Agapie (1998b). Consider the threshold accepting (TA) algorithm (Dueck and

---

Algorithm 1 Threshold accepting

initialize individual $X_0 \in \mathbb{B}^n$
set $t = 0$
repeat
 $Y_t = X_t \oplus B_t$
 if $f(Y_t) \le f(X_t) + T$ **then**
  $X_{t+1} = Y_t$
 else
  $X_{t+1} = X_t$
 end if
 increment t
until stopping criterion fulfilled

Scheuer 1990) as shown in ❯ *Algorithm 1*, where $B_0, B_1, \ldots$ is an i.i.d. sequence of Bernoulli random vectors (i.e., each component is an independent Bernoulli random variable with parameter $0 < p_m < 1$), $T > 0$ is a positive threshold and $\oplus$ denotes the bitwise exclusive-or operation. Notice that TA would be exactly an *elitist* $(1+1)$-EA if $T = 0$.

Evidently, TA can reach every point in the search space in one step, it accepts every improvement and also worse points, provided the difference does not exceed the threshold $T > 0$. Since worse points may be accepted, the method is not elitist. If this algorithm is applied to the minimization of the objective function

$$f(x) = \begin{cases} 0, & \text{if } x = 0 \\ 1 + \sum_{i=1}^{n} x_i, & \text{otherwise} \end{cases}$$

with threshold $T = 1$, then it is easily seen that TA may move to states with worse objective function value. TA may cycle freely through all states/solutions, but as soon as it hits the optimum at $x = 0$, then this state cannot be left since the difference in objective function values to all other solutions is larger than $T = 1$. Thus, only the optimal state 0 is recurrent whereas all other states are transient. Now, ❯ Theorem 9 ensures convergence to the optimum without elitism. □

The rigorous analysis in terms of Markov chain theory and the insight gained thereby has led to the conclusion that powerful sufficient conditions concerning properties and combinations of variation and selection operators can be derived already via simple basic probabilistic arguments (Rudolph 1998b) without using results from Markov chain theory: Let $(x_1, x_2, \ldots, x_\mu) \in \mathscr{X}^\mu$ denote the population of μ parents. An offspring is produced as follows: At first, ρ parents are selected to serve as mates for the recombination process. This operation is denoted by

$$\texttt{mat} : \mathscr{X}^\mu \to \mathscr{X}^\rho$$

where $2 \le \rho \le \mu$. These individuals are then recombined by the procedure

$$\texttt{reco} : \mathscr{X}^\rho \to \mathscr{X}$$

yielding a preliminary offspring. Finally, a mutation via

$$\texttt{mut} : \mathscr{X} \to \mathscr{X}$$

yields the complete offspring. After all λ offspring have been produced in this manner, the selection procedure

$$\texttt{sel} : \mathscr{X}^k \to X^\mu$$

decides which offspring and possibly parents ($k \geq \mu$) will serve as the new parents in the next iteration. Thus, a single iteration of the evolutionary algorithm can be described as follows:

$$\forall i \in \{1, \ldots, \lambda\} : x'_i = \texttt{mut}(\texttt{reco}(\texttt{mat}(x_1, \ldots, x_\lambda)))$$

$$(y_1, \ldots, y_\mu) = \begin{cases} \texttt{sel}(x_{\pi(1)}, \ldots, x_{\pi(q)}, x'_1, \ldots, x'_\lambda) & \text{(parents and offspring)} \\ \texttt{sel}(x'_1, \ldots, x'_\lambda) & \text{(only offspring)} \end{cases}$$

where $1 \leq q \leq \mu$ and $\pi(1), \ldots, \pi(\mu)$ is a permutation of the indices $1, \ldots, n$ such that $f(x_{\pi(1)}) \leq f(x_{\pi(2)}) \leq \ldots \leq f(x_{\pi(\mu)})$. This formulation includes selection methods that choose from the offspring and a subset of parents under the restriction that the best parent is a member of this subset.

After this operational description of evolutionary algorithms, one is in the position of defining some assumptions about the properties of the variation and selection operators:

(A1) $\forall x \in (x_1, \ldots, x_\mu) : \mathsf{P}\{x \in \texttt{reco}(\texttt{mat}(x_1, \ldots, x_\mu))\} \geq \delta_r > 0$.

(A2) For every pair $x, y \in \mathscr{X}$ there exists a finite path $x_1, x_2, \ldots, x_\tau$ of pairwise distinct points with $x_1 = x$ and $x_\tau = y$ such that $\mathsf{P}\{x_{i+1} = \texttt{mut}(x_i)\} \geq \delta_m > 0$ for all $i = 1, \ldots, \tau - 1$.

(A'2) For every pair $x, y \in \mathscr{X}$ holds $\mathsf{P}\{y = \texttt{mut}(x)\} \geq \delta_m > 0$.

(A3) $\forall x \in (x_1, \ldots, x_k) : \mathsf{P}\{x \in \texttt{sel}(x_1, \ldots, x_k))\} \geq \delta_s > 0$.

(A4) Let $v^*_k(x_1, \ldots, x_k) = \max\{f(x_i) : i = 1, \ldots, k\}$ denote the best fitness value within a population of k individuals ($k \geq \mu$). The selection method fulfills the condition

$$\mathsf{P}\{v^*_\mu(\texttt{sel}(x_1, \ldots, x_k)) = v^*_k(x_1, \ldots, x_k)\} = 1 .$$

Assumption (A1) means that every parent may be selected for mating and is not altered by recombination with minimum probability $\delta_r > 0$. Assumption (A2) ensures that every individual can be changed to an arbitrary other individual by a finite number of successive mutations, whereas assumption (A'2) asserts the same but within a single mutation. Assumption (A3) guarantees that every individual competing for survival may survive with minimum probability $\delta_s > 0$, whereas assumption (A4) makes sure that the best individual among the competitors in the selection process will survive with probability one.

These assumptions lead to a nonzero lower bound $\delta_r \cdot \delta_m \cdot \delta_s > 0$ on the probability to move from some nonoptimal state to some other state on the finite path of length τ to the optimum. Repetitions of this argument lead to the lower bound $\delta = (\delta_r \cdot \delta_m \cdot \delta_s)^\tau > 0$ to reach an optimal state within τ iterations. Consequently, the probability that the optimal state has not been found after $t \geq \tau$ iterations is at most $(1 - \delta)^{\lfloor t/\tau \rfloor}$ converging to zero geometrically fast as $t \to \infty$. This proves complete and mean convergence to the optimum.

Theorem 14 (Rudolph 1998b) *If either assumption (A'2) or the assumptions (A1), (A2), and (A3) are valid then the evolutionary algorithm visits the global optimum after a finite number of iterations with probability one, regardless of the initialization. If assumption (A4) is valid additionally and the selection method chooses from parents as well as offspring then the evolutionary algorithm converges completely and in mean to the global optimum regardless of the initialization.* □

Finally, note that there are a number of general convergence conditions for EAs (see, e.g., He and Kang 1999 or He and Yu 2001) that are only of limited value since they are not further refined to properties of the variation and selection operations. It is not clear if such a refinement can lead to an extension of the results known currently.

4.1.2 Innumerable State Space

The convergence theory of probabilistic optimization methods resembling a $(1+1)$-EA with $\mathscr{X} = \mathbb{R}^n$ and time-homogeneous transitions was established in Devroye (1976), Oppel and Hohenbichler (1978), Born (1978), Solis and Wets (1981), Pintér (1984), and others between the mid-1970s and mid-1980s. The proofs in each of these publications exploited the algorithms' property that the parent of the next generation cannot be worse than the current one, that is, it is guaranteed by the construction of the algorithms that the stochastic sequence $(Z_t : t \geq 0)$ is *monotonically* decreasing. As a consequence, $(Z_t : t \geq 0)$ is a nonnegative supermartingale that converges w.p.1 to a finite limit. Before summarizing the convergence results, some regularity conditions must be proclaimed: For all objective functions $f : \mathbb{R}^n \to \mathbb{R}$ holds

(R_1) $f^* > -\infty$ and
(R_2) the set $\mathscr{X}_\varepsilon^* = \bigcup_{x^* \in \mathscr{X}^*} \mathscr{V}_\varepsilon(x^*)$ has nonzero measure for all $\varepsilon > 0$,

where $\mathscr{V}_\varepsilon(x^*) = \{x \in \mathscr{X} : \|x - x^*\| \leq \varepsilon\}$ denotes the ε-vicinity around $x^* \in \mathscr{X}$. The first condition ensures the existence of an optimum whereas the second condition makes sure that at least one optimum is not an isolated point. The set $\mathscr{X}_\varepsilon^*$ is the set of ε-optimal solutions whereas $E_\varepsilon^* = \{x \in E \mid \exists i = 1, \dots, \mu : x_i \in \mathscr{X}_\varepsilon^*\}$ is the set of ε-optimal states.

Let $F^*(x) = \min\{f(x_1), f(x_2), \dots, f(x_\mu)\}$ denote the best objective function value in a state / population $x \in E$ consisting of μ individuals. Then the set $B(y) = \{x \in E : F^*(x) < F^*(y)\}$ is simply the set of states (populations) that contain an individual better than the best individual in population $y \in E$.

Theorem 15 (Rudolph 1997a, p. 201) *Let $X_0 \in E$ be the initial population of some elitist EA and let $P_c(x,A)$, $P_m(x,A)$ and $P_s(x,A)$ denote the transition probability functions of the crossover, mutation and selection operator, respectively, with $x \in E$ and $A \subseteq E$. If the conditions*

(a) $\exists \delta_c > 0 : \forall x \in E : P_c(x, B(x)) \geq \delta_c$,
(b) $\exists \delta_m > 0 : \forall x \in B(X_0) : P_m(x, E_\varepsilon^*) \geq \delta_m$ *and*
(c) $\exists \delta_s > 0 : \forall x \in E : P_s(x, B(x)) \geq \delta_s$

hold simultaneously, then for every $\varepsilon > 0$, there exists a $\delta > 0$ such that $P_{cms}(x, A_\varepsilon) \geq \delta > 0$ for every $x \in B(X_0)$ and the EA converges completely and in mean to the optimum. □

The next result refines ❯ Theorem 15 to specific variation and selection operators.

Theorem 16 (Rudolph 1997a, p. 204f) *A population-based EA with elitism that uses*

(a) *multipoint, parametrized uniform, parametrized intermediate, averaging, gene pool, or parametrized intermediate gene pool recombination (with replacement);*

(b) *a mutation distribution with support $\mathbb{R}^n$ and (possibly varying) positive definite covariance matrix whose spectrum stays in a fixed interval $[a, b] \subset \mathbb{R}_+$;*

(c) *standard proportional, proportional SUS, q–ary tournament, q–fold binary tournament or top μ selection*

converges completely and in mean to the global minimum of an objective function $f: \mathbb{R}^n \to \mathbb{R}$ from the set $\{f \in F : f(x) \to \infty \text{ as } \|x\| \to \infty\}$. □

EAs with search space $\mathscr{X} = \mathbb{R}^n$ can converge to the optimum even without elitism – but only for special classes of problems (Rudolph 1994a, 1997b,a) (e.g., (L, Q)-convex functions) and where the mutation distribution is scaled proportionally to the length of the gradient at the current position. If this kind of information is available, then it is possible to apply general convergence results for supermartingales like ❷ Theorem 4, which is specialized to EAs below.

Theorem 17 (Rudolph 1997b) *Let $(X_t : t \geq 0)$ be the sequence of populations generated by some evolutionary algorithm and let F_t^* denote the best objective function value of the population at generation $t \geq 0$. If $E[Z_t] < \infty$ and*

$$E[Z_{t+1} \mid X_t, X_{t-1}, \dots, X_0] \leq c_t Z_t \quad w.p.1 \tag{9}$$

where $Z_t = F_t^* - f^*$ and $c_t \in [0,1]$ for all $t \geq 0$ such that the infinite product of the c_t converges to zero, then the evolutionary algorithm converges in mean and with probability 1 to the global minimum of the objective function $f(\cdot)$. □

Convergence properties of EAs applied to constrained "corridor model" function in the context of Markov chains can be found in Agapie and Agapie (2007).

4.2 Time-Inhomogeneous Transitions

4.2.1 Finite Search Space

Apparently, the development of EAs with time-inhomogeneous transitions was motivated by the observation that, owing to ❷ Theorem 11, many popular EAs do not converge but there existed many convergence proofs for the simulated annealing (SA) optimization algorithm (Hajek 1988; Aarts and Korst 1989; Haario and Saksman 1991) that is endowed with time-inhomogeneous transitions. As a consequence, the EAs were modified to meet the convergence conditions of the SA convergence theory (Davis 1991; Mahfoud and Goldberg 1992, 1995; Adler 1993). A more general point of view was the starting point in Suzuki (1998) and Schmitt et al. (1998) and Schmitt (2001, 2004). They exploited the theory of time-inhomogeneous Markov chains. Before summarizing their results, the case of elitist selection and time-dependent variation operators is considered.

Theorem 18 *Let $(X_t)_{t \geq 0}$ be the stochastic sequence generated by an EA with elitist selection and a transition probability function $P_t(x, A)$ with $A \subseteq E$, that only describes the sequence of variation operations. If $P_t(x, E^*) \geq \delta_t$ for all $x \in E \backslash E^*$ and*

$$\sum_{t=0}^{\infty} \delta_t = \infty$$

then the EA converges w.p.1 and in mean to the optimum as $t \to \infty$.

Proof Since the selection operation is elitist, it suffices to show that the probability of not transitioning to some optimal state in E^* converges to zero as $t \to \infty$. This probability can be bounded via

$$P\{X_t \notin E^*\} \leq \prod_{k=1}^{t}(1 - \delta_k) \tag{10}$$

In order to prove that the product above converges to zero, the following equivalence (see ❯ Sect. 2.3.1 in Rudolph (1999)) is useful:

$$\prod_{t=1}^{\infty}(1 - \delta_t) \to 0 \quad \Leftrightarrow \quad \sum_{t=1}^{\infty}\log\left(\frac{1}{1 - \delta_t}\right) \to \infty$$

From the series expansion of the logarithm, it is clear that

$$\log\left(\frac{1}{1 - z}\right) > z$$

for all $z \in (0,1)$. As a consequence, since

$$\sum_{t=1}^{\infty}\log\left(\frac{1}{1 - \delta_t}\right) > \sum_{t=1}^{\infty}\delta_t = \infty$$

is fulfilled by the precondition of the theorem, it has been proven that the product in ❯ Eq. 10 converges to zero. This proves convergence in probability. Since selection is elitist, the associated sequence of best function values $(Z_t)_{t \geq 0}$ is a nonnegative supermartingale that converges w.p.1 to some finite limit. Since the limits of w.p.1 convergence and convergence in probability must be unique, one obtains w.p.1 convergence to the optimum. Moreover, elitism implies that $E[Z_t] \leq E[Z_0]$ for all $t \geq 0$, which in turn implies convergence in mean. ☐

It remains to show how the preconditions of the theorem can be fulfilled. For example, use an arbitrary crossover operation and the usual bit-flipping mutation with time-variant mutation probability $p_m(t)$ with $p_m(t) \to 0$ as $t \to \infty$. It suffices to bound the probability that just one individual becomes optimal. Evidently, if $p_m(t) \leq 1/2$ then

$$\delta_t \geq \min\{p_m^h(t) \cdot (1 - p_m(t))^{n-h} : h = 0, 1, \ldots, n\} = p_m^n(t)$$

The inequality below

$$\sum_{t=1}^{\infty}\delta_t \geq \sum_{t=1}^{\infty}p_m^n(t) = \infty$$

is fulfilled if, for example, $p_m(t) = (t + 1)^{-1/n}$. Thus, the mutation probability must not decrease too fast.

If elitism is dropped from the list of preconditions, then at least the selection operation must become time-dependent. The detailed studies in Schmitt et al. (1998) and Schmitt (2001, 2004) contain numerous results in this direction, which are dared to be summarized as follows:

Theorem 19 (Schmitt 2004) *An EA that uses*

(a) *scaled mutations with $p_m(t) \sim t^{-\kappa_m/n}$ where $\kappa_m \in (0,1]$,*

(b) *gene-lottery or regular pair-wise crossover with $p_c(t) \sim p_m^{\kappa_c}(t)$ where $\kappa_c \in (0,1]$,*
(c) *proportional selection using exponentiation $f^{g(t)}$ with $g(t) \sim \log(t+1)$*

converges w.p.1 and in mean to the optimum. □

4.2.2 Innumerable Search Space

Time-inhomogeneous transitions in innumerable search spaces have not been studied for EAs, apparently. If crossover is arbitrary, mutation is chosen as in ❯ Theorem 16(b), and selection is done as in simulated annealing, then the SA convergence theory can be applied directly (Bélisle 1992). Moreover, there is a simple analogon to the result with elitist selection in finite state space.

Corollary 1 (to ❯ Theorem 18) *Let $(X_t)_{t\geq 0}$ be the stochastic sequence generated by an EA with elitist selection and a transition probability function $P_t(x,A)$ with $A \subseteq E$, that only describes the sequence of variation operations. If $P_t(x,E_\varepsilon^*) \geq \delta_t$ for all $x \in E\backslash E_\varepsilon^*$ and*

$$\sum_{t=0}^{\infty} \delta_t = \infty$$

then the EA converges w.p.1 and in mean to the optimum as $t \to \infty$.

Proof Set $\mathcal{X} = \mathbb{R}^n$, $E = \mathbb{R}^{n \times \mu}$, replace E by E_ε^* in the statement and proof of ❯ Theorem 18. □

4.3 Self-Adaptive Transitions

Self-adaptation in EAs may appear in various forms. Here, the focus is on the special case where each individual consists of the position in the search space $\mathcal{X} = \mathbb{R}^n$ and a number of (strategy) parameters that determine the shape of the mutation distribution.

In the simplest case, an individual might be described by the pair (x,σ) with $x \in \mathcal{X}$ and $\sigma > 0$. A mutation of an individual can be realized by $y = x + \sigma Z$ where Z is a standard multinormally distributed random vector. If σ is altered depending on the outcome y, then σ has been self-adapted. Evidently, in the context of Markov chains, this is the time-homogeneous case – but the state space becomes larger since the strategy parameters also must be represented in the state space. This fact makes the analysis considerably more complicated. Therefore, the presentation of results for self-adaptive transitions have been separated from the case with fixed, externally set strategy parameters (like mutation and crossover probability).

General results known by now are rare. One may apply ❯ Theorem 16 to ensure convergence to the optimum for any kind of self-adaptive mutation mechanisms as long as it is guaranteed that the support is $\mathbb{R}^n$ and that the eigenvalues of the covariance matrices stay in a fixed compact interval in $\mathbb{R}_+$. In this case, one can derive a lower positive bound on the probability to hit $\mathcal{X}_\varepsilon^*$ in every iteration, which implies convergence to the optimum. The situation changes if the spectrum of the eigenvalues has no restrictions, that is, the smallest eigenvalue may approach zero and the largest eigenvalue may grow arbitrarily large. This may happen for Rechenberg's $\frac{1}{5}$-rule. It has been shown (Rudolph 1999, 2001b) that the probability to escape from local, non-global optima under the $\frac{1}{5}$-rule is strictly less than 1. As a consequence, convergence to the optimum is not guaranteed. But if the objective function only has a single local and therefore global optimum, then the $\frac{1}{5}$-rule leads to convergence w. p.1 where the approximation error is decreased geometrically fast (Jägersküpper 2006, 2007). If selection only accepts points whose objective function value is at least $\varepsilon > 0$ better than the parent, and if the $\frac{1}{5}$-rule is changed such that the variance is increased for success probabilities far below $\frac{1}{5}$ (say, below $\frac{1}{20}$), then convergence to the optimum can be asserted (Greenwood and Zhu 2001).

In case of mutative self-adaptation of the variance of the mutation distributions, only unimodal or spherical objective functions have been analyzed. Whereas the proof in Semenov and Terkel (2003) via stochastic Lyapunov functions/supermartingales has a theoretical gap (a lemma is "proven" by experiments), the proof in Bienvenüe and Francois (2003) and Auger (2004) via Markov processes ensure convergence to the optimum for objective function $f(x) = x^2$ with $x \in \mathbb{R}$.

5 Further Reading and Open Questions

The previous section tacitly assumed that EAs are applied to optimization problems with a single objective function. But optimization under multiple objectives is the *de facto* standard for practical applications nowadays. Unfortunately, the convergence theory cannot easily be transferred from single- to multi-objective EAs. First, the solution in multi-objective optimization (via the *a posteriori* approach) is not a single point but a set of points (the Pareto set) whose objective function values (the Pareto front) are only partially ordered. Second, it is unclear how to assess the quality of the Pareto front approximation: all points should be close to the true Pareto front, the points should represent the entire Pareto front, and they should be "somehow uniformly distributed" over the Pareto front approximation. Early work (Rudolph 1998a, c) proved that at least a single individual approaches the Pareto front (convergence of the scalar Euclidean distance between point and set). Subsequent work (Hanne 1999; Rudolph and Agapie 2000; Rudolph 2001a) proved that the entire population approaches the Pareto front (convergence of the distance between sets). But nothing was shown with respect to the spread of the individuals and their distribution: it was possible that the entire population converges to a single point of the Pareto front. The problem was that the quality of an approximation is described by three objectives. As a consequence, achieving a good Pareto front approximation is itself a multi-objective problem with possibly incomparable approximations. A good compromise was a scalar measure termed *dominated hypervolume* or *S-metric* (Zitzler and Thiele 1998), as it rewards closeness to the Pareto front, it rewards a good spread, and it rewards a certain distributions of solutions (the more curved the Pareto front the more solutions). Although this scalar measure is appealing for formulating convergence results, it seems to be quite difficult to accomplish this goal. Moreover, it is quite likely that only those multi-objective EAs can reach an approximation with best S-metric value that deploy the S-metric indicator in the selection process for deciding which individuals should be integrated or discarded from the approximation. For such an EA, it was shown that it can converge to an approximation with best S-metric value if the true Pareto front is linear (Beume et al. 2009) but there are counter-examples for nonlinear Pareto fronts (Zitzler et al. 2008). Another approach was initiated in Schütze et al. (2007) by proving approximation properties of the Pareto archive of multi-objective EAs.

Summing up, the convergence theory for self-adaptive transitions and for multi-objective EAs is not complete yet. Challenging open questions that await answers are the convergence properties of model-assisted EAs, the covariance matrix adaptation evolution strategy (CMA-ES), and other related methods.

References

Aarts EHL, Korst J (1989) Simulated annealing and Boltzman machines: a stochastic approach to combinatorial optimization and neural computing. Wiley, Chichester

Adler D (1993) Genetic algorithms and simulated annealing: A marriage proposal. In: IEEE international conference on neural networks, San Francisco, CA, March–April 1993. IEEE Press, Piscataway, NJ, pp 1104–1109

Agapie A (1998a) Genetic algorithms: minimal conditions for convergence. In: Hao JK, Lutton E, Ronald E, Schoenauer M, Snyers D (eds) AE'97: Artificial evolution: Third European conference; selected papers, Nimes, France, October 1997. Springer, Berlin, pp 183–193

Agapie A (1998b) Modelling genetic algorithms: from Markov chains to dependance with complete connections. In: Bäck T, Eiben AE, Schoenauer M, Schwefel HP (eds) Parallel problem solving from nature – PPSN V. Springer, Berlin, pp 3–12

Agapie A (2007) Evolutionary algorithms: modeling and convergence. Editura Academiei Române, Bucarest, Romania

Agapie A, Agapie M (2007) Transition functions for evolutionary algorithms on continuous state-space. J Math Model Algorithms 6(2):297–315

Auger A (2004) Convergence results for the $(1,\lambda)$-ES using the theory of ϕ-irreducible Markov chains. Theor Comput Sci 334(1–3):181–231

Bélisle CJP (1992) Convergence theorems for a class of simulated annealing algorithms on $\mathbb{R}^d$. J Appl Probability 29:885–895

Beume N, Naujoks B, Preuss M, Rudolph G, Wagner T (2009) Effects of 1-greedy S-metric-selection on innumerably large pareto fronts. In: Ehrgott M et al. (eds) Proceedings of 5th international conference on evolutionary multi-criterion optimization (EMO 2009), Nantes, France, April 2009. Springer, Berlin, pp 21–35

Bienvenüe A, Francois O (2003) Global convergence for evolution strategies in spherical problems: some simple proofs and difficulties. Theor Comput Sci 306(1–3):269–289

Born J (1978) Evolutionsstrategien zur numerischen Lösung von Adaptationsaufgaben. Dissertation A, Humboldt-Universität, Berlin

Bucy R (1965) Stability and positive supermartingales. J Differ Equations 1(2):151–155

Bucy R, Joseph P (1968) Filtering for stochastic processes with applications to guidance. Interscience Publishers, New York

Davis T (1991) Toward an extrapolation of the simulated annealing convergence theory onto the simple genetic algorithm. PhD thesis, University of Florida, Gainesville

Davis T, Principe J (1993) A Markov chain framework for the simple genetic algorithm. Evol Comput 1(3):269–288

Devroye LP (1976) On the convergence of statistical search. IEEE Trans Syst Man Cybern 6(1):46–56

Doob J (1967) Stochastic processes, 7th edn. Wiley, New York

Dueck G, Scheuer T (1990) Threshold accepting: a general purpose optimization algorithm superior to simulated annealing. J Comput Phys 90(1):161–175

Eiben AE, Aarts EHL, van Hee KM (1991) Global convergence of genetic algorithms: a Markov chain analysis. In: Schwefel HP, Männer R (eds) Parallel problem solving from nature. Springer, Berlin, pp 4–12

Fogel D (1994) Asymptotic convergence properties of genetic algorithms and evolutionary programming: analysis and experiments. Cybern Syst 25(3):389–407

Greenwood G, Zhu Q (2001) Convergence in evolutionary programs with self-adaptation. Evol Comput 9(2):147–157

Haario H, Saksman E (1991) Simulated annealing process in general state space. Adv Appl Probability 23:866–893

Hajek B (1988) Cooling schedules for optimal annealing. Math Oper Res 13(2):311–329

Hanne T (1999) On the convergence of multiobjective evolutionary algorithms. Eur J Oper Res 117(3):553–564

He J, Kang K (1999) On the convergence rates of genetic algorithms. Theor Comput Sci 229(1–2):23–39

He J, Yu X (2001) Conditions for the convergence of evolutionary algorithms. J Syst Archit 47(7):601–612

Iosifescu M (1980) Finite Markov processes and their applications. Wiley, Chichester

Iosifescu M, Grigorescu S (1990) Dependence with complete connections and its applications. Cambridge University Press, Cambridge

Jägersküpper J (2006) How the (1+1) ES using isotropic mutations minimizes positive definite quadratic forms. Theor Comput Sci 361(1):38–56

Jägersküpper J (2007) Algorithmic analysis of a basic evolutionary algorithm for continuous optimization. Theor Comput Sci 379(3):329–347

Lukacs E (1975) Stochastic convergence, 2nd edn. Academic, New York

Mahfoud SW, Goldberg DE (1992) A genetic algorithm for parallel simulated annealing. In: Männer R, Manderick B (eds) Parallel problem solving from nature, vol 2. North Holland, Amsterdam, pp 301–310

Mahfoud SW, Goldberg DE (1995) Parallel recombinative simulated annealing: a genetic algorithm. Parallel Comput 21:1–28

Meyn S, Tweedie R (1993) Markov chains and stochastic stability. Springer, London

Muhammad A, Bargiela A, King G (1999) Fine-grained parallel genetic algorithm: a global convergence criterion. Int J Comput Math 73(2):139–155

Neveu J (1975) Discrete-parameter martingales. North Holland, Amsterdam

Nix A, Vose M (1992) Modeling genetic algorithms with Markov chains. Ann Math Artif Intell 5:79–88

Nummelin E (1984) General irreducible Markov chains and non-negative operators. Cambridge University Press, Cambridge

Oppel U, Hohenbichler M (1978) Auf der Zufallssuche basierende Evolutionsprozesse. In: Schneider B, Ranft U (eds) Simulationsmethoden in der Medizin und Biologie. Springer, Berlin, pp 130–155

Pintér J (1984) Convergence properties of stochastic optimization procedures. Math Oper Stat Ser Optimization 15:405–427

Rudolph G (1994a) Convergence of non-elitist strategies. In: Proceedings of the first IEEE conference on evolutionary computation, vol 1. Orlando, FL, June 1994. IEEE Press, Piscataway, NJ, pp 63–66

Rudolph G (1994b) Convergence properties of canonical genetic algorithms. IEEE Trans Neural Netw 5(1):96–101

Rudolph G (1994c) An evolutionary algorithm for integer programming. In: Davidor Y, Schwefel HP, Männer R (eds) Parallel problem solving from nature, vol 3. Springer, Berlin, pp 139–148

Rudolph G (1996) Convergence of evolutionary algorithms in general search spaces. In: Proceedings of the third IEEE conference on evolutionary computation, Nagoya, Japan, May 1996. IEEE Press, Piscataway, NJ, pp 50–54

Rudolph G (1997a) Convergence properties of evolutionary algorithms. Kovač, Hamburg

Rudolph G (1997b) Convergence rates of evolutionary algorithms for a class of convex objective functions. Control Cybern 26(3):375–390

Rudolph G (1998a) Evolutionary search for minimal elements in partially ordered finite sets. In: Porto VW, Saravanan N, Waagen D, Eiben AE (eds) Evolutionary programming VII, Proceedings of the 7th annual conference on evolutionary programming, San Diego, CA, March 1998. Springer, Berlin, pp 345–353

Rudolph G (1998b) Finite Markov chain results in evolutionary computation: a tour d'horizon. Fund Inform 35(1–4):67–89

Rudolph G (1998c) On a multi-objective evolutionary algorithm and its convergence to the Pareto set. In: Proceedings of the 1998 IEEE international conference on evolutionary computation, Anchorage, AK, May 1998. IEEE Press, Piscataway, NJ, pp 511–516

Rudolph G (1999) Global convergence and self-adaptation: A counter-example. In: CEC'99: Proceedings of the 1999 congress of evolutionary computation, vol 1. Washington, DC, July 1999. IEEE Press, Piscataway, NJ, pp 646–651

Rudolph G (2001a) Evolutionary search under partially ordered fitness sets. In: Sebaaly MF (ed) ISI 2001: Proceedings of the international NAISO congress on information science innovations, Dubai, UAE, March 2001. ICSC Academic Press, Millet and Sliedrecht, pp 818–822

Rudolph G (2001b) Self-adaptive mutations may lead to premature convergence. IEEE Trans Evol Comput 5(4):410–414

Rudolph G (2005) Analysis of a non-generational mutationless evolutionary algorithm for separable fitness functions. Int J Comput Intell Res 1(1):77–84

Rudolph G, Agapie A (2000) Convergence properties of some multi-objective evolutionary algorithms. In: Zalzala A, Fonseca C, Kim JH, Smith A, Yao X (eds) CEC 2000: Proceedings of the 2000 congress on evolutionary computation, vol 2. La Jolla, CA, July 2000. IEEE Press, Piscataway, NJ, pp 1010–1016

Schmitt L (2001) Theory of genetic algorithms. Theor Comput Sci 259(1–2):1–61

Schmitt L (2004) Theory of genetic algorithms II: models for genetic operators over the string-tensor representation of populations and convergence to global optima for arbitrary fitness functions under scaling. Theor Comput Sci 310(1–3):181–231

Schmitt LM, Nehaniv CL, Fujii RH (1998) Linear analysis of genetic algorithms. Theor Comput Sci 200 (1–2):101–134

Schütze O, Laumanns M, Tantar E, Coello Coello C, Talbi EG (2007) Convergence of stochastic search algorithms to gap-free Pareto front approximations. In: GECCO 2007: Proceedings of the 9th annual conference on genetic and evolutionary computation, London, July 2007. ACM, New York, pp 892–901

Semenov M, Terkel D (2003) Analysis of convergence of an evolutionary algorithm with self-adaptation using a stochastic Lyapunov function. Evol Comput 11(4):363–379

Seneta E (1981) Non-negative matrices and Markov chains, 2nd edn. Springer, New York

Solis FJ, Wets RJB (1981) Minimization by random search techniques. Math Oper Res 6(1):19–30

Suzuki J (1993) A Markov chain analysis on a genetic algorithm. In: Forrest S (ed) Proceedings of the fifth international conference on genetic algorithms, Urbana-Champaign, IL, June 1993. Morgan Kaufmann, San Mateo, CA, pp 146–153

Suzuki J (1995) A Markov chain analysis on simple genetic algorithms. IEEE Trans Syst Man Cybern 25(4):655–659

Suzuki J (1998) A further result on the Markov chain model of genetic algorithm and its application to a simulated annealing-like strategy. IEEE Trans Syst Man Cybern B 28(1):95–102

Williams D (1991) Probability with martingales. Cambridge University Press, Cambridge

Zitzler E, Thiele L (1998) Multiobjective optimization using evolutionary algorithms: comparative case study. In: Eiben A et al. (eds) Parallel problem solving from nature (PPSN V). Springer, Berlin, pp 292–301

Zitzler E, Thiele L, Bader J (2008) On set-based multi-objective optimization. Technical Report 300, Computer Engineering and Networks Laboratory, ETH Zurich

28 Evolutionary Multiobjective Optimization

Eckart Zitzler
PHBern – University of Teacher Education, Institute for Continuing
Professional Education, Weltistrasse 40, CH-3006, Bern, Switzerland
eckart.zitzler@phbern.ch
eckart.zitzler@tik.ee.ethz.ch

G. Rozenberg et al. (eds.), *Handbook of Natural Computing*, DOI 10.1007/978-3-540-92910-9_28,

Abstract

This chapter provides an overview of the branch of evolutionary computation that is dedicated to solving optimization problems with multiple objective functions. On the one hand, it sketches the foundations of multiobjective optimization and discusses general approaches to deal with multiple optimization criteria. On the other hand, it summarizes algorithmic concepts that are employed when designing corresponding search methods and briefly comments on the issue of performance assessment. This chapter concludes with a summary of the main application areas of evolutionary multiobjective optimization, namely, learning/decision making and multiobjectivization.

1   Introduction

The term *evolutionary multiobjective optimization* – EMO for short – refers to the employment of evolutionary algorithms to search problems involving multiple optimization criteria. This type of problem arises naturally in practical applications, simply because there is usually no such thing as a 'free lunch': in embedded system design high performance comes along with high cost, in structural engineering the level of stability is related to the weight of the construction, and in machine learning the accuracy of a model needs to be traded off against the model complexity. There exist countless other examples, many of which are not restricted to a pair of counteracting criteria but a multitude of competing objectives. The design of a car, for instance, requires many aspects to be taken into account, be it driving characteristics, cost, environmental sustainability, or more subjective criteria such as comfort and appearance. One may argue that the multiobjective case represents the *normal* situation in reality, while the single-objective case is rather an exception.

Example 1   Consider the following knapsack problem, a classical combinatorial optimization problem. Given is a set $\{1, \ldots, l\}$ of items and with each item i a corresponding weight $w_i \in \mathbb{R}^+$ and a corresponding profit $p_i \in \mathbb{R}^+$ is associated. The goal is to identify a subset S of the items that (i) maximizes the overall profit $f_p(S)$ and (ii) minimizes the overall weight $f_w(S)$ where

$$f_p(S) = \sum_{i \in S} p_i \tag{1}$$

$$f_w(S) = \sum_{i \in S} w_i \tag{2}$$

are the corresponding objective functions. ❷ *Figure 1* illustrates a problem instance with four items.

It is obvious that with the knapsack problem, in general, both objectives cannot be achieved at the same time: maximum profit means all items need to be selected, minimum weight is obtained by choosing none of the items. In between these extremes, several compromise solutions may emerge, each of them representing a particular trade-off of profit versus weight. Classically, this problem is approached by transforming the weight objective into a constraint, that is,

$$\begin{aligned} &argmax_{S \subseteq \{1,\ldots,l\}} \ f_p(S) \\ &subject\ to \qquad f_w(S) \leq w_{\max} \end{aligned} \tag{3}$$

◻ Fig. 1

Example of a knapsack problem instance with four items: (1) a camera with profit $p_1 = 5$ and weight $w_1 = 750$, (2) a pocket knife with profit $p_2 = 7$ and weight $w_2 = 300$, (3) a thermos flask with profit $p_3 = 8$ and weight $w_3 = 1{,}500$, and (4) a book with profit $p_4 = 3$ and weight $w_4 = 1{,}000$. All possible solutions are plotted with respect to their objective function values.

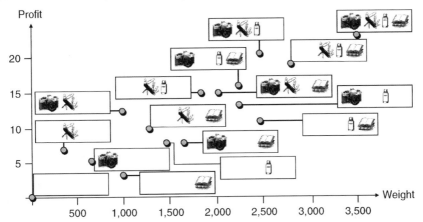

where w_{max} is a fixed weight limit. This is one way to handle scenarios with multiple objectives, but various others exist.

The field of *Multiple Criteria Decision Making (MCDM)* explicitly addresses these issues and can be defined as (Steuer and Miettinen 2008)

▶ the study of methods and procedures by which concerns about multiple conflicting criteria can be formally incorporated into the management planning process.

In this sense, EMO can be regarded as a branch of MCDM (as well as it can be considered a branch of evolutionary computation) where evolutionary algorithms and other randomized search heuristics are the methods of choice. From a historical perspective, though, MCDM has been – until recently – mainly devoted to exact optimization techniques such as linear programming, while the first EMO studies pursued a specific MCDM approach, namely to find a *set* of suitable compromise solutions. Meanwhile, the boundaries between these communities have blurred, but still the large majority of papers dedicated to EMO is related to approximating the set of optimal compromise solutions, as will be detailed later.

This chapter is an attempt to summarize 25 years of EMO research from a unified point of view. The focus is on general concepts, and therefore the following overview does not lay claim to completeness, with respect to both specific research topics and publications in the field. Readers interested in a detailed discussion of specific algorithms, a complete historical overview, and an extensive collection of references may consult, for example, the textbooks by Deb (2001) and Coello Coello et al. (2007).

2 Fundamental Concepts

Before different ways on how to approach scenarios with multiple optimization criteria and how to design evolutionary algorithms for this purpose are discussed, some basic notions and reflections on multiobjective optimization will be presented in general.

2.1   Multiobjective Optimization Problems

Roughly speaking, optimization stands for the task of identifying within a set of solutions one solution that is best (or close to best). The set of potential solutions is denoted as *decision space*, in the following represented by the symbol X, and its elements are denoted as decision vectors – or simply solutions. Sometimes, the terms "decision space" and "search space" are used interchangeably, although here a subtle differentiation will be made: the search space is the space a concrete evolutionary algorithm operates on; it can be different from the decision space, which is part of the formal description of the optimization problem.

In order to define what "best" means, all solutions are evaluated by means of objective functions on the basis of which they can be compared. Without loss of generality, here it is assumed that an objective function f returns a real value, that is, $f: X \mapsto \mathbb{R}$, although the codomain could be any set. In single-objective optimization, there is just one objective function, while in the multiobjective case there are several ones, say $f_1, \ldots, f_n$. These functions form a vector function $\mathbf{f} = (f_1, \ldots, f_n)$ that assigns each decision vector a real vector, or *objective vector* in general. The objective vector contains the function values for all n objective functions and is an element of the so-called objective space Z, here $Z = \mathbb{R}^n$. Therefore, $\mathbf{f}: X \mapsto Z$ provides the mapping from the decision space to the objective space.

Finally, a binary relation $\preccurlyeq$ specifies an order on the objective space, which completes the formal description of the optimization problem. The minimum requirement for this relation is usually that it is a partial order, that is, reflexive, antisymmetric, and transitive. (A binary relation $\preceq$ over a set Y is (1) reflexive iff $a \preceq a$ for all $a \in Y$, (2) antisymmetric iff $(a \preceq b \wedge b \preceq a) \Rightarrow (a = b)$ for all $a, b \in Y$, and (3) transitive iff $(a \preceq b \wedge b \preceq c) \Rightarrow (a \preceq c)$ holds for all $a, b,$ and c in Y.) In the single-objective case, usually the relations $\leq$ (minimization) or $\geq$ (maximization) are employed, in the multiobjective case, the choice is less obvious and will be discussed later in ❷ Sect. 2.2. Now given $\preccurlyeq$, two solutions $\mathbf{a}, \mathbf{b} \in X$ can be compared by first determining the objective vectors $\mathbf{f}(\mathbf{a}), \mathbf{f}(\mathbf{b})$ and then checking whether $\mathbf{f}(\mathbf{a}) \preccurlyeq \mathbf{f}(\mathbf{b})$ or vice versa. The goal is to find a solution $\mathbf{a}$ that is best, that is, there is no better solution; formally

$$\forall \mathbf{b} \in X : (\mathbf{f}(\mathbf{b}) \preccurlyeq \mathbf{f}(\mathbf{a})) \Rightarrow (\mathbf{f}(\mathbf{a}) \preccurlyeq \mathbf{f}(\mathbf{b})) \tag{4}$$

Furthermore, constraints may be involved, each of which can be regarded as a special case of an objective function in combination with a prespecified threshold. Fulfilling a constraint means the given threshold must not be exceeded; here it is assumed that a constraint is given by a constraint function $g: X \mapsto \mathbb{R}$, which needs to be equal to or less than 0 to be met. Several such constraint functions may be considered simultaneously, and a decision vector fulfilling the constraints is called a *feasible solution*; the entirety of the feasible solutions constitutes the *feasible set*. Hence, the optimization goal can be restated as finding the best among the feasible solutions.

Example 2   In the classical formulation of the knapsack problem according to ❷ Eq. 3, the decision space is $X = \{0, 1\}^l$, the objective space is $Z = \mathbb{R}$, the objective function is $\mathbf{f} = (f_1)$ with $f_1(a_1, \ldots, a_l) = \sum_{1 \leq i \leq l} a_i \cdot p_i$, the considered relation is $\geq$, and the weight is included using a constraint function $g(a_1, \ldots, a_l) = \left(\sum_{1 \leq i \leq l} a_i \cdot w_i \right) - w_{max}$ where w_{max} is the weight limit. For the example in ❷ Fig. 1, the feasible set is $X_f = \{(0, 0, 0, 0), (0, 1, 0, 0), (1, 0, 0, 0)\}$ when setting the weight limit to $w_{max} = 800$.

In summary, a (multiobjective) optimization scenario can be described as follows:

▶ A *multiobjective optimization problem* is defined by a 5-tuple $(X, Z, \mathbf{f}, \mathbf{g}, \preceq)$ where
  - X is the decision space,
  - $Z = \mathbb{R}^n$ is the objective space,
  - $\mathbf{f} = (f_1, \ldots, f_n)$ is a vector-valued function consisting of n objective functions $f_i \colon X \mapsto \mathbb{R}$,
  - $\mathbf{g} = (g_1, \ldots, g_m)$ is a vector-valued function consisting of m constraint functions $g_i \colon X \mapsto \mathbb{R}$, and
  - $\preceq \subseteq Z \times Z$ is a binary relation on the objective space.

  The goal is to identify a decision vector $\mathbf{a} \in X$ such that (i) for all $1 \leq i \leq m$ holds $g_i(\mathbf{a}) \leq 0$ and (ii) for all $\mathbf{b} \in X$ holds $\mathbf{f}(\mathbf{b}) \preceq \mathbf{f}(\mathbf{a}) \Rightarrow \mathbf{f}(\mathbf{a}) \preceq \mathbf{f}(\mathbf{b})$.

The question of how to define the relation $\preceq$ in a multiobjective context has remained open so far – it will be addressed next.

2.2 Preference Relations and Optimality

Consider the biobjective knapsack problem from ❷ Example 1. Given two item sets S and T and the corresponding objective vectors (p_S, w_S) and (p_T, w_T), when can S be said to be superior to T? Clearly, if S is better in both objectives, that is, it provides higher profit at lower weight, then it can be called overall superior. Even if S is only better in one objective, while there is no difference in the other, for example, $p_S > p_T \wedge w_S = w_T$, then one may make the same statement. That means, in general, that a solution is better than another if the former is (1) not worse in any objective and (2) better in at least one objective. These considerations lead to a concept known as *(weak) Pareto dominance*.

▶ An objective vector $\mathbf{u} = (u_1, \ldots, u_n)$ *weakly Pareto dominates* an objective vector $\mathbf{v} = (v_1, \ldots, v_n)$, written as $\mathbf{u} \preceq_{par} \mathbf{v}$, iff

$$\forall 1 \leq i \leq n : u_i \leq v_i \tag{5}$$

assuming without loss of generality that all objectives are to be minimized. Furthermore, iff

$$\mathbf{u} \preceq_{par} \mathbf{v} \wedge \mathbf{v} \npreceq_{par} \mathbf{u} \tag{6}$$

then $\mathbf{u}$ is said to *(Pareto) dominate* $\mathbf{v}$.

Using this concept, an objective vector is said to be better than another if the former weakly Pareto dominates the latter, while the latter does not weakly Pareto dominate the former. Pareto dominance represents the most commonly accepted notion of superiority in a multiobjective setting since it is a canonical generalization of the single-objective case. However, dominance may be defined in other ways as well, see ❷ *Fig. 2*. One example is the so-called (additive) epsilon dominance (Helbig and Pateva 1994; Laumanns et al. 2002; Zitzler et al. 2003) defined as

$$\mathbf{u} \preceq_{\varepsilon} \mathbf{v} :\Leftrightarrow \forall 1 \leq i \leq n : u_i \leq v_i + \varepsilon \tag{7}$$

which is a relaxation of Pareto dominance using an additive term $\varepsilon \in \mathbb{R}$; another possibility is to use ordering cones (Miettinen 1999; Ehrgott 2005).

The dominance relation $\preceq$ structures the objective space and thereby implicitly induces a corresponding *preference relation* $\preccurlyeq$ on the decision space with

$$\mathbf{a} \preccurlyeq \mathbf{b} :\Leftrightarrow \mathbf{f}(\mathbf{a}) \preceq \mathbf{f}(\mathbf{b}) \tag{8}$$

□ **Fig. 2**

Illustration of three dominance concepts based on the example shown in ❷ *Fig. 1*: Pareto dominance, epsilon dominance, and dominance using a pointed convex cone. The figure shows for a particular objective vector the area that is dominated in the objective space. Note that the considered knapsack problem is a mixed minimization/maximization problem and that, therefore, the definitions of the dominance relations change accordingly.

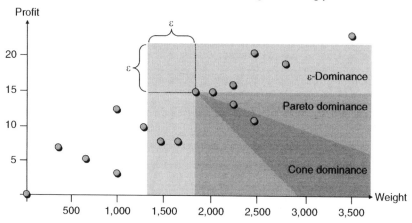

In this sense, the triple $(\mathbf{f}, Z, \leqslant)$ can be regarded as an indirect means to construct an order on X. In principle, there are also other ways to define this order and therefore the term preference relation will be used in a general notion:

▶ A *preference relation* $\preccurlyeq \subseteq X \times X$ is a binary relation on the decision space that is both reflexive and transitive, that is, $\preccurlyeq$ is a preorder.

Note that there is an important difference between the relation $\leqslant \subseteq Z \times Z$ and the corresponding preference relation $\preccurlyeq \subseteq X \times X$. Since two different decision vectors may be mapped to the same objective vector, antisymmetry is not guaranteed, that is, from $\mathbf{a} \preccurlyeq \mathbf{b} \wedge \mathbf{b} \preccurlyeq \mathbf{a}$ it cannot be inferred that $\mathbf{a} = \mathbf{b}$. Hence, $\preccurlyeq$ is usually only a preorder, while $\leqslant$ is a partial order.

The above considerations hold for the single-objective as well as the multiobjective case. What is now the difference between these cases? To illustrate this, a graphical representation of a preference relation is helpful. The idea is to interpret the pair $(X, \preccurlyeq)$ as a graph where X stands for the nodes and $\preccurlyeq \subseteq X \times X$ defines the edges: there is a directed edge from node $\mathbf{a} \in X$ to node $\mathbf{b} \in X$ iff $\mathbf{a} \preccurlyeq \mathbf{b}$. Such a graph will be denoted as a *relation graph*. For better visualization and to make the representation more compact, both reflexive edges and edges that can be inferred through the transitivity property will be omitted in the graphical rendering in the following.

Example 3 ❷ *Figure 3* depicts three relation graphs for the knapsack problem instance defined by ❷ *Fig. 1*. For all graphs, weak Pareto dominance is the underlying dominance relation, but each time a different set of objectives is considered; the resulting preference relations are denoted as $\preccurlyeq_{par}^{f_w}$ (only weight is taken into account), $\preccurlyeq_{par}^{f_p}$ (only profit is taken into account), and $\preccurlyeq_{par}^{f_p, f_w}$ (biobjective case).

■ Fig. 3
Relation graphs for the knapsack problem considering (a) $(X, \preccurlyeq_{par}^{f_w})$, **(b)** $(X, \preccurlyeq_{par}^{f_p})$, **and (c)** $(X, \preccurlyeq_{par}^{f_p, f_w})$.
The optimal solutions are marked by dotted circles.

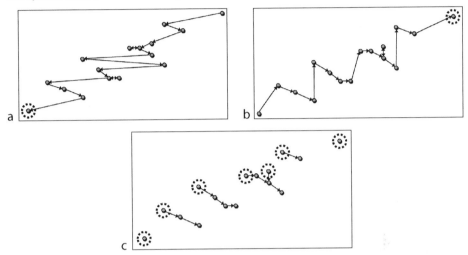

As can be seen from ❯ *Fig. 3*, in the case of $n = 1$ (only a single objective is considered), the resulting graph represents a chain – a chain means that any two solutions $\mathbf{a}, \mathbf{b} \in X$ are *comparable*, that is, it holds either $\mathbf{a} \preccurlyeq \mathbf{b}$ or $\mathbf{b} \preccurlyeq \mathbf{a}$ (or both). In other words: the preference relation $\preccurlyeq$ is a total preorder. (A relation $\leqq \subseteq Y \times Y$ is total if for all $a, b \in Y$ holds $(a \leqq b) \vee (b \leqq a)$.) When $n > 1$, then there may be some solution pairs $\mathbf{a}, \mathbf{b}$ for which neither $\mathbf{a} \preccurlyeq \mathbf{b}$ nor $\mathbf{b} \preccurlyeq \mathbf{a}$ holds, see ❯ *Fig. 3c*; in such a situation, $\mathbf{a}$ and $\mathbf{b}$ are denoted as *incomparable*. Among a set of incomparable solutions – a so-called *antichain* – no solution is preferable a priori unless further preference information is provided.

If several optimal solutions exist that are incomparable to each other, then multiple optima emerge in the objective space – a situation that is particular to multiobjective settings. In general, every solution for which no better solution exists is an optimal solution, cf. ❯ Sect. 2.1; in order theory, such a solution is denoted as a minimal element.

▶ The *minimal set* of a preordered set $(Y, \leqq)$ is defined as

$$Min(Y, \leqq) := \{a \in Y \mid \forall b \in Y : b \leqq a \Rightarrow a \leqq b\} \qquad (9)$$

The elements of $Min(Y, \leqq)$ are denoted as *minimal elements*.

If we denote the situation $\mathbf{a} \preccurlyeq \mathbf{b} \wedge \mathbf{b} \not\preccurlyeq \mathbf{a}$ as "a dominates b," then the minimal elements are those not dominated by others, that is, the *nondominated* ones. Whenever weak Pareto dominance is the underlying dominance relation, then the minimal elements of $(X, \preccurlyeq_{par})$ are called *Pareto optimal* and the entirety of the minimal elements is also known as *Pareto-optimal set*. The corresponding elements of the objective space are also called *Pareto-optimal*, while the set of Pareto-optimal objective vectors constitutes the *Pareto-optimal front*. The Pareto-optimal front may contain fewer elements than the Pareto-optimal set since decision vectors can be mapped to identical objective vectors.

In the example shown in ❯ *Fig. 3*, the Pareto-optimal sets for the single-objective scenarios (a) and (b) contain one solution, while in the biobjective setting (c) seven incomparable solutions are included in the Pareto-optimal set. Note that, in principle, also in the case of a single objective, multiple Pareto optima may exist; however, they would not be incomparable and all Pareto-optimal solutions would be assigned the same objective function value. In the presence of multiple objectives, the Pareto-optimal set usually contains an antichain that reflects different trade-offs among the objectives.

2.3 Properties of the Pareto-Optimal Set

The Pareto-optimal set has different characteristics, each of which can have an impact on the optimization and the decision-making process. The most important aspects are the following:

Range: The ideal and the nadir objective vectors represent lower and upper bounds, respectively, on the objective function values of the Pareto-optimal set; knowledge of these points is not only helpful for decision making, but can also be important for interactive optimization procedures (Deb et al. 2006).

▶ Let X_p be the Pareto-optimal set of a multiobjective optimization problem. The *ideal point* $\underline{z} = (\underline{z}_1, \ldots, \underline{z}_n) \in \mathbb{R}^n$ of this problem is defined as

$$\underline{z}_i := \min_{\mathbf{a} \in X_p} f_i(\mathbf{a}) \tag{10}$$

and the *nadir point* $\bar{z} = (\bar{z}_1, \ldots, \bar{z}_n) \in \mathbb{R}^n$ is given by

$$\bar{z}_i := \max_{\mathbf{a} \in X_p} f_i(\mathbf{a}) \tag{11}$$

where all objectives are without loss of generality to be minimized.

❯ *Figure 4* illustrates these concepts. If ideal point and nadir point are identical, then there is no conflict among the objectives and the Pareto-optimal front consists solely of the ideal point.

Shape: The image of the Pareto optima in the objective space can have different shapes that reflect how strongly objectives are in agreement or disagreement. For instance, the curvature of the front may be convex (❯ *Fig. 4*, left), concave (❯ *Fig. 4*, right), or contain convex and non-convex parts. If the decision space is continuous, then connectedness is another property that characterizes the front shape, see Ehrgott (2005). Roughly speaking, connectedness means that there are no gaps in the Pareto-optimal front; in the biobjective case, this implies that the Pareto-optimal objective vectors constitute a line. Pareto-optimal sets that are not contiguous in the objective space can cause difficulties with respect to both decision making and search.

Size: Whenever the underlying optimization problem is discrete, that is, the feasible set is finite as with the knapsack problem from ❯ Example 1, the number of the Pareto-optimal solutions can become a crucial issue. If the size of the minimal set is rather small, one may aim at generating the entire Pareto-optimal set using appropriate optimization methods. The larger the set of Pareto optima becomes, the less feasible is this approach. If the number of Pareto-optimal solutions is even exponential in the input size, for example, exponential in the number of items with the knapsack problem, then the problem of identifying the Pareto-optimal set becomes intractable in general. This is the case with many multiobjective variants of combinatorial optimization problems like the multiobjective shortest path problem and the traveling salesperson problem (Ehrgott 2005).

◻ **Fig. 4**

Two hypothetical minimization problems with two objectives each. The ideal points and the nadir points are identical, whereas the shapes of the Pareto-optimal fronts – visually emphasized by the auxiliary dotted lines – are different. The conflict between the objectives is weaker on the left side.

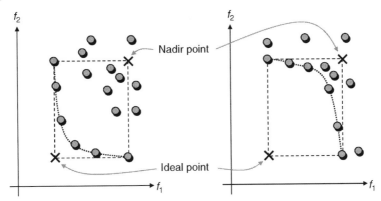

It is interesting to note that the size is related to the number of objectives involved. Winkler (1985) proved that the number of incomparable solution pairs increases, if further randomly generated objectives are added. Thereby, on the one hand the Pareto-optimal front may become larger and on the other hand the power of the dominance relation to guide the search may diminish – these are the main arguments that various researchers (Fonseca and Fleming 1995; Horn 1997; Deb 2001; Coello Coello et al. 2002; Fleming et al. 2005; Wagner et al. 2007), list in favor of the assumption that the search becomes harder the more objectives are involved. In the extreme case, when all solution pairs are incomparable, the Pareto-optimal set equals the decision space.

Computational complexity: The difficulty of combinatorial optimization problems can be expressed in terms of complexity classes. For instance, an NP-hard problem is assumed to be not solvable generally in polynomial time. In this context, it is remarkable that a polynomially solvable single-objective problem can become NP-hard when considering the multiobjective variant of it. One example is the shortest path problem: already in the bicriterion case it is, in general, not possible to identify *any* Pareto-optimal solution in polynomial time, while the extremes of the Pareto-optimal front can be computed efficiently, see Ehrgott (2005).

Several other properties could be listed here, for example, necessary and sufficient conditions for Pareto optimality. Optimality conditions are not discussed here, since the focus is on black-box optimization; the interested reader is referred to Miettinen (1999) instead.

3 Approaches to Multiobjective Optimization

In solving an optimization problem, one is interested in choosing the best solution for the application at hand. However, in the presence of multiple objectives the problem is usually underspecified. Whenever there is a conflict between the objectives and different trade-offs emerge, the choice of a Pareto-optimal decision vector is not arbitrary. Otherwise, it would be sufficient to consider one of the objectives only and to generate an extreme point. It matters which Pareto-optimum is selected, and what is missing is the specification of this choice. But what is then the use of formulating a problem in terms of multiple optimization criteria?

The answer is that a multiobjective model can be regarded as an intermediate step. A *decision-making process* is necessary to add the information that is not contained in the model yet. To this end, a *decision maker*, a person or a group of persons responsible for the selection, expresses his or her preferences based on experience, expert knowledge, or further insights into the application. The preference information is then used to carry out a problem transformation that – usually – leads to a fully specified model where a total preorder on the decision space is induced. In principle, two types of problem transformations can be distinguished: (1) solution-oriented ones that leave the decision space unchanged, and (2) set-oriented ones where the approximation of the entire Pareto-optimal set is the focus.

3.1 Solution-Oriented Problem Transformations

Taking the sum of the objective function values is a commonly known method to reformulate a multiobjective problem in terms of a single objective problem. It is one example for handling multiple objectives by means of *scalarization* where a scalarizing function represents the new optimization criterion.

▸ A *scalarizing function s* is a function $s : Z \mapsto \mathbb{R}$ that maps each objective vector $(u_1, \dots, u_n) \in Z$ to a real value $s(u_1, \dots, u_n) \in \mathbb{R}$.

Typically, each scalarization relies on individual parameters by which the trade-off between the original objectives can be specified. The transformed problem is then given by $(X, \mathbb{R}, s \circ \mathbf{f}, \mathbf{g}, \leq)$, assuming that the scalarizing function is to be minimized. Three selected functions will be briefly presented here, more detailed discussions and other techniques can be found in Miettinen (1999) and Ehrgott (2005).

Weighted Sum Method: The scalarizing function is given by

$$s_W(u_1, \dots, u_n) := \sum_{1 \leq i \leq n} w_i u_i \qquad (12)$$

where $w_1, \dots, w_n \in \mathbb{R}^+$ are the scalarization parameters that assign each objective a positive weight; typically, the weights sum up to 1. If all weights are greater than 0, then each optimal solution of the transformed problem is also a Pareto-optimal solution of the original multiobjective problem. The opposite does not hold in general, that is, in certain cases, there exists no weight combination such that a specific Pareto-optimal solution is also an optimal solution of the transformed problem (Ehrgott 2005).

Tchebycheff Method: This technique considers the distance to a reference point $(z_1, \dots, z_k) \in Z$ using an l_∞-norm

$$s_T(u_1, \dots, u_n) := \max_{1 \leq i \leq n} w_i(u_i - z_i) \qquad (13)$$

where $w_1, \dots, w_n \in \mathbb{R}^+$ are the weights as before with $\sum_{1 \leq i \leq n} w_i = 1$. Here, for every Pareto-optimal solution a corresponding weight combination exists, but not all minimal elements of the transformed problem are Pareto-optimal in the common sense. The reference point stands for an objective vector that the decision maker would like to achieve and that may represent an ideal choice.

ε-Constraint Method: Here, one of the objective functions is selected for optimization, without loss of generality the last, leading to the scalarizing function

$$s_C(u_1, \ldots, u_n) := u_n \tag{14}$$

while the remaining objectives are converted into constraints, that is, $n - 1$ additional constraints $g_{m+i}(\mathbf{a}) = f_i(\mathbf{a}) - \varepsilon_i$ for $1 \le i < n$ are created where $\varepsilon_i \in \mathbb{R}$ represents the threshold for the ith objective. (The constraints are added to the existing ones, that is, $\mathbf{g} = (g_1, \ldots, g_m, g_{m+1}, \ldots, g_{m+n-1})$.) The optimal solution to an ε-constraint problem is not necessarily Pareto-optimal with respect to the original problem, but any Pareto-optimal solution can be found by solving multiple ε-constraint transformations (Miettinen 1999; Laumanns et al. 2006).

Besides scalarization techniques, another possibility to totally order the decision space is to exchange the relation $\le$ of the multiobjective problem $(X, Z, \mathbf{f}, \mathbf{g}, \le)$ under consideration. A lexicographic order $\le_{lex}$, for instance, is a total order that encodes a ranking of the objectives

$$(u_1, \ldots, u_n) \le_{lex} (v_1, \ldots, v_n) :\Leftrightarrow$$
$$\exists 1 \le i \le n : (\forall 1 \le j \le i : u_j = v_j) \wedge (i = n \ \vee \ u_{i+1} < v_{i+1}) \tag{15}$$

with $(u_1, \ldots, u_n), (v_1, \ldots, v_n) \in \mathbb{R}^n$. The principle is that the objectives are considered one by one according to the ranking; the first objective where the objective values differ between two solutions decides which solution is better. Note that by renumbering the objectives, different rankings can be generated.

Independently of how exactly a solution-oriented problem transformation is carried out, the general idea is to refine the preference relation $\prec$ on the decision space such that all incomparable solutions become comparable, respectively, that $\prec$ becomes total. When considering the relation graph for $(X, \prec)$, this mainly means that edges are inserted until a complete chain emerges; the edges represent additional knowledge or preferences that are specified by weights, constraint thresholds, objective rankings, etc., as sketched above. The resulting chain again should be a preference relation, $\prec_{ref}$ which should, to a certain extent, preserve the original preference relation. This can be formulated as follows:

▶ A relation $\le_{ref} \subseteq Y \times Y$ is a *refinement* of another relation $\le \subseteq Y \times Y$ iff

$$\forall a, b \in Y : a \le b \wedge b \not\le a \Rightarrow a \le_{ref} b \wedge b \not\le_{ref} a \tag{16}$$

and iff

$$\forall a, b \in Y : a \le b \wedge b \not\le a \Rightarrow a \le_{ref} b \tag{17}$$

then $\le_{ref}$ is a *weak refinement* of $\le$.

For instance, the preference relations induced by the weighted sum approach, the Tchebycheff method, and lexicographic ordering are refinements of the Pareto dominance relation $\prec_{par}$. That means if a $\mathbf{a} \in X$ dominates $\mathbf{b} \in X$, then $\mathbf{a}$ is also better than $\mathbf{b}$ with respect to the scalarizing function – in other words: dominance is preserved after the problem transformation. The ε-constraint method only represents a weak refinement of $\prec_{par}$; this guarantees that dominance is not violated, that is, whenever $\mathbf{a}$ dominates $\mathbf{b}$, then $\mathbf{a}$ is not worse than $\mathbf{b}$ regarding the chosen objective. Other terms like (weak) *compatibility* (Hansen and Jaszkiewicz 1998), (weak) *Pareto compliance* (Zitzler et al. 2008a), or (strict) *monotonicity* (Miettinen 1999) are used as well in the literature to denote a (weak) refinement.

3.2   Set-Oriented Problem Transformations

Often, it is difficult to provide the preference information necessary for a solution-oriented problem transformation. Instead of specifying in advance which Pareto optimum is sought, an alternative is to generate the Pareto-optimal set and make a decision afterward on the basis of this knowledge. Since with most applications the identification of all Pareto-optimal solutions is infeasible, one is rather interested in approximating the Pareto-optimal set, that is, finding a suitable *set* of compromise solutions. In this case, one deals with a set problem where the decision space consists of all possible sets of solutions.

▶ For a multiobjective optimization problem $(X, Z, \mathbf{f}, \mathbf{g}, \leqslant)$, the associated *set problem* is given by $(\Psi, \Omega, F, \mathbf{G}, \preccurlyeq)$ where
  - $\Psi = 2^X$ is the space of decision vector sets, that is, the powerset of X,
  - $\Omega = 2^Z$ is the space of objective vector sets, that is, the powerset of Z,
  - F is the extension of $\mathbf{f}$ to sets, that is, $F(A) := \{\mathbf{f}(\mathbf{a}) : \mathbf{a} \in A\}$ for $A \in \Psi$,
  - $\mathbf{G} = (G_1, \ldots, G_m)$ is the extension of $\mathbf{g}$ to sets, that is, $G_i(A) := \max \{g_i(\mathbf{a}) : \mathbf{a} \in A\}$ for $1 \leq i \leq m$ and $A \in \Psi$, and
  - $\preccurlyeq$ extends $\leqslant$ to sets where $A \preccurlyeq B : \Leftrightarrow \forall \mathbf{b} \in B \; \exists \, \mathbf{a} \in A : \mathbf{a} \leqslant \mathbf{b}$.

The dominance relation $\preccurlyeq$ on sets represents a natural extension of a dominance relation $\leqslant$ on solutions. It induces a corresponding *set preference relation* $\precsim$ on the space Ψ of solution sets and therefore the terms comparability, incomparability, etc., also apply here. In the case of $\preccurlyeq_{par}$ and the corresponding set preference relation $\precsim_{par}$, a solution set is also denoted as *Pareto set approximation*, and an optimal Pareto set approximation is one that contains for each Pareto-optimal objective vector a corresponding decision vector.

Example 4   Consider the knapsack problem and the four solution sets depicted in ❷ *Fig. 5*: both A and B dominate C, D is incomparable to A, B, and C, and A and B are *indifferent*, that is, $A \precsim B \wedge B \precsim A$, with respect to Pareto dominance. Although A contains only Pareto-optimal solutions, it does not represent an optimal Pareto set approximation as the entire Pareto-optimal front is not covered.

It is important to note that the set problem resulting from the transformation is still underspecified – similarly to the original multiobjective problem – as usually incomparable Pareto set approximations exist. Again, preference information is required to impose a total preorder on Ψ; however, the preferences are weaker and easier to specify than the ones needed for solution-oriented problem transformations. This is also the motivation for making the problem more complex by considering sets instead of single solutions (the size of Ψ is exponential in the size of X).

A common way to define a total preorder on Pareto set approximation relies on *quality indicators* – also denoted as performance measures or performance metrics – which represent functions that assign a tuple of solution sets a real value. Most popular are unary quality indicators:

▶ A (unary) *quality indicator* I is a function $I : \Psi \mapsto \mathbb{R}$ that assigns a Pareto set approximation a real value.

According to this definition, a unary quality indicator I can be regarded as a set quality measure or an objective function on solution sets. Therefore, it can be used to transform the

□ **Fig. 5**
Four Pareto set approximations for the knapsack problem instance from ❷ Fig. 1.

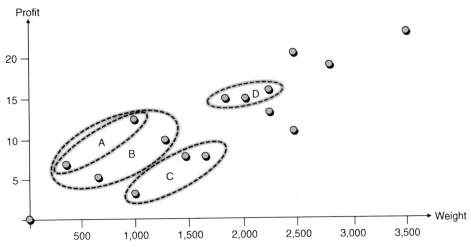

set problem $(\Psi, \Omega, F, \mathbf{G}, \lessdot)$ into a single-objective problem $(\Psi, \mathbb{R}, I, \mathbf{G}, \leq)$ assuming that the indicator values are to be minimized.

Various quality indicators have been proposed in the literature, see Hansen and Jaszkiewicz (1998), Knowles and Corne (2002), and Zitzler et al. (2003); three selected unary indicators are briefly discussed here:

Hypervolume indicator: The *hypervolume indicator* I_H (Zitzler and Thiele 1998) gives the volume of the portion of the objective space that is weakly dominated by a specific Pareto set approximation. It can be formally defined as

$$I_H(A) := \lambda(H(A, R)) \tag{18}$$

where $R \subseteq Z$ is a given reference set of objective vectors,

$$H(A, R) := \{z \in Z \;:\; \exists a \in A \,\exists r \in R : f(a) \leq z \leq r\} \tag{19}$$

denotes the set of objective vectors that are enclosed by the front $F(A)$ given by A and the reference set R, and λ is the Lebesgue measure with $\lambda(H(A, R)) = \int_{\mathbb{R}^n} 1_{H(A,R)}(z) dz$ and $1_{H(A,R)}$ being the characteristic function of $H(A, R)$. The hypervolume indicator is to be maximized. In ❷ Fig. 6 on the left, the gray shaded area represents the set $H(A, R)$.

Epsilon indicator: The *epsilon indicator* I_ε (Zitzler et al. 2003) measures the objective-wise distance to a given reference set in the objective space; it is based on the concept of epsilon dominance

$$I_\varepsilon(A) := \inf_\varepsilon \; F(A) \lessdot_\varepsilon R \tag{20}$$

where $\lessdot_\varepsilon$ is the natural extension of epsilon dominance from ❷ Eq. 7 to solution sets with $A \lessdot_\varepsilon B : \Leftrightarrow \forall b \in B \,\exists a \in A : a \lessdot_\varepsilon b$. In other words, it provides the minimum ε such that the reference set is weakly Pareto dominated as illustrated in ❷ Fig. 6 on the right; the smaller the value, the better the solution set.

◘ **Fig. 6**

Illustration of the hypervolume indicator (*left*) and the (additive) epsilon indicator (*right*) for a minimization problems with two objectives each.

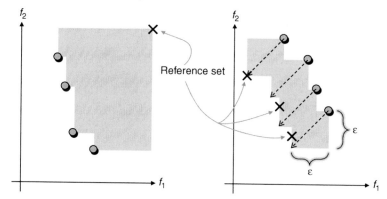

R indicators: The *R indicators* proposed in Hansen and Jaszkiewicz (1998) can be used to assess Pareto set approximations on the basis of scalarizing functions. The following quality indicator I_{RT} relies on the Tchebycheff method and requires multiple scalarizing functions $s_T^{(1)}, \ldots, s_T^{(l)}$, each using a different combination of weights and the same reference point $(z_1, \ldots, z_n)$

$$I_{RT}(A) := \frac{1}{l} \sum_{1 \le i \le l} \min_{\mathbf{a} \in A} s_T^{(i)}(\mathbf{a}) \tag{21}$$

It assumes the scalarizing functions to be minimized and simply takes the average over the best values per scalarizing function, which is to be minimized as well.

Often, it is beneficial to use a combination of quality indicators instead of a single indicator. One possibility is to define a sequence of indicators $(I_1, \ldots, I_l)$ and to apply a lexicographic order $\le_{lex}$ on $\mathbb{R}^l$, see Zitzler et al. (2008b). Thereby, the set problem $(\Psi, \Omega, F, \mathbf{G}, \prec)$ is further transformed into the optimization problem $(\Psi, \mathbb{R}^l, (I_1, \ldots, I_l), \mathbf{G}, \le_{lex})$. Such a hierarchy of indicators is especially useful for search as will be discussed in the next section.

Finally, note that the same considerations as with scalarizing functions apply to quality indicators: the induced set preference relation should be in accordance with the original set preference relation $\prec$. The relation $\prec_H$ associated with the hypervolume indicator refines the set Pareto dominance relation $\prec_{par}$, while the relations $\prec_\varepsilon$ and $\prec_{RT}$ associated with I_ε and I_{RT}, respectively, only weakly refine $\prec_{par}$.

4 Search Algorithm Design

When designing a search algorithm for a multiobjective optimization problem, one either has to deal with a solution-oriented scenario or a set-oriented scenario – depending on the type of problem transformation chosen. In the first case, usually a single-objective problem emerges and the standard components of a black-box search method can be used. The latter case requires specialized methods and by far most of the studies in the area of evolutionary multiobjective optimization focus on approximating the Pareto-optimal set. Therefore, in the following, a set-oriented problem transformation will be assumed and the corresponding algorithmic concepts that have been developed for this purpose will be discussed.

Note that the presentation focuses on evolutionary algorithms, although also other natural computing methods such as ant colony optimization and particle swarm optimization have been successfully applied to multiobjective scenarios. The general concepts are more or less the same, but the latter techniques may employ specialized procedures tailored to the algorithmic framework used.

4.1 Types of Algorithms

One can distinguish between two types of search strategies to identify a suitable set of compromise solutions: the algorithms of the first category perform multiple single-objective runs, while the methods of the second category require only a single run. As to the multiple runs approach, consider, for example, the I_{RT} quality indicator from ❷ Eq. 21 in combination with l Tchebycheff scalarizing functions $s_T^{(1)}, \ldots, s_T^{(l)}$. For each scalarizing function $s_T^{(i)}$, a separate optimization run is carried out where the objective function f is $f = s_T^{(i)}$, that is, overall l runs are performed. Afterward, the best solutions found per run are combined to form a Pareto set approximation. A similar strategy can be devised on the basis of the ε-constraint method. First, the thresholds ε_i for the constraints that represent converted objective functions are set to ∞, which means only the chosen objective is optimized. The best solution found is used to adapt the constraints, so that it falls into the infeasible region, and the next optimization run is carried out. This process is iterated until the feasible set equals the empty set. As before, the union of the best solutions per run form the approximation of the Pareto-optimal set. While for $n = 2$ this scheme is simple to implement as depicted in ❷ Fig. 7, it becomes much more complex, if the number of optimization criteria is arbitrary, see Laumanns et al. (2006).

Most multiobjective evolutionary algorithms proposed in the literature are of the second type. Here, the population as a whole represents a Pareto set approximation and therefore a single run is sufficient. Taking this line of thought further, the successive application of mating selection, variation, and environmental selection aims at generating a new, better Pareto set

◻ **Fig. 7**
Visualization of the multiple run strategy to approximate the Pareto-optimal set using the ε-constraint method. The figure shows for the knapsack problem and for two successive runs the current weight constraint and the best solution found.

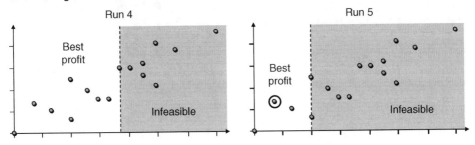

approximation and therefore can be regarded as a multi-stage mutation operator on solution sets. From this perspective, a common multiobjective evolutionary algorithm implements a single-objective $(1, 1)$-strategy on solution sets with a clever mutation operator; this applies to almost all search strategies currently available. The idea of operating on a population of solution sets, that is, employing an evolutionary algorithm where each individual is a Pareto set approximation on its own and where possibly solution sets are recombined, has not gained much attention yet (end of 2008): there only exists a preliminary study (Bader et al. 2009) that fully addresses this issue, while some papers in the context of parallelism partially exploit this idea (Poloni 1995; Baita et al. 1995; Lee and Hajela 1996; Aherne et al. 1997; Sawai and Adachi 2000; Branke et al. 2004b; Mezmaz et al. 2006; Coello Coello et al. 2007). Thus, the potential of this multi-solution set strategy still needs to be explored.

4.2   Basic Algorithmic Concepts

The main difference between a single-objective evolutionary optimizer and a multiobjective one approximating the Pareto-optimal set lies in fitness assignment and selection, while variation has been treated similarly (and therefore is not discussed in the following). (Note that most recently (March 2009), specific set variation operators have been proposed (Bader et al. 2009; Voß et al. 2009).) The key question is how a better Pareto set approximation can be generated from the current population in a randomized fashion – independently of whether the search algorithm works with a single or multiple solution sets.

4.2.1   Fitness Assignment

One issue is to quantify the usefulness of a particular individual with regard to the entire population. The general idea behind most prevalent concepts is to assign fitness values respectively to rank the individuals according to the loss in quality of the Pareto set approximation that can be accounted to the removal of a particular individual. More specifically, the fitness of an individual $\mathbf{a} \in X$ is determined by taking the difference between the quality of the entire population P and the quality of the population without $\mathbf{a}$, for example, $fit_{\mathbf{a}} = I(P) - I(P \setminus \{\mathbf{a}\})$ where P is a multiset of decision vectors and I a unary quality indicator used for optimization. (Note that for reasons of simplicity, here individuals and decision vectors are not distinguished, although in practice usually a representation is required to encode decision vectors appropriately.)

A difficulty that arises is the fact that the set preference relation or the quality indicator used are usually insensitive to dominated population members. That means the quality of the population solely depends on the nondominated individuals. The consequence is that all dominated individuals are assigned a fitness of 0 when employing the above principle. For instance, the hypervolume indicator, which has become a popular measure for fitness assignment, does not change its value when dominated solutions are added to or removed from the solution set. The standard solution to this problem is to partition the population into nonoverlapping dominance classes that are organized in a hierarchy. Each class contains only mutually nondominating solutions, and the fitness values are then computed per dominance class. Furthermore, the individuals in the first class are considered superior to the second class, which in turn is better than the third class and so forth.

■ Fig. 8
Two different partitionings into dominance classes for a biobjective minimization problem. The numbers refer to the hierarchy of the dominance classes.

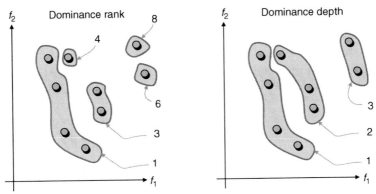

There exist different ways to create such a partitioning, ❯ *Fig. 8* illustrates two popular schemes. The method of *dominance rank* (Fonseca and Fleming 1993) counts for each individual the number of population members that dominate it; this number gives the so-called dominance rank. All individuals with dominance rank 0 form the first dominance class, the second class consists of individuals with dominance rank 1, and so forth. Another method originally proposed in Goldberg (1989) and for the first time integrated in a multiobjective evolutionary algorithm by Deb and colleagues (Srinivas and Deb 1994; Deb et al. 2002a) uses the *dominance depth* – a strategy also known as nondominated sorting. Here, the minimal elements of the population form the first dominance class, that is, $Min(P, \preccurlyeq_{par})$; the next class is represented by the minimal elements of the remaining population members, that is, $Min(P \setminus Min(P, \preccurlyeq_{par}), \preccurlyeq_{par})$, and the other classes are defined accordingly. Partitioning in terms of dominance depth has in comparison to dominance rank the additional property that classes with lower numbers dominate classes with higher numbers in the hierarchy.

As mentioned above, the fitness values, respectively, the ranking within each dominance class are determined using the underlying set measure or set preference relation. Density-based measures that resemble niching techniques like fitness sharing in single-objective evolutionary computation have been the most popular, particularly in the first decade of EMO. This approach usually relies on Euclidean distance in the objective space: it first computes the distances between all population members and on this basis then estimates the density around each individual. One example is the kth nearest neighbor method where k is a prespecified constant which is often set to 2. Here, the distance to the kth closest individual is taken as the density estimate for each population member and the individuals in each dominance class are then ranked in descending order of the density estimates (Zitzler et al. 2002). Many other techniques exist (Fonseca and Fleming 1993; Knowles and Corne 2000b; Deb et al. 2002a). All these techniques can be regarded as implementations of a specific quality indicator, I_D, which is based on Euclidean distance in the objective space. However, the problem with this type of indicator is that, in general, it does not represent a proper preorder as transitivity is violated (Knowles and Corne 2002; Zitzler et al. 2003, 2008b). This theoretical consideration can also be observed in practice: Euclidean distance-based multiobjective evolutionary algorithms tend to exhibit cyclic behavior, which means they do not progress after a certain point in time

(Laumanns et al. 2002). This can cause problems especially with applications involving many objective functions (Wagner et al. 2007).

The alternative to density-based fitness assignment is to directly use set preference relations or quality indicators that are preorders and (weak) refinements. In particular, the hypervolume indicator has been employed in this context since it provides a refinement of weak Pareto dominance and hence allows a fine-grained ranking of the individuals. Several hypervolume-based algorithms have been proposed (Emmerich et al. 2005; Igel et al. 2007; Zitzler et al. 2007) that all rely on the same principle: the population is first partitioned into dominance classes based on dominance depth and then for each dominance class the individuals are ranked according to their contribution to the overall hypervolume of the class. Precisely, the hypervolume contribution $h_\mathbf{a}$ of an individual $\mathbf{a}$ with respect to its dominance class A is given by

$$h_\mathbf{a} = I_H(A) - I_H(A \setminus \{\mathbf{a}\}) \tag{22}$$

which equals the loss in hypervolume resulting from the removal of $\mathbf{a}$.

Example 5   Consider the population in ❯ *Fig. 9* consisting of five individuals. When using the hypervolume-based fitness assignment described above, the ranking of the individuals is $\mathbf{a}$, $\mathbf{c}$, $\mathbf{b}$, $\mathbf{d}$, $\mathbf{e}$. The individuals $\mathbf{a}$, $\mathbf{b}$, $\mathbf{c}$ fall in the first dominance class A_1 and it holds $h_\mathbf{a} > h_\mathbf{c} > h_\mathbf{b}$. The second dominance class A_2 contains the remaining population members where $\mathbf{d}$ contributes more to the overall hypervolume of A_2 than $\mathbf{e}$, that is, $h_\mathbf{d} > h_\mathbf{e}$.

Unfortunately, the exact computation of the hypervolume values is computationally expensive (Bringmann and Friedrich 2008) and the available algorithms have a worst-case runtime complexity that is exponential in the number of objectives (Fonseca et al. 2006; Beume and Rudolph 2006). Therefore, different methods to speed up this computation have

◻ **Fig. 9**

Hypervolume-based fitness assignment for a hypothetical population for the biobjective knapsack problem. The boxed areas represent the hypervolume contributions attributed to the specific individuals.

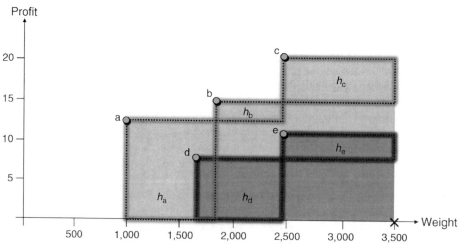

been proposed (Bradstreet et al. 2008; Bader et al. 2010) that enable hypervolume-based search also in high-dimensional objective spaces.

The majority of the proposed fitness assignment methods follows the principle sketched above, a combination of dominance-based partitioning and a quality indicator. It is interesting to note in this context that this scheme can be expressed in terms of a sequence of indicators as outlined in ❷ Sect. 3.2. Regarding the example with the hypervolume indicator, let $I_H^{(i)}$ denote the indicator value of the ith dominance class A_i of a population P, that is, $I_H^{(i)}(P) := I_H(A_i)$. Then the fitness assignment procedure corresponds to a set preference relation induced by the sequence $(I_H^{(1)}, I_H^{(2)}, \ldots, I_H^{(N)})$ of quality indicators where N is the population size. This idea can be extended to arbitrary set preference relations as proposed in Zitzler et al. (2008b).

Finally, also other fitness assignment variants exist that do not rely on a dominance-based partitioning (Schaffer 1985; Horn et al. 1994; Czyzak and Jaskiewicz 1998; Zitzler and Künzli 2004; Zitzler et al. 2008b). When, for instance, the quality indicator used is composed of scalarizing functions, the fitness values may be directly inferred from the scalarizing functions without further dominance checks.

4.2.2 Selection

The fitness values and the ranking of the individuals, respectively, provide the basis for mating selection and environmental selection.

Mating selection is often implemented in the same way as in single-objective evolutionary algorithms, although some concepts have been proposed that are different. For instance, Schaffer's vector evaluated genetic algorithm (Schaffer 1985) switches between the objectives during the mating selection phase, the niched pareto genetic algorithm by Horn et al. (1994) carries out dominance-based tournaments, and other approaches based on the weighted-sum method randomly choose weight combinations to determine the members of the mating pool (Ishibuchi and Murata 1996). These techniques can be regarded as randomized fitness assignment schemes; however, they are rather seldom used.

Environmental selection plays a much more important role in multiobjective search than mating selection. The first evolutionary methods for multiobjective optimization (Schaffer 1985; Fonseca and Fleming 1993; Srinivas and Deb 1994; Horn et al. 1994) used the standard genetic algorithm scheme where the offspring population entirely replaces the parent population (a comma strategy). However, when empirical studies demonstrated that elitism is of high importance when approximating the Pareto-optimal set (Zitzler and Thiele 1999; Knowles and Corne 2000b), the picture changed. Nowadays, most multiobjective search algorithms implement a plus strategy, that is, the best solutions from the union P^+ of parent population and offspring population are chosen for survival. This leads to a subset selection problem: from the N parents and M children a subset of N solutions needs to be determined that constitutes the next population. In the case of the hypervolume indicator for instance, one searches for a subset A with $|A| = N$ that has the largest hypervolume value among all subsets of size N, that is,

$$\forall B \subseteq P^+ : \ |B| = N \Rightarrow I_H(B) \leq I_H(A) \tag{23}$$

Determining the best subset can be computationally expensive, and often heuristic procedures are used instead. One strategy iteratively removes the worst solution from P^+ and updates the fitness values until the desired population size is reached (see Zitzler et al. 2002). Alternatively, one may remove the M worst solutions in a single step without intermediate fitness updates as, for example, in Deb et al. (2002a); this strategy is faster, but also less accurate than the iterative approach.

The choice of the environmental selection scheme is also important for the convergence properties of a multiobjective optimizer as has been shown in Rudolph (1998, 2001), Rudolph and Agapie (2000), Laumanns et al. (2002), and Zitzler et al. (2008b). Therefore, sometimes a separate archive is maintained that stores the nondominated solutions (Knowles and Corne 2003a; Fieldsend et al. 2003). The archive may use a different selection scheme than the population; for instance, one can employ a strategy that ensures certain quality bounds (Laumanns et al. 2002).

4.3 Advanced Algorithmic Concepts

In the following, advanced concepts to integrate constraints, user preferences, and other types of optimizers are dealt with. The presentation is not exhaustive; for instance, other topics like uncertainty and robustness, which are of high practical relevance are not treated here. On the one hand, these issues can be addressed in the same manner as in single-objective optimization where various techniques have been developed (Jin and Branke 2005). On the other hand, the methodologies that have been specifically proposed in a multiobjective context (Hughes 2001; Teich 2001; Deb and Gupta 2006; Basseur and Zitzler 2006; Gaspar-Cunha and Covas 2008) form a relatively new area of research that is still maturing and does not provide a clear picture yet of potential standard concepts. Furthermore, specialized data structures and algorithms as well as parallelization schemes to speed up fitness assignment and selection are important when implementing a multiobjective optimizer (see Habenicht 1983; Mostaghim et al. 2002; Jensen 2003; Talbi et al. 2008). Such implementation issues are not covered in the following.

4.3.1 Constraint Handling

In principle, constraints can be handled in the same manner as in single-objective optimization, for example, by using repair methods, by not accepting infeasible solutions, or by employing penalty functions. However, multiobjective optimization offers new opportunities where the constraint functions are treated as objective functions. The idea is to modify the underlying preference relation such that the constraints are included and

1. Among feasible solutions, the original dominance relation holds,
2. Feasible solutions dominate infeasible solutions, and
3. Among infeasible solutions, the one with the lower constraint violation is preferred.

There are different ways how this principle can be implemented in practice, see Coello et al. (2007) for an extensive list of references. The following scheme reflects the notion proposed in Fonseca and Fleming (1998):

▶ Let $(X, Z, \mathbf{f}, \mathbf{g}, \preccurlyeq)$ be a multiobjective optimization problem. The *constraint-integrating preference relation* $\preccurlyeq_{con}$ is defined as

$$\begin{aligned}
\mathbf{a} \preccurlyeq_{con} \mathbf{b} :\Leftrightarrow & (\mathbf{g}(\mathbf{a}) \leq \mathbf{0} \ \wedge \ \mathbf{g}(\mathbf{b}) \leq \mathbf{0} \ \wedge \ \mathbf{f}(\mathbf{a}) \preccurlyeq \mathbf{f}(\mathbf{b})) \ \vee \\
& (\mathbf{g}(\mathbf{a}) \leq \mathbf{0} \ \wedge \ \mathbf{g}(\mathbf{b}) \not\leq \mathbf{0}) \ \vee \\
& (\mathbf{g}(\mathbf{a}) \not\leq \mathbf{0} \ \wedge \ \mathbf{g}(\mathbf{b}) \not\leq \mathbf{0} \ \wedge \ \mathbf{g}(\mathbf{a}) \preccurlyeq_{par} \mathbf{g}(\mathbf{b}))
\end{aligned} \tag{24}$$

for all $\mathbf{a}, \mathbf{b} \in X$.

This scheme compares infeasible solutions using weak Pareto dominance; alternatively, one may consider the overall constraint violation, for example, the sum of the individual violations, and only compare this scalar value.

The constraint-integrating preference relation can be used in the same way as described in ❯ Sect. 4.2.1 to partition the population in dominance classes. The construction ensures that (1) a class contains either only feasible or only infeasible solutions, and (2) infeasible classes follow the feasible classes in the hierarchy of dominance classes.

Example 6 Consider the single-objective variant of the knapsack problem defined in ❯ Eq. 3. To handle the weight constraint, a preference relation $\preccurlyeq_{con}$ can be formulated as follows:

$$S \preccurlyeq_{con} T :\Leftrightarrow \left(f_p(S) \leq f_p(T) \ \wedge \ f_w(S) \leq w_{\max}\right) \ \vee \ f_w(S) \leq f_w(T) \qquad (25)$$

where S, T are arbitrary set of items. To illustrate how this preference relation can be used for partitioning the population, recall the knapsack problem instance from ❯ Fig. 1. For a hypothetical population of ten individuals, the resulting dominance classes are depicted in ❯ Fig. 10 assuming that dominance depth is used for partitioning.

4.3.2 Preference Articulation

Often, the decision maker is not interested in approximating the entire Pareto-optimal set, but has further preferences that specify certain regions of interest. Such preferences can be in the form of constraints, priorities, reference points, etc. To integrate such information in the search process is especially helpful when using the evolutionary algorithm in an interactive fashion (Jaszkiewicz and Branke 2008): the optimizer first presents a general approximation set, for example, according to the hypervolume indicator, and afterward, the decision maker

◻ **Fig. 10**
Partitioning of a population into dominance classes using dominance depth and a constraint-integrating preference relation. The problem is taken from ❯ Fig. 1 with a weight threshold of $w_{\max} = 2,000$.

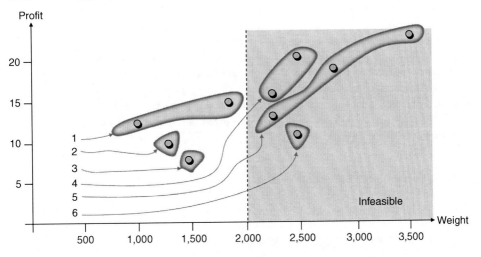

restricts respectively refines the search goal by adding further preferences; these two steps are iterated until the decision maker arrives at a final solution.

There basically exist two methodologies to guide the search on the basis of additional preference information: one can modify either the preference relation on solutions or the set preference relation. The first possibility corresponds to the aforementioned strategy to include constraints. For instance, the approach proposed in Fonseca and Fleming (1998) allows us to combine goals, which correspond to desired objective functions values, and priorities, which allow us to define a hierarchy of objectives and constraints. A further option is to change the dominance relation (Branke et al. 2001; Molina et al. 2009), for example, to reflect trade-off information or to incorporate reference points. Several other solution-oriented preference handling techniques exist (Cvetković and Parmee 2002; Rachmawati and Srinivasan 2006).

The second possibility consists in adapting the set preference relation resp. the underlying quality indicator according to the user preferences. Often, a reference point is used that specifies a region in the objective space that is of particular interest. In density-based fitness assignment procedures, this information can be incorporated by introducing a bias in the distance metric that is directly affected by the reference point (Branke and Deb 2004; Deb and Sundar 2006). Alternatively, one may consider an extension of the hypervolume indicator where a weight function puts emphasis on certain regions of the objective space (Zitzler et al. 2007); thereby, it is possible to concentrate the Pareto set approximation around one or several reference points. A similar approach can be implemented based on the epsilon indicator or the I_{RT} indicator (Zitzler et al. 2008b). Besides reference points, it is also possible to include directions: with the I_{RT} indicator, for instance, the chosen weight combinations determine which parts of the front are most important.

4.3.3 Hybridization

A promising approach to exploit knowledge about the problem structure in single-objective optimization is to combine an evolutionary algorithm with other techniques such as local search strategies or exact optimization methods. The same principle can be highly beneficial in multiobjective optimization (Knowles and Corne 2000a; Jaszkiewicz 2002), but some difficulties arise here: when for instance using a local search algorithm, this strategy focuses on the improvement of a single solution, while all population members together constitute a Pareto set approximation. Therefore, a search direction needs to be specified that corresponds to the population as a whole. In Jaszkiewicz (2002), for instance, hybrid multiobjective evolutionary algorithms based on scalarizing functions with weights have been investigated. Each time the local search procedure is invoked, a weight combination is chosen at random and the corresponding scalarizing function is taken as objective function for the local search strategy. Another approach does not use scalarizing functions, but relies on dominance to decide whether a newly generated solution replaces the current solution of the local search algorithm (Knowles and Corne 2000a).

4.4 Performance Assessment

When designing a multiobjective evolutionary algorithm or specific algorithmic concepts, it is crucial to compare the effectiveness of the new method to existing methods. To this end, both benchmark problems and comparison methodologies are needed.

4.4.1 Test Problems

Various suites of test functions have been proposed in the literature and many of them are widely used in the community, for example, the ZDT test functions (Zitzler et al. 2000), the DTLZ test functions (Deb et al. 2002b, 2005), the framework presented in Okabe et al. (2004), and the WFG suite (Huband et al. 2006). To ensure a consistent nomenclature, the proposed functions are often named according to the authors of the corresponding papers: the first letters of the authors' last names form the acronym, which may be extended by a number distinguishing specific functions of the same suite.

Artificial test functions can be designed and used to evaluate algorithms with respect to selected problem characteristics such as the shape of the front or the neighborhood structure in the decision space. Combinatorial optimization problems, in particular multiobjective versions of well-known NP hard problems, usually do not offer this flexibility, but they are often closer to reality since they can be found in various application areas. Although the problem formulation may be simple, a combinatorial problem as such can be hard to solve. Prominent examples are the multiobjective knapsack problem (Zitzler and Thiele 1999) and the multiobjective quadratic assignment problem (Knowles and Corne 2003b), which have been widely used in the EMO community.

Nevertheless, the ultimate benchmarks are real-world applications themselves (Künzli et al. 2004; Hughes 2007; Siegfried et al. 2009). The difficulty with these is that implementations need to be available in order to make them accessible for the research community. For instance, the PISA framework offers a methodology to facilitate the exchange of program code for performance assessment (Bleuler et al. 2003; Bader et al. 2008). PISA stands for "A Platform and Programming Language Independent Interface for Search Algorithms" and basically defines a text-based interface to separate general optimization principles (fitness assignment and selection) from application-specific aspects (representation, variation, and objective function evaluation). The PISA Web site (Bader et al. 2008) contains implementations of various multiobjective optimizers and benchmark problems that can be arbitrarily combined, independently of the computing platform.

4.4.2 Comparison Methodologies

Comparing stochastic algorithms for approximating the Pareto-optimal set involves comparing samples of solution sets. One can distinguish between three possibilities, how this can be accomplished: on the basis of (1) quality indicators, (2) set preference relations, and (3) attainment functions, which will be introduced later.

If the optimization goal is given in terms of a quality indicator, then the outcomes that have been obtained by running the different optimizers multiple times can be converted into samples of indicator values. Afterward, standard statistical testing procedures such as the Mann–Whitney rank sum test or Fisher's permutation test can be applied to make statistical statements about whether some algorithms generate higher quality solution sets than others, see Conover (1999). Note that multiple testing issues need to be taken into account when comparing more than two search methods (Zitzler et al. 2008a).

In the case that not a single indicator, but a more complex set preference relation is considered – possibly composed of multiple indicators – the solution sets under consideration can be compared pairwisely using this relation. On the basis of these comparisons, a standard

Mann–Whitney rank sum test can be applied, provided that the set preference relation is a total preorder. If totality is not given, for example, when directly using the weak Pareto dominance set relation, then a slightly modified testing procedure is necessary as described in Knowles et al. (2006).

The use of quality indicators or set preference relations in general implies that the outcome of the comparison is always specific to the encoded preference information. This means that Algorithm 1 can be better than Algorithm 2 when quality indicator I_1 is used, while the opposite can hold when another indicator I_2 is considered. This preference dependency can be circumvented by performing a comparison based on the weak Pareto dominance relation only. One possibility is the aforementioned methodology for set preference relations. Another possibility is the attainment function approach, which allows a more detailed analysis, but at the same time is only applicable in scenarios with few objectives.

The *attainment function method* proposed in Fonseca and Fleming (1996) and Grunert da Fonseca et al. (2001) summarizes a sample of Pareto set approximations in terms of a so-called empirical attainment function. To explain the underlying idea, suppose that a certain stochastic multiobjective optimizer is run once on a specific problem. For each objective vector $\mathbf{u} \in Z$, there is a certain probability p that the resulting Pareto set approximation contains a solution $\mathbf{a}$ such that $\mathbf{f}(\mathbf{a})$ weakly dominates $\mathbf{u}$. We say p is the probability that $\mathbf{u}$ is *attained* by the optimizer. The *attainment function* gives for each objective vector $\mathbf{u} \in Z$ the probability that $\mathbf{u}$ is attained in one optimization run of the considered algorithm. The true attainment function is usually unknown, but it can be estimated on the basis of a sample of solution sets: one simply counts the number of Pareto set approximations by which each objective vector is attained and normalizes the resulting number with the overall sample size. The attainment function is a first-order moment measure, meaning that it estimates the probability that $\mathbf{u}$ is attained in one optimization run of the considered algorithm *independently* of attaining any other $\mathbf{u}$; however, also higher-order attainment functions may be considered (Grunert da Fonseca et al. 2001).

Example 7 Consider ❯ *Fig. 11*. For the scenario on the left, the four Pareto front approximations cut the objective space into five regions: the upper right region is attained in all of the runs and therefore is assigned a relative frequency of 1, the lower left region is attained in none of the runs, and the remaining three regions are assigned relative frequencies of $1/4$, $2/4$, and $3/4$ because they are attained, respectively, in one, two, and three of the four runs. In the scenario on the right, the objective space is partitioned into nine regions; the relative frequencies are determined analogously as shown in the figure.

The estimated attainment functions can be used to compare two optimizers by performing a corresponding statistical test as proposed in (Grunert da Fonseca et al. 2001). The test reveals in which regions in the objective space there are significant differences between the attainment functions. In addition, the estimated attainment functions can also be used for visualizing the outcomes of multiple runs of an optimizer. For instance, one may be interested in plotting all the goals that have been attained (independently) in 50% of the runs. The 50%-attainment surface represents the border of this subspace; roughly speaking, the 50%-attainment surface divides the objective space in two parts: the goals that have been attained and the goals that have not been attained with a frequency of at least 50%, see ❯ *Fig. 11*.

⬛ Fig. 11

Hypothetical outcomes of four runs for two different stochastic optimizers (*left* and *right*). The numbers in the figures give the relative frequencies according to which the distinct regions in the objective space are attained. The thick gray lines represent the 50%-attainment surfaces.

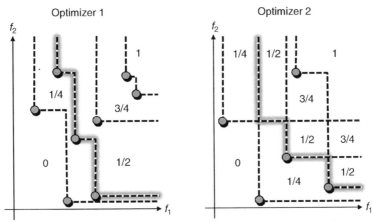

5   Applications

As outlined in the previous sections, the main use of evolutionary algorithms in multiobjective optimization scenarios lies in identifying a suitable set of compromise solutions, that is, in approximating the Pareto-optimal set. In the following, it will be briefly discussed why this type of optimization is beneficial.

5.1   Learning and Decision Making

The main reason for formulating an optimization problem in terms of a multiple criteria model is the lack of knowledge: too little is known about the problem structure and the influence of the different optimization criteria; the same holds for user preferences, and it may even be that sufficient preferences are not available or cannot be quantified in a mathematical manner. Therefore, multiobjective optimization focuses on "modeling in progress" where the model is iteratively refined to best suit (the relevant aspects of) reality. Since evolutionary algorithms as black-box optimizers also provide high flexibility with regard to the model, evolutionary multiobjective optimization offers an excellent framework for successively learning about a problem. The workflow of this model-centered problem-solving strategy is depicted in ❷ *Fig. 12.*

Learning can have different meanings. First of all, an approximation of the Pareto-optimal set allows us to perform a trade-off analysis. Starting with a particular compromise solution, one may ask how much deterioration is necessary to obtain a certain improvement in another objective. Such information is basically given by the shape of the front. Non-convex regions of the front stand for large trade-offs, while convex parts represent low trade-offs.

☐ **Fig. 12**

Model-centered problem solving using evolutionary multiobjective optimization.

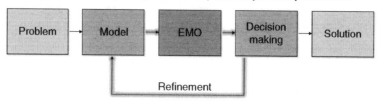

Example 8 In Siegfried et al. (2009), a multiobjective groundwater management problem is tackled using an evolutionary algorithm. The task is to optimize the placement and the operation of pumping facilities over time, while considering multiple neighboring regions that are economically independent. For each region a separate cost objective function is considered, which reflects how much money needs to be spent for that region to ensure sufficient water supply for a prespecified time horizon.

❯ *Figure 13* shows the approximated trade-off front for a three-region scenario. It is remarkable that there exists one solution which represents a so-called knee point (Branke et al. 2004a): it can be regarded as a decision vector that epsilon dominates the other decision vectors in the approximation set with a relatively small value for ε, cf. ❯ Eq. 7. Such information is highly useful for decision making, as it shows that – given the current estimate for the front – only small trade-offs need to be taken into account.

Another form of learning is to better understand the relationships between the considered objectives. Interesting questions in this context can be: how strongly are the objectives conflicting, can certain objectives be omitted without losing too much information, and which are the most important objectives? These are important aspects in particular from the perspective of decision making. It can become quickly infeasible to analyze a Pareto set approximation when the number of objectives increases. Therefore, an automatic dimensionality reduction of the objective space can be a valuable tool for a human user. Some studies addressed this problem and proposed methods based on (1) classical dimensionality reduction techniques such as principal component analysis (Deb and Saxena 2006; Saxena and Deb 2007), and (2) dedicated dominance-oriented reduction techniques (Brockhoff and Zitzler 2007, 2006). A remarkable result is that the notion of objective conflict always applies to groups of objectives and not to objective pairs only. Objective conflict can be defined in different ways (Deb 2001; Purshouse and Fleming 2003; Tan et al. 2005; Brockhoff and Zitzler 2007); the following notion is the most general (Brockhoff and Zitzler 2009):

▶ Let $(X, Z, \mathbf{f} = (f_1, \ldots, f_n), \mathbf{g}, \leq_{par})$ be a multiobjective optimization problem. A set $\mathscr{F}_1 \subseteq \{f_1, \ldots, f_n\}$ of objectives is denoted as *conflicting* with another set $\mathscr{F}_2 \subseteq \{f_1, \ldots, f_n\}$ iff

$$\leq_{par}^{\mathscr{F}_1} \neq \leq_{par}^{\mathscr{F}_2} \tag{26}$$

where the relation $\leq_{par}^{\mathscr{F}}$ restricts the weak Pareto dominance relation to the objectives in $\mathscr{F}$, that is,

$$\mathbf{u} \leq_{par}^{\mathscr{F}} \mathbf{v} :\Leftrightarrow \forall f \in \mathscr{F} : f(\mathbf{u}) \leq f(\mathbf{v}) \tag{27}$$

for all objective vectors $\mathbf{u}, \mathbf{v} \in Z$.

⬛ **Fig. 13**

Pareto set approximation obtained for the groundwater management application (Siegfried et al. 2009). Each axis corresponds to one cost function, all three are to be minimized. The annotated points represent the minima per objectives as well as the objective vectors providing the minimum resp. maximum unweighted sum over the objective components.

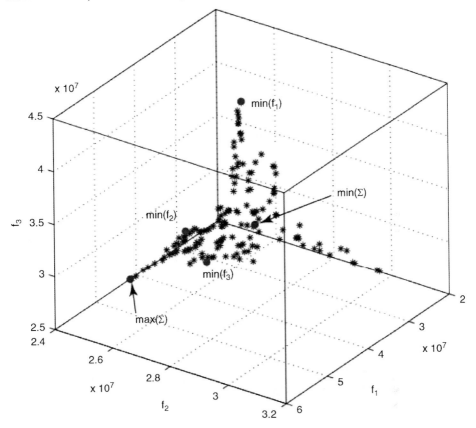

That means two groups of objectives are nonconflicting if they induce the same preference relation, which in turn implies that the Pareto-optimal sets are identical.

Example 9 ❷ *Figure 14* shows the parallel coordinates plot, cf. Purshouse and Fleming (2003), of three solutions **a** (solid line), **b** (dotted), and **c** (dashed) that are pairwisely incomparable. Assuming that X consists of only these three elements, the objective sets $\{f_1\}$, $\{f_2\}$, and $\{f_3\}$ are mutually conflicting, that is, all objectives are pairwisely conflicting, and also conflicting with the set $\{f_1, f_2, f_3\}$, that is, no objective alone can preserve the dominance relationships. However, the objective subset $\{f_1, f_3\}$ is not conflicting with $\{f_1, f_2, f_3\}$, which implies that the second objective could be omitted while preserving the dominance structure.

The above example shows that objectives may be removed without changing the dominance structure. Experimental results indicate that up to 50% of the objectives may be omitted

■ **Fig. 14**

Parallel coordinates plot for three solutions and three objectives. Each solution is represented by a line that connects the function values per objective.

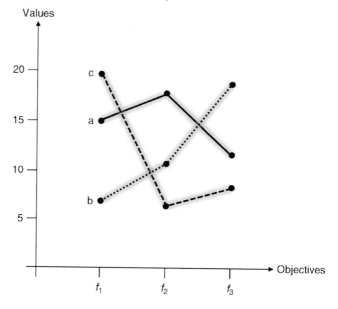

when focusing on a subset $A \subseteq X$ of the decision space, for example, a Pareto set approximation (Brockhoff and Zitzler 2009). The same techniques can also be used to quantify the degree of conflict and to determine on this basis how strongly certain objectives are conflicting (Brockhoff and Zitzler 2006). Thereby, one can reveal which objectives are most influential and can explain the shape of the front best.

Thirdly, learning may include a combined decision space/objective space analysis of a Pareto set approximation in order to identify structural patterns. It is valuable information to reveal that certain regions of the Pareto-optimal front are correlated to and characterized by specific decision variable settings. However, to identify interesting structural patterns is itself a challenging optimization problem closely related to issues of data mining. Researchers have developed and applied dedicated clustering techniques for this purpose (Ulrich et al. 2008; Aittokoski et al. 2009). Such a clustering aims at finding groups of solutions (within a Pareto set approximation) that shows similarities in both decision space and objective space. The clustering results can later be used for a dimensionality reduction of the decision space, possibly in combination with the objective space.

5.2 Multiobjectivization

The term *multiobjectivization* has been coined by Knowles et al. (2001) and originally refers to reformulating a single-objective problem by means of a multiobjective model, which is then tackled using a set-based problem transformation. The idea is that the resulting problem is easier to solve or can be better handled than the original problem formulation – a principle that applies not only to single-objective problems, but also to multiobjective ones.

One can distinguish between three types of multiobjectivization, or more precisely ways to increase the number of objectives, to make a problem easier: (1) disaggregation, (2) decomposition, and (3) addition. The first type means that an originally multiobjective model, which was tackled using a solution-oriented problem transformation is now approached on the basis of a set-oriented problem transformation. The motivation for disaggregation is given by empirical and theoretical results. Different studies (Zitzler and Thiele 1999; Zitzler et al. 2000) have shown that a set-based multiobjective optimizer can be more effective than, for example, a weighted-sum single-objective optimizer for certain ranges of weight combinations. These empirical findings are supported by theoretical studies where a similar picture emerges (Laumanns et al. 2004a, b).

Decomposition, the second type of multiobjectivization, refers to reformulating an objective function in a different way in terms of several objectives. This reformulation can have the effect that the optimizer has more information available and therefore can search more efficiently. Scharnow et al. (2002) showed for the single-source shortest-paths problem and Neumann and Wegener (2006) demonstrated for the minimum spanning tree problem that a multiobjective approach is indeed faster in finding the (single-objective) optimum than a single-objective approach; these results are based on running time analyses. Practical evidence for these findings are provided in Knowles et al. (2001) and Handl et al. (2008).

As to the third type of multiobjectivization, adding objectives, there is a prominent application area, namely, the phenomenon of bloat in genetic programming, which was addressed first by different research groups simultaneously (Bleuler et al. 2001; Ekárt and Németh 2001; de Jong et al. 2001) and later further investigated in several follow-up studies (de Jong and Pollack 2003; Panait and Luke 2004; Bernstein et al. 2004; Kotanchek et al. 2006; Bleuler et al. 2007). The idea is to avoid useless growth of trees and the resulting decrease of search efficiency by considering the tree size as an additional objective. For different test problems, it was shown that a set-based multiobjective optimizer can find better and more compact solutions than a standard genetic programming approach and other solution-oriented approaches that combine quality and size in a single objective. In addition, there are theoretical results available that prove that a problem can become easier when adding an objective function (Brockhoff et al. 2007). Brockhoff et al. (2007) showed for a particular problem that optimizing each objective separately requires more steps in terms of running time complexity than treating the combined biobjective problem – although in the latter case a set of solutions has to be found. Clearly, the addition of objectives often makes a problem harder, as outlined in ❷ Sect. 2.3, but if additional knowledge is integrated into the model using a further objective this can be beneficial for the search process.

References

Aherne FJ, Thacker NA, Rockett PI (1997) Optimising object recognition parameters using a parallel multiobjective genetic algorithm. In: Proceedings of the 2nd IEE/IEEE international conference on genetic algorithms in engineering systems: innovations and applications (GALESIA'97). IEEE, Prague, pp 1–6

Aittokoski T, Ayramo S, Miettinen K (2009) Clustering aided approach for decision making in computationally expensive multiobjective optimization.

Optimization Meth Softw 24(2):157–174. doi: http://dx.doi.org/10.1080/10556780802525331

Bader J, Bleuler S, Künzli S, Laumanns M, Thiele L, Zitzler E (2008) PISA Website. http://www.tik.ee.ethz.ch/sop/pisa/

Bader J, Brockhoff D, Welten S, Zitzler E (2009) On using populations of sets in multiobjective optimization. In: Ehrgott M et al. (eds) Conference on evolutionary multi-criterion optimization (EMO 2009), LNCS, vol 5467. Springer, pp 140–154

Bader J, Deb K, Zitzler E (2010) Faster hypervolume-based search using Monte Carlo sampling. In: Ehrgott M et al. (eds) Conference on multiple criteria decision making (MCDM 2010), LNEMS, vol 634. Springer, Heidelberg, Germany, pp 313–326

Baita F, Mason F, Poloni C, Ukovich W (1995) Genetic algorithm with redundancies for the vehicle scheduling problem. In: Biethahn J, Nissen V (eds) Evolutionary algorithms in management applications. Springer, Berlin, Germany, pp 341–353

Basseur M, Zitzler E (2006) Handling uncertainty in indicator-based multiobjective optimization. Int J Comput Intell Res 2(3):255–272

Bernstein Y, Li X, Ciesielski V, Song A (2004) Multi-objective parsimony enforcement for superior generalisation performance. In: Congress on evolutionary computation (CEC 2004). IEEE Press, Piscataway, NJ, pp 83–89

Beume N, Rudolph G (2006) Faster S-metric calculation by considering dominated hypervolume as Klee's measure problem. Tech. Rep. CI-216/06, Sonderforschungsbereich 531 Computational Intelligence, Universität Dortmund, shorter version published at IASTED International Conference on Computational Intelligence (CI 2006)

Bleuler S, Bader J, Zitzler E (2007) Reducing bloat in GP with multiple objectives. In: Knowles J, Corne D, Deb K (eds) Multi-objective problem solving from nature: from concepts to applications. Springer, Heidelberg, pp 177–200

Bleuler S, Brack M, Thiele L, Zitzler E (2001) Multi-objective genetic programming: reducing bloat by using SPEA2. In: Congress on evolutionary computation (CEC 2001). IEEE, Piscataway, NJ, pp 536–543

Bleuler S, Laumanns M, Thiele L, Zitzler E (2003) PISA—A platform and programming language independent interface for search algorithms. In: Fonseca CM, Fleming PJ, Zitzler E, Deb K, Thiele L (eds) Conference on evolutionary multi-criterion optimization (EMO 2003), LNCS, vol 2632. Springer, Berlin, pp 494–508

Bradstreet L, While L, Barone L (2008) A fast incremental hypervolume algorithm. IEEE Trans Evolut Comput 12(6):714–723

Branke J, Deb K (2004) Integrating user preferences into evolutionary multi-objective optimization. Tech. Rep. 2004004, Indian Institute of Technology, Kanpur, India, also published as book chapter in Jin Y (ed): Knowledge incorporation in evolutionary computation. Springer, Berlin, pp 461–477

Branke J, Deb K, Dierolf H, Osswald M (2004a) Finding knees in multi-objective optimization. In: Runarsson TP et al. (eds) Conference on parallel problem solving from nature (PPSN VIII), LNCS, vol 3242. Springer, Berlin, pp 722–731

Branke J, Schmeck H, Deb K, Reddy M (2004b) Parallelizing multi-objective evolutionary algorithms: cone separation. In: Congress on evolutionary computation (CEC 2004), vol 2. IEEE Service Center, Portland, OR, pp 1952–1957

Branke J, Kaußler T, Schmeck H (2001) Guidance in evolutionary multi-objective optimization. Adv Eng Softw 32:499–507

Bringmann K, Friedrich T (2008) Approximating the volume of unions and intersections of high-dimensional geometric objects. In: Proceedings of the 19th international symposium on algorithms and computation (ISAAC 2008). Springer, Berlin, Germany

Brockhoff D, Zitzler E (2006) Are all objectives necessary? On dimensionality reduction in evolutionary multiobjective optimization. In: Runarsson T et al. (eds) Conference on parallel problem solving from nature (PPSN IX), vol 4193. Springer, Berlin, Germany, LNCS, pp 533–542

Brockhoff D, Zitzler E (2007) Dimensionality reduction in multiobjective optimization: the minimum objective subset problem. In: Waldmann KH, Stocker UM (eds) Operations Research Proceedings 2006. Springer, Karlsruhe, pp 423–429

Brockhoff D, Zitzler E (2009) Objective reduction in evolutionary multiobjective optimization: theory and applications. Evolut Comput 17(2):135–166

Brockhoff D, Friedrich T, Hebbinghaus N, Klein C, Neumann F, Zitzler E (2007) Do additional objectives make a problem harder? In: Thierens D et al. (eds) Genetic and evolutionary computation conference (GECCO 2007). ACM Press, New York, NY, pp 765–772

Coello Coello CA, Lamont GB, Van Veldhuizen DA (2007) Evolutionary algorithms for solving multiobjective problems, 2nd edn. Springer, Berlin, Germany

Coello Coello CA, Van Veldhuizen DA, Lamont GB (2002) Evolutionary algorithms for solving multiobjective problems. Kluwer, Norwell

Conover WJ (1999) Practical nonparametric statistics, 3rd edn. Wiley, New York

Cvetković D, Parmee IC (2002) Preferences and their application in evolutionary multiobjective optimisation. IEEE Trans Evolut Comput 6(1):42–57

Czyzak P, Jaskiewicz A (1998) Pareto simulated annealing—a metaheuristic for multiobjective combinatorial optimization. Multi-criteria Decis Anal 7:34–47

de Jong ED, Pollack JB (2003) Multi-objective methods for tree size control. Genet Programming Evol Mach 4:211–233

de Jong ED, Watson RA, Pollack JB (2001) Reducing bloat and promoting diversity using multi-objective methods. In: Spector L et al. (eds) Genetic and evolutionary computation conference (GECCO

2001). Morgan Kaufmann, San Francisco, CA, pp 11–18

Deb K (2001) Multi-objective optimization using evolutionary algorithms. Wiley, Chichester, UK

Deb K, Gupta H (2006) Introducing robustness in multi-objective optimization. Evolut Comput 14 (4):463–494

Deb K, Saxena D (2006) Searching for pareto-optimal solutions through dimensionality reduction for certain large-dimensional multi-objective optimization problems. In: Congress on evolutionary computation (CEC 2006). IEEE Press, Seattle, WJ, pp 3352–3360

Deb K, Sundar J (2006) Reference point based multiobjective optimization using evolutionary algorithms. In: Keijzer M et al. (eds) Conference on genetic and evolutionary computation (GECCO 2006). ACM, New York, pp 635–642

Deb K, Chaudhuri S, Miettinen K (2006) Towards estimating nadir objective vector using evolutionary approaches. In: Keijzer M et al. (eds) Conference on genetic and evolutionary computation (GECCO 2006). ACM, New York, pp 643–650

Deb K, Pratap A, Agarwal S, Meyarivan T (2002a) A fast and elitist multiobjective genetic algorithm: NSGA-II. IEEE Trans Evolut Comput 6(2):182–197

Deb K, Thiele L, Laumanns M, Zitzler E (2002b) Scalable multi-objective optimization test problems. In: Congress on evolutionary computation (CEC 2002). IEEE Press, Honolulu, pp 825–830

Deb K, Thiele L, Laumanns M, Zitzler E (2005) Scalable test problems for evolutionary multi-objective optimization. In: Abraham A, Jain R, Goldberg R (eds) Evolutionary multiobjective optimization: theoretical advances and applications. Springer, chap 6, Berlin, pp 105–145

Ehrgott M (2005) Multicriteria optimization, 2nd edn. Springer, Berlin, Germany

Ekárt A, Németh SZ (2001) Selection based on the pareto nondomination criterion for controlling code growth in genetic programming. Genet Programming Evol Mach 2:61–73

Emmerich M, Beume N, Naujoks B (2005) An EMO algorithm using the hypervolume measure as selection criterion. In: Conference on evolutionary multi-criterion optimization (EMO 2005), LNCS, vol 3410. Springer, Berlin, pp 62–76

Fieldsend JE, Everson RE, Singh S (2003) Using unconstrained elite archives for multiobjective optimization. IEEE Trans Evolut Comput 7(3):305–323

Fleming PJ, Purshouse RC, Lygoe RJ (2005) Many-objective optimization: an engineering design perspective. In: Coello Coello CA et al. (eds) Conference on evolutionary multi-criterion optimization (EMO 2005), LNCS, vol 3410. Springer, Berlin, Germany, pp 14–32

Fonseca CM, Fleming PJ (1993) Genetic algorithms for multiobjective optimization: formulation, discussion and generalization. In: Forrest S (ed) Conference on genetic algorithms. Morgan Kaufmann, San Mateo, CA, pp 416–423

Fonseca CM, Fleming PJ (1995) An overview of evolutionary algorithms in multiobjective optimization. Evolut Comput 3(1):1–16

Fonseca CM, Fleming PJ (1996) On the performance assessment and comparison of stochastic multiobjective optimizers. In: Parallel problem solving from nature (PPSN IV). Springer, Berlin, Germany, pp 584–593

Fonseca CM, Fleming PJ (1998) Multiobjective optimization and multiple constraint handling with evolutionary algorithms—Part I: a unified formulation. IEEE Trans Syst Man Cybern 28(1):26–37

Fonseca CM, Paquete L, López-Ibáñez M (2006) An improved dimension-sweep algorithm for the hypervolume indicator. In: Congress on evolutionary computation (CEC 2006), Vancouver. IEEE Press, pp 1157–1163

Gaspar-Cunha A, Covas JA (2008) Robustness in multiobjective optimization using evolutionary algorithms. Comput Optimization Appl 39(1):75–96

Goldberg DE (1989) Genetic algorithms in search, optimization, and machine learning. Addison-Wesley, Reading, MA

Grunert da Fonseca V, Fonseca CM, Hall AO (2001) Inferential performance assessment of stochastic optimisers and the attainment function. In: Zitzler E et al. (eds) Conference on evolutionary multi-criterion optimization (EMO 2001), LNCS, vol 1993. Springer, Zurich, pp 213–225

Habenicht W (1983) Quad trees: a datastructure for discrete vector optimization problems. In: Hansen P (ed) Essays and surveys on multiple criteria decision making, LNEMS, vol 209. Springer, Berlin, pp 136–145

Handl J, Lovell SC, Knowles J (2008) Multiobjectivization by decomposition of scalar cost functions. In: Conference on parallel problem solving from nature (PPSN X). Springer, Berlin, pp 31–40

Hansen MP, Jaszkiewicz A (1998) Evaluating the quality of approximations of the non-dominated set. Tech. rep., Institute of Mathematical Modeling, Technical University of Denmark, iMM Technical Report IMM-REP-1998-7

Helbig S, Pateva D (1994) On several concepts for ε-efficiency. OR Spektrum 16(3):179–186

Horn J (1997) Multicriterion decision making. In: Bäck T, Fogel DB, Michalewicz Z (eds) Handbook of evolutionary computation, IOP Publishing and Oxford Universtity Press, Bristol, UK

Horn J, Nafpliotis N, Goldberg DE (1994) A niched pareto genetic algorithm for multiobjective optimization. In: Congress on evolutionary computation (CEC 1994). IEEE Press, Piscataway, pp 82–87

Huband S, Hingston P, Barone L, While L (2006) A review of multiobjective test problems and a scalable test problem toolkit. IEEE Trans Evolut Comput 10(5):477–506

Hughes E (2001) Evolutionary multi-objective ranking with uncertainty and noise. In: Evolutionary multi-criterion optimization, Lecture notes in computer science. Springer, Berlin, pp 329–343

Hughes EJ (2007) Radar waveform optimization as a many-objective application benchmark. In: Conference on evolutionary multi-criterion optimization (EMO 2007), LNCS, vol 4403. Springer, Heidelberg, pp 700–714

Igel C, Hansen N, Roth S (2007) Covariance matrix adaptation for multi-objective optimization. Evolut Comput 15(1):1–28

Ishibuchi H, Murata T (1996) Multi-objective genetic local search algorithm. In: IEEE (ed) Proceedings of the 1996 International conference on evolutionary computation. Nagoya, Japan, pp 119–124

Jaszkiewicz A (2002) On the performance of multiple-objective genetic local search on the 0/1 knapsack problem—a comparative experiment. IEEE Trans Evolut Comput 6(4):402–412

Jaszkiewicz A, Branke J (2008) Interactive multiobjective evolutionary algorithms. In: Branke J, Deb K, Miettinen K, Slowinski R (eds) Multiobjective optimization: interactive and evolutionary approaches. Springer, Heidelberg, pp 179–193

Jensen MT (2003) Reducing the run-time complexity of multiobjective EAs: the NSGA-II and other algorithms. IEEE Trans Evolut Comput 7(5):503–515

Jin Y, Branke J (2005) Evolutionary optimization in uncertain environments—a survey. IEEE Trans Evolut Comput 9(3):303–317

Knowles J, Corne D (2000a) M-PAES: a memetic algorithm for multiobjective optimization. In: Congress on evolutionary computation (CEC 2000). IEEE Press, Piscataway, NJ, pp 325–332

Knowles JD, Corne DW (2000b) Approximating the non-dominated front using the pareto archived evolution strategy. Evolut Comput 8(2):149–172

Knowles J, Corne D (2002) On metrics for comparing non-dominated sets. In: Congress on evolutionary computation (CEC 2002). IEEE Press, Piscataway, NJ, pp 711–716

Knowles J, Corne D (2003a) Properties of an adaptive archiving algorithm for storing nondominated vectors. IEEE Trans Evolut Comput 7(2):100–116

Knowles JD, Corne DW (2003b) Instance generators and test suites for the multiobjective quadratic assignment problem. In: Fonseca CM, Fleming PJ, Zitzler E, Deb K, Thiele L (eds) Evolutionary

multi-criterion optimization (EMO 2003), LNCS, vol 2632. Springer, Berlin, pp 295–310

Knowles J, Thiele L, Zitzler E (2006) A tutorial on the performance assessment of stochastic multiobjective optimizers. TIK Report 214, Computer Engineering and Networks Laboratory (TIK), ETH Zurich

Knowles JD, Watson RA, Corne DW (2001) Reducing local optima in single-objective problems by multi-objectivization. In: Zitzler E et al. (eds) Conference on evolutionary multi-criterion optimization (EMO 2001), LNCS, vol 1993. Springer, Berlin, pp 269–283

Kotanchek M, Smits G, Vladislavleva E (2006) Pursuing the pareto paradigm tournaments, algorithm variations & ordinal optimization. In: Riolo RL, Soule T, Worzel B (eds) Genetic programming theory and practice IV, genetic and evolutionary computation, vol 5. Springer, chap 3

Künzli S, Bleuler S, Thiele L, Zitzler E (2004) A computer engineering benchmark application for multiobjective optimizers. In: Coello CAC, Lamont G (eds) Applications of multi-objective evolutionary algorithms. World Scientific, Singapore, pp 269–294

Laumanns M, Thiele L, Deb K, Zitzler E (2002) Combining convergence and diversity in evolutionary multiobjective optimization. Evolut Comput 10(3):263–282

Laumanns M, Thiele L, Zitzler E (2004a) Running time analysis of evolutionary algorithms on a simplified multiobjective knapsack problem. Nat Comput 3(1):37–51

Laumanns M, Thiele L, Zitzler E (2004b) Running time analysis of multiobjective evolutionary algorithms on pseudo-Boolean functions. IEEE Trans Evolut Comput 8(2):170–182

Laumanns M, Thiele L, Zitzler E (2006) An efficient, adaptive parameter variation scheme for metaheuristics based on the epsilon-constraint method. Eur J Oper Res 169(3):932–942

Lee J, Hajela P (1996) Parallel genetic algorithm implementation in multidisciplinary rotor blade design. J Aircraft 33(5):962–969

Mezmaz M, Melab N, Talbi E-G (2006) Using the multi-start and island models for parallel multi-objective optimization on the computational grid. In: eScience. IEEE Computer Society, Washington, DC, p 112

Miettinen K (1999) Nonlinear multiobjective optimization. Kluwer, Boston, MA

Molina J, Santana LV, Hernández-Díaz A, Coello Coello CA, Caballero R (2009) G-dominance: reference point based dominance for multiobjective metaheuristics. Eur J Oper Res 197(2):685–692. doi:10.1016/j.ejor. 2008.07.015, http://www.sciencedirect.com/science/article/B6VCT-4T2M5WY-1/2/498e5b5a39d874c7aee 53e01a5557910

Mostaghim S, Teich J, Tyagi A (2002) Comparison of data structures for storing pareto-sets in MOEAs.

In: Congress on evolutionary computation (CEC 2002). IEEE Press, Piscataway, NJ, pp 843–848

Neumann F, Wegener I (2006) Minimum spanning trees made easier via multi-objective optimization. Nat Comput 5(3):305–319, conference version in Beyer H-G et al. (eds.) Genetic and evolutionary computation conference, GECCO 2005, Volume 1. ACM Press, New York, pp 763–770

Okabe T, Jin Y, Olhofer M, Sendhoff B (2004) On test functions for evolutionary multi-objective optimization. In: Parallel problem solving from nature (PPSN VIII). Springer, Berlin, pp 792–802

Panait L, Luke S (2004) Alternative bloat control methods. In: Genetic and evolutionary computation conference (GECCO 2004), LNCS. Springer, pp 630–641

Poloni C (1995) Hybrid GA for multi-objective aerodynamic shape optimization. In: Winter G, Periaux J, Galan M, Cuesta P (eds) Genetic algorithms in engineering and computer science. Wiley, Chichester, UK, pp 397–416

Purshouse RC, Fleming PJ (2003) Conflict, harmony, and independence: relationships in evolutionary multi-criterion optimisation. In: Conference on evolutionary multi-criterion optimization (EMO 2003), LNCS, vol 2632. Springer, Berlin, pp 16–30

Rachmawati L, Srinivasan D (2006) Preference incorporation in multi-objective evolutionary algorithms: a survey. In: Congress on evolutionary computation (CEC 2006). IEEE Press, pp 962–968

Rudolph G (1998) On a multi-objective evolutionary algorithm and its convergence to the pareto set. In: Proceedings of the IEEE International conference on evolutionary computation. IEEE Press, Piscataway, pp 511–516

Rudolph G (2001) Some theoretical properties of evolutionary algorithms under partially ordered fitness values. In: Evolutionary algorithms workshop (EAW-2001). Bucharest, pp 9–22

Rudolph G, Agapie A (2000) Convergence properties of some multi-objective evolutionary algorithms. In: Zalzala A, Eberhart R (eds) Congress on evolutionary computation (CEC 2000), vol 2. IEEE Press, New York, pp 1010–1016

Sawai H, Adachi S (2000) Effects of hierarchical migration in a parallel distributed parameter-free GA. In: Congress on evolutionary computation (CEC 2000). IEEE Press, Piscataway, NJ, pp 1117–1124

Saxena DK, Deb K (2007) Non-linear dimensionality reduction procedures for certain large-dimensional multi-objective optimization problems: employing correntropy and a novel maximum variance unfolding. In: Conference on evolutionary multi-criterion optimization (EMO 2007), LNCS, vol 4403. Springer, Berlin, pp 772–787

Schaffer JD (1985) Multiple objective optimization with vector evaluated genetic algorithms. In: Grefenstette JJ (ed) Conference on genetic algorithms and their applications. Pittsburgh, PA, pp 93–100

Scharnow J, Tinnefeld K, Wegener I (2002) Fitness landscapes based on sorting and shortest paths problems. In: Conference on parallel problem solving from nature (PPSN VII), LNCS, vol 2439. Springer, Berlin, pp 54–63

Siegfried T, Bleuler S, Laumanns M, Zitzler E, Kinzelbach W (2009) Multi-objective groundwater management using evolutionary algorithms. IEEE Trans Evolut Comput 13(2):229–242

Srinivas N, Deb K (1994) Multiobjective optimization using nondominated sorting in genetic algorithms. Evolut Comput 2(3):221–248

Steuer R, Miettinen K (2008) International Society on Multiple Criteria Decision Making. http://www.terry.uga.edu/mcdm/

Talbi EG, Mostaghim S, Okabe T, Ishibuchi H, Rudolph G, Coello Coello CA (2008) Parallel approaches for multiobjective optimization. In: Branke J et al. (eds) Multiobjective optimization: interactive and evolutionary approaches. Springer, Heidelberg, pp 349–372

Tan KC, Khor EF, Lee TH (2005) Multiobjective evolutionary algorithms and applications. Springer, London

Teich J (2001) Pareto-front exploration with uncertain objectives. In: Conference on evolutionary multi-criterion optimization (EMO 2001). In: Zitzler E, Deb K, Thiele L, Coello Coello CA, Corne D, LNCS, vol 1993. Springer, pp 314–328

Ulrich T, Brockhoff D, Zitzler E (2008) Pattern identification in pareto-set approximations. In: Keijzer M et al. (eds) Genetic and evolutionary computation conference (GECCO 2008). ACM, pp 737–744

Voß T, Hansen N, Igel C (2009) Recombination for learning strategy parameters in the MO-CMA-ES. In: Ehrgott M et al. (eds) Evolutionary multi-criterion optimization. Lecture notes in computer science, vol 5467. Springer, pp 155–168

Wagner T, Beume N, Naujoks B (2007) Pareto-, aggregation-, and indicator-based methods in many-objective optimization. In: Obayashi S et al. (eds) Conference on evolutionary multi-criterion optimization (EMO 2007), LNCS, vol 4403. Springer, pp 742–756

Winkler P (1985) Random orders. Order 1(1985):317–331

Zitzler E, Künzli S (2004) Indicator-based selection in multiobjective search. In: Conference on parallel problem solving from nature (PPSN VIII), LNCS, vol 3242. Springer, Heidelberg, pp 832–842

Zitzler E, Thiele L (1998) Multiobjective optimization using evolutionary algorithms—a comparative case study. In: Conference on parallel problem solving from Nature (PPSN V). Amsterdam, pp 292–301

Zitzler E, Thiele L (1999) Multiobjective evolutionary algorithms: a comparative case study and the strength pareto approach. IEEE Trans Evolut Comput 3(4):257–271

Zitzler E, Brockhoff D, Thiele L (2007) The hypervolume indicator revisited: on the design of pareto-compliant indicators via weighted integration. In: Obayashi S et al. (eds) Conference on evolutionary multi-criterion optimization (EMO 2007), LNCS, vol 4403. Springer, Berlin, pp 862–876

Zitzler E, Deb K, Thiele L (2000) Comparison of multiobjective evolutionary algorithms: empirical results. Evolut Comput 8(2):173–195

Zitzler E, Knowles J, Thiele L (2008a) Quality assessment of pareto set approximations. In: Branke J, Deb K Miettinen K, Slowinski R (eds) Multiobjective optimization: interactive and evolutionary approaches. Springer, Berlin, pp 373–404

Zitzler E, Laumanns M, Thiele L (2002) SPEA2: Improving the strength pareto evolutionary algorithm for multiobjective optimization. In: Giannakoglou K et al. (eds) Evolutionary methods for design, optimisation and control with application to industrial problems (EUROGEN 2001), International Center for Numerical Methods in Engineering (CIMNE). Athens, Greece, pp 95–100

Zitzler E, Thiele L, Laumanns M, Fonseca CM, Grunert da Fonseca V (2003) Performance assessment of multiobjective optimizers: an analysis and review. IEEE Trans Evolut Comput 7(2):117–132

Zitzler E, Thiele L, Bader J (2008b) On set-based multiobjective optimization (Revised Version). TIK Report 300, Computer Engineering and Networks Laboratory (TIK), ETH Zurich

29 Memetic Algorithms

Natalio Krasnogor
Interdisciplinary Optimisation Laboratory, The Automated Scheduling,
Optimisation and Planning Research Group, School of Computer
Science, University of Nottingham, UK
natalio.krasnogor@nottingham.ac.uk

G. Rozenberg et al. (eds.), *Handbook of Natural Computing*, DOI 10.1007/978-3-540-92910-9_29,
© Springer-Verlag Berlin Heidelberg 2012

Abstract

Memetic algorithms (MA) has become one of the key methodologies behind solvers that are capable of tackling very large, real-world, optimization problems. They are being actively investigated in research institutions as well as broadly applied in industry. This chapter provides a pragmatic guide on the key design issues underpinning memetic algorithms (MA) engineering. It begins with a brief contextual introduction to memetic algorithms and then moves on to define a pattern language for MAs. For each pattern, an associated design issue is tackled and illustrated with examples from the literature. The last section of this chapter "fast forwards" to the future and mentions what, in our mind, are the key challenges that scientists and practitioners will need to face if memetic algorithms are to remain a relevant technology in the next 20 years.

1   Introduction

Memetic algorithms (MAs) was the name given by Moscato (1989) to a class of stochastic global search techniques that, broadly speaking, combine within the framework of evolutionary algorithms (EAs) the benefits of problem-specific local search heuristics and multi-agent systems. MAs have been successfully applied to a wide range of domains that cover problems in combinatorial optimization (Reeves 1996; Burke et al. 1996; Fleurent and Ferland 1997; He and Mort 2000; Cheng and Gen 1997; Carr et al. 2002; Tang and Yao 2007; Gutin et al. 2007), continuous optimization (Hart 1994; Niesse and Mayne 1996; Morris et al. 1998), dynamic optimization (Vavak and Fogarty 1996; Caponio et al. 2007; Wang et al. 2008), multi-objective optimization (Liu et al. 2007; Ishibuchi and Kaige 2004; Jaszkiewicz 2002), etc. It could be argued that, unlike other nature-inspired algorithms such as ant colony optimization (ACO) (Dorigo et al. 1997), simulated annealing (Kirkpatrick et al. 1983), neural networks (McCulloch and Pitts 1943), evolutionary algorithms (Holland 1976), etc., memetic algorithms lack, at their core, a clear natural metaphor. Whether this is a strength or weakness of the paradigm is a question for another time and this chapter focuses first on a pragmatic software engineering presentation of this remarkably malleable search technology and then, toward the end of the chapter, argues that there is indeed a potentially powerful nature-inspired metaphor that could lead to important new breakthroughs in the field.

Early in the history of the application of EAs to real-world problems, it became apparent that a *canonical GA*, namely, one using a simple binary representation, n-point crossover, bitwise mutation, and fitness proportionate selection, could not possibly compete with tailor-made algorithms. This empirical observation resonated well with theoretical (and experimental) studies on the so called "Baldwin effect" and on "Lamarckian evolution" (Hinton and Nowlan 1987; Bull et al. 2000; Houck et al. 1997; Mayley 1996; Turney 1996; Whitley et al. 1994; Whitley and Gruau 1993) that focused on how learning could affect the process of evolution. Thus, it became apparent that the global search dynamics of EAs ought to be complemented with local search refinement provided by a suitable hybridization using problem-specific solvers including heuristics, approximate, and exact algorithms. Moreover, further theoretical results (Wolpert and Macready 1997) (and similar subsequent work) debunked the idea that effective and efficient "black box" general problem solvers were attainable, and hence gave further impetus to the school of thought that supported, as an essential methodological component, the incorporation of problem (or domain)-specific information in EAs. (Please note that everything said so far about evolutionary algorithms can also be applied to other search frameworks such as tabu search, simulated annealing or ant colony optimization, etc.) Domain-specific knowledge was thus added to the EA framework by means of specialized crossover and mutation operators, sophisticated problem-specific representations, smart population initialization, complex fitness functions (closer to the spirit of MAs), local search heuristics and, when available, approximate and exact methods. More recently, R. Dawkins' concept of "memes" (Dawkins 1976) has been gathering pace within the memetic algorithms literature as they can be thought of as representing "evolvable" strategies for problem solving, thus breaking the mould of a fixed and static domain knowledge captured once during the design of the MAs and left untouched afterward. Thus Dawkins' memes, and their extensions (Cavalli-Sforza and Feldman 1981; Durham 1991; Gabora 1993; Blackmore 1999) as evolvable search strategies (Krasnogor 1999, 2002, 2004b; Krasnogor and Gustafson 2004; Smith 2003; Burke et al. 2007a) provide a critical link to the possibility of open-ended combinatorial and/or continuous problem solving.

Software development is a process of knowledge acquisition (Armour 2007) and the development of successful memetic algorithms is no different. The popularity behind MAs is more closely related to the relative ease by which a reasonably good solver can be implemented than to any fundamental advantage over other optimization techniques such as tabu search or simulated annealing (to name but two). Indeed, any successful nature-inspired search method owes its popularity not to an intrinsic problem-solving feature, which might be absent from a competing method, but rather to the fact that, in spite of obvious design flaws (e.g., large number of parameters, lack of operational theory for their use, etc.), they help to structure around them a healthy research and practice milieu. That is, nature-inspired search methods are computational "research programs," or "research paradigms," in their own right (Kuhn 1962). Thus the question of what are the key components of the memetic algorithms research paradigm takes center stage. The literature has a large number of papers in which a variety of methods are classified as memetic algorithms. Thus, although the large majority of memetic algorithms are instances of evolutionary algorithms-local search hybrids, numerous MAs are derived from other metaheuristics, for example, ant colony optimization (ACO) (Lee and Lee 2005), particle swarm (Liu et al. 2005), artificial immune systems (AIS) (Yanga et al. 2008), etc. What all of these implementations have in common is a *carefully choreographed interplay between (stochastic) global search and local search strategies*. The remaining parts of this chapter consider some of the key implementation strategies that, over the course of the years, have (re)appeared in the form of tried and tested algorithmic design solutions to the ubiquitous problem of how to successfully orchestrate global and local search methods in complex search spaces.

2   A Pattern Language for Memetic Algorithms

Alexander et al. (1977) introduced, within the context of architecture and urban planning, the concept of "Patterns" and "Pattern Languages":

▶ In this book, we present one possible pattern language,... The elements of this language are entities called patterns. Each pattern describes a problem which occurs over and over again in our environment, and then describes the core of the solution to that problem, in such a way that you can use this solution a million times over, without ever doing it the same way twice.

A pattern language is then defined as a collection of interrelated patterns, with each and every one of them expressed in a concise, clear, and uniform format. The content of a pattern includes at least the following elements (Gamma et al. 1995):

- *The pattern name*, which is a concise handler to refer to both a specific problem and a tried and tested solution. By having a carefully selected set of pattern names (e.g., selection mechanism, crossover strategy, exploration strategy, diversification plan, etc.) the pattern language, that is, the vocabulary of, in our case, nature-inspired paradigms – for example, memetic algorithms – is enriched. Critically, a small increase in the size of a vocabulary creates a rich and expressive combinatorial explosion of patterns' compositions, thus opening the road for substantial research and practical experimentation. The importance of this observation will become apparent once we describe self-generating memetic algorithms.
- *The problem statement* depicting the situation in which the pattern is best applied, that is, the problem that the pattern attempts to provide a solution to, for example, maintaining pareto front diversity, etc. The problem statement might also contain a set of constraints describing situations where the pattern should not be applied or, symmetrically, conditions that must be fulfilled before the pattern can be used.
- *The solution*, in turn, provides a template on how to approach the solution of the problem to which the pattern is applied. The description in this section is not prescriptive but rather qualitative. As emphasized by Alexander et al. (1977), one might reuse a solution under myriads of different shapes, yet the essential core of all those implementations should be easily distinguishable and invariant.
- *The consequences* of applying the pattern. There are no free lunches, hence, even when a pattern might be the best (or perhaps only) solution to a given problem, its application might lead to a series of trade-offs. The more explicit and clearly stated these are, the more clear and precise the pattern language as a whole will be. For example, a pattern that calls for the reinitialization of a population due to a diversity crisis might carry with it, as obvious collateral damage, certain loss of information. Thus, by employing a reinitialization pattern, one might be forced to utilize a pattern that safeguards partial solutions.
- *Representative examples* briefly mentioning cases where the pattern has been used.

Thus, a collection of well-defined patterns, that is, a rich pattern language, substantially enhances one's ability to communicate solutions to recurring problems without the need to discuss specific implementation details. The pattern language thus serves the dual purpose of being both a taxonomy of problems and a catalog of solutions. This chapter provides a series of patterns that will be defined as per the tableaux that appears above. A reader interested in, for example, finding out about diversity handling strategies for MAs, irrespective of which underlying framework (e.g., evolutionary, ant colony, artificial immune system) the MA is

◘ **Fig. 1**

In **(a)** the pseudocode (reproduced from Bacardit and Krasnogor (2009)) for a Memetic Learning Classifier System, **(b)** the pseudocode (reproduced from Merz (2003)) for an Estimation of Distribution-Like Memetic Algorithm, **(c)** a Memetic Particle Swarm Optimization pseudocode (reproduced from Petalas et al. (2007)), **(d)** and **(e)** pseudocode of a representative Ant Colony Optimization metaheuristic (reproduced from Cordon et al. (2002)) with a solution refining strategy, through local search, for Ant Colony Optimization (ACO) and **(f)** an Artificial Immune System (AIS) inspired memetic algorithm flowchart (reproduced from Yanga et al. 2008).

```
Procedure GA Cycle
Population = Initialize population
Evaluate (Population)
For it = 1 to NumIterations
  Selection (Population)
  Offspring = CrossOver (Population)
  Mutation(Offspring)
  Localsearch (Offspring)
  Population = Replacement (Population, Offspring)
  Evaluate (Population)
EndFor
Output : Best individual from Population


Procedure LocalSerch
Input : Population
ForEach individual in Population
  If rand(0,1) < probLocalSearch
    Apply Rule-wise local search operators to individual
  Endif
EndForEach

Output : Population
```
a

```
function cMA(psize, gens, rec, ub) : X

begin
  for i = 1 to l do p[i] : = 0.5;
  X : = generate (p);
  X : = localSearch(X);
  X* : = X;
  i : = 1;
  repeat
    X : = generate(p);
    X : = localSearch(X);
    if rec then X' = recombine (X*, X);
    else X' : = generate (p);
    X' : = localSearch (X');
    if f(X') < f(X) then Swap (X, X');
    if f(X') > f(X*) then X* : = X';
    if up then update (p, x', x, psize);
    else update (p, x', x, psize);
    i : = i + 1;
  until (i > gens);
  return X*;
end
```
b

```
Input : N, X, c_1, c_2, x_min, x_max (lower & upper bounds), F (objective function),
Set t = 0.
Initialize x_i^{(t)}, v_i^{(t)} ∈ [x_min, x_max], p_i^{(t)} ← x_i^{(t)}, i = 1,..., N.
Evaluate F(x_i^{(t)})
Determine the indices g_r, i = 1,..., N.
While (stopping criterion is not satisfied) Do
  Update the velocities v_i^{(t+1)}(, i = 1,..., N, according to (1).
  Set x_i^{(t+1)} = v_i^{(t+1)}, i = 1,...,N.
  Constrain each particle x_i in [x_min, x_max].
  Evaluate F(x_i^{(t+1)}), i = 1,...,N.
  If F(x_i^{(t+1)}) < F(p_i^{(t)}) Then p_i^{(t+1)} ← p_i^{(t+1)}
  Else p_i^{(t+1)} ← p_i^{(t)}.
  Update the indices g_r
  When (local search is applied) Do
    Choose (according to one of the Schemata 1–3) p_q^{(t+1)}, q ∈ (1,...,N).
    Apply local search on p_q^{(t+1)} and obtain a new solution, y.
    If F(y) < F(p_q^{(t+1)}) Then p_q^{(t+1)} ← y.
  End When
  Set t = t+1.
End while
```
c

```
1  Procedure daemon actions
2    for each S_k do local_search(S_k) {optional}
3    rank (S_1,...,S_m) in decreasing order of solution
       quality into (S'_1,...,S'_m)
4    if (best (S'_1.S_global-best))
5      S_global-best = S'_1
6    end if
7    for μ = 1 to (σ − 1) do
8      for each edge a_rs ∈ S'_μ do
9        τ_rs = τ_rs + (σ − μ) . f(C(S'_μ))
10     end for
11   end for
12   for each edge a_rs ∈ S_global-best do
13     τ_rs = τ_rs + σ . f(C(S_global-best))
14   end for
15 end procedure
```
d

```
1  Procedure ACO_Metaheuristic
2    parameter_initialization
3    while (termination_criterion not_satisfied)
4      schedule_activities
5        ants_generation_and_activity()
6        pheromone_evaporation()
7        daemon_actions() {optional}
8      end schedule_activities
9    end while
10 end Procedure


1  Procedure ants_generation_and_activity()
2    repeat in parallel for k=1 to m (number_of_ants)
3      new_ant(k)
4    end repeat in parallel
5  end Procedure


1  Procedure new_ant (ant.id)
2    initialize_ant (ant_id)
3    L = update_ant_memory()
4    while (current_state ≠ target_state)
5      P = compute_transition_probabilities (A,L,Ω)
6      next_state = apply_ant_decision_policy(P,Ω)
7      move_to_next_state(next_state)
       if (on_line_step_by_step_pheromone_update)
8        deposit_pheromone_on_the_visited_edge()
       end if
9      L = update_internal_state()
10   end while
     if (online.delayed_pheromone_update)
11     for each visited edge
12       deposit_pheromone_on_the_visited_edge()
13     end for
     end if
14   release_ant_resources (ant_id)
15 end Procedure
```
e

f

being implemented in, can quickly scan the various patterns in the pattern language catalog, identify the one related to diversity strategies and rapidly gain an idea of the tried and tested approaches, the pattern's motivation and consequences. Furthermore, the reader could then refer to the mentioned literature for concrete, detailed codes and methods.

It has been argued (Cooper 2000) that design patterns are seldom created but rather they are discovered through a process of datamining the source code base and literature base for reusable solutions to recurring problems. In this spirit, and before describing a specific pattern language for memetic algorithms, one sees the pseudocodes and flowcharts of some representative MA instances. These algorithms are reported in exactly the same form as found in the original publication so as to emphasize both the *invariants* in their architecture as well as the variety of "decorations" found in the many implementations of memetic algorithms. Other examples of "in the wild" MAs can be found in Krasnogor and Smith (2005).

Memetic improvements have been used in, for example, learning classifier systems (Bacardit and Krasnogor 2009) with the top level pseudocode shown in ❷ *Fig. 1a.* An estimation of distribution (EDA) like MA, a compact memetic algorithm (Merz 2003) is shown in ❷ *Fig. 1b* (for a more recent EDA-MA see Duque et al. (2008)) while a memetic particle swarm optimization (PSO) (Petalas et al. 2007) pseudocode is depicted in ❷ *Fig. 1c.* ❷ *Figure 1d* and *e* show a generic pseudocode of an ant colony optimization-based MA (Cordon et al. 2002) and its explicit use of solution-refining strategies (in the form of local search methods), respectively. An example of an immune system-inspired memetic algorithm's (Yanga et al. 2008) flowchart is shown in ❷ *Fig. 1f.* The key invariant property that is present in the architecture of all these memetic algorithms is the combination of a search mechanism operating over (in principle) the entire search space with other search operators focusing on local regions of these search spaces. This key invariant holds true regardless of the nature-inspired paradigm the memetic algorithm is derived from or whether it is meant to solve an NP-hard combinatorial problem or a highly complex (e.g., multimodal, nonlinear, and multidimensional) continuous one.

We argue that the key problem that is addressed by memetic algorithms is the balance between global and local search, in other words, the strategy that different nature-inspired paradigms (e.g., ACO, AIS, etc.) might need to implement so as to benefit from a successful tradeoff between exploration and exploitation. Thus, the first top-level pattern can be defined.

Name: Memetic Algorithm Pattern (MAP) [1]

- *Problem statement:* A memetic algorithm provides solution patterns for the ubiquitous problem of how to successfully orchestrate a balanced tradeoff between exploring a search space and exploiting available (partial) solutions. It provides a suitable solution to complex problems where standard and efficient (i.e., approximation algorithms, exact algorithms, etc.) methods do not exist. The memetic algorithm is said to explore the search space through a "global" search technique, while exploitation is achieved through "local" search.
- *The solution:* This pattern relies on finding, for a given problem domain, an adequate instantiation of exploration and exploitation. Exploration is performed by "global search" methods usually implemented by means of a population-based nature-inspired method such as evolutionary algorithms, ant colony optimization, artificial immune systems, etc. Exploitation is commonly done through the use of local search methods and domain-specific heuristics. The global scale exploration is achieved by, for example, keeping track of multiple solutions or by virtue of specific "jump" operators that are able to connect

distant regions of the search space. The local scale exploitation focuses the search on the vicinity of a given candidate solution.

- *The consequences:* Hybridizing a global search method of any kind with local search and/ or domain-specific heuristics usually results in better end-results, but this comes at the expense of increased computational time. The correct tradeoff between exploration and exploitation must be such that, were the global searcher given the same total CPU "budget" as the memetic algorithm, then, its solutions would still be worse than those derived from the MA. Needless to say, if the local searcher by itself, or through a naive multi-start shell could achieve the same quality of results as the memetic algorithm then the global searcher becomes irrelevant. Another likely consequence is that local search and domain-specific heuristics usually result in premature convergence (also called diversity crisis).
- *Examples:* Papers (Cordon et al. 2002; Yanga et al. 2008; Petalas et al. 2007; Duque et al. 2008; Bacardit and Krasnogor 2009) report on the use of memetic algorithms under a variety of nature-inspired incarnations.

The memetic algorithm patterns (MAP) can be refined through a variety of template patterns (TP) (Gamma et al. 1995), which allow for the definition of the *skeleton* of an algorithm, method, or protocol, through deferring problem-specific details to subclasses. In this way, through a judicious use of the template pattern, one can have a very generic and reusable recipe for implementing solutions to a range of, perhaps very different, problems.

❯ *Figure 2* shows a class diagram capturing a template method pattern. The abstract class defines a template method that provides the algorithmic skeleton for a specific functionality. To achieve its functionality the template method calls one or more primitive operations (defined abstractly in the abstract class) that can be redefined in more specialized subclasses (concreteClass in the figure). ❯ *Figures 1a-f* are examples of template method patterns for ACO, AIS, particle swarm optimization (PSO), EDA and LCS based memetic algorithms

◼ **Fig. 2**
A UML class diagram sketching the structure for implementing Template Method patterns. (Adapted from Gamma et al. (1995).)

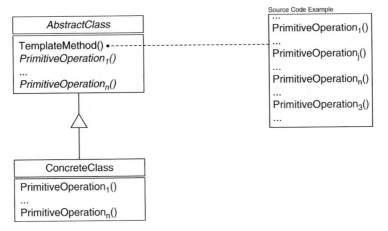

Template method pattern

respectively. Each TP captures the invariant properties of memetic algorithms when it is implemented from the perspective of a specific nature-inspired algorithm. It also captures the invariant features of MAs regardless of which nature-inspired route is used to implement it, for example, the entwining of global and local search procedures. The following sections will discuss other design patterns that are critical to the implementation of competent memetic algorithms thus extending the pattern language for MAs.

2.1   A Template Pattern for an Evolutionary Memetic Algorithm

As the design issues and critical considerations behind the implementation of MAs are almost identical across the various possible nature-inspired algorithms that could be used as templates, and as evolutionary memetic algorithms are the best known MAs, the focus of the following discussions will be on an evolutionary algorithm template pattern for MAs. The pattern language that will emerge, however, will be valid for other nature-inspired paradigms as well.

Name: Evolutionary Memetic Algorithm Template Pattern (EMATP) [2]

- *Problem statement:* the EMATP provides a viable route for solving the problem of how best to coordinate global and local search methods from within an evolutionary algorithms paradigm. Although in principle the simultaneous exploration of the search space by the evolutionary process and the exploitation (by refinement) of candidate solutions should result on an improved algorithm, this is not always the case. The expectation is that a hybridization of an EA would result in a synergistic net effect that would productively balance local and global search.

- *The solution:* the (almost) standard evolutionary cycle composed of *Initiate* → *Evaluate* → *Mate* → *Mutate* → *Select* → *RepeatUntilDone* is expanded with domain-specific operators that can refine this cycle. The refinement processes could vary for the different elements of the core EA's pipeline and could be implemented through exact, approximated or heuristic (e.g., local search) methods. Domain-specific operations are often implemented in the form of smart initializations, local search procedures, etc., that refine the input solutions to the mate and/or mutate processes as well as (more often) their outputs. Refinements could also be applied to the selection, initialization, and population management processes through, for example, fitness sharing, crowding, population structuring (e.g., cellular/lattice structures, demes, islands), and diversity management strategies.

- *The consequences:* paradoxically, although the key motivation for using memetic algorithms has been to refine solutions and converge toward good local optima (or ideally global optima) fast, sometimes the balance of local and global search is poorly implemented resulting in an untimely diversity crisis because the algorithm converges too fast. Another direct consequence of using memetic algorithms is that often, refining solutions by problem-specific strategies incurs in a substantial additional CPU commitment. Thus, it becomes essential to validate whether the exploration mechanisms of the core EA's pipeline when augmented with refining strategies do not end up producing worse solutions than having run the pure EAs with the same total CPU commitment (or the refinement strategies alone with an equivalent total CPU budget).

- *Examples:* ❯ *Figure 3* shows a concrete example of an evolutionary memetic algorithm. A detailed description and analysis of several successful implementations of the EMATP pattern are described in Krasnogor and Smith (2005). The paper also provides a taxonomy for EMAs as well as a discussion of design issues.

❑ **Fig. 3**

An evolutionary algorithm-based memetic algorithm template. The figure highlights the EA's core operators as well as the hotspots where the algorithm can be refined, that is, where "memetic" operators might come into play.

```
/* Variable and constants definitions:
/* t is generation number
/* IPi stands for intermediate population ith
/* P(t) stands for the population at time t
/* H stands for historical population archive
*/
Begin
t ← 0;
```

Initialize P(t);

```
Evaluate P(t);
Repeat Until (termination conditions = True) do
    /* Mating FineGrainedScheduler, fSR, coordinates the application of      */
    /* Crossover and Refinement Operators to population P(t)                 */

    IP1 ← fSR(Crossover, Refinement, P(t));

    /* Mutation FineGrainedScheduler, fSM, coordinates the application of     */
    /* Mutation and Refinement Operators to population IP1                    */

    IP2 ← fSM(Mutation, Refinement, IP1);

    /* Selection CoarseGrainedScheduler, cS, coordinates the application of   */
    /* Selection and Refinement Operators to population IP2                   */

    IP3 ← cS(Selection, Refinement, IP2);

    t ← t+1;
    P(t) ← IP3;

    /* MetaScheduler, mS, coordinates the application of                      */
    /* Refinement Operators to populations H ⊇ P(i) with i ∈ [0..t]           */

    H ← mS(Refinement, IP3);

End;
End;
```

Evolutionary algorithm backbone

Refinement hooks

❯ *Figure 3* shows a concrete example of the EMATP. The figure identifies the key EA's components as the "backbone" and the potential places where refinement might take place as refinement hooks. Each hook is represented by a "scheduler" operator that manages the flow of information (e.g., partial/candidate solutions, time-dependent parameters, etc.) between each of the core EA's component and the refinement strategies associated with them. A formal notation for these schedulers can be found in Krasnogor and Smith (2005). A detailed study of the literature reveals that three types of schedulers can be abstracted from the various templates that are implemented. These schedulers are: a *fine-grain scheduler* for coordinating the operation of the genetic operators mutation and crossover with refinement methods, a *coarse-grain scheduler* for overseeing the interplay between population management strategies (e.g., selection) and refinement strategies and *meta scheduler* that orchestrate refinement procedures over longer timescales and larger spatial scales (e.g., islands, population structures, pareto archives). These schedulers receive their names because they operate at different level of granularities and timeframes. The fine-grain schedulers (represented by fS_R, fS_M in ❯ *Fig. 3*) have access to a very limited number of solutions and hence they have a very localized view of the current state of the search and thus the decision they can take in terms of parameter and operators adaptation is confined to small regions of the search space. The course-grain scheduler (represented by cS in ❯ *Fig. 3*), on the other hand, has access to a complete population and, perhaps, even to intermediate populations that could have been created in the main EA's pipeline. Thus it has a more global view of the search state that might be used to strategically guide the application of further refining methods to (parts of) the population it has access to. The meta scheduler (represented by mS in ❯ *Fig. 3*), on the other hand, has access to potentially the power set of all previously visited solutions and hence it has an even broader information base to decide the appropriate balance between exploration and exploitation. These schedulers also operate at different time-scales. Clearly, fS_* can operate very frequently indeed, potentially each time that a crossover or mutation event takes place. The coarse-grain scheduler operates once per generation and the mS, potentially, even less often. These differing spatial (i.e., access to solutions) and temporal (frequency of operation) features result in constraints on the complexity of the algorithmic strategies that these schedulers can implement and have a direct impact in several design issues such as whether or not a surrogate fitness function (Zhou 2004; Zhou et al. 2007) should be used, etc. Needless to say, EMATP can be implemented through a series of (object-oriented) class hierarchies. The EMATP pattern in ❯ *Fig. 3* can be encapsulated into an abstract class from which subclasses implement specific versions of the pattern. One could thus imagine a family of evolutionary memetic algorithms where one or more features are either removed from the pattern or added to it simply by overriding the behavior of the scheduler methods.

2.2   Strategy Patterns for Memetic Algorithms Design Issues

The memetic algorithms pattern language is extended by describing a series of strategy patterns associated with design issues arising from attempting to implement effective instances of the EMATP. The evolutionary algorithm's processes such as crossover and mutation, as well as the refinement strategies and the schedulers that coordinate their operations, are best thought of as strategy patterns (Gamma et al. 1995). These patterns are useful for defining a family of interchangeable algorithms.

■ Fig. 4

A UML class diagram sketching the structure for implementing strategy patterns. (Adapted from Gamma et al. (1995).)

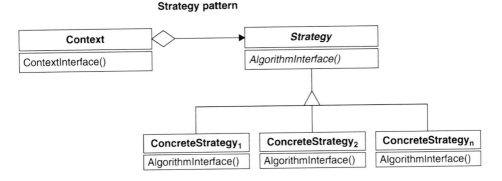

The idea is that the client that employs a member of this family of algorithms, for example, the fine-grained scheduler coordinating a local search strategy with a crossover operator, should be able to use any of the available local search and/or crossover strategies without need for recoding. A most frequent example is provided when testing different, for example, crossover operators such as one-point crossover, two-point crossover and uniform crossover without the need to rewrite the EA's backbone. Similarly, one should be able to change the local search used without affecting the pattern in which it is used. That is, a strategy pattern should be used each time that a program, in this case a memetic algorithm, needs a specific service or function and when there are, in fact, several ways of executing these functions. To promote reusability, strategy patterns follow a scheme similar to the one depicted in ❯ *Fig. 4*. The choice of which strategy to implement as to perform the required function is problem- and instance-specific. Thus by using strategy patterns, one can change algorithms without causing a chain reaction of changes in the code that uses those algorithms. As argued below, there is a direct link between critical memetic algorithms design issues and strategy patterns.

2.2.1 Refinement Strategies

It has been argued that an EMATP provides conceptual solutions to the problem of balancing local and global search. The global search is performed through the implementation of some form of evolutionary algorithm while the local search is meant to be implemented by some form of refinement operator. The choice of operator refinement is a critical design decision in memetic algorithms engineering as it tremendously impacts the synergy between the EA's backbone and the refinement methods of choice.

Name: Refinement Strategy Pattern (RSP) [3]
- *Problem statement:* this pattern attempts to address the following key questions: (1) what refinement operator should be used and, ultimately, (2) what fitness landscape will the MA be exploring? While the EMATP helps define the algorithmic skeleton of an

evolutionary-based memetic algorithm, it leaves to the strategy patterns the problem of how to decide what strategy would be implemented at the various refinement stages. Indeed, both theoretical and experimental work has analyzed the effectiveness of different local search strategies within memetic algorithms.

- *The solution:* the key design issue related to refinement strategies is how they prevent getting trapped in (poor) local optima, that is, search stagnation, while performing an efficient and fast local search. The answer to questions (1) and (2) above is problem, instance, and time dependent, hence a solution can only be provided at an abstract engineering level that allows for the seamless adaptation of the EMATP to different situations. More precisely, for a given set of low-level move operators, for example, n-exchange, n-swaps, bit-flip, etc., a variety of navigation rules have been implemented. Navigation rules specify how the search landscape induced by a given move operator is traversed. Some of the alternatives are breadth first search, depth first search, variable depth, etc. Besides the navigation method, the refinement strategy also needs to specify an acceptance criteria, for example, greedy, Monte Carlo, Great-Deluge (Dueck 1993; Burke et al. 2004), etc. Finally, the most sophisticated refinement strategies employ multiple move operators and acceptance criteria, for example, multimeme algorithms (Krasnogor 2004a; Krasnogor and Smith 2001), variable-neighborhood search (Hansen and Mladenovic 1998, 2001), etc. Indeed, a refinement strategy pattern might use at its core a full memetic algorithm pattern, see Romero-Campero et al. (2008) for an example. Thus, through the recursive nesting of MAP and RSP, it is possible to obtain extremely complex and versatile search methodologies.

- *The consequences:* the refinement strategy pattern comes with some "strings attached" to it. Its very flexibility means that a given memetic algorithm might be exploring not one but several search landscapes simultaneously. The precise nature of this combined search space should, ideally, be adequately analyzed either experimentally or theoretically (ideally in both ways). Fitness landscapes statistics were utilized in, for example, Kallel et al. (2001) and Merz and Freisleben (1999). Krasnogor and Smith (2008) resorted to polynomial local search theory to show the impact of various combinations of local search and genetic operators. The chapter provides a worst case analyzing for the complexity of heuristics applied to a well-known problem, the two-dimensional and Euclidean TSP. By way of simple arguments, the paper shows the PLS-completeness of a family of memetic algorithms for the TSP as well as graph partitioning and maximum network flow (both with important practical applications). It was shown that potentially very long paths to local optima exists even when the neighborhood used by some MAs are of polynomial size. Interestingly, the chapter's arguments are also valid for a variable-neighborhood search, rather than evolutionary based, template pattern for global search. A similar result is also provided in Sudholt (2007) for MinCut, MaxSAT, and Knapsack problems.

- *Examples:* intelligent refinement strategy patterns can be seen in Jakob (2006) and Neri et al. (2007a).

2.2.2 Exact and Approximate Hybridization

The previous section presented a refinement strategy pattern that dealt with the general question of how to design a navigable fitness landscape for memetic algorithms through a judicious selection of refinement strategies. One complementary and related design issue is that of the hybridization of evolutionary algorithms with exact and approximate methods for which a new strategy pattern is presented.

Name: Exact and Approximate Hybridization Strategy Pattern (EAHSP) [4]

- *Problem statement:* The integration of exact and approximation method into an evolutionary memetic algorithm (EMA) pattern requires the consideration of specific design issues that are distinct than those of other (e.g., heuristic) refinement strategies. More specifically, the development of exact methods requires considerably more effort than that of a set of relatively simple local searchers. The majority of the effort goes into obtaining a tight formulation of the problem for the exact method or a suitable relaxation. Moreover, exact and approximation methods might take a substantial amount of CPU effort to run, sometimes dwarfing that is required by the EMA. Thus, how to synergistically exploit both approaches for obtaining a better solver remains a critical problem.
- *The solution:* It has been argued that there are essentially two main ways of integrating exact and EMA methods. The first approach is based on running the EMA and the exact method in tandem, that is, the EMA provides solutions to the exact method (for example, a branch and bound method) that could, in turn, use the evolved solutions as good bounds, thus helping reduce the branch and bound runtime. Symmetrically, the exact method might be used to seed EMA populations with good candidate solutions. In this loose integration, each algorithm runs essentially independent from each other but exchanging information from time to time. On the second implementation strategy the branch and bound calls the EMA from time to time and thus uses the EMA as a heuristic to produce bounds. It might call it only once as to provide initial solutions or many times over its execution, thus forming a more tightly coupled pipeline. Gallardo et al. (2007) calls these two strategies coercive and cooperative, respectively. Sometimes, when the problem is excessively large, for example, it does not fit in memory and there is no possibility of a distributed computation. Branch and bound is heuristically modified and converted into a beam search. Although this approximation loses performance, it does allow for much larger problems to be tackled (Gallardo et al. 2007). In Pirkwieser et al. (2008), the authors interleave the execution of an EMA with a Lagrangian relaxation method based on a loose coupling strategy.
- *The consequences:* the application of this strategy pattern requires a detailed modeling of the problem at hand and a realistic evaluation of the exact method and memetic algorithm relative runtimes. It should be clear that this strategy should be used if the added benefit of guaranteed results outweigh the additional efforts both in engineering the system and in running it.
- *Examples:* Mezmaz et al. (2007) explores the application of both tandem and pipeline strategies for the bi-objective flowshop scheduling problem. Raidl and Puchinger (2008) provide a detailed analysis of how to combine integer and linear programming methods with metaheuristics. Many of the strategies reported are directly applicable to EMAP.

2.2.3 Population Diversity Handling Strategies

A key characteristic of memetic algorithms is that, they combine search through a population of solutions with search focused around specific promising ones. The introduction of

■ **Fig. 5**

The mapping from genotype (i.e., solution encoding) to phenotype (i.e., encoding interpretation) to fitness (i.e., a concrete measure of a solution's worth) can be very complex.

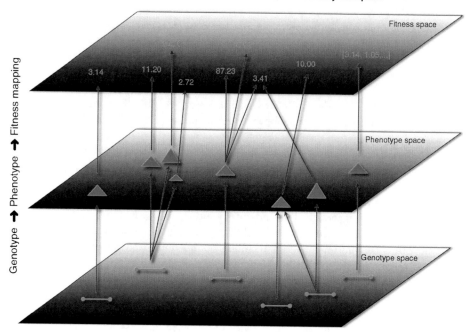

refinement strategies to improve over a subset of solutions often produces additional selection that, in turn, brings a premature population diversity crisis. A population's lack of diversity causes the MA to allocate repeated reproductive, mutation, and refinement trials to the same individuals, thus wasting precious CPU resources. Balancing population diversity is a critical issue, that in itself necessitates a detailed analysis of what is meant by "diversity." The complexity of this task is represented in ❯ *Fig. 5*. Only in the most idealized problem, the mapping from genotypes to phenotypes and then to fitness values is a one-to-one direct mapping. In real-world applications these mappings are highly complex. In some cases, the interpretation of a given solution representation (i.e., genotype) into a phenotype (i.e., actual solution) is nonlinear and/or stochastic and results in potentially different phenotypes for the same genotype. Similarly, fitness assignment to phenotypes might also be nonlinear and/or stochastic thus producing, for a given phenotype, potentially different fitness values. The existence of nonlinearities and stochasticity in the genotype → phenotype → fitness mappings gets compounded when multi-objective problems are tackled. Hence, the definition of population diversity used by diversity strategies must take these complex mappings into account.

Name: Population Diversity Handling Strategy Pattern (PDHSP) [5]

- *Problem statement:* the goal of this pattern is to provide interchangeable strategies for the robust management of population diversity within memetic algorithms.
- *The solution:* sustaining population diversity in memetic algorithms is a difficult task. It is usually tackled by a combination of strategies operating at various levels. First,

the population is usually initialized in an intelligent manner as to avoid resampling and nonrepresentative coverage of the search space at the beginning of the search. Examples of initialization mechanisms can be found in Burke et al. (1998) and Kretwski (2008). Later, during the search, population diversity maintenance can be implemented at various levels. For example Krasnogor et al. (2002) uses a memory of solution features during the mating process as to avoid generating phenotypes with over-represented features. In this work population diversity is controlled at the phenotype level. An example of a diversity preservation strategy at the fitness level is adopted by Kononova et al. (2007), where the selection of which local search to use is guided by the range of fitness in the population. Similarly, Krasnogor and Smith (2000) uses an adaptation in the local searcher to monitor and affect fitness distributions. In the context of AIS-based templates for memetic algorithms, the aging operator eliminates the oldest candidate solutions from a population so as to contribute to maintaining diversity and thus avoiding (poor) local optima. AIS and differential evolution (DE)-based solvers for multidimensional problems, where diversity handling and refinement strategies are compared, are reported in Cutello et al. (2007). Keeping track of the "age" of the solutions is quite an ubiquitous strategy used on several multi-solution templates. Sorensen and Sevaux (2006) propose several population management strategies based on measuring diversity in the solution space for multidimensional knapsack problem and weighted tardiness single-machine scheduling problems. For continuous problems, diversity is sometimes preserved through structuring the population using a cellular memetic algorithm (Nguyen et al. 2007).

- *The consequences:* Care should be taken in designing the diversity measure to use. In Burke et al. (2002, 2003) a variety of diversity measures for problems have been analyzed where mappings such as those depicted in ❷ *Fig. 5* take place. Although the examples used in this chapter are based on genetic programming, the lessons learnt are universal for memetic algorithms, namely, (1) increased diversity does not always positively correlate with improved performance and (2) whether it leads to improved performance depends on the complex mapping between genotypes, phenotypes, and fitness. Thus population diversity handling strategies must be implemented in such a way that they should be easy to change and benchmark as to continuously assess their performance in specific problems.

- *Examples:* Neri et al. present a comparative study of several fitness diversity-based adaptation schemes for multimeme algorithms in Neri et al. (2007b). Landa Silva and Burke (2004) address the issue of the interplay between diversity and multi-objective optimization.

2.2.4 Surrogate Strategies

Evolutionary memetic algorithms are usually used to solve very complex and hard problems. Oftentimes, these problems also give rise to uncertainties in the assignment of the fitness values of individuals upon which selection can be applied. On the other hand, refinement strategies used within memetic algorithms are usually computationally expensive, there is no explicit knowledge of a fitness function (e.g., in interactive evolutionary design problems) and hence fitness must be "reverse-engineered." Moreover, in some cases the objective function is extremely multi-modal and a smoothing criteria is required. The above considerations lead directly to an important design issue in memetic algorithms engineering, namely, the use of surrogate objective functions.

Name: Surrogate Objective Function Strategy Pattern (SOFSP) [6]

- *Problem statement:* Memetic algorithms are computationally intensive methods and one of the most important computational bottlenecks that must be considered when engineering an MA in the calculation of the objective (i.e., fitness) function. As shown in ❯ *Fig. 5*, many problems are inherently noisy, that is, the same candidate solution might be assigned a different fitness value due to stochasticity or nonlinearities in their evaluation. In other cases, several promising solutions might be repeatedly improved by a refinement strategy pattern and this refinement, almost invariably, leads to a large number of additional objective function's evaluations. In optimal design problems, one often finds that no explicit knowledge of fitness is available and the quality of a solution must be inferred from samples given by experts. The aim of the surrogate objective function strategy pattern is to provide effective ways of replacing an expensive, noisy, or unknown fitness function with a suitably defined approximation. The goal is to ameliorate the cost of highly expensive fitness function evaluations while preserving fitness assignment quality by providing a "proxy" to the original objective function that is faster than the former, but of sufficient quality so that it can provide good quality estimations of the true fitness value of a candidate solution.

- *The solution:* Surrogate fitness functions have also been called metamodels, local models, and partial objective functions in the literature. Although the use of effective strategies for solving the surrogacy problem are also important to other branches of natural computation, it is particularly poignant in memetic algorithm's design. This is so because MAs should be able to cope with uncertainty in the genotype $\rightarrow$ phenotype $\rightarrow$ fitness mappings, which could lead to a noisy fitness assignment and ambiguity in the very definition of fitness but also MAs must contend with and balance the additional computational effort introduced by the refinement strategies. Explicit averaging has been used to ascertain the fitness of candidate solutions under uncertainty. This is exemplified by the reduction of variance techniques (e.g., latin hypercube sampling), reevaluation of distinct individuals (e.g., those with high fitness or variance), weighted histories (Branke 1998, 2001, 2002), etc. Fitness inheritance (Llora et al. 2007), artificial neural networks and related approaches (Ong et al. 2008), models (Bull 1999), and metamodels (Bhattacharya 2007) as well as design of experiments (Sacks et al. 1989) have also been used for surrogate fitness implementation. In Jin (2005) and Paenke and Jin (2006) the authors present a detailed analysis of surrogate methods for evolutionary computation.

- *The consequences:* The use of a surrogate objective function strategy is, in many real-world scenarios, unavoidable. However, its use brings forth a series of other design decisions the memetic algorithms engineer will need to take into account, for example, at what level will the approximation be more productive? the fitness level or the problem itself? Is it possible to use global models or should a collection of local approximations be preferred, etc.

- *Examples:* Zhou et al. (2007) provide specific recommendations for integrating approximate fitness models within memetic algorithms.

2.2.5 Continuous Problems

The utilization of EMATP for continuous problems presents, in addition to the previous design considerations, new opportunities and challenges. This section only focuses on the issue of integrating refinement strategies for continuous optimization within a memetic algorithm framework. The reader must note that there is a large body of literature on the use of evolutionary computation for continuous domains, which is not dealt with here. Unlike combinatorial optimization, where it is always possible to detect when a given point is a local optimum, in the continuous case it is, in general, not possible and hence one must settle for an a priori decision on what solution precisions are acceptable, or provide a decision-making mechanism to the MA itself. This becomes even more critical when facing continuous design optimization problems with multiple optimality criteria (e.g., Siepmann et al. 2007). The following section describes a refinement strategy pattern for continuous problems that addresses some of the design issues reported in Hart et al. (2004).

Name: Continuous Problems Refinement Strategy Pattern (CPRSP) [7]

- *Problem statement:* Effective search requires an appropriate determination of search scales, which in continuous optimization is not always possible, for example, it is often impossible to determine, given a candidate solution, whether it represents a local optimum or not. The lack of explicit local scale might also result on very long searches, that when combined with lack of gradient information pose a very difficult problem for optimization techniques. Thus, whether one uses as a refinement strategy, a derivative-free method such as, for example, Nelder–Mead simplex (Nelder and Mead 1965) or Lagrangian interpolation (Berrut and Trefethen 2004), or methods that use first or second order information, for example, Gram–Schmidt orthogonalization or Broyden–Fletcher–Goldfarb–Shanno (Jongen et al. 2004; Hoshino 1971); in general, it is not advisable to rely on detecting a local optimum to stop a local search within a continuous problem.

- *The solution:* Schwefel (1993) surveys several continuous optimization methods, some of which use derivatives information and some which do not. Any of these methods can be integrated within an EMATP through a judicious balance of the amount of effort allocated to the local searchers. As in general it is not possible to run these standard methods all the way to stationary points or local optima, one relies on either truncated searches or a selective application of the local continuous optimization technique to some of the potential solutions in the population. Local search frequency and intensity parameters defining the proportion of individuals in the population that will undergo refinement and how much effort each one will be allocated are introduced. As mentioned in Nguyen et al. (2007), the values for these parameters and the decision on which of the above-mentioned refinement methods are to be used are very much problem dependent. As it is the case for discrete optimization problems, MAs applied to continuous domains also, if improperly engineered, might suffer from premature convergence through a rapid loss of diversity in the population. The use of infrequent or incomplete searches contributes to maintaining diversity but it has been shown, for example, Quang et al. (2009), that a lattice-type population structure promotes a more diverse search.

- *The consequences:* The difficulty of a priori deciding step sizes, local search probability, refinement strategy, etc., for each new problem suggests that, unless one knows beforehand the problem's characteristic features, an adaptive mechanism must be used to decide on the fly on the various choices available. Similarly to the combinatorial case (Krasnogor et al. 2002), Ong and Keane propose to use multimeme algorithms for continuous optimization problems (Ong and Keane 2004).

- *Examples:* Other examples on intelligent hybridization strategies for continuous problems could be seen in Lozano et al. (2004) and Molina et al. (2008). In Hart (2003, 2005) it is argued that many of the difficulties in guaranteeing an efficient combination of local-global search resulting in adequate convergence properties might be tackled by discretizing the decision space. These papers provide both theoretical as well as practical guidance on how to achieve this.

2.2.6 Multimeme Strategy Pattern

In previous sections, it was argued that, for the evolutionary memetic algorithm Template Pattern to be of practical use, it should be further decorated with a series of strategy patterns. Most of these strategy patterns are conceptually linked to considerations about MA's design issues. Chief among these issues is the selection for a given problem instance, at a given point in time during the search process and for a given potential solution (out of a potentially large population), which refinement strategy to use. Deciding which refinement strategy to use can be further extended to decide not only among heuristic methods but also among approximate and exact methods in both the discrete and continuous case. The following strategy pattern addresses this design issue.

Name: Multimeme Strategy Pattern (MSP) [8]
- *Problem statement:* Deciding which type of refinement or complementary search strategy to use at any given time for a given problem instance and a set of candidate solutions is a far from trivial matter. This is particularly difficult in the absence of abundant statistical information on an algorithm's behavior in relation to problem instances' features.
- *The solution:* Rather than deciding a priori on a particular search strategy (meme) by which to complement an EMATP, one can incorporate adaptive or self-adaptive techniques that would allow the MA to dynamically change the refinement strategy accordingly to how search is progressing. These methods have been called multimeme algorithms. Several relatively simple schemes have been shown to be extremely successful both for discrete and continuous optimization. For example, Krasnogor and Smith (2001) Krasnogor et al. (2002), Krasnogor and Pelta (2002), Carr et al. (2002), and Krasnogor (2004a) assign to each available local search method-meme a label which, during crossover, is inherited by an offspring from the fittest parent. Thus, the most effective local search methods are adaptively used during the search process and the fittest problem-solving memes survive over generations. As to maintain local search diversity, the mutation operator can also override an inherited local search label and assign a new one, thus ensuring that search is also done in the heuristic (memetic) space. Similarly, for continuous optimization, Ong and Keane (2004) present multimeme algorithms that employ on-line reinforcement performance information to adaptively change the memes used. Also, a roulette wheel decision mechanism can be exploited for sampling memetic space, that is, stochastically selecting (with bias) from the set of available local searchers the one to use next. An extensive discussion on how to "schedule" various local searchers-memes within an EMATP, and the levels at which search performance information can be incorporated into the decision making process behind the selection of each refinement strategy, is available in Krasnogor and Smith (2005, 2008).
- *The consequences:* The on-line adaptation of the local search choice requires some kind of bookkeeping mechanism as to ascertain the usefulness of the various refinement memes during the search. Besides the simple mechanisms for scheduling multimeme algorithms

appearing in the references mentioned above, Smith (2007) discusses other ways of assigning credit to a potentially varying set of local searchers. In principle, any reinforcement-learning (Kaelbling et al. 1996) type of mechanism could be used. A related family of algorithms, called hyperheuristics (Burke et al. 2007c; Dowsland et al. 2007), has been proposed as an attempt to raise the level of generality in problem solving by intelligently managing the application of collections of "low level" heuristics. The learning mechanisms behind hyper-heuristics could also be used within multimeme algorithms (and viceversa).

- *Examples:* Other examples of multimeme strategies appear in Jakob (2006) and Neri et al. (2007a). Multimeme strategies are closely related to adaptive and self-adaptive methods, hence the reader should also refer to the vast literature on the subject. Some pointers are Smith and Smuda (1995), Smith (2001, 2007), and Landa-Silva and Le (2008).

2.2.7 Self-Generating Strategies

Before describing a strategy pattern for self-generating strategies, the etymology of "memetic" in memetic algorithms is revisited and it is argued that there indeed is a promising nature-inspired research direction that could lead to important new algorithmic variants. Memetic theory started as such with the introduction by R. Dawkins of the concept of a meme in Dawkins (1976) and later refined in Dawkins (1982):

▶ I think that a new kind of replicator has recently emerged on this very planet. It is staring us in the face. It is still in its infancy, still drifting clumsily about in its primeval soup, but already it is achieving evolutionary change at a rate that leaves the old gene panting far behind. The new soup is the soup of human culture. We need a name for the new replicator, a noun that conveys the idea of a unit of cultural transmission, or a unit of imitation. "Mimeme" comes from a suitable Greek root, but I want a monosyllable that sounds a bit like "gene". I hope my classicist friends will forgive me if I abbreviate mimeme to meme. If it is any consolation, it could alternatively be thought of as being related to "memory", or to the French word "même". It should be pronounced to rhyme with "cream". Examples of memes are tunes, ideas, catch-phrases, clothes fashions, ways of making pots or of building arches. Just as genes propagate themselves in the gene pool by leaping from body to body via sperms or eggs, so memes propagate themselves in the meme pool by leaping from brain to brain via a process which, in the broad sense, can be called imitation.

The fact that cultural phenomena could, in certain cases, be modeled or understood as some sort of evolutionary process was not a new idea, for example, the elementary unit of cultural change and/or transmission was sometimes called *m-culture* and *i-culture* (Cloak 1975), *culture-type* (Richerson and Boyd 1978), etc. Durham (1991) provides quite a compelling case for the coevolution of genes and culture in H. Sapiens. The fundamental innovation of memetic theory is the recognition that a dual system of inheritance, by means of the existence of two distinct replicators, moulds human culture. Moreover, these two replicators interact and co-evolve shaping each other's environment. As a consequence, evolutionary changes at the gene level are expected to influence the second replicator, the memes. Symmetrically, evolutionary changes in the meme pool can have consequences for the genes. From a computer science perspective, this is a very appealing nature-inspired computation paradigm as it defines a possible two-scale learning and adaptation mechanism, one at the level of solutions to problems in exactly the same way in which EAs (or any other population-based metaheuristic) attempts to solve problems and a second level in which the solving methods themselves evolve, or more generally, change. The distinction between standard EAs and a meme-gene evolutionary process can be seen in ❷ *Fig. 6a* and ❷ *b*, respectively.

In ❷ *Fig. 6a* a hypothetical population of individuals is represented at two different points in time, generation 1 (G1) and at a later generation (G2). In the lower line, G_i for $i = 1, 2$ represents the distribution of genotypes in the population. In the upper line, P_i represents the distribution of phenotypes at the given time. Transformations, T_A, account for epigenetic phenomena, for example, interactions with the environment, in-migration and out-migrations, individual development, etc., all of them affecting the distribution of phenotypes and producing a change in the distribution of genotypes during this generation. On the other hand, transformations, T_B, account for the Mendelian principles that govern genetic inheritance and transform a distribution of genotypes G_1' into another one G_2. Evolutionary computation endeavors to concentrate on the study and assessment of many different ways the cycle depicted in ❷ *Fig. 6a* can be implemented. This evolutionary cycle implicitly assumes the existence of only one replicator, namely genes, representing solutions to hard and complex problems. On the other hand, ❷ *Fig. 6b* reflects a coevolutionary system where two replicators

◻ **Fig. 6**

In **(a)** evolutionary genetic cycle (adapted from Durham (1991) p 114), **(b)** Coevolutionary memetic-genetic cycle (adapted from Durham (1991) p 186).

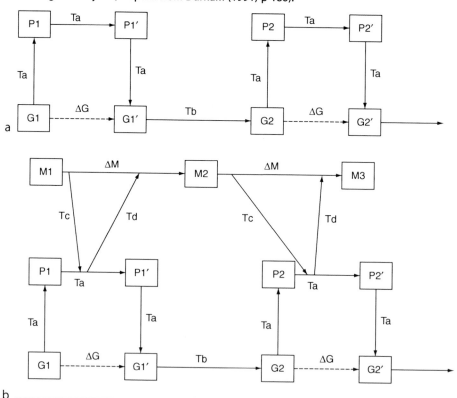

of a different nature interact. In the context of memetic algorithms, memes represent instructions to self-improve. That is, memes specify sets of rules, programs, heuristics, strategies, behaviors, etc., defining the very essence of the underlying search algorithm. Cavalli-Sforza and Feldman (1981) suggest that memetic-genetic search processes might be driven by:

▶ the balance of several evolutionary forces: (1) *mutation*, which is both purposive (innovation) and random (copy error); (2) *transmission*, which is not as inert as in biology [i.e., conveyance may also be horizontal and oblique]; (3) *cultural drift* (sampling fluctuations); (4) *cultural selection* (decisions by individuals); and (5) *natural selection* (the consequences at the level of Darwinian fitness)

❯ *Figure 6b* shows the type of transformations that are possible under a coevolutionary setting, namely, the usual transformations in terms of 'genes' and phenotypes distributions but also meme-phenotypes and memes-memes processes. There are mainly two transformations for memes that are depicted, T_C and T_D. Transformations T_C represents the various ways in which "cultural" instructions can reshape phenotypes distributions, for example, individuals learn, adopt, or imitate certain memes or modify other memes. T_D, on the other hand, reflects the changes in memetic distribution that can be expected from changes in phenotypic distributions, for example, those attributed to teaching, preaching, etc. Thus, Dawkin's memes, and their extensions (Cavalli-Sforza and Feldman 1981; Durham 1991; Gabora 1993; Blackmore 1999) as evolvable search strategies (Krasnogor 1999, 2002, 2004b; Krasnogor and Gustafson 2004; Smith 2003; Burke et al. 2007a) provide a critical link to the possibility of open-ended combinatorial and/or continuous problem solving. Next, a self-generating strategy pattern is illustrated that captures some of the solutions for implementing the above concepts.

Name: Self-Generating Strategy Pattern (SGSP) [9]

- *Problem statement:* The essential problem these strategies tackle can be succinctly put as "how to implement a search mechanism that learns how to search?" or "how to optimize through learning in a reusable manner?" (Please note that this does not refer to the well-known fact that many learning processes can be recast as optimization ones.) This problem statement is not about adaptation or self-adaptation mechanisms that mainly deal with parameter learning within fixed algorithmic templates, but rather how to infer, on the fly, new algorithmic templates that could be useful at a later time.

- *The solution:* Needless to say, this is one of the more recent cutting-edge developments in MA and the proposed solutions are still in their infancy. Self-generating MAs (sometimes called coevolutionary MAs) were first proposed as coevolutionary MAs in Krasnogor (1999) and these first ideas were explained in some more detail in Krasnogor (2002). The first experimental confirmation of the potential behind these strategies came with Krasnogor and Gustafson (2002), Smith (2002a, b), Krasnogor and Gustafson (2003, 2004), and Krasnogor (2004b). The key concept behind these new algorithms was the capturing of algorithmic building blocks through a formal grammar. Through a process not unlike grammatical evolution (O'Neill and Ryan 2003), sentences in the language induced by the grammar were evolved. These evolved sentences represented entire search methods that were self-generated on the fly while attempting to solve a given problem. The above algorithms were used for a range of difficult problems such as polynomial time solvable with low epistasis NK landscapes, polynomial time solvable with high epistasis NK landscapes all the way up to NP hard NK landscapes with both low and high epistasis as well as in randomly generated benchmarks, simplified models of proteins structure prediction and on a real-world bioinformatics application, namely, protein structure comparison.

- *The consequences:* It is evident that a self-generating strategy can only be used in a setting where the problem is either sufficiently hard, that it is worth paying the additional cost of searching an algorithmic design space or for which no good heuristics are available and one is interested in evolving them. Interestingly, after one has evolved a set of refinement strategies, these can be readily used (without need to rediscover them) in new instances or problems, under a multimeme strategy setting. Thus, the self-generating mechanisms described in the papers mentioned above provide a concrete implementation for the nature-inspired gene-meme coevolutionary process depicted in ❷ *Fig. 6b*.
- *Examples:* Recent papers exploiting GP-type learning processes for evolving problem solvers are Geiger et al. (2006), Burke et al. (2006, 2007a, b), Pappa and Freitas (2007), Bader-El-Din and Poli (2007, 2008), Fukunaga (2008), Tabacman et al. (2008), and Tay and Ho (2008). It is important to note that from this set of examples, only Fukunaga (2008) evolve algorithms, the others evolve rules or evaluation functions that guide other heuristics. Thus Krasnogor and Gustafson (2002, 2003, 2004), Smith (2002a, b), Krasnogor (2004b), and Fukunaga (2008) are the closest examples available of a coupled optimization-learning process such as the one depicted in ❷ *Fig. 6b*.

3 A Pragmatic Guide to Fitting It All Together

In previous sections, the core nine terms defining a pattern language for memetic algorithms were described. From an analysis of the literature and software available, the following patterns define our language.

At the top of the conceptual hierarchy, the memetic algorithm pattern (MAP) appears. Its aim is to provide organizational principles for effective global-local search on hard problems. MAP can be implemented through a number of available (successful) templates some of which are based on particle swarm optimization, ant colony optimization, evolutionary algorithms, etc. Each of these gives rise to new terms in the language, namely, the EMATP (for the evolutionary memetic algorithm algorithmic template), the ACOMATP (for the ant colony memetic algorithm template), the SAMATP (for the simulated annealing memetic algorithm template), etc. Each of these nature-inspired paradigm templates have their own "idiosyncrasies." However, at the time of implementing them as memetic algorithms, the following common features are captured in new design patterns, namely, the refinement strategy pattern (RSP), with focus on heuristic and local search methods, the exact and approximate hybridization strategy pattern (EAHSP) that defines ways in which expensive exact (or approximate) algorithms are integrated with a memetic algorithms template pattern under any nature-inspired realization. Two other terms in our pattern language are the population diversity handling strategy pattern (PDHSP), dealing with ways to preserve – in the face of aggressive refinement strategies – global diversity and the surrogate objective function strategy pattern (SOFSP) whose role is to define methods for dealing within a memetic algorithm with very expensive/noisy/undetermined objective functions. Finally, the pattern language presented also includes terms for defining the "generation" of memetic algorithm one is trying to implement. A memetic algorithm of the first generation in which only one refinement strategy is used, is called a simple strategy pattern (SSP) and that is the canonical MAs (e.g., see ❷ *Fig. 3*). The multimeme strategy pattern (MSP) and the self-generating strategy pattern (SGSP), which provide methods for utilizing several refinement strategies simultaneously and for synthesizing new refinement strategies on the fly respectively, are second and third generation MAs. Thus the resulting pattern language has at least the following terms: {MAP, EMATP, PSOMATP, ACOMATP, SAMATP, ... , AISMATP, RSP, EASHP, CPRSP, PDHSP, SOFSP, SSP, MSP, SGSP}.

These patterns are functionally related as depicted in ❷ *Fig. 7* that represents the series of design decisions involved in the implementation of memetic algorithms. One could choose to implement an MA by following, for example, an ant colony optimization template or – more often – an evolutionary algorithm template or any of the other template patterns. Each one of these nature-inspired template patterns will have their own algorithmic peculiarities, for example, crossovers and mutations for EAs, pheromones updating rules for ACOs, etc., but all of them when taken as a memetic algorithm will need to address the issues of population diversity, refinement strategies, exact algorithms, surrogate objective functions, single meme versus multimeme versus self-generation, etc. In order to instantiate code for any of these design patterns, one can simply look at the description of the design pattern provided in this chapter and refer to any of the literature references given within the pattern description. Thus, ❷ *Fig. 7* provides a reference handbook for MAs engineering.

◰ **Fig. 7**

Integrative view of the design patterns for the memetic algorithms patterns language. A path through the graph represents a series of design decision that an algorithms engineer would need to take while implementing an MA. Each design pattern provides solutions to specific problems the pattern addresses. By indexing through the patterns' acronyms into the main text in this chapter, the reader can have access to examples from the literature where the issues have been solved to satisfaction.

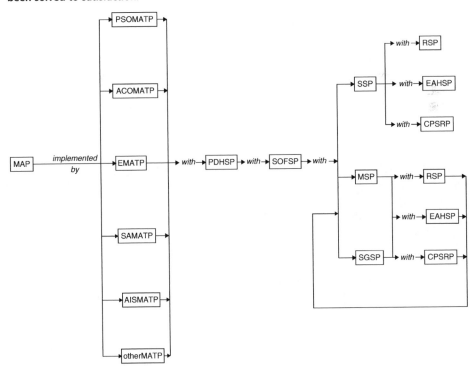

4 Brief Notes on Theoretical Aspects

Memetic algorithms' notorious successes in practical applications notwithstanding, no systematic effort has been made to try to understand them from a theoretical viewpoint. Memetic algorithms do not adhere to pure EA theory, not even those applied to continuous problems for which convergence analysis are somewhat easier. Moreover, MAs are also much more complex than other heuristic methods such as greedy-like algorithms for which precise theoretical knowledge is possible. Thus, while the body of empirical knowledge is steadily growing, their theoretical underpinnings remain poorly understood. A collection of disjoint, but only partial, analyses were published in, e.g., Krasnogor and Smith (2005), Sudholt (2007) for MAs applied to combinatorial optimization and Hart (2003, 2005) for continuous optimization. Two emerging methodologies to assess heuristic performance that might in the future lead to important insights for MAs are PLS-theory (Johnson et al. 1988; Yannakakis 1997) and Domination analysis (Glover and Punnen 1997; Gutin and Yeo 2006). Recent results based on both of these theoretical approaches to heuristic analysis are compiled in Michiels et al. (2007). For an organizational perspective of the field, the reader is also referred to the papers by Krasnogor and Smith (2005) and Ong et al. (2006) in which several best practices are described and research lines proposed.

5 The Future: Toward Nature-Inspired Living Software Systems

Memetic algorithms have gone through three major transitions in their evolution. The first transition was coincidental with their introduction and the bulk of the research and applications produced was based on the use of, mainly, an evolutionary algorithm with one specialized local searcher. The second transition brought the concept of multimeme algorithms that expanded the repertoire of both the key skeleton, that is, EAs were no longer the only global search template used but rather any global-(population) based search could be employed, and gave way to the constraint on having a single local searcher. The third major transition is taking place right now and involves the incorporation of explicit learning mechanisms that can, while optimizing, "distill" new refinement operators. These new operators, in turn, can be reused in a later application of the MA, thus amortizing the cost invested in discovering them. This third transition is also producing more frequent and more confident use of exact methods in tandem or hybridized with MAs. This third major transition is likely to produce very exciting results over the next few years and indeed much research remains to be done in the theory and practice of memetic algorithms.

The question that needs to be briefly addressed is "what's next?," "where do memetic algorithms go from here?"

Memetic algorithms in particular and optimization algorithms in general are but one class of the millions of different software systems humans create to solve problems. The vast majority – if not all – of software systems development follows, to various degrees of formalism and details, a process of requirement collection, specification, design, implementation, testing, deployment, and long term maintenance. There are, indeed, a variety of software engineering techniques that advocate the concurrent realization or elimination altogether of (some of) these steps as a way to achieve a more agile software development. The unrelenting scaling up of available computational, storage, and communication power is bringing closer the day when software will cease to be created following such a process and will, instead, be "planted/seeded" within a production environment (e.g., a factory, university, bank, etc.) and then "grown" from the bottom-up into a fully developed functional software system. Although within the remit of optimization problems, self-generating techniques such as Krasnogor and Gustafson (2002), Smith (2002a, b), Krasnogor and Gustafson (2003, 2004), Krasnogor (2004b), Burke et al. (2006, 2007a, b), Pappa and Freitas (2007), Bader-El-Den and Poli (2008), Fukunaga (2008), and Tabacman et al. (2008) are moving incrementally in this direction, perhaps an even more radical scenario can be imagined. Consider, for example, a given production environment, a factory, where a series of optimization problems must be solved routinely, such as personnel rostering (Π_{pr}), job-shop schedules (Π_{jss}), path planning (Π_{pp}), inventory management (Π_{in}), etc. Each of these problems, during long periods of time (weeks, months, years, and even decades) give rise to large collections of problem instances (I_{Π_j} with $j \in \{\Pi_{pr}, \Pi_{jss}, \Pi_{pp}, \Pi_{im}\}$). Current practice dictates that for each Π, a tailor-made solver is used that in practice works well for the associated $I(\Pi)$. A key observation worth making is that the distribution of instances for each of the problems arising from a given factory (or bank, university, supermarket, airport, etc.) is not random but rather strongly dependent on physical, legal, commercial, and societal constraints operating over the factory. That is, the instances derived from each one of these problems, for a given factory, will have strong common features and similarities. This, in turn, provides a tremendous opportunity for a more organic and autonomic problem solving. What we have in mind can best be described as a new discipline of "pluripotential problem solvers (PPS)." The idea behind pluripotential problem solvers is that a minimum set of self-generating features would be hardcoded into a given pluripotential solver cell (PSC), then this software cell would be immersed (in multiple copies) into the factory environment and bombarded with problem instances over a period of time. After a while, some of the PSC would differentiate along problem instance lines (very much as a stem cell differentiates along cell type lines). Once differentiated, a mature PSC (or solver cell for simplicity) would only be sensitive to new instances of the same problem. Furthermore, as time progresses, the solver cell would be able to incorporate new problem-specific and instance-distribution-specific problem solving capabilities (e.g., perhaps using some refinement of the technologies described in the self-generating strategy pattern) thus further specializing. In the end, the set of initially planted PSC would have been developing along very specialized lines within a specific production environment. Needless to say, if one were to transplant a specialized solver cell operating on, let's say, Π_{jss} in factory 1 as a solver for Π_{jss} in factory 2, the performance of the solver will be limited because, unless the distributions of instances generated in both factories are very similar, the specialized solver cell would not be able to cope with the new ones as it would have lost its capacity to specialize. What one would rather do is seed again in factory 2 a new set of PSC and let them mature within their production environment. The scenario described above is predicated under two key assumptions. The first one is that one does not expect to have a cutting-edge solver "out of the tin" (for which arguably one would be willing to pay considerable amounts of money) but rather than one would pay less (or nothing) for a PSC and be willing to wait until the software grows into a very specialized, niche-specific, cutting-edge solver. The second assumption is that computational power, communication bandwidth, and storage capacity will keep increasing while the price per floating point operation and gigabytes transmitted/stored will go on decreasing. The advent of pluripotential problem solvers will require a concerted, creative, and unorthodox research on the amalgamation of optimization algorithms, data mining, and machine learning as well as more in-depth studies of nature-inspired principles that – complementing evolutionary approaches – so far have not been tapped for optimization, namely, the creative power of self-assembly and self-organization (Krasnogor et al. 2008).

6 Conclusions

This chapter provided an unorthodox introduction to memetic algorithms. It analyzed memetic algorithms as they appear in the literature and are used in practice, rather than conforming to a top down definition of what a memetic algorithm is and then, based on a definition, prescribe how to implement them. The approach taken here allows for a pragmatic view of the field and provides a recipe for creating memetic algorithms by resorting to a shared pattern language. This emerging pattern language for memetic algorithms serves as a catalog of parts (or concerns) that an algorithms engineer might want to resort to for solving specific design issues arising from the need to interlink global search and local search for hard complex problems. Due to space and time constraints, design patterns for multiobjective (Burke and Landa-Silva 2004; Li and Landa-Silva 2008) problems, parallelization strategies (Alba and Tomassini 2002; Nebro et al. 2007), etc., have not been considered, and how these important new components of the pattern language would interact with the patterns described here is also not discussed. It is hoped that the pattern language for MAs presented in this chapter will, in the future, be expanded and refined by researchers and practitioners alike.

Acknowledgments

The author would like to acknowledge the many friends and colleagues with whom he has collaborated over the years. Their ideas, scientific rigor and enthusiasm for memetic algorithms has been a continuous source of inspiration and challenges. The author would also like to thank Jonathan Blakes and James Smaldon for their valuable comments during the preparation of this paper. The author wishes to acknowledge funding from the EPSRC for projects EP/D061571/1 and EP/C523385/1. Finally, the editors of this book are thanked for giving the author an opportunity to contribute.

References

Alba E, Tomassini M (2002) Parallelism and evolutionary algorithms. IEEE Trans Evolut Comput 6:443–462

Alexander C, Ishikawa S, Silverstein M, Jacobson M, Fiksdahl-King I, Angel S (1977) A pattern language - towns, buildings, construction. Oxford University Press, New York

Armour P (2007) The conservation of uncertainty, exploring different units for measuring software. Commun ACM 50:25–28

Bacardit J, Krasnogor N (2009) Performance and efficiency of memetic Pittsburgh learning classifier systems. Evolut Comput 17(3)

Bader-El-Den M, Poli R (2008) Evolving heuristics with genetic programming. In: GECCO '08: Proceedings of the 10th annual conference on genetic and evolutionary computation, ACM, New York, pp 601–602, doi: http://doi.acm.org/10.1145/1389095.1389212

Bader-El-Din MB, Poli R (2007) Generating SAT local-search heuristics using a GP hyper-heuristic framework. In: LNCS 4926. Proceedings of the 8th international conference on artifcial evolution, Honolulu, pp 37–49

Berrut JP, Trefethen L (2004) Barycentric Lagrange interpolation. SIAM Rev 46(3):501–517

Bhattacharya M (2007) Surrogate based EA for expensive optimization problems. In: Proceedings for the IEEE congress on evolutionary computation (CEC), Singapore, pp 3847–3854

Blackmore S (1999) The meme machine. Oxford University Press, Oxford

Branke J (1998) Creating robust solutions by means of an evolutionary algorithm. In: Parallel problem solving from nature PPSN V, Amsterdam, pp 119–128

Branke J (2001) Reducing the sampling variance when searching for robust solutions. In: Spector L, et al. (eds) Proceedings of the genetic and evolutionary computation conference, Kluwer, San Francisco, CA, pp 235–242

Branke J (2002) Evolutionary Optimization in Dynamic Environments. Kluwer, Boston, MA

Bull L (1999) On model-based evolutionary computation. J Soft Comput Fus Found Methodol Appl 3: 76–82

Bull L, Holland O, Blackmore S (2000) On meme–gene coevolution. Artif Life 6:227–235

Burke E, Landa-Silva J (2004) The design of memetic algorithms for scheduling and timetabling problems. In: Hart W, Krasnogor N, Smith J (eds) Recent advances in memetic algorithms. Springer, pp 289–312

Burke E, Newall J, Weare R (1996) A memetic algorithm for university exam timetabling. In: Burke E, Ross P (eds) The practice and theory of automated timetabling, Lecture notes in computer science, vol 1153. Springer, Berlin, pp 241–250

Burke E, Newall J, Weare R (1998) Initialization strategies and diversity in evolutionary timetabling. Evolut Comput 6:81–103

Burke E, Bykov Y, Newall J, Petrovic S (2004) A time-predefined local search approach to exam timetabling problems. IIE Trans 36:509–528

Burke E, Gustafson S, Kendall G, Krasnogor N (2002) Advanced population diversity measures in genetic programming. In: Guervos JM, Adamidis P, Beyer H, Fernandez-Villacanas J, Schwefel H (eds) 7th International conference parallel problem solving from nature, PPSN, Springer, Granada, Spain, Lecture notes in computer science, vol 2439. Springer, New York, pp 341–350

Burke E, Gustafson S, Kendall G, Krasnogor N (2003) Is increased diversity beneficial in genetic programming: an analysis of the effects on fitness. In: IEEE congress on evolutionary computation, CEC, IEEE, Canberra, pp 1398–1405

Burke E, Hyde M, Kendall G (2006) Evolving bin packing heuristics with genetic programming. In: Runarsson T, Beyer HG, Burke E, Merelo-Guervos J, Whitley D, Yao X (eds) Proceedings of the 9th International conference on parallel problem solving from nature (PPSN 2006), LNCS 4193. Springer, pp 860–869

Burke E, Hyde M, Kendall G, Woodward J (2007a) Automatic heuristic generation with genetic programming: evolving a jack-of-all-trades or a master of one. In: Proceedings of the genetic and evolutionary computation conference (GECCO 2007), ACM, London, pp 1559–1565

Burke E, Hyde M, Kendall G, Woodward J (2007b) Scalability of evolved on line bin packing heuristics. In: Proceedings of the congress on evolutionary computation (CEC 2007). Singapore, pp 2530–2537

Burke E, McCollum B, Meisels A, Petrovic S, Qu R (2007c) A graph-based hyper-heuristic for timetabling problems. Eur J Oper Res 176:177–192

Caponio A, Cascella G, Neri F, Salvatore N, Sumner M (2007) A fast adaptive memetic algorithm for on-line and off-line control design of PMSM drives. IEEE Trans Syst Man Cybern Part B 37:28–41

Carr R, Hart W, Krasnogor N, Burke E, Hirst J, Smith J (2002) Alignment of protein structures with a memetic evolutionary algorithm. In: Langdon W, Cantu-Paz E, Mathias K, Roy R, Davis D, Poli R, Balakrishnan K, Honavar V, Rudolph G, Wegener J, Bull L, Potter M, Schultz A, Miller J, Burke E, Jonoska N (eds) GECCO-2002: Proceedings of the genetic and evolutionary computation conference, Morgan Kaufmann, San Mateo, CA

Cavalli-Sforza L, Feldman M (1981) Cultural transmission and evolution: a quantitative approach. Princeton University Press, Princeton, NJ

Cheng R, Gen M (1997) Parallel machine scheduling problems using memetic algorithms. Comput Ind Eng 33(3–4):761–764

Cloak F (1975) Is a cultural ethology possible. Hum Ecol 3:161–182

Cooper J (2000) Java design patterns: a tutorial. Addison-Wesley, Boston, MA

Cordon O, Herrera F, Stutzle T (2002) A review on the ant colony optimization metaheuristic: basis, models and new trends. Mathware Soft Comput 9:141–175

Cutello V, Krasnogor N, Nicosia G, Pavone M (2007) Immune algorithm versus differential evolution: a comparative case study using high dimensional function optimization. In: International conference on adaptive and natural computing algorithms, ICANNGA 2007. LNCS, Springer, Berlin, pp 93–101

Dawkins R (1976) The selfish gene. Oxford University Press, New York

Dawkins R (1982) The extended phenotype. Freeman, Oxford

Dorigo M, Gambardela L (1997) Ant colony system: a cooperative learning approach to the travelling salesman problem. IEEE Trans Evolut Comput 1(1):53–66

Dowsland K, Soubeiga E, Burke EK (2007) A simulated annealing hyper-heuristic for determining shipper sizes. Eur J Oper Res 179:759–774

Dueck G (1993) New optimisation heuristics. the Great Deluge algorithm and record-to-record travel. J Comput Phys 104:86–92

Duque T, Goldberg D, Sastry K (2008) Improving the efficiency of the extended compact genetic algorithm. In: GECCO '08: Proceedings of the 10th annual conference on genetic and evolutionary computation, ACM, New York, pp 467–468. doi: http://doi.acm.org/10.1145/1389095.1389181

Durham W (1991) Coevolution: genes, culture and human diversity. Stanford University Press, Stanford, CA

Fleurent C, Ferland J (1997) Genetic and hybrid algorithms for graph coloring. Ann Oper Res 63:437–461

Fukunaga A (2008) Automated discovery of local search heuristics for satisfiability testing. Evolut Comput 16(1):31–61, doi: 10.1162/evco.2008.16.1.31, URL http://www.mitpressjournals.org/doi/abs/10.1162/evco.2008.16.%1.31, pMID: 18386995, http://www.mitpressjournals.org/doi/pdf/10.1162/evco.2008.16.1.31

Gabora L (1993) Meme and variations: a computational model of cultural evolution. In: L Nadel, Stein D (eds) 1993 Lectures in complex systems. Addison-Wesley, Boston, MA, pp 471–494

Gallardo J, Cotta C, Fernandez A (2007) On the hybridization of memetic algorithms with branch-and-bound techniques. Syst Man Cybern Part B IEEE Trans 37(1):77–83. doi: 10.1109/TSMCB.2006.883266

Gamma E, Helm R, Johnson R, Vlissides J (1995) Design patterns, elements of reusable object-oriented software. Addison-Wesley, Reading, MA

Geiger CD, Uzsoy R, Aytug H (2006) Rapid modeling and discovery of priority dispatching rules: an autonomous learning approach. J Scheduling 9(1):7–34

Glover F, Punnen A (1997) The traveling salesman problem: new solvable cases and linkages with the development of approximation algorithms. J Oper Res Soc 48:502–510

Gutin G, Yeo A (2006) Domination analysis of combinatorial optimization algorithms and problems. In: Graph theory, combinatorics and algorithms, operations research/computer science interfaces, vol 34. Springer, New York, pp 145–171

Gutin G, Karapetyan D, Krasnogor N (2007) Memetic algorithm for the generalized asymmetric traveling salesman problem. In: Pavone M, Nicosia G, Pelta D, Krasnogor N (eds) Proceedings of the 2007 workshop on nature inspired cooperative strategies for optimisation. Studies in computational intelligence. Springer, Berlin

Hansen P, Mladenovic N (1998) Variable neighborhood search for the *p*-median. Location Sci 5(4):207–226

Hansen P, Mladenovic N (2001) Variable neighborhood search: principles and applications. Eur J Oper Res (130):449–467

Hart W (2003) Locally-adaptive and memetic evolutionary pattern search algorithms. Evolut Comput 11:29–52

Hart W (2005) Rethinking the design of real-coded evolutionary algorithms: making discrete choices in continuous search domains. J Soft Comput Fus Found Methodol Appl 9:225–235

Hart W, Krasnogor N, Smith J (2004) Recent advances in memetic algorithms, studies in fuzziness and soft computing, vol 166, Springer, Berlin/Heidelberg/New York, chap Memetic Evolutionary Algorithms, pp 3–27

Hart WE (1994) Adaptive global optimization with local search. Ph.D. thesis, University of California, San Diego, CA

He L, Mort N (2000) Hybrid genetic algorithms for telecommunications network back-up routing. BT Technol J 18(4):42–50

Hinton G, Nowlan S (1987) How learning can guide evolution. Complex Syst 1:495–502

Holland JH (1976) Adaptation in natural and artificial systems. The University of Michigan Press, New York

Hoshino S (1971) On Davies, Swann and Campey minimisation process. Comput J 14:426

Houck C, Joines J, Kay M, Wilson J (1997) Empirical investigation of the benefits of partial lamarckianism. Evolut Comput 5(1):31–60

Ishibuchi H, Kaige S (2004) Implementation of simple multiobjective memetic algorithms and its application to knapsack problems. Int J Hybrid Intell Syst 1(1–2):22–35

Jakob W (2006) Towards an adaptive multimeme algorithm for parameter optimisation suiting the engineers needs. In: Runarsson TP, et al. (eds) Proceedings of the IX parallel problem solving from nature conference (PPSN IX). Lecture notes in computer science 4193. Springer, Berlin, pp 132–141

Jaszkiewicz A (2002) Genetic local search for multiobjective combinatorial optimization. Eur J Oper Res 137

Jin Y (2005) A comprehensive survey of fitness approximation in evolutionary computation. Soft Comput Fus Found Methodol Appl 9:3–12

Johnson D, Papadimitriou C, Yannakakis M (1988) How easy is local search. J Comput Syst Sci 37:79–100

Jongen H, Meer K, Triesch E (2004) Optimization theory. Springer, New York

Kaelbling L, Littman M, Moore A (1996) Reinforcement learning: a survey. J Artif Intell Res 4:237–285

Kallel L, Naudts B, Reeves C (2001) Properties of fitness functions and search landscapes. In: Kallel L, Naudts B, Rogers A (eds) Theoretical aspects of evolutionary computing. Springer, Berlin, pp 175–206

Kirkpatrick S, Gelatt C, Vecchi MP (1983) Optimization by simulated annealing. Science 220:671–680

Kononova A, Hughes K, Pourkashanian M, Ingham D (2007) Fitness diversity based adaptive memetic algorithm for solving inverse problems of chemical kinetics. In: IEEE Congress on Evolutionary Computation (CEC), IEEE. Singapore, pp 2366–2373

Krasnogor N (1999) Coevolution of genes and memes in memetic algorithms. In: Wu A (ed) Proceedings of the 1999 genetic and evolutionary computation conference, Graduate students workshop program, San Francisco, CA, http://www.cs.nott.ac.uk/ nxk/ PAPERS/memetic.pdf, (poster)

Krasnogor N (2002) Studies on the theory and design space of memetic algorithms. Ph.D. thesis, University of the West of England, Bristol, http://www.cs. nott.ac.uk/ nxk/PAPERS/thesis.pdf

Krasnogor N (2004a) Recent advances in memetic algorithms, Studies in fuzziness and soft computing, vol 166, Springer, Berlin, Heidelberg New York, chap Towards robust memetic algorithms, pp 185–207

Krasnogor N (2004b) Self-generating metaheuristics in bioinformatics: the protein structure comparison case. Genet Programming Evol Mach 5(2):181–201

Krasnogor N, Gustafson S (2002) Toward truly "memetic" memetic algorithms: discussion and proof of concepts. In: Corne D, Fogel G, Hart W, Knowles J, Krasnogor N, Roy R, Smith JE, Tiwari A (eds) Advances in nature-inspired computation: the PPSN VII workshops, PEDAL (Parallel, Emergent and Distributed Architectures Lab). University of Reading, UK ISBN 0-9543481-0-9

Krasnogor N, Gustafson S (2003) The local searcher as a supplier of building blocks in self-generating memetic algorithms. In: Hart JS WE, Krasnogor N (eds) Fourth international workshop on memetic algorithms (WOMA4), In GECCO 2003 workshop proceedings. Chicago, IL

Krasnogor N, Gustafson S (2004) A study on the use of "self-generation" in memetic algorithms. Nat Comput 3(1):53–76

Krasnogor N, Pelta D (2002) Fuzzy memes in multimeme algorithms: a fuzzy-evolutionary hybrid. In: Verdegay J (ed) Fuzzy sets based heuristics for optimization. Springer, Berlin

Krasnogor N, Smith J (2000) A memetic algorithm with self-adaptive local search: TSP as a case study. In: Whitley D, Goldberg D, Cantu-Paz E, Spector L, Parmee I, Beyer HG (eds) GECCO 2000: Proceedings of the 2000 genetic and evolutionary computation conference, Morgan Kaufmann, San Francisco, CA

Krasnogor N, Smith J (2001) Emergence of profitable search strategies based on a simple inheritance mechanism. In: Spector L, Goodman E, Wu A,

Langdon W, Voigt H, Gen M, Sen S, Dorigo M, Pezeshj S, Garzon M, Burke E (eds) GECCO 2001: Proceedings of the 2001 genetic and evolutionary computation conference, Morgan Kaufmann, San Francisco, CA

Krasnogor N, Smith J (2005) A tutorial for competent memetic algorithms: Model, taxonomy and design issues. IEEE Trans Evolut Algorithms 9(5):474–488

Krasnogor N, Smith J (2008) Memetic algorithms: the polynomial local search complexity theory perspective. J Math Model Algorithms 7:3–24

Krasnogor N, Blackburne B, Hirst J, Burke E (2002) Multimeme algorithms for protein structure prediction. In: Guervos JM, Adamidis P, Beyer H, Fernandez-Villacanas J, Schwefel H (eds) 7th International conference parallel problem solving from nature, PPSN, Springer, Berlin/Heidelberg, Granada, Spain, Lecture notes in computer science, vol 2439. Springer, pp 769–778

Krasnogor N, Gustafson S, Pelta D, Verdegay J (eds) (2008) Systems self-assembly: multidisciplinary snapshots, Studies in multidisciplinarity, vol 5. Elsevier, Spain

Kretwski M (2008) A memetic algorithm for global induction of decision trees. In: Proceedings of SOFSEM: theory and practice of computer science. Lecture notes in computer science, Springer, New York, pp 531–540

Kuhn T (1962) The structure of scientific revolution. University of Chicago Press, Chicago, IL

Landa-Silva D, Le KN (2008) A simple evolutionary algorithm with self-adaptation for multi-objective optimisation. Springer, Berlin, pp 133–155

Landa Silva J, Burke EK (2004) Using diversity to guide the search in multi-objective optimization. World Scientific, Singapore, pp 727–751

Lee Z, Lee C (2005) A hybrid search algorithm with heuristics for resource allocation problem. Inf Sci 173:155–167

Li H, Landa-Silva D (2008) Evolutionary multi-objective simulated annealing with adaptive and competitive search direction. In: Proceedings of the 2008 IEEE congress on evolutionary computation (CEC 2008). IEEE Press, Piscataway, NJ, pp 3310–3317

Liu BF, Chen HM, Chen JH, Hwang SF, Ho SY (2005) Meswarm: memetic particle swarm optimization. In: GECCO '05: Proceedings of the 2005 conference on Genetic and evolutionary computation, ACM, New York, pp 267–268. doi: http://doi.acm.org/ 10.1145/1068009.1068049

Liu D, Tan KC, Goh CK, Ho WK (2007) A multiobjective memetic algorithm based on particle swarm optimization. Syst Man Cybern Part B IEEE Trans 37 (1):42–50. doi: 10.1109/TSMCB.2006.883270

Llora X, Sastry K, Yu T, Goldberg D (2007) Do not match, inherit: fitness surrogates for genetics-based machine learning techniques. In: Proceedings of the 9th annual conference on genetic and evolutionary computation. ACM, San Mateo, CA, pp 1798–1805

Lozano M, Herrera F, Krasnogor N, Molina D (2004) Real-coded memetic algorithms with crossover hill-climbing. Evolut Comput 12(3):273–302

Mayley G (1996) Landscapes, learning costs and genetic assimilation. Evolut Comput 4(3):213–234

McCulloch W, Pitts W (1943) A logical calculus of the ideas immanent in nervous system. Bull Math Biophys 5:115–133

Merz P (2003) The compact memetic algorithm. In: Proceedings of the IV International workshop on memetic algorithms (WOMA IV). GECCO 2003, Chicago, IL. http://w210.ub.uni-tuebingen.de/portal/woma4/

Merz P, Freisleben B (1999) Fitness landscapes and memetic algorithm design. In: New ideas in optimization, McGraw-Hill, Maidenhead, pp 245–260

Mezmaz M, Melab N, Talbi EG (2007) Combining metaheuristics and exact methods for solving exactly multi-objective problems on the grid. J Math Model Algorithms 6:393–409

Michiels W, Aarts E, Korst J (2007) Theoretical aspects of local search. Monographs in theoretical computer science. Springer, New York

Molina D, Lozano M, Garcia-Martines C, Herrera F (2008) Memetic algorithm for intense local search methods using local search chains. In: Hybrid metaheuristics: 5th international workshop. Lecture notes in computer science. Springer, Berlin/Heidelberg/New York, pp 58–71

Morris GM, Goodsell DS, Halliday RS, Huey R, Hart WE, Belew RK, Olson AJ (1998) Automated docking using a lamarkian genetic algorithm and an empirical binding free energy function. J Comp Chem 14:1639–1662

Moscato P (1989) On evolution, search, optimization, genetic algorithms and martial arts: towards memetic algorithms. Tech. Rep. Caltech Concurrent Computation Program, Report. 826, California Institute of Technology, Pasadena, CA

Nebro A, Alba E, Luna F (2007) Multi-objective optimization using grid computing. Soft Comput 11:531–540

Nelder J, Mead R (1965) A simplex method for function minimization. Comput J 7(4):308–313. doi:10.1093/comjnl/7.4.308

Neri F, Jari T, Cascella G, Ong Y (2007a) An adaptive multimeme algorithm for designing HIV multidrug therapies. IEEE/ACM Trans Comput Biol Bioinformatics 4(2):264–278

Neri F, Tirronen V, Karkkainen T, Rossi T (2007b) Fitness diversity based adaptation in multimeme algorithms: a comparative study. In: Proceedings of the IEEE congress on evolutionary computation. IEEE, Singapore, pp 2374–2381

Nguyen QH, Ong YS, Lim MH, Krasnogor N (2007) A comprehensive study on the design issues of memetic algorithm. In: Proceedings of the 2007 IEEE congress on evolutionary computation. IEEE, Singapore, pp 2390–2397

Niesse J, Mayne H (Sep. 15, 1996) Global geometry optimization of atomic clusters using a modified genetic algorithm in space-fixed coordinates. J Chem Phys 105(11):4700–4706

O'Neill M, Ryan C (2003) Grammatical evolution: evolutionary automatic programming in an arbitrary language. Genetic Programming, vol 4. Springer, Essex

Ong Y, Keane A (2004) Meta-lamarckian learning in memetic algorithms. IEEE Trans Evolut Comput 8:99–110

Ong Y, Lim M, Zhu N, Wong KW (2006) Classification of adaptive memetic algorithms: a comparative study. IEEE Trans Syst Man Cybern Part B 36:141–152

Ong Y, Lum K, Nair P (2008) Hybrid evolutionary algorithm with hermite radial basis function interpolants for computationally expensive adjoint solvers. Comput Opt Appl 39:97–119

Paenke I, Jin J (2006) Efficient search for robust solutions by means of evolutionary algorithms and fitness approximation. IEEE Trans Evolut Comput 10:405–420

Pappa G, Freitas A (2007) Discovering new rule induction algorithms with grammar-based genetic programming. In: Maimon O, Rokach L (eds) Soft computing for knowledge discovery and data mining. Springer, New York, pp 133–152

Petalas Y, Parsopoulos K, Vrahatis M (2007) Memetic particle swarm optimisation. Ann Oper Res 156:99–127

Pirkwieser S, Raidl GR, Puchinger J (2008) A Lagrangian decomposition/evolutionary algorithm hybrid for the knapsack constrained maximum spanning tree problem. In: Cotta C, van Hemert J (eds) Recent advances in evolutionary computation for combinatorial optimization. Springer, Valencia, pp 69–85

Quang Q, Ong Y, Lim M, Krasnogor N (2009) Adaptive cellular memetic algorithm. Evolut Comput 17(2):231–256

Raidl GR, Puchinger J (2008) Combining (integer) linear programming techniques and metaheuristics for combinatorial optimization. In: Blum C, et al. (eds) Hybrid Metaheuristics - an emergent approach for combinatorial optimization. Springer, Berlin/Heidelberg/New York, pp 31–62

Reeves C (1996) Hybrid genetic algorithms for bin-packing and related problems. Ann Oper Res 63:371–396

Richerson P, Boyd R (1978) A dual inheritance model of the human evolutionary process: I. Basic postulates and a simple model. J Soc Biol Struct I:127–154

Romero-Campero F, Cao H, Camara M, Krasnogor N (2008) Structure and parameter estimation for cell systems biology models. In: Keijzer, M et al. (eds) Proceedings of the genetic and evolutionary computation conference (GECCO-2008), ACM, Seattle, WA, pp 331–338

Sacks J, Welch W, Mitchell T, Wynn H (1989) Design and analysis of computer experiments. Stat Sci 4:409–435

Schwefel H (1993) Evolution and optimum seeking: the sixth generation. Wiley, New York, NY

Siepmann P, Martin C, Vancea I, Moriarty P, Krasnogor N (2007) A genetic algorithm approach to probing the evolution of self-organised nanostructured systems. Nano Lett 7(7):1985–1990

Smith J (2001) Modelling GAs with self adaptive mutation rates. In: GECCO-2001: Proceedings of the genetic and evolutionary computation conference. Morgan Kaufmann, San Francisco, CA

Smith J (2002a) Co-evolution of memetic algorithms: Initial results. In: Merelo, Adamitis, Beyer, Fernandez-Villacans, Schwefel (eds) Parallel problem solving from nature – PPSN VII, LNCS 2439. Springer, Spain, pp 537–548

Smith J (2002b) Co-evolution of memetic algorithms for protein structure prediction. In: Hart K, Smith J (eds) Proceedings of the third international workshop on memetic algorithms, New York

Smith J (2003) Co-evolving memetic algorithms: A learning approach to robust scalable optimisation. In: Proceedings of the 2003 congress on evolutionary computation. Canberra, pp 498–505

Smith JE (2007) Credit assignment in adaptive memetic algorithms. In: GECCO '07: Proceedings of the 9th annual conference on genetic and evolutionary computation, ACM, New York, pp 1412–1419. doi: http://doi.acm.org/10.1145/1276958.1277219

Smith R, Smuda E (1995) Adaptively resizing populations: algorithms, analysis and first results. Complex Syst 1(9):47–72

Sorensen K, Sevaux M (2006) MA:PM: memetic algorithms with population management. Comput Oper Res 33:1214–1225

Sudholt D (2007) Memetic algorithms with variable-depth search to overcome local optima. In: Proceedings of the 2007 conference on genetic and evolutionary computation (GECCO), ACM, New York, pp 787–794

Tabacman M, Bacardit J, Loiseau I, Krasnogor N (2008) Learning classifier systems in optimisation problems: a case study on fractal travelling salesman problems. In: Proceedings of the international workshop on learning classifier systems, Lecture notes in computer science, Springer, New York, URL http://www.cs.nott.ac.uk/ nxk/PAPERS/maxi.pdf

Tang M, Yao X (2007) A memetic algorithm for VLSI floorplanning. Syst Man Cybern Part B IEEE Trans 37(1):62–69. doi: 10.1109/TSMCB.2006.883268

Tay JC, Ho NB (2008) Evolving dispatching rules using genetic programming for solving multi-objective flexible job-shop problems. Comput Ind Eng 54(3):453–473

Turney P (1996) How to shift bias: lessons from the Baldwin effect. Evolut Comput 4(3):271–295

Vavak F, Fogarty T (1996) Comparison of steady state and generational genetic algorithms for use in non-stationary environments. In: Proceedings of the 1996 IEEE conference on evolutionary computation, Japan, pp 192–195

Wang H, Wang D, Yang S (2009) A memetic algorithm with adaptive hill climbing strategy for dynamic optimization problems. Soft Comput 13(8–9)

Whitley L, Gruau F (1993) Adding learning to the cellular development of neural networks: evolution and the Baldwin effect. Evolut Comput 1:213–233

Whitley L, Gordon S, Mathias K (1994) Lamarkian evolution, the Baldwin effect, and function optimisation. In: Davidor Y, Schwefel HP, Männer R (eds) PPSN, Lecture notes in computer science, vol 866. Springer, Berlin, pp 6–15

Wolpert D, Macready W (1997) No free lunch theorems for optimisation. IEEE Trans Evolut Comput 1(1):67–82

Yanga J, Suna L, Leeb H, Qiand Y, Liang Y (2008) Clonal selection based memetic algorithm for job shop scheduling problems. J Bionic Eng 5:111–119

Yannakakis M (1997) Computational complexity. In: Aarts E, Lenstra J (eds) Local search in combinatorial optimization. Wiley, New York, pp 19–55

Zhou Z (2004) Hierarchical surrogate-assisted evolutionary optimization framework. In: Congress on evolutionary computation, 2004. CEC 2004. Portland, pp 1586–1593

Zhou Z, Ong Y, Lim M, Lee B (2007) Memetic algorithm using multi-surrogates for computationally expensive optimization problems. Soft Comput Fus Found Methodol Appl 11:957–971

30 Genetics-Based Machine Learning

Tim Kovacs
Department of Computer Science, University of Bristol, UK
kovacs@cs.bris.ac.uk

G. Rozenberg et al. (eds.), *Handbook of Natural Computing*, DOI 10.1007/978-3-540-92910-9_30,

Abstract

This is a survey of the field of genetics-based machine learning (GBML): the application of evolutionary algorithms (ES) to machine learning. We assume readers are familiar with evolutionary algorithms and their application to optimization problems, but not necessarily with machine learning. We briefly outline the scope of machine learning, introduce the more specific area of supervised learning, contrast it with optimization and present arguments for and against GBML. Next we introduce a framework for GBML, which includes ways of classifying GBML algorithms and a discussion of the interaction between learning and evolution. We then review the following areas with emphasis on their evolutionary aspects: GBML for subproblems of learning, genetic programming, evolving ensembles, evolving neural networks, learning classifier systems, and genetic fuzzy systems.

1 Introduction

Genetics-based machine learning (GBML) is the application of evolutionary algorithms (EAs) to machine learning. We assume readers are familiar with EAs, which are well documented elsewhere, and their application to optimization problems. In this introductory section we outline the scope of machine learning, introduce the more specific area of supervised learning, and contrast it with optimization. However, the treatment is necessarily brief and readers who desire to work in GBML are strongly advised to first gain a solid foundation in non-evolutionary approaches to machine learning. ❷ Sect. 2 describes a framework for GBML, which includes ways of classifying GBML algorithms and a discussion of the interaction between learning and evolution. ❷ Sect. 3 reviews the work of a number of GBML communities with emphasis on their evolutionary aspects. Finally, ❷ Sect. 4 concludes the chapter.

What is Missing

Given the breadth of the field and the volume of the literature, the coverage herein is necessarily somewhat arbitrary and misses a number of significant subjects. These include a general introduction to machine learning including the structure of learning problems and their fitness landscapes (which we must exploit in order to learn efficiently), non-evolutionary algorithms (which constitute the majority of machine learning methods, and include both simple and effective methods), and theoretical limitations of learning (such as the *no free lunch* theorem for supervised learning (Wolpert 1996) and the *conservation law of generalization* (Schaffer 1994)). Also missing is coverage of GBML for clustering, reinforcement learning, Bayesian networks, artificial immune systems, artificial life, and application areas. Finally, some areas which have been touched on have been given an undeservedly cursory treatment, including EAs for data preparation (e.g., feature selection), coevolution, and comparisons between GBML and non-evolutionary alternatives. However, Freitas (2002a) contains good treatments of GBML for, among others, clustering and data preparation.

1.1 Machine Learning

Machine learning is concerned with machines that improve with experience and reason inductively or abductively in order to optimize, approximate, summarize, generalize from specific examples to general rules, classify, make predictions, find associations, propose explanations, and propose ways of grouping things. For simplicity, we will restrict ourselves to classification and optimization problems.

Inductive Generalization

Inductive generalization refers to the inference of unknown values from known values. Induction differs from deduction in that the unknown values are in fact *unknowable*, which gives rise to fundamental limitations in what can be learned. (If at a later time new data makes all such values known the problem ceases to be inductive.) Given that the unknown values are unknowable, we *assume* they are correlated with the known values and we seek to learn the correlations. We formulate our objective as maximizing a function of the unknown values. In evolutionary computation this objective is called the fitness function, whereas in other areas the analogous feedback signal may be known as the error function, or by other names. There is *no need for induction* if: (i) all values are known and (ii) there is enough time to process them. We consider two inductive problems: function optimization and learning. We will not deal with abduction.

1-Max: A Typical Optimization Problem

The 1-max problem is to maximize the number of 1s in a binary string of length n. The optimal solution is trivial for humans although it is less so for EAs. The *representation* of this problem follows. Input: none. Output: bit strings of length n. *Data generation*: we can generate as many output strings as time allows, up to the point where we have enumerated the search space (in which case the problem ceases to be inductive). *Training*: the fitness of a string is the number of 1s it contains. We can evaluate a learning method on this task by determining how close it gets to the known optimal solution. In more realistic problems, the optimum is not known and we may not even know the maximum possible fitness. Nonetheless, for both toy and realistic problems, we can evaluate how much training was needed to reach a certain fitness and how a learning method compares to others.

Classification of Mushrooms: A Typical Learning Problem

Suppose we want to classify mushroom species as poisonous or edible given some training data consisting of features of each species (color, size and so on) including edibility. Our task is to learn a hypothesis, which will classify new species whose edibility is unknown. *Representation*: the input is a set of nominal attributes and the output is a binary label indicating edibility. *Data generation*: a fixed dataset of input/output examples derived from a book. Typically the dataset is far, far smaller than the set of possible inputs, and we partition it into train and test sets. *Training*: induce a hypothesis which maximizes classification accuracy on the train set. *Evaluation*: evaluate the accuracy of the induced hypothesis on the test set, which we take as an indication of how well a newly encountered species might be classified.

Terminology in Supervised Learning

Although many others exist, we focus on the primary machine learning paradigm: standard supervised learning (SL), of which the preceding mushroom classification task is a good example. In SL we have a dataset of labeled input/output pairs. Inputs are typically called instances or exemplars and are factored into attributes (also called features), while outputs are called classes (for classification tasks) or the output is called the dependent variable (for regression tasks).

Comparison of Supervised Learning and Optimization

In SL we typically have limited training data and it is crucial to find a good inductive bias for later use on new data. Consequently, we *must* evaluate the generalization of the induced hypothesis from the train set to the previously unused test set. In contrast, in optimization we can typically generate as much data as time allows and we can typically evaluate any output. We are concerned with finding the optimum output in minimum time, and, specifically, inducing which output to evaluate next. As a result no test set is needed.

Issues in Supervised Learning

A great many issues arise in SL including overfitting, underfitting, producing human readable results, dealing with class imbalances in the training data, asymmetric cost functions, noisy and nonstationary data, online learning, stream mining, learning from particularly small datasets, learning when there are very many attributes, learning from positive instances only, incorporating bias and prior knowledge, handling structured data, and using additional unlabeled data for training. None of these will be dealt with here.

1.2   Arguments For and Against GBML

GBML methods are a niche approach to machine learning and much less well-known than the main non-evolutionary methods, but there are many good reasons to consider them.

Accuracy

Importantly, the classification accuracy of the best evolutionary and non-evolutionary methods are comparable (Freitas 2002a, Sect. 12.1.1).

Synergy of Learning and Evolution

GBML methods exploit the synergy of learning and evolution, combining global and local search and benefitting from the Baldwin effect's smoothing of the fitness landscape ❯ Sect. 2.3.

Epistasis

There is some evidence that the accuracy of GBML methods may not suffer from epistasis as much as typical non-evolutionary greedy search (Freitas 2002a, Sect. 12.1.1).

Integrated Feature Selection and Learning

GBML methods can combine feature selection and learning in one process. For instance, feature selection is intrinsic in LCS methods ❯ Sect. 3.5.

Adapting Bias

GBML methods are well-suited to adapting inductive bias. We can adapt representational bias by, for example, selecting rule condition shapes ❯ Sect. 3.5.3, and algorithmic bias by, for example, evolving learning rules ❯ Sect. 3.4.

Exploiting Diversity

We can exploit the diversity of a population of solutions to combine and improve predictions (the ensemble approach, ❯ Sect. 3.3) and to generate Pareto sets for multi-objective problems.

Dynamic Adaptation

All the above can be done dynamically to improve accuracy, to deal with nonstationarity, and to minimize population size. This last is of interest in order to reduce overfitting, improve run-time, and improve readability.

Universality

Evolution can be used as a wrapper for *any* learner.

Parallelization

Population-based search is easily parallelized.

Suitable Problem Characteristics

From an optimization perspective, learning problems are typically large, non-differentiable, noisy, epistatic, deceptive, and multimodal (Miller et al. 1989). To this list we could add high-dimensional and highly constrained. EAs are a good choice for such problems. See Cantú-Paz and Kamath (2003) and ❷ Sect. 3.4 for more arguments in favor of and against GBML.

Algorithmic Complexity

GBML algorithms are typically more complex than their non-evolutionary alternatives. This makes them harder to implement and harder to analyze, which means there is less theory to guide parameterization and development of new algorithms.

Increased Run-time

GBML methods are generally much slower than the non-evolutionary alternatives.

Suitability for a Given Problem

No single learning method is a good choice for all problems. For one thing the bias of a given GBML method may be inappropriate for a given problem. Problems to which GBML methods are particularly prone include prohibitive run-time (or set-up time) and that simpler and/or faster methods may suffice. Furthermore, even where GBML methods perform better, the improvements may be marginal. See the strengths, weaknesses, opportunities, threats (SWOT) analysis of GBML in Orriols-Puig et al. (2008b) for more.

2 A Framework for GBML

The aim of the framework presented in this section is to structure the range of GBML systems into more specific categories about which we can make more specific observations than we could about GBML systems as a whole. We present two categorizations. In the first (❷ Sect. 2.1), GBML systems are classified by their role in learning; specifically their applications to i) subproblems of machine learning, ii) learning itself, or iii) meta-learning. In the second categorization (❷ Sect. 2.2), GBML systems are classified by their high-level algorithmic approach as either Pittsburgh or Michigan systems. Following this, in ❷ Sect. 2.3 we briefly review ways in which learning and evolution interact and in ❷ Sect. 2.4 we consider various models of GBML not covered earlier.

Before proceeding, we note that evolution can output a huge range of phenotypes, from scalar values to complex learning agents, and that agents can be more or less plastic (independent of evolution). For example, if evolution outputs a fixed hypothesis, that hypothesis has no plasticity. In contrast, evolution can output a neural net which, when trained with backpropagation, can learn much. (In the latter approach, evolution may specify the network structure while backpropagation adapts the network weights.)

Structure of GBML Systems

We can divide any evolutionary (meta)-learning system into the following parts: (i) *Representation*, which consists of the genotype (the learner's genes) and phenotype (the learner itself, built according to its genes). In simple cases, the genotype and phenotype may be identical, for example with the simple ternary LCS rules of ❱ Sect. 3.5.2. In other cases, the two are very different and the phenotype may be derived through a complex developmental process (as in nature); see ❱ Sect. 3.4 on developmental encodings for neural networks. (ii) *Feedback*, which consists of the learner's objective function (e.g., the error function in supervised learning) and the fitness function which guides evolution. (iii) *The production system*, which applies the phenotypes to the learning problem. (iv) *The evolutionary system*, which adapts genes.

2.1 Classifying GBML Systems by Role

In order to contrast learning and meta-learning, we define learning as a process which outputs a fixed hypothesis. Accordingly, when evolution adapts hypotheses it is a learner and when it adapts learners it is a meta-learner. However, this distinction between learning and meta-learning should not be overemphasized; if evolution outputs a learner with little plasticity then evolution may be largely responsible for the final hypothesis, and in this case plays both a learning and a meta-learning role. Furthermore, both contribute to the ultimate goal of adaptation, and in ❱ Sect. 2.3 we will see ways in which they interact.

Evolution as learning is illustrated in the left of ❱ *Fig. 1*, which shows a GBML agent interacting directly with the learning problem. In contrast, the right of the figure shows GBML as meta-learning: the learner (or a set of learners) is the output of evolution, and the learner interacts directly with the learning problem while evolution interacts with it only through learners. At time step 1 of each generation, evolution outputs a learning agent and at the generation's final step, T, it receives an evaluation of the learner's fitness. During the intervening time steps the learner interacts with the problem. This approach to meta-learning is *universal* as any learner can be augmented by GBML, and is related to the wrapper approach to feature selection in ❱ Sect. 3.1.

Meta-learning is a broad term with different interpretations but the essential idea is *learning about learning*. A meta-learner may optimize parameters of a learner, learn which learner to apply to a given input or a given problem, learn which representation(s) to use, optimize the update rules used to train learners, learn an algorithm which solves the problem, evolve an ecosystem of learners, and potentially be open ended. See Vilalta and Drissi (2002) and Giraud-Carrier and Keller (2002) on non-evolutionary meta-learning and Burke et al. (2003), Krasnogor (2004), Krasnogor and Gustafson (2004), and Burke and Kendall (2005) on the hyperheuristics (*heuristics to learn heuristics*) approach, of which a subset is evolutionary.

◻ Fig. 1
(*Left*) GBML as learner. Input, Output, and Fitness shown. (*Right*) GBML as meta-learner.
Subscripts denote generation and time step (1...T).

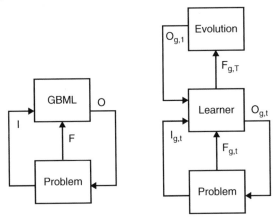

A third role for evolution is application to various subproblems of learning including feature selection, feature construction, and other optimization roles within learning agents. In these cases evolution neither outputs the final hypothesis nor outputs a learning agent which does so. ❷ Section 3.1 deals with such applications.

2.2 Classifying GBML Systems Algorithmically

In the Pittsburgh (Pitt) approach, one chromosome encodes one solution. We assume that fitness is assigned to chromosomes, so in Pitt systems it is assigned to solutions. This leaves a credit assignment problem: how did the chromosome's component genes contribute to the observed fitness of the chromosome? This is left to evolution as this is what EAs are designed to deal with. In the Michigan approach, one solution is (typically) represented by many chromosomes and so fitness is assigned to partial solutions. Credit assignment differs from the Pitt case as chromosomes not only compete for reproduction but may also complement and cooperate with each other. This gives rise to the issues of how to encourage cooperation, complementarity, and coverage of the inputs, all of which makes designing an effective fitness function more complex than in Pitt systems. In Michigan systems, the credit assignment problem is how to measure a chromosome's contributions to the overall solution, as reflected in the various aspects of fitness just mentioned. To sum up the difficulty in Michigan systems: the best set of chromosomes may not be the set of best (i.e., fittest) chromosomes (Freitas 2002a). To illustrate, ❷ *Fig. 2* depicts the representation used by Pitt and Michigan versions of the rule-based systems called learning classifier systems (LCS) (see ❷ Sect. 3.5). In a Pittsburgh LCS, a chromosome is a variable-length *set* of rules, while in a Michigan LCS, a chromosome is a single fixed-length rule.

Although the Pittsburgh and Michigan approaches are generally presented as two discrete cases, some hybrids exist (e.g., Wilcox (1995)).

■ **Fig. 2**
Michigan and Pittsburgh rule-based systems compared. The F:x associated with each chromosome indicates its fitness.

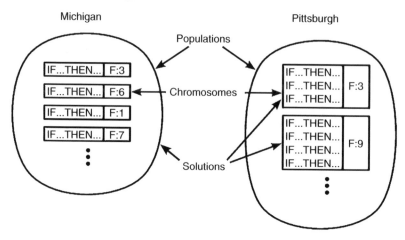

Pittsburgh and Michigan Compared

Pittsburgh systems (especially naive implementations) are slower, since they evolve more complex structures and they assign credit at a less specific (and hence less informative) level. (See, however, ❷ Sect. 3.5.5 on the windowing approach to improving run-time and Bacardit et al. (2009a) for an approach which avoids performing matching operations between rule conditions and irrelevant features.) Additionally, their chromosomes and their genetic operators are more complex. On the other hand they face less complex credit assignment problems and hence are more robust, that is, more likely to adapt successfully. Michigan systems use a finer grain of credit assignment than the Pittsburgh approach, which means bad partial solutions can be deleted without restarting from scratch. This makes them more efficient and also more suitable for incremental learning. However, credit assignment is more complex in Michigan systems. Since the solution is a *set* of chromosomes: (i) the population must not converge fully, and (ii) as noted, the best set of chromosomes may not be the set of best chromosomes.

The two approaches also tend to be applied in different ways. Pitt systems are typically used offline and are algorithm-driven; the main loop processes each chromosome in turn and seeks out data to evaluate them (which is how a standard genetic algorithm (GA) works, although fitness evaluation is typically simpler in a GA). In contrast, Michigan systems are typically used online and are data-driven; the main loop processes each data input in turn and seeks out applicable chromosomes (see ❷ *Fig. 3*). As a result, Michigan systems are more often used as learners (though not necessarily more often as meta-learners) for reinforcement learning, which is almost always online. The Michigan approach has mainly been used with LCS. See Greene and Smith (1993), Janikow (1993), Wilcox (1995), Freitas (2002b) and Kovacs (2004) for comparison of the approaches.

Iterative Rule Learning

IRL is a variation on the Michigan approach in which, as usual, one solution is represented by many chromosomes, but only the single best chromosome is selected after each run, which

☐ **Fig. 3**
A basic Michigan algorithm.

On each time step:
1. Identify match set: subset of population which match current input
2. Compute support in match set for each class
3. Select class
4. Identify action set: subset of match set which advocate selected class
5. Update action set based on feedback
6. Optionally alter population

alters the coevolutionary dynamics of the system. The output of multiple runs is combined to produce the solution. The approach originated with SIA (Supervised Inductive Algorithm) (Venturini 1993; Juan Liu and Tin-Yau Kwok 2000), a supervised genetic rule learner.

Genetic Cooperative-Competitive Learning

GCCL is another Michigan approach in which each generation is ranked by fitness and a *coverage-based filter* then allocates inputs to the first rule which correctly covers them. Inputs are only allocated to one rule per generation and rules which have no inputs allocated die at the end of a generation. The collective accuracy of the remaining rules is compared to the previous best generation, which is stored offline. If the new generation is more accurate (or the same but has fewer rules) it replaces the previous best. Examples include COGIN (Greene and Smith 1993, 1994), REGAL (Giordana and Neri 1995), and LOGENPRO (Wong and Leung 2000).

2.3 The Interaction of Learning and Evolution

This section briefly touches on the rich interactions between evolution and learning.

Memetic Learning

We can characterize evolution as a form of global search, which is good at finding good basins of attraction, but poor at finding the optimum of those basins. In contrast, many learning methods are forms of local search and have the opposite characteristics. We can get the best of both by combining them, which generally outperforms either alone (Yao 1999). For example, evolving the initial weights of a neural network and then training them with gradient descent can be two orders of magnitude faster than using random initial weights (Floreano et al. 2008). Methods which combine global and local search are called *memetic* algorithms (Hart et al. 2004, 2005; Ong et al. 2006, 2007; Smith 2007; Ong et al. 2009; Rozenberg et al. 2012). See Krasnogor and Smith (2005) for a self-contained tutorial.

Darwinian and Lamarckian Evolution

In Lamarckian evolution/inheritance, learning during an individual's lifetime directly alters the genes passed to offspring, so offspring inherit the result of their parents' learning. This does not occur in nature but can in computers and has the potential to be more efficient than Darwinian evolution since the results of learning are not thrown away. Indeed, Ackley and

Littman (1992) showed Lamarckian evolution was much faster on stationary learning tasks but Saski and Tokoro (1997) showed Darwinian evolution was generally better on nonstationary tasks. See also Whitley et al. (1994), Yamasaki and Sekiguchi (2000), Pereira and Costa (2001), and Whiteson and Stone (2006).

The Baldwin Effect

The Baldwin effect is a two-part dynamic between learning and evolution which depends on *Phenotypic Plasticity* (PP): the ability to adapt (e.g., learn) during an individual's lifetime. The first aspect is this. Suppose a mutation would have no benefit except for PP. Without PP, the mutation does not increase fitness, but with PP it does. Thus PP helps evolution to adopt beneficial mutations; it effectively smooths the fitness landscape. A possible example from nature is lactose tolerance in human adults. At a recent point in human evolution a mutation occurred, which allowed adult humans to digest milk. Subsequently, humans learned to keep animals for milk, which in turn made the mutation more likely to spread. The smoothing effect on the fitness landscape depends on PP; the greater the PP the more potential there is for smoothing. All GBML methods exploit the Baldwin effect to the extent that they have PP. See Whiteson and Stone (2006, Sect. 7.2) for a short review of the Baldwin effect in reinforcement learning.

The second aspect of the Baldwin effect is genetic assimilation. Suppose PP has a cost (e.g., learning involves making mistakes). If PP can be replaced by new genes, it will be; for instance a learned behavior can become instinctive. This allows learned behaviors to become inherited without Lamarckian inheritance.

Turney (1996) has connected the Baldwin effect to inductive bias. All inductive algorithms have a bias and the Baldwin effect can be seen as a shift from weak to strong bias. When bias is weak, agents rely on learning; when bias is strong, agents rely on instinctive behavior.

Evaluating Evolutionary Search by Evaluating Accuracy

Kovacs and Kerber (2004) point out that high classification accuracy does not imply effective genetic search. To illustrate, they initialized XCS (Wilson 1995) with random condition/action rules and disabled evolutionary search. Updates to estimates of rule utility, however, were made as usual. They found the system was still able to achieve very high training set accuracy on the widely used 6 and 11 multiplexer tasks since ineffective rules were simply given low weight in decision making, though neither removed nor replaced. Care is therefore warranted when attributing good accuracy to genetic search. A limitation of this work is that test set accuracy was not evaluated.

2.4 Other GBML Models

This section covers some models which are orthogonal to those discussed earlier.

Online Evolutionary Computation

In many problems, especially sequential ones, feedback is very noisy and needs averaging. Whiteson and Stone (2006) allocated trials to chromosomes in proportion to their fitness with the following procedure. At each new generation, each chromosome is evaluated once only. Subsequent evaluations are allocated using a softmax distribution based on the initial fitnesses and the average fitness of a chromosome is recalculated after each evaluation. In nonstationary problems a recency-weighted average of fitness samples is used. This approach is called *online evolutionary computation*. Its advantages are that less time is wasted evaluating weaker chromosomes, and in cases where mistakes matter, fewer mistakes are made by agents during fitness evaluations. However, the improvement is only on average; worst case performance is not improved. This is related to other work on optimizing noisy fitness functions (Stagge 1998; Beielstein and Markon 2002), except that they do not reduce online mistakes.

Steady-State EAs

Whereas standard generational EAs replace the entire population each generation, steady state EAs replace a subset (e.g., only two in XCS). This approach is standard in Michigan LCS because they minimize disruption to the population, which is useful for online learning. Steady state EAs introduce selection for deletion as well as reproduction and this is typically biased toward lower fitness chromosomes or to reduce crowding.

Co-evolving Learners and Problems

Another possibility not mentioned in our earlier classifications is to coevolve both learners and problems. When successful, this allows learners to gradually solve harder problems rather than tackling the most difficult problems from the start. It also allows us to search the space of problems to find those which are harder for a given learner, and to explore the dynamics between learners and problems.

3   GBML Areas

This section covers the main GBML research communities. These communities are more disjoint than the methods they use and the lines between them are increasingly blurring. For example, LCS often evolve neural networks and fuzzy rules, and some are powered by genetic programming. Nonetheless, the differences between the communities and their approaches is such that it seemed most useful to structure this section by community and not, for example, by phenotype or learning paradigm; such integrated surveys of GBML are left to the future. Many communities have reinvented the same ideas, yet each has its own focus and strengths and so each has much to learn from the others.

3.1   GBML for Subproblems of Learning

This section briefly reviews ways in which evolution has been used not for the primary task of learning – generating hypotheses – but for subproblems including data preparation and optimization within other learning methods.

Evolutionary Feature Selection

Some attributes (features) of the input are of little or no use in classification. We can simplify and speed learning by selecting only useful attributes to work with, especially when there are very many attributes and many contribute little. EAs are widely used in the wrapper approach to feature selection (John et al. 1994) in which the base learner (the one which generates hypotheses) is treated as a black box to be optimized by a search algorithm. In this, EAs usually give good results compared to non-evolutionary methods (Jain and Zongker 1997;

Sharpe and Glover 1999; Kudo and Skalansky 2000) but there are exceptions (Jain and Zongker 1997). In Cantú-Paz (2002) Estimation of Distribution Algorithms were found to give similar accuracy but run more slowly than a GA. More generally we can weight features (instead of making an all-or-nothing selection) and some learners can use weights directly, for example, weighted k-nearest neighbors (Raymer et al. 2000). The main drawback of EAs for feature selection is their slowness compared to non-evolutionary methods. See Martin-Bautista and Vila (1999) and Freitas (2002a, b) for overviews, and Stout et al. (2008), Bacardit et al. (2009b) for some recent real-world applications.

Evolutionary Feature Construction

Some features are not very useful by themselves but can be when combined with others. We can leave the base learner to discover this or we can preprocess data to construct informative new features by combining existing ones, for example new feature $f_{new} = f_1$ AND f_3 AND f_8. This is also called constructive induction and there are different approaches. GP has been used to construct features out of the original attributes, for example, Hu (1998), Krawiec (2002), and Smith and Bull (2005). The original features have also been linearly transformed by evolving a vector of coefficients (Kelly and Davis 1991; Punch et al. 1993). Simultaneous feature transformation and selection has had good results (Raymer et al. 2000).

Other Subproblems of Learning

EAs have been used in a variety of other ways. One is training set optimization in which we can partition the data into training sets (Romaniuk 1994), select the most useful training inputs (Ishibuchi and Nakashima 2000), and even generate synthetic inputs (Zhang and Veenker 1991; Cho and Cha 1996). EAs have also been used for optimization within a learner, for example, Kelly and Davis (1991) optimized weighted k-nearest neighbors with a GA, Cantú-Paz and Kamath (2003) optimized decision tree tests using a GA and an evolution strategy (ES) and Thompson (1998, 1999) optimized voting weights in an ensemble. Janikow (1993) replaced beam search in AQ with a genetic algorithm and similarly Tameddoni-Nezhad and Muggleton (2000, 2003), Divina and Marchiori (2002), and Divina et al. (2002, 2003) have investigated inductive logic programming driven by a GA.

3.2 Genetic Programming

Genetic Programming (GP) is a major evolutionary paradigm which evolves programs (Vanneschi and Poli 2012). The differences between GP and GAs are not precise but typically GP evolves variable-length structures, typically trees, in which genes can be functions. See Woodward (2003) for discussion. Freitas (2002a) discusses differences between GAs and GP which arise because GP representations are more complex. Among the pros of GP: (i) it is easier to represent complex languages, such as first-order logic in GP, (ii) it is easier to represent complex concepts compactly, and (iii) GP is good at finding novel and complex patterns overlooked by other methods. Among the cons of GP: (i) expressive representations have large search spaces, (ii) GP tends to overfit / does not generalize well, and (iii) variable-length representations suffer from bloat (see, e.g., Poli et al. (2008)).

While GAs are typically applied to function optimization, GP is widely applied to learning. To illustrate, there are many learning problems among the set of "typical GP problems" defined by Koza (1992), which have become more-or-less agreed benchmarks for the

GP community (Vanneschi and Poli 2012), there are many learning problems. These include the multiplexer and parity Boolean functions, symbolic regression of mathematical functions and the intertwined spirals problem, which involves the classification of two-dimensional points as belonging to one of the two spirals. GP usually follows the Pittsburgh approach. We cover the two representations most widely used for learning with GP: GP trees and decision trees.

3.2.1 GP Trees

GP Trees for Classification
❯ *Figure 4* shows the 3 multiplexer Boolean function as a truth table on the left and as a GP tree on the right. To classify an input with the GP tree: (i) instantiate the leaf variables with the input values, (ii) propagate values upward from leaves though the functions in the non-leaf nodes and (iii) output the value of the root (top) node as the classification.

GP Trees for Regression
In regression problems leaves may be constants or variables and non-leaves are mathematical functions. ❯ *Figure 5* shows a real-valued function as an algebraic expression on the left and as

◻ **Fig. 4**
Two representations of the 3 multiplexer function: truth table (*left*) and GP tree (*right*).

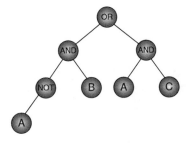

A	B	C	Class
0	0	0	0
0	0	1	0
0	1	0	1
0	1	1	1
1	0	0	0
1	0	1	1
1	1	0	0
1	1	1	1

◻ **Fig. 5**
Two representations of a real-valued function.

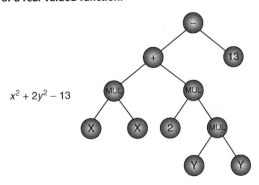

$$x^2 + 2y^2 - 13$$

a GP tree on the right. (Note that $x^2 + 2y^2 - 13 = ((x*x) + (2*(y*y))) - 13$.) The output of the tree is computed in the same way as in the preceding classification example.

3.2.2 Decision Trees

❯ *Figure 6* shows the 3 multiplexer as a truth table and as a decision tree. To classify an input in such a tree: (i) start at the root (top) of tree, (ii) follow the branch corresponding to the value of the attribute in the input, (iii) repeat until a leaf is reached, and (iv) output the value of the leaf as the classification of the input.

Evolving First-Order Trees
First-order trees use both propositional and first-order internal nodes. Rouwhorst and Engelbrecht (2000) found first-order logic made trees more expressive and allowed much smaller solutions than found by the rule learner CN2 or the tree learner C4.5, with similar accuracy.

Oblique (Linear) Trees
Whereas conventional tree algorithms learn axis-parallel decision boundaries, oblique trees make tests on a linear combination of attributes. The resulting trees are more expressive but have a larger search space. See Bot and Langdon (2000).

Evolving Individual Nodes in Decision Trees
In most GP-based tree evolvers, an individual is a complete tree but in Marmelstein and Lamont (1998) each individual is a tree node. The tree is built incrementally: one GP run is made for each node. This is similar to IRL in ❯ Sect. 2.2 but results are added to a tree structure rather than a list.

3.2.3 Extensions to GP

Ensemble Methods and GP
Ensemble ideas have been used in two ways. First, to reduce fitness computation time and memory requirements by training on subsamples of the data. The bagging approach has been used in Folino et al. (2003) and Iba (1999) and the boosting approach in Song et al. (2005).

◻ **Fig. 6**
Two representations of the 3 multiplexer function: truth table (*left*) and decision tree (*right*).

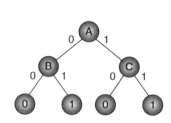

A B C	Class
0 0 0	0
0 0 1	0
0 1 0	1
0 1 1	1
1 0 0	0
1 0 1	1
1 1 0	0
1 1 1	1

Although not an ensemble technique, the limited error fitness (LEF) method introduced in Gathercole and Ross (1997) as a way of reducing GP run-time works in a similar manner: in LEF, the proportion of the training set used to evaluate fitness depends on the individual's performance. The second ensemble approach improves accuracy by building an ensemble of GP trees. In Keijzer and Babovic (2000) and Paris et al. (2001) each run adds one tree to the ensemble and weights are computed with standard boosting.

GP Hyperheuristics

Schmidhuber (1987) proposed a meta-GP system evolving evolutionary operators as a way of expanding the power of GP's evolutionary search. Instead of evolving decision rules Krasnogor proposes applying GP to the much harder task of evolving classification algorithms, represented using grammars (Krasnogor 2002, 2004; Krasnogor and Gustafson 2004). Freitas (2002a, Sect. 12.2.3) sketches a similar approach, which he calls *algorithm induction*, while (Pappa and Freitas 2010) goes into the subject in much more detail. Burke et al. (2009) also deal with GP hyperheuristics.

3.2.4 Conclusions

Lack of Test Sets in GP

GP terminology follows a convention in the GA field since at least (Holland 1986) in which *brittleness* refers to overfitting or poor generalization to unseen cases, and *robustness* refers to good generalization. A feature of the GP literature is that GP is usually evaluated only on the training set (Kushchu 2002; Vanneschi and Poli 2012). Kushchu has also criticized the way in which test sets have been used (Kushchu 2002). Nonetheless GP has the same need for test sets to evaluate generalization as other methods (Kuschu 2002) and as a result, the ability of GP to perform inductive generalization is one of the open issues for GP identified in Vanneschi and Poli (2012). See Kushchu (2002) and Vanneschi and Poli (2012) for methods which have been used to encourage generalization in GP, many of which can be applied to other methods.

Reading

See Koza's 1994 book (Koza 1994) for the basics of evolving decision trees with GP, Wong and Leung's 2000 book on data mining with grammar-based GP (Wong and Leung 2000), Freitas' 2002 book (Freitas 2002a) for a good introduction to GP, decision trees and both evolutionary and non-evolutionary learning, Poli, Langdon, and McPhee's free 2008 GP book (Poli et al. 2008), and Vanneschi and Poli's chapter on GP in this volume (Vanneschi and Poli 2012). The GP bibliography has over 5,000 entries (Langdon et al. 2009).

3.3 Evolving Ensembles

Ensembles, also called *multiple classifier systems* and *committee machines*, is the field which studies how to combine predictions from multiple sources. Ensemble methods are widely applicable to evolutionary systems where a population intrinsically provides multiple predictors. Ensemble techniques can be used with any learning method, although they are most useful for unstable learners, whose hypotheses are sensitive to small changes in their training. Ensembles can be heterogeneous (composed of different types of predictors) in which case they are called *hybrid* ensembles. Relatively few studies of hybrid systems exist (Brown et al. 2005) but see for example Woods et al. (1997), Cho and Park (2001), and Chandra and Yao (2006). Ensembles enjoy good theoretical foundations (Brown et al. 1996; Tumer and Ghosh 1996), perform very well in practice (Caruana and Niculescu-Mizil 2006) and were identified by Dietterich as one of the four current directions for machine learning in 1998 (Dietterich 1998). While the key advantage of using ensembles is better test-set generalization, there are others: ensembles can perform more complex tasks than individual members, the overall system can be easier to understand and modify and ensembles are more robust/degrade more gracefully than individual predictors (Sharkey 1996).

Working with an ensemble raises a number of issues. How to create or select ensemble members? How many members are needed? When to remove ensemble members? How to combine their predictions? How to encourage diversity in members? There are many approaches to these issues, among the best known of which are bagging (Breiman 1996, 1998) and boosting (Chandra and Yao 2006a; Meir and Rätsch 2003).

Creating a good ensemble is an inherently multi-objective problem (Valentini and Masulli 2002). In addition to maximizing accuracy we also want to maximize diversity in errors; after all, having multiple identical predictors provides no advantage. On the other hand, an ensemble of predictors which make different errors is very useful since we can combine their predictions so that the ensemble output is at least as good on the training set as the average predictor (Krogh and Vedelsby 1995). Hence we want to create accurate predictors with diverse errors (Dietterich 1998; Hansen and Salamon 1990; Krogh and Vedelsby 1995; Opitz and Maclin 1999; Opitz and Slavlik 1996). In addition we may want to minimize ensemble size in order to reduce run-time and to make the ensemble easier to understand. Finally, evolving variable-length chromosomes without pressure toward parsimony results in bloat (Poli et al. 2008), in which case we have a reason to minimize the size of individual members.

3.3.1 Evolutionary Ensembles

Although most ensembles are non-evolutionary, evolution has many applications within ensembles. (i) *Classifier creation and adaptation*: providing the ensemble with a set of candidate members. (ii) *Voting*: Lam and Suen (1995), Thompson (1998, 1999), and Cho (1999) evolve weights for the votes of ensemble members. (iii) *Classifier selection*: the winners of evolutionary competition are added to the ensemble. (iv) *Feature selection*: generating diverse classifiers by training them on different features (see ❷ Sect. 3.1 and Kuncheva (2004, Sect. 8.1.4)). (v) *Data selection*: generating diverse classifiers by training on different data (see ❷ Sect. 3.1). All these approaches have non-evolutionary alternatives. We now go into more detail on two of the above applications.

Classifier Creation and Adaptation

Single-objective evolution is common in evolving ensembles. For example, Liu and Yao (1999) combines accuracy and diversity into a single objective. In comparison, multiobjective evolutionary ensembles are rare (Chandra and Yao 2006) but they are starting to appear for example Abbass (2003) and Chandra and Yao (2006a, b). In addition to upgrading GBML to multi-objective GBML, other measures can be taken to evolve diversity, for example, fitness sharing (Liu et al. 2000) and the coevolutionary fitness method we describe next. Gagné et al. (2007) compare boosting and coevolution of learners and problems – both gradually focus on cases which are harder to learn – and argue that coevolution is less likely to overfit noise. Their coevolution-inspired fitness works as follows: Let Q be a set of reference classifiers. The hardness of a training input, x_i, is based on how many members of Q misclassify it. The fitness of a classifier is the sum of hardnesses of the inputs, x_i, it classifies correctly. This method results in accurate, yet error-diverse classifiers, since both are required to obtain high fitness. Gagné et al. exploit the population of classifiers to provide Q. They also introduce a greedy margin-based scheme for selection of ensemble members. They find that a simpler off-line version of their evolving ensemble learning (EEL) approach dominates their online version as the latter lacks a way to remove bad classifiers. Good results were obtained compared to Adaboost on six UCI (Asuncion and Newman 2009) datasets.

Evolutionary Selection of Members

There are two extremes. Usually each run produces one member of the ensemble and many runs are needed. Sometimes, however, the entire population is eligible to join the ensemble, in which case only one run is needed. The latter does not resolve the *ensemble selection problem*: which candidates to use? There are many combinations possible from just a small pool of candidates, and, as with selecting a solution set from a Michigan population, the set of best individuals may not be the best set of individuals (that is, the best ensemble). The selection problem is formally equivalent to the feature selection problem (Gagné et al. 2007) (❷ Sect. 3.2). See, for example, Sirlantzis et al. (2001) and Ruta and Gabrys (2001) for evolutionary approaches.

3.3.2   Conclusions

Research directions for evolutionary ensembles include multiobjective evolution (Chandra and Yao 2006a), hybrid ensembles (Chandra and Yao 2006b), and minimizing ensemble complexity (Liu et al. 2000).

Reading

Key works include Opitz and Shavlik's classic 1996 paper on evolving NN ensembles (Opitz and Shavlik (1996), Kuncheva's 2004 book on ensembles (Kuncheva 2004), Chandra and Yao's 2006 discussion of multiobjective evolution of ensembles (Chandra and Yao 2006a), Yao and Islam's 2008 review of evolving NN ensembles (Yao and Islam 2008) and Brown's 2005 and 2010 surveys of ensembles (Brown et al. 2005; Brown 2010). We cover evolving NN ensembles in ❷ Sect. 3.4.

3.4   Evolving Neural Networks

The study of neural networks (NNs) is a large and interdisciplinary area. The term artificial neural network (ANN) is often used to distinguish simulations from biological NNs, but having noted this we shall refer simply to NNs. When evolution is involved such systems may be called evolving artificial neural networks (EANNs) Yao (1999) or evolving connectionist systems (ECoSs) (Kasabov 2007).

A neural network consists of a set of nodes, a set of directed connections between a subset of nodes, and a set of weights on the connections. The connections specify inputs and outputs to and from nodes and there are three forms of nodes: input nodes (for input to the network from the outside world), output nodes, and hidden nodes, which only connect to other nodes.

Nodes are typically arranged in layers: the input layer, hidden layer(s), and output layer. Nodes compute by integrating their inputs using an activation function and passing on their activation as output. Connection weights modulate the activation they pass on and in the simplest form of learning weights are modified while all else remains fixed. The most common approach to learning weights is to use a gradient descent-based learning rule such as back-propagation. The *architecture* of a NN refers to the set of nodes, connections, activation functions, and the plasticity of nodes (that is, whether they can be updated or not). Most often all nodes use the same activation function and in virtually all cases all nodes can be updated. Evolution has been applied at three levels: weights, architecture, and learning rules. In terms of architecture, evolution has been used to determine connectivity, select activation functions, and determine plasticity.

Representations

Three forms of representations have been used: (i) direct encoding (Yao 1999; Floreano et al. 2008) in which all details (connections and nodes) are specified, (ii) indirect encoding (Yao 1999; Floreano et al. 2008) in which general features are specified (e.g., number of hidden layers and nodes) and a learning process determines the details, and (iii) developmental encoding (Floreano et al. 2008) in which a developmental process is genetically encoded (Kitano 1990; Gruau 1995; Nolfi et al. 1994; Husbands et al. 1994; Pal and Bhandari 1994; Sziranyi 1996). Implicit and developmental representations are more flexible and tend to be used for evolving architectures, while direct representations tend to be used for evolving weights alone.

Credit Assignment

Evolving NNs virtually always use the Pittsburgh approach although there are a few Michigan systems (Andersen and Tsoi 1993; Smith and Cribbs 1994; Smith and Cribbs 1997). In Michigan systems, each chromosome specifies only one hidden node, which raises issues. How should the architecture be defined? A simple method is to fix it in advance. How can we make nodes specialize? Two options are to encourage diversity during evolution, for example with fitness sharing, or, after evolution, by pruning redundant nodes (Andersen and Tsoi 1993).

Adapting Weights

Most NN learning rules are based on gradient descent, including the best known: back-propagation (BP). BP has many successful applications, but gradient descent-based methods require a continuous and differentiable error function and often get trapped in local minima (Sutton 1986; Whitley et al. 1990).

An alternative is to evolve the weights which has the advantages that EAs do not rely on gradients and can work on discrete fitness functions. Another advantage of evolving weights is that the same evolutionary method can be used for different types of network (feedforward, recurrent, and higher order), which is a great convenience for the engineer (Yao 1999). Consequently, much research has been done on evolution of weights. Unsurprisingly fitness functions penalize NN error but they also typically penalize network complexity (number of hidden nodes) in order to control overfitting. The expressive power of a NN depends on the number of hidden nodes: fewer nodes = less expressive = fits training data less, while more nodes = more expressive = fits data better. As a result, if a NN has too few nodes it underfits, while with too many nodes it overfits. In terms of training rate, there is no clear winner between evolution and gradient descent; which is better depending on the problem

(Yao 1999). However, Yao (1999) states that evolving weights and architecture is better than evolving weights alone and that evolution seems better for reinforcement learning and recurrent networks. Floreano (2008) suggests evolution is better for dynamic networks. Happily we do not have to choose between the two approaches.

Evolving and Learning Weights

Evolution is good at finding a good basin of attraction but poor at finding the optimum of the basin. In contrast, gradient descent has the opposite characteristics. To get the best of both (Yao 1999) we should evolve initial weights and then train them with gradient descent. Floreano (2008) claims that this can be two orders of magnitude faster than beginning with random initial weights.

Evolving Architectures

Architecture has an important impact on performance and can determine whether a NN under- or over-fits. Designing architectures by hand is a tedious, expert, trial-and-error process. Alternatives include constructive NNs, which grow from a minimal network and destructive NNs, which shrink from a maximal network. Unfortunately, both can become stuck in local optima and can only generate certain architectures (Angeline et al. 1994). Another alternative is to evolve architectures. Miller et al. (1989) make the following suggestions (quoted from Yao (1999)) as to why EAs should be suitable for searching the space of architectures.

▶ The surface is infinitely large since the number of possible nodes and connections is unbounded.

▶ The surface is nondifferentiable since changes in the number of nodes or connections are discrete and can have a discontinuous effect on EANN's performance.

▶ The surface is complex and noisy, since the mapping from an architecture to its performance is indirect, strongly epistatic, and dependent on the evaluation method used.

▶ The surface is deceptive since similar architectures may have quite different performance.

▶ The surface is multimodal since different architectures may have similar performance.

There are good reasons to evolve architectures and weights simultaneously. If we learn with gradient descent there is a many-to-one mapping from NN genotypes to phenotypes (Yao and Liu 1997). Random initial weights and stochastic learning lead to different outcomes, which makes fitness evaluation noisy, and necessitates averaging over multiple runs, which means the process is slow. On the other hand, if we evolve architectures and weights simultaneously we have a one-to-one genotype to phenotype mapping, which avoids the problem above and results in faster learning. Furthermore, we can co-optimize other parameters of the network (Floreano 2008) at the same time. For example, Belew et al. (1992) found the best networks had a very high learning rate which may have been optimal due to many factors such as initial weights, training order, and amount of training. Without co-optimizing, architecture and weights evolution would not have been able to take all factors into account at the same time.

Evolving Learning Rules (Yao 1999)

There is no one best learning rule for all architectures or problems. Selecting rules by hand is difficult and if we evolve the architecture then we do not know a priori what it will be.

A way to deal with this is to evolve the learning rule, but we must be careful: the architectures and problems used in learning the rules must be representative of those to which it will eventually be applied. To get general rules, we should train on general problems and architectures, not just one kind. On the other hand, to obtain a training rule specialized for a specific architecture or problem type, we should train just on that architecture or problem.

One approach is to evolve only learning rule parameters (Yao 1999) such as the learning rate and momentum in backpropagation. This has the effect of adapting a standard learning rule to the architecture or problem at hand. Non-evolutionary methods of adapting training rules also exist. Castillo et al. (2007), working with multi-layer perceptrons, found evolving the architecture, initial weights, and rule parameters together as good or better than evolving only the first two or the third.

We can also evolve new learning rules (Yao 1999; Radi and Poli 2003). Open-ended evolution of rules was initially considered impractical and instead Chalmers (1990) specified a generic, abstract form of update and evolved its parameters to produce different concrete rules. The generic update was a linear function of ten terms, each of which had an associated evolved real-valued weight. Four of the terms represented local information for the node being updated while the other six terms were the pairwise products of the first four. Using this method Chalmers was able to rediscover the delta rule and some of its variants. This approach has been used by a number of others and has been reported to outperform human-designed rules (Dasdan and Oflazer 1993). More recently, GP was used to evolve novel types of rules from a set of mathematical functions and the best new rules consistently outperformed standard backpropagation (Radi and Poli 2003). Whereas architectures are fixed, rules could potentially change over their lifetime (e.g., their learning rate could change) but evolving dynamic rules would naturally be much more complex than evolving static ones.

3.4.1 Ensembles of NNs

Most methods output a single NN (Yao and Islam 2008) but a population of evolving NNs is naturally treated as an ensemble and recent work has begun to do so. Evolving NNs is inherently multiobjective: we want accurate yet simple and diverse networks. Some works combine these objectives into one fitness function while others are explicitly multiobjective.

Single-Objective Ensembles
Yao and Liu (1998) used EPNet's (Yao and Liu 1997) population as an ensemble without modifying the evolutionary process. By treating the population as an ensemble, the result outperformed the population's best individual. Liu and Yao (1990) pursued accuracy and diversity in two ways. The first was to modify backpropagation to minimize error and maximize diversity using an approach they call negative correlation learning (NCL) in which the errors of members become negatively correlated and hence diverse. The second method was to combine accuracy and diversity in a single objective. Evolutionary ensembles for NCL (EENCL) (Liu et al. 2000) automatically determines the size of an ensemble. It encourages diversity with fitness sharing and NCL, and it deals with the ensemble member selection problem (❷ Sect. 3.3.1) with a cluster-and-select method (see Jin and Sendhoff (2004)). First we cluster candidates, based on their errors, on the training set so that clusters of

candidates make similar errors. Then we select the most accurate in each cluster to join the ensemble; the result is the ensemble can be much smaller than the population. Cooperative neural net ensembles (CNNE) (Islam et al. 2003) used a constructive approach to determine the number of individuals and how many hidden nodes each has. Both contribute to the expressive power of the ensemble and CNNE was able to balance the two to obtain suitable ensembles. Unsurprisingly, it was found that more complex problems needed larger ensembles.

Multiobjective Ensembles

Memetic pareto artificial NN (MPANN) (Abbass 2003) was the first ensemble of NNs to use multiobjective evolution. It also uses gradient-based local search to optimize network complexity and error. Diverse and accurate ensembles (DIVACE) (Chandra and Yao 2006a) uses multi-objective evolution to maximize accuracy and diversity. Evolutionary selection is based on non-dominated sorting (Srinivas and Deb 1994), a cluster-and-select approach is used to form the ensemble, and search is provided by simulated annealing and a variant of differential evolution (Storn and Price 1996). DIVACE-II (Chandra and Yao 2006) is a heterogeneous multi-objective Michigan approach using NNs, support vector machines, and radial basis function nets. The role of crossover and mutation is played by bagging (Breiman 1996) and boosting (Freund and Schapire 1996), which produce accurate and diverse candidates. Each generation bagging and boosting make candidate ensemble members and only dominated members are replaced. The accuracy of DIVACE-II was very good compared to 25 other learners on the Australian credit card and diabetes datasets and it outperformed the original DIVACE.

3.4.2 Yao's Framework for Evolving NNs

❯ *Figure 7* shows Yao's framework for evolving architectures, training rules, and weights as nested processes (Yao 1999). Weight evolution is the innermost as it occurs at the fastest time scale, while either rule or architecture evolution is outermost. If we have prior knowledge, or are interested in a specific class of either rule or architecture, this constrains the search space and Yao suggests the outermost should be the one which constrains it most. The framework can be thought of as a three-dimensional space of evolutionary NNs where 0 on each axis represents one-shot search and infinity represents exhaustive search. If we remove references to EAs and NNs, it becomes a general framework for adaptive systems.

3.4.3 Conclusions

Evolution is widely used with NNs, indeed according to Floreano et al. (2008) most studies of neural robots in real environments use some form of evolution. Floreano et al. go on to claim that evolving NNs can be used to study "brain development and dynamics because it can encompass multiple temporal and spatial scales along which an organism evolves, such as genetic, developmental, learning, and behavioral phenomena. The possibility to coevolve both the neural system and the morphological properties of agents . . . adds an additional valuable perspective to the evolutionary approach that cannot be matched by any other approach." (Floreano et al. 2008, p. 59).

◘ Fig. 7

Yao's framework for evolving architectures, training rules, and weights.

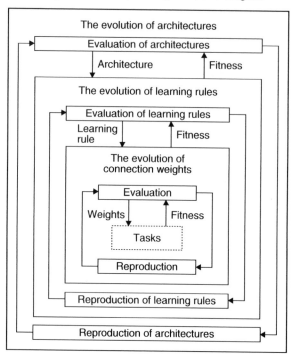

Reading

Key reading on evolving NNs includes Yao's classic 1999 survey (Yao 1999), Kasabov 2007 book (Kasabov 2007), Floreano, Dürr, and Mattiussi's 2008 survey (Floreano et al. 2008), which includes reviews of evolving dynamic and neuromodulatory NNs, and Yao and Islam's 2008 survey of evolving NN ensembles (Yao and Islam 2008).

3.5   Learning Classifier Systems (LCS)

Learning classifier systems (LCS) originated in the GA community as a way of applying GAs to learning problems. The LCS field is one of the oldest, largest, and most active areas of GBML. The majority of LCS research is currently carried out on XCS (Wilson 1995; Butz and Wilson 2001) and its derivatives XCSF (Wilson 2001b, 2002a) for regression/function approximation, and UCS (Bernadó-Mansilla and Garrell-Guiu 2003; Orriols-Puig and Bernadó-Mansilla 2008) for supervised learning.

The Game of the Name

Terminology has been contentious in this area (Heitkötter and Beasley 2001). LCS are also widely simply called classifier systems (abbreviated CS or CFS) and sometimes evolutionary (learning) classifier systems. At one time GBML referred exclusively to LCS. None of these names is very satisfactory but the field appears to have settled on LCS.

The difficulty in naming the field relates in part to the difficulty in defining what an LCS is (Smith 1992; Holland et al. 2000). In practice, what is accepted as an LCS has become more inclusive over the years. A reasonable definition of an LCS would be an evolutionary rule-based system – except that a significant minority of LCS are not evolutionary! On the other hand, most non-evolutionary rule-based systems are not considered LCS, so the boundaries of the field are defined more by convention than principle. Even EA practitioners are far from unanimous; work continues to be published which some would definitely consider forms of LCS, but which make no reference to the term and which contain few or no LCS references.

(L)CS has at times been taken to refer to Michigan systems only (see, e.g., Greene and Smith (1993)) but it now generally includes Pitt systems as well, as implied by the name and content of the international workshop on learning classifier systems (IWLCS) which includes both Pitt and Michigan, evolutionary and non-evolutionary systems. As a final terminological note, rules in LCS are often referred to as "classifiers".

3.5.1 Production Systems and Rule(Set) Parameters

LCS evolve condition-action (IF–THEN) rules. Recall from ❷ Sect. 2.2 and ❷ *Fig. 2* that in Michigan rule-based systems a chromosome is a single rule, while in Pittsburgh systems a chromosome is a variable-length *set* of rules. Pittsburgh, Michigan, IRL, and GCCL are all used. Michigan systems are rare elsewhere but are the most common form of LCS. Within LCS, IRL is most common with fuzzy systems, but see Aguilar-Ruiz et al. (2003) for a non-fuzzy version. In LCS, we typically evolve rule conditions and actions although non-evolutionary operators may act upon them. In addition, each phenotype has parameters associated with it and these parameters are typically learned rather than evolved using the Widrow–Hoff update or similar (see Lanzi et al. (2006b) for examples). In Michigan LCS parameters are associated with each rule but in Pittsburgh systems they are associated with each ruleset. For example, in UCS the parameters are: fitness, mean action set size (to bias a deletion process, which seeks to balance action set sizes), and experience (a count of the number of times a rule has been applied, in order to estimate confidence in its fitness). In GAssist (a supervised Pittsburgh system) the only parameter is fitness. Variations of the above exist; in some cases, rules predict the next input or read and write to memory.

3.5.2 Representing Conditions and Actions

The most common representation in LCS uses fixed-length strings with binary inputs and outputs and ternary conditions. In a simple Michigan version (see, e.g., Wilson (1995)), each rule has one action and one condition from $\{0, 1, \#\}$ where # is a wildcard, matching both 0 and 1 in inputs. For example, the condition 01# matches two inputs: 010 and 011. Similar representations were used almost exclusively prior to approximately 2000 and are inherited from GAs and their preference for minimal alphabets. (Indeed, ternary conditions have an interesting parallel with ternary schemata (Reeves and Rowe 2002) for binary GAs.) Such rules individually have limited expressive power (Schuurmans and Schaeffer 1989) (but see also Booker (1991)), which necessitates that solutions are sets of rules. More insidiously, the lack of individual expressiveness can be a factor in pathological credit assignment (strong/fit overgenerals

(Kovacs 2004)). Various extensions to the simple scheme described above have been studied (see (Kovacs 2004, Sect. 2.2.2)).

Real-Valued Intervals

Following Bacardit (2004, p. 84) we distinguish two approaches to real-valued interval representation in conditions. The first is representations based on discretization: HIDER* uses *natural coding* (Giraldez et al. 2003), ECL clusters attribute values and evolves constraints on them (Divina et al. 2003) while GAssist uses *adaptive discretization intervals* (Bacardit 2004). The second approach is to handle real values directly. In HIDER genes specify a lower and upper bound (where lower is always less than upper) (Aguilar-Ruiz et al. 2003). In Corcoran and Sen (1994) a variation of HIDER's scheme is used where the attribute is ignored when the upper bound is less than the lower. Interval representations are also used in Wilson (2001a) and Stone and Bull (2003). Finally, Wilson (2000) specifies bounds using centre and spread genes.

Default/Exception Structures

Various forms of default/exception rule structures have been used with LCS. It has been argued that they should increase the number of solutions possible without increasing the search space and should allow gradual refinement of knowledge by adding exceptions (Holland 1986). However, the space of *combinations* of rules is much larger than the set of rules and the evolutionary dynamics of default/exception rule combinations have proved difficult to manage in Michigan systems. Nonetheless, default rules can significantly reduce the number of rules needed for a solution (Valenzuela-Rendón 1989) and there have been some successes. ❯ *Figure 8* illustrates three representations for a Boolean function. The leftmost is a truth table, which lists all possible inputs and their outputs. The middle representation is the ternary language commonly used by LCS, which requires only four rules to represent the eight input/output pairs, thanks to the generalization provided by the # symbol. Finally, on the right, a default rule (### → 1) has been added to the ternary representation. This rule matches all inputs and states that the output is always 1. This rule is incorrect by itself, but the two rules above it provide exceptions and, taken together, the three accurately represent the function using one less rule than the middle representation. One difficulty in evolving such

◻ **Fig. 8**

Three representations for the 3 multiplexer function.

Truth table			
A	B	C	Output
0	0	0	0
0	0	1	0
0	1	0	1
0	1	1	1
1	0	0	0
1	0	1	1
1	1	0	0
1	1	1	1

Ternary rules
0 0 # → 0
0 1 # → 1
1 # 0 → 0
1 # 1 → 1

Default rule
0 0 # → 0
1 # 0 → 0
→ 1

default/exception structures lies in identifying which rules are the defaults and which the exceptions; a simple solution is to maintain the population in some order and make earlier rules exceptions to later ones (as in a decision list (Rivest 1987)). This is straightforward in Pitt systems in which individual rulesets are static but is more complex in Michigan populations in which individual rules are created and deleted dynamically. The other issue is how to assign credit to the overall multi-rule structure. In Pittsburgh systems this is again straightforward since fitness is assigned only at the level of rulesets, but in Michigan systems each rule has a fitness, and it is not obvious how to credit the three rules in the default/exception structure in a way which recognizes their cooperation.

The Pittsburgh GABIL (De Jong and Spears 1991) and GAssist (Bacardit 2004) use decision lists and often evolve default rules spontaneously (e.g., a fully general last rule). Bacardit found that enforcing a fully general last rule in each ruleset in GAssist (and allowing evolution to select the most useful class for such rules) was effective (Bacardit 2004).

In Michigan systems, default/exception structures are called default hierarchies. Rule specificity has been used as the criterion for determining which rules are exceptions and accordingly conflict resolution methods have been biased according to specificity. There are, however, many problems with this approach (Smith and Goldberg 1991). It is difficult for evolution to produce these structures since they depend on cooperation between otherwise competing rules. The structures are unstable since they are interdependent; unfortunate deletion of one member alters the utility of the entire structure. As noted, they complicate credit assignment and conflict resolution since exception rules must override defaults (Wilson 1989; Smith and Goldberg 1991). There are also problems with the use of specificity to prioritize rules. For one, having fewer #s does not mean a rule actually matches fewer inputs; counting #s is a purely syntactic measure of generality. For another, there is no reason why exception rules should be more specific. The consequence of these difficulties is that there has not been much interest in Michigan default hierarchies since the early 1990s (but see Vallim et al. (2003)) and indeed not all Michigan LCS support them (e.g., ZCS (Wilson 1995), XCS/XCSF and UCS do not). Nonetheless, the idea should perhaps be revisited and an ensembles perspective might prove useful.

Other Representations for Conditions and Actions

A great range of other representations has been used, particularly in recent years. These include VL_1 logic (Michalski et al. 1986) as used in GIL (Janikow 1993), first-order logic (Mellor 2005a, b, 2008), decision lists as used in GABIL (De Jong and Spears 1991) and GAssist (Bacardit 2004), messy encoding (Lanzi 1999a), ellipses (Butz 2005) and hyperellipses (Butz et al. 2006), hyperspheres (Marshall and Kovacs 2006), convex hulls (Lanzi and Wilson 2006a), tile coding (Lanzi et al. 2006) and a closely related hyperplane coding (Booker 2005a, b), GP trees (Ahluwalia and Bull 1999; Lanzi 1999b, 2001), Boolean networks defined by GP (Bull 2009), support vectors (Loiacono et al. 2007), edges of an augmented transition network (Landau et al. 2005), gene expression programming (Wilson 2008), fuzzy rules (see ❯ Sect. 3.6), and neural networks (Smith and Cribbs 1994; Cribbs and Smith 1996; Smith and Cribbs 1997; Bull and O'Hara 2002; O'Hara and Bull 2005; Dam et al. 2008; Howard et al. 2008; Howard and Bull 2008). GALE (Llorá and Garrell 2001; Llorá 2002; Llorá and Wilson 2004) has used particularly complex representations, including the use of GP to evolve trees defining axis-parallel and oblique hyper-rectangles (Llorá and Wilson 2004), and evolved prototypes, which are used with a k-nearest-neighbor classifier. The prototypes need not be fully specified; some attributes can be left undefined. This representation has also been used in

GAssist (Bacardit 2004). There has been limited work with alternative action representations including computed actions (Tran et al. 2007; Lanzi and Loiacono 2007) and continuous actions (Wilson 2007).

3.5.3 Evolutionary Selection of Representations

As we have seen, there are many representations to choose from. Unfortunately, it is generally not clear which might be best for a given problem or part of a problem. One approach is to let evolution make these choices. This can be seen as a form of meta-learning in which evolution adapts the inductive bias of the learner. In Bacardit (2004) and Bacardit et al. (2007), evolution was used to select default actions in decision lists in GAssist. GAssist's initial population was seeded with last rules, which together advocated all possible classes and over time evolution selected the most suitable of these rules. To obtain good results, it was necessary to encourage diversity in default actions. In GALE (Llorá 2002), evolution selects both classification algorithms and representations. GALE has elements of cellular automata and artificial life: individuals are distributed on a two-dimensional grid. Only neighbors within r cells interact: two neighbors may perform crossover, an individual may be cloned and copied to a neighboring cell, and an individual may die if its neighbors are fitter. A number of representations have been used: rule sets, prototypes, and decision trees (orthogonal, oblique, and multivariate based on nearest neighbor). Decision trees are evolved using GP while prototypes are used by a k-nearest-neighbor algorithm to select outputs. An individual uses a particular representation and classification algorithm and hence evolutionary selection operates on both. Populations may be homogeneous or heterogeneous and in Llorá and Wilson (2004) GALE was modified to interbreed orthogonal and oblique trees.

In the representational ecology approach (Marshall and Kovacs 2006) condition shapes were selected by evolution. Two Boolean classification tasks were used: a plane function, which is easy to describe with hyperplanes, but hard with hyperspheres, and a sphere function, which has the opposite characteristics. Three versions of XCS were used: with hyperplane conditions (XCS-planes), with hyperspheres (XCS-spheres), and both (XCS-both). XCS was otherwise unchanged, but in XCS-both the representations compete due to XCS's pressure against overlapping rules (Kovacs 2004). In XCS-both planes and spheres do not interbreed; they constitute genetically independent populations, that is, species. In terms of classification accuracy XCS-planes did well on the plane function and poorly on the sphere function while XCS-sphere showed the opposite results. XCS-both performed well on both; there was no significant difference in accuracy compared to the better single-representation version on each problem. Furthermore, XCS-both selected the more appropriate representation for each function. In terms of the amount of training needed, XCS-both was similar to XCS-sphere on the sphere function but was significantly slower than XCS-plane on the plane function.

Selecting Discretization Methods and Cut Points

GAssist's Adaptive Discretization Intervals (ADI) approach has two parts (Bacardit 2004). The first consists of adapting interval sizes. To begin, a discretization algorithm proposes cut points for each attribute and this defines the finest discretization possible, said to be composed of micro-intervals. Evolution can merge and split macro-intervals, which are composed of micro-intervals, and each individual can have different macro-intervals. The second part consists of selecting discretization algorithms. Evolution is allowed to select discretization algorithms for each attribute or rule from a pool including uniform width, uniform frequency, ID3, Fayyad and Irani, Màntaras, USD, ChiMerge, and random. Unfortunately, evolving the best discretizers was found to be difficult and the use of ADI resulted in only small improvements in accuracy. However, further work was suggested.

3.5.4 Optimization of Population Representation: Macroclassifiers

Wilson (1995) introduced an optimization for Michigan populations called macroclassifiers. He noted that as an XCS population converges on the solution set, many identical copies of this set accumulate. A macroclassifier is simply a rule with an additional *numerosity* parameter, which indicates how many identical virtual rules it represents. Using macroclassifiers saves a great deal of run-time compared to processing a large number of identical rules. Furthermore, macroclassifiers provide interesting statistics on evolutionary dynamics. Empirically, macroclassifiers perform essentially as the equivalent "micro" classifiers (Kovacs 1996). ❯ *Figure 9* illustrates how the rules m and m' in the top can be represented by m alone in the bottom by adding a numerosity parameter.

3.5.5 Rule Discovery

LCS are interesting from an evolutionary perspective, particularly Michigan systems in which evolutionary dynamics are more complex than in Pittsburgh systems. Where Pittsburgh systems face two objectives (evolving accurate and parsimonious rulesets) Michigan systems face a third: coverage of the input (or input/output) space. Furthermore, Michigan systems have coevolutionary dynamics as rules both cooperate and compete. Since the level of selection (rules) is lower than the level of solutions (rulesets) researchers have attempted to coax better results by modifying rule fitness calculations with various methods. Fitness sharing and crowding have been used to encourage diversity and, hence, coverage. Fitness sharing is naturally based on inputs (see ZCS) while crowding has been implemented by making deletion probability proportional to the degree of overlap with other rules (as in XCS). Finally, restricted mating as implemented by a niche GA plays an important role in XCS and UCS (see ❯ Sect. 3.5.5).

☐ **Fig. 9**

A population of microclassifiers (*top*) and the equivalent macroclassifiers (*bottom*).

Rule	Cond.	Action	Strength
m	# # 0 0 1 1	1	200.0
m'	# # 0 0 1 1	1	220.0
n	# # 0 0 1 1	0	100.0
o	0 0 1 1 1 0	1	100.0

Rule	Cond.	Action	Strength	Numerosity
m	# # 0 0 1 1	1	200.0	2
n	# # 0 0 1 1	0	100.0	1
o	0 0 1 1 1 0	1	100.0	1

Windowing in Pittsburgh LSS

As noted in ❷ Sect. 2.2 naive implementations of the Pittsburgh approach are slower than Michigan systems, which are themselves slow compared to non-evolutionary methods. The naive Pitt approach is to evaluate each individual on the entire data set, but much can be done to improve this. Instead, windowing methods (Fürnkranz 1998) learn on subsets of the data to improve runtime. Windowing has been used in Pitt LCS since, at least, ADAM (Greene and Smith 1987). More recently GAssist used incremental learning by alternating strata (ILAS) (Bacardit 2004) which partitions the data into *n* strata, each with the same class distribution as the entire set. A different stratum is used for fitness evaluation each generation. On larger data sets speed-up can be an order of magnitude. Windowing has become a standard part of recent Pittsburgh systems applied to real-world problems (e.g., Bacardit and Krasnogor (2008) and Bacardit et al. (2008, 2009)).

Many ensemble methods improve classification accuracy by sampling data in similar ways to windowing techniques, which suggests the potential for both improved accuracy and runtime, but this has not been investigated in LCS.

Michigan Rule Discovery

Most rule discovery work focuses on Michigan LCS as they are more common and their evolutionary dynamics are more complex. The rest of this section deals with Michigan systems although many ideas, such as self-adaptive mutation, could be applied to Pitt systems. Michigan LCS use the steady state GAs introduced in ❷ Sect. 2.4 as they minimize disruption to the rule population during on-line learning. An unusual feature of Michigan LCS is the emphasis placed on minimizing population size, for which various techniques are used: niche GAs, the addition of a generalization term in fitness, subsumption deletion, condensation and various compaction methods.

Niche GAs

Whereas in a standard panmictic GA all rules are eligible for reproduction, in a niche GA mating is restricted to rules in the same action set (which is considered a niche). (See ❷ Fig. 3 on action sets.) The input spaces of rules in an action set overlap and their actions agree, which suggests their predictions will tend to be related. Consequently, mating these related rules is more effective, on average, than mating rules drawn from the entire population. This is a form of speciation since it creates non-interbreeding sub-populations. However, the niche GA has many other effects (Wilson 2001). First, a strong bias toward general rules, since they match more inputs and hence appear in more action sets. Second, pressure against overlapping rules, since they compete for reproduction (Kovacs 2004). Third, complete coverage of the input space, since competition occurs for each input. The niche GA was introduced in Booker (1989) and originally operated in the match set but was later further restricted to the action set (Wilson 1998). It is used in XCS and UCS and is related to *universal suffrage* (Giordana and Saitta 1994).

EDAs Instead of GAs

Recently Butz et al. (2005, 2006) and Butz and Pelikan (2006) replaced XCS's usual crossover with an estimation of distribution algorithm (EDA)-based method to improve solving of difficult hierarchical problems, while (Llorà et al. 2005a, b) introduced CCS: a Pitt LCS based on compact GAs (a simple form of EDA).

Subsumption Deletion

A rule *x* logically subsumes a rule *y* when *x* matches a superset of the inputs *y* matches and they have the same action. For example, 00#→0 subsumes 000→0 and 001→0. In XCS *x* is allowed to subsume *y* if: (i) *x* logically subsumes *y*, (ii) *x* is sufficiently accurate and (iii) *x* is sufficiently experienced (has been evaluated sufficiently) so we can have confidence in its accuracy. Subsumption deletion was introduced in XCS (see Butz and Wilson (2001)) and takes two forms. In *GA subsumption*, when a child is created, we check to see if its parents subsume it, which constrains accurate parents to only produce more general children. In *action set subsumption*, the most general of the sufficiently accurate and experienced rules in the action set is given the opportunity to subsume the others. This removes redundant, specific rules from the population but is too aggressive for some problems.

Michigan Evolutionary Dynamics

Michigan LCS have interesting evolutionary dynamics and plotting macroclassifiers is a useful way to monitor population convergence and parsimony. ❯ *Figure 10* illustrates by showing XCS learning the 11 multiplexer function. The performance curve is a moving average of the proportion of the last 50 inputs which were classified correctly, %[O] shows the proportion of the minimal set of 16 ternary rules XCS needs to represent this function (indicated by the straight line labeled "Size of minimal DNF solution" in the figure) and macroclassifiers were explained in ❯ Sect. 3.5.4. In this experiment, the population was initially empty and was seeded by covering ❯ Sect. 3.5.5. "Cycles" refers to the number of inputs presented, inputs were drawn uniform randomly from the input space, the population size limit was 800, all input/output pairs were in both the train and test sets, GA subsumption was used but action set subsumption was not and curves are the average of 10 runs. Other settings are as in Wilson (1995).

◼ **Fig. 10**
Evolutionary dynamics of XCS on the 11 multiplexer.

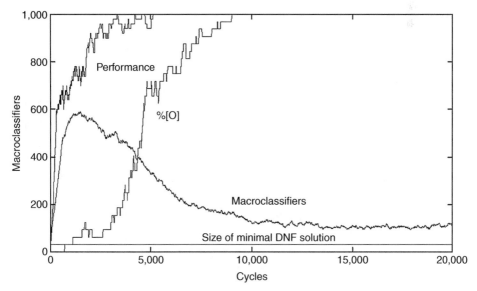

Note that XCS continues to refine its solution (population) after 100% performance is reached and that it finds the minimal representation (at the point where %[O] reaches the top of the figure) but that continued crossover and mutation generate extra transient rules, which make the population much larger.

Condensation

As illustrated by ❯ *Fig. 10*, an evolved population normally contains many redundant and low-fitness rules. These rules are typically transient, but more are generated while the GA continues to run. Condensation (Wilson 1998; Kovacs 1996) is a very simple technique to remove such rules, which consists of running the system with crossover and mutation turned off; we only clone and delete existing rules. ❯ *Figure 11* repeats the experiment from ❯ *Fig. 10* but switches after 15,000 cycles to condensation after which the population quickly converges to the minimal solution. Other methods of compacting the population have been investigated (Kovacs 1997; Wilson 2002b; Dixon et al. 2002).

Tuning Evolutionary Search

XCS is robust to class imbalances (Orriols-Puig and Bernadó-Mansilla 2006) but for very high imbalances tuning the GA based on a facetwise model improved performance (Orriols-Puig and Bernadó-Mansilla 2006; Orriols-Puig et al. 2007b). Self-tuning evolutionary search has also been studied. The mutation rate can be adapted during evolution for example Hurst and Bull (2003, 2004), Howard et al. (2008), and Butz et al. (2008), while De Jong et al. (1993) dynamically controls use of two generalization operators: each has a control bit specifying whether it can be used and control bits evolve with the rest of the genotypes.

◘ **Fig. 11**
XCS with condensation on the 11 multiplexer.

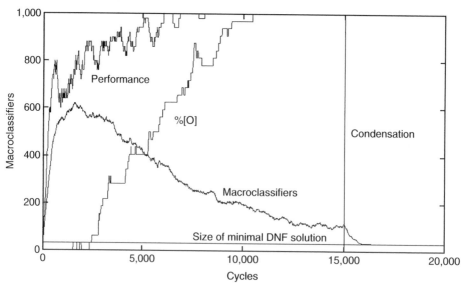

Non-evolutionary Rule Discovery

Evolution has been supplemented by heuristics in various ways. Covering, first suggested in Holland (1976), creates a rule to match an unmatched input. It can be used to create ("seed") the initial population (Venturini 1993; Wilson 1995; Hekanaho 1995) or to supplement the GA throughout evolution (Wilson 1995). Kovacs (2004, p. 42) found covering each action set was preferable to covering the match set when applying XCS to sequential tasks. Most covering/seeding is done as needed but instead (Juan Liu and Tin-Yau Kwok 2000) selects inputs at the center of same-class clusters. For other non-evolutionary operators see (Booker 1989; Riolo 1987), the work on corporations of rules (Wilson and Goldberg 1989; Smith 1994; Tomlinson and Bull 1998, 2002; Tomlinson 1999) and the work on non-evolutionary LCS.

Non-evolutionary LCS

Although LCS were originally conceived as a way of applying GAs to learning problems (Holland and Reitman 1978), not all LCS include a GA. Various heuristics have been used to create and refine rules in for example YACS (Gerard et al. 2002) and MACS (Gérard and Sigaud 2003). A number of systems have been inspired by psychological models of learning. ACS (Stolzmann 1996; Butz 2002) and ACS2 (Butz 2002a) are examples, although ACS was also later supplemented by a GA (Butz et al. 2000a, b). Another is AgentP, a specialized LCS for maze tasks (Zatuchna 2004, 2005).

3.5.6 LCS Credit Assignment

While credit assignment in Pittsburgh LCS is a straightforward matter of multiobjective fitness evaluation, as noted in ❷ Sect. 3.5.5 it is far more complex in Michigan systems with their more complex evolutionary dynamics. Credit assignment is also more complex in some learning paradigms, particularly reinforcement learning, which we will not cover here. Within supervised learning, credit assignment is more complex in regression tasks than in classification. These difficulties have been the major issue for Michigan LCS and have occupied a considerable part of the literature, particularly prior to the development of XCS, which provided a reasonable solution for both supervised and reinforcement learning.

Strength and Accuracy in Michigan LCS

Although we are not covering reinforcement learning, Michigan LCS have traditionally been designed for such problems. XCS/XCSF are reinforcement learning systems, but since supervised learning can be formulated as simplified reinforcement learning, they have been applied to SL tasks. Consequently, we now very briefly outline the difference between the two major forms of Michigan reinforcement learning LCS.

In older (pre-1995) reinforcement learning, LCS fitness is proportional to the magnitude of reward and is called *strength*. Strength is used both for conflict resolution and as fitness in the GA (see e.g., ZCS (Wilson 1994)). Such LCS are referred to as strength-based and they suffer from many difficulties with credit assignment (Kovacs 2004), the analysis of which is quite complex. Although some strength-based systems incorporate accuracy as a component of fitness, their fitness is still proportional to reward. In contrast, the main feature of XCS is that it adds a prediction parameter, which estimates the reward to be obtained if the action advocated by a rule is taken. Rule fitness is proportional to the accuracy of reward prediction and not to its magnitude, which avoids many problems strength-based systems have with credit assignment. In XCS, accuracy is estimated from the variance in reward and since overgeneral rules have high variance they have low fitness. Although XCS has proved robust in a range of applications, a major limitation is that the accuracy estimate conflates several things: (i) overgenerality in rules, (ii) noise in the training data, and (iii) stochasticity in the transition function in sequential problems. In contrast, strength-based systems may be less affected by noise and stochasticity since they are little affected by reward variance. See Kovacs (2004) for analysis of the two approaches.

Prediction Updates

To update rule predictions while training, the basic XCS system (Wilson 1995; Butz and Wilson 2001) uses the Widrow–Hoff update for nonsequential problems and the Q-learning update for sequential ones. Various alternatives have been used: average rewards (Tharakannel and Goldberg 2002; Lanzi and Loiacono 2006), gradient descent (Butz et al. 2005b; Lanzi et al. 2007), and eligibility traces (Drugowitsch and Barry 2005). The basic XCSF uses NLMS (linear piecewise) prediction (Wilson 2000b, 2002a) but Lanzi et al. (2006b) has compared various alternative classical parameter estimation (RLS and Kalman filter) and gain adaptation algorithms (K1, K2, IDBD, and IDD). He found that Kalman filter and RLS have significantly better accuracy than the others and that Kalman filter produces more compact solutions than RLS. There has also been recent work on other systems; UCS is essentially a supervised version of XCS and the main difference is its prediction update. Bull has also studied simplified LCS (Bull 2005).

Evolutionary Selection of Prediction Functions

In Lanzi et al. (2008), Lanzi selects prediction functions in XCSFHP (XCSF with Heterogeneous Predictors) in a way similar to the selection of condition types in the representational ecology approach in ❯ Sect. 3.5.3. Polynomial functions (linear, quadratic, and cubic) and constant, linear and NN predictors were available. XCSFHP selected the most suitable predictor for regression and sequential tasks and performed almost as well as XCSF using the best single predictor.

Theoretical Results

Among the notable theoretical works on LCS, Lanzi (2002a) demonstrates that XCS without generalization implements tabular Q-learning, Butz et al. (2005a) investigate the computational complexity of XCS in a probably approximately correct (PAC) setting, and Wada et al. (2005a, b, c, 2007) analyze credit assignment and relate LCS to mainstream reinforcement learning methods. Kovacs (2004) identifies pathological rule types: strong overgeneral and fit overgeneral rules, which are overgeneral yet stronger/fitter than not-overgeneral competitors. Fortunately, such rules are only possible under specific circumstances. A number of papers seek to characterize problems which are hard for LCS (Goldberg et al. 1992; Kovacs 2000, 2001, 2004; Bernadó-Mansilla and Ho 2005; Bagnall and Zatuchna 2005) while others model evolutionary dynamics (Butz et al. 2004a, b, 2007; Butz 2006; Orriols-Puig 2007c, d) and yet others attempt to reconstruct LCS from first principles using probabilistic models (Drugowitsch 2007, 2008; Edakunni et al. 2009).

Hierarchies and Ensembles of LCS

Hierarchical LCS have been studied for some time and Barry (1996) reviews early work. Dorigo and Colombetti (1998), Donnart and Meyer (1996a, b), and Donnart (1998) apply hierarchical LCS to robot control while Barry (2000) uses hierarchical XCSs to learn long sequences of actions. The ensembles field (❯ Sect. 3.3) studies how to combine predictions (Kuncheva 2004) and all the above could be reformulated as ensembles of LCS. Some recent work has taken this approach (Dam et al. 2005; Bull et al. 2007).

3.5.7 Conclusions

LCS face certain inherent difficulties; Michigan systems face complex credit assignment problems while in Pittsburgh systems, run-time can be a major issue. The same is true for all GBML systems, but the Michigan approach has been explored far more extensively within LCS than elsewhere. Recently, there has been much integration with mainstream machine learning and much research on representations and credit assignment algorithms. Most recent applications have been to data mining and function approximation, although some work continues on reinforcement learning. Future directions are likely to include exposing more of the LCS to evolution and further integration with machine learning, ensembles, memetic algorithms, and multiobjective optimization.

Reading

No general up-to-date introduction to LCS exists. For the basics see Goldberg (1989) and the introductory parts of Kovacs (2004) or Butz (2006). For a good introduction to representations and operators see chapter 6 of Freitas (2002a). For a review of early LCS see Barry (2000). For reviews of LCS research see Wilson and Goldberg (1989), Lanzi and Riolo (2000), and Lanzi (2008). For a review of state-of-the-art GBML and empirical comparison to non-evolutionary pattern recognition methods see Orriols-Puig et al. (2008b). For other comparisons with non-evolutionary methods see Bonelli and Alexandre (1991), Greenyer (2000), Saxon and Barry (2000), Wilson (2000), Bernadó et al. (2002), and Bernadó-Mansilla and Garrell-Guiu (2003). Finally, the LCS bibliography (Kovacs 2009) has over 900 references.

3.6 Genetic Fuzzy Systems

Following the section on LCS, this section covers a second actively developing approach to evolving rule-based systems. We will see that the two areas overlap considerably and that the distinction between them is somewhat arbitrary. Nonetheless, the two communities and their literatures are somewhat disjoint.

Fuzzy logic is a major paradigm in soft computing which provides a means of approximate reasoning not found in traditional crisp logic. Genetic fuzzy systems (GFS) apply evolution to fuzzy learning systems in various ways: GAs, GP, and evolution strategies have all been used. We will cover a particular form of GFS called genetic fuzzy rule-based systems (FRBS), which are also known as learning fuzzy classifier systems (LFCS) (Bonarini 2000) or referred to as, for example, "genetic learning of fuzzy rules" and (for reinforcement learning tasks) "fuzzy Q-learning". Like other LCS, FRBS evolve if-then rules but in FRBS the rules are fuzzy. Most systems are Pittsburgh, but there are many Michigan examples (Valenzuela-Rendón 1991, 1998; Geyer-Schulz 1997; Bonarini 2000; Orriols-Puig et al. 2007a, 2008a; Casillas et al. 2007). In addition to FRBS, we briefly cover genetic fuzzy NNs, but we do not cover genetic fuzzy clustering (see Oscar-Cordón et al. (2001)).

In the terminology of fuzzy logic, ordinary scalar values are called *crisp* values. A *membership function* defines the degree of match between crisp values and a set of fuzzy linguistic *terms.* The set of terms is a *fuzzy set.* The following figure shows a membership function for the set {cold, warm, and hot}.

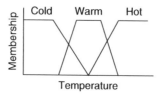

Each crisp value matches *each* term to some degree in the interval [0,1], so, for example, a membership function might define 5° as 0.8 cold, 0.3 warm and 0.0 hot. The process of computing the membership of each term is called fuzzification and can be considered a form of discretization. Conversely, defuzzification refers to computing a crisp value from fuzzy values.

Fuzzy rules are condition/action (IF–THEN) rules composed of a set of linguistic variables (e.g., temperature, humidity), which can each take on linguistic terms (e.g., cold, warm, and hot). For example:

IF temperature IS cold AND humidity IS high THEN heater IS high
IF temperature IS warm AND humidity IS low THEN heater IS medium

As illustrated in ❷ *Fig. 12,* a fuzzy rule-based system consists of:

- A rule base (RB) of fuzzy rules
- A data base (DB) of linguistic terms and their membership functions
- Together the RB and DB are the *knowledge base* (KB)
- A fuzzy inference system which maps from fuzzy inputs to a fuzzy output
- Fuzzification and defuzzification processes

◘ Fig. 12

Components and information flow in a fuzzy rule-based system. (Adapted from Hekanaho 1995.)

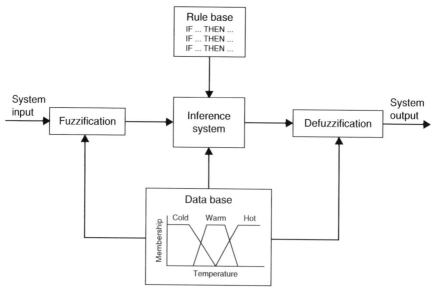

3.6.1 Evolution of FRBSs

We distinguish (i) genetic tuning and (ii) genetic learning of DB, RB, or inference engine parameters.

Genetic Tuning

The concept behind genetic tuning is to first train a hand-crafted FRBS and then to evolve the DB (linguistic terms and membership functions) to improve performance. In other words, we do not alter the hand-crafted rule base but only tune its parameters. Specifically, we can adjust the shape of the membership functions and the parameterized expressions in the (adaptive) inference system and adapt defuzzification methods.

Genetic Learning

The concept of genetic learning is to evolve the DB, RB, or inference engine parameters. There are a number of approaches. In *genetic rule learning*, we usually predefine the DB by hand and evolve the RB. In *genetic rule selection* we use the GA to remove irrelevant, redundant, incorrect, or conflicting rules. This is a similar role to condensation in LCS (see ❯ Sect. 3.5.5). In *genetic KB learning* we learn both the DB and RB. We either learn the DB first and then learn the RB or we iteratively learn a series of DBs and evaluate each one by learning an RB using it.

It is also possible to learn components simultaneously, which may produce better results though the larger search space makes it slower and more difficult than adapting components independently. As examples, Morimoto et al. (1997) learn the DB and RB simultaneously while Homaitar and McCormick (1995) learn KB components and inference engine parameters simultaneously.

Recently, Sánchez and Couso (2007) claimed that all existing GFS have been applied to crisp data and that with such data the benefits of GFS compared to other learning methods are limited to linguistic interpretability. However, GFS has the potential to outperform other methods on fuzzy data and they identify three cases (Sánchez and Couso 2007, p. 558):

1. Crisp data with hand-added fuzziness
2. Transformations of data based on semantic interpretations of fuzzy sets
3. Inherently fuzzy data

They argue that GFS should use fuzzy fitness functions in such cases to deal directly with the uncertainty in the data and propose such systems as a new class of GFS to add to the taxonomy of Herrera (2008).

3.6.2 Genetic Neuro-fuzzy Systems

A neuro-fuzzy system (NFS) or fuzzy neural network (FNN) is any combination of fuzzy logic and neural networks. Among the many examples of such systems, Liangjie and Yanda (1996) use a GA to minimize the error of the NN, Hanebeck and Schmidt (1996) use both a GA and backpropagation to minimize error, Perneel and Themlin (1995) optimize a fuzzy expert system using a GA and NN, and Morimoto et al. (1997) use a NN to approximate the fitness function for a GA, which adapts membership functions and controls rules. See Cordón et al.

(2001) for an introduction to NFS, Linkens and Nyongesa (1996) for a review of EAs, NNs, and fuzzy logic from the perspective of intelligent control, and He et al. (1999) for a discussion of combining the three. Kolman and Margaliot (2009) introduce fuzzy all-permutations rule-bases (FARBs), which are mathematically equivalent to NNs.

3.6.3 Conclusions

Herrera (2008, p. 38) lists the following active areas within GFS:

1. Multi-objective genetic learning of FRBSs: interpretability–precision trade-off
2. GA-based techniques for mining fuzzy association rules and novel data mining approaches
3. Learning genetic models based on low quality data (e.g., noisy data)
4. Genetic learning of fuzzy partitions and context adaptation
5. Genetic adaptation of inference engine components
6. Revisiting the Michigan-style GFSs

Herrera also lists (p. 42) current issues for GFS:

1. Human readability
2. New data mining tasks: frequent and interesting pattern mining, mining data streams
3. Dealing with high dimensional data

Reading

There is a substantial GFS literature. Notable works include the four seminal 1991 papers on genetic tuning of the DB (Karr 1991), the Michigan approach (Valenzuela-Rendón 1989), the Pittsburgh approach (Thrift 1991), and relational matrix-based FRBS (Pham and Karaboga 1991). Subsequent work includes Geyer-Schulz's 1997 book on Michigan fuzzy LCS learning RBs with GP (Geyer-Schulz 1997), Bonarini's 2000 introductory chapter from an LCS perspective (Bonarini 2000), Mitra and Hayashi's 2000 survey of neuro-fuzzy rule generation methods (Mitra and Hayashi 2000), Cordon et al.'s 2001 book on genetic fuzzy systems in general (Cordón et al. 2001), Angelov's 2002 book on evolving FRBS (Angelov 2002), chapter 10 of Freitas' 2002 book on evolutionary data mining (Freitas 2002a), Herrera's 2008 survey article on GFS (Herrera 2008) (which lists further key reading), and finally Kolman and Margaliot's 2009 book on the neuro-fuzzy FARB approach (Kolman and Margaliot 2009).

4 Conclusions

The reader should need no convincing that GBML is a very diverse and active area. Although much integration with mainstream machine learning has taken place in the last 10 years, more is needed. The use of multiobjective EAs in GBML is spreading. Integration with ensembles is natural given the population-based nature of EAs but it is only just beginning. Other areas which need attention are memetics, meta-learning, hyperheuristics, and estimation of distribution algorithms. In addition to further integration with other areas, the constituent areas of GBML need more interaction with each other.

Two persistent difficulties for GBML are worth highlighting. First, run-time speed remains an issue as EAs are much slower than most other methods. While this sometimes matters little (e.g., in off-line learning with modest datasets), it is sometimes equally critical (e.g., in stream mining). Various methods to speed up GBML exist (see e.g., Freitas (2002a, Sect. 12.1.3) and more research is warranted, but this may simply remain a weakness. The second difficulty is theory. EA theory is notoriously difficult and when coupled with other processes becomes even less tractable. Nonetheless, substantial progress has been made in the past 10 years, most notably with LCS.

Other active research directions will no doubt include meta-learning such as the evolution of bias (e.g., selection of representation), evolving heuristics and learning rules for specific problem classes, and other forms of self-adaptation. In the area of data preparation, Freitas (2002a, Sect. 12.2.1) argues that attribute construction is a promising area for GBML and that filter methods for feature selection are faster than wrappers and deserve more GBML research. Finally, many specialized learning problems (not to mention specific applications) remain little or unexplored with GBML, including ranking, semi-supervised learning, transductive learning, inductive transfer, learning to learn, stream mining, and no doubt others that have not yet been formulated.

Acknowledgments

Thanks to my editor Thomas Bäck for his patience and encouragement, and to Larry Bull, John R. Woodward, Natalio Krasnogor, Gavin Brown and Arjun Chandra for comments.

Glossary

EA	Evolutionary Algorithm
FRBS	Fuzzy Rule-Based System
GA	Genetic Algorithm
GBML	Genetics-Based Machine Learning
GFS	Genetic Fuzzy System
GP	Genetic Programming
LCS	Learning Classifier System
NN	Neural Network
SL	Supervised Learning

References

Abbass HA (2003) Speeding up backpropagation using multiobjective evolutionary algorithms. Neural Comput, 15(11):2705–2726

Ackley DH, Littman ML (1992) Interactions between learning and evolution. In: Langton C, Taylor C, Rasmussen S, Farmer J (eds) Artificial life II: Santa Fe institute studies in the sciences of Complexity, vol 10. Addison-Wesley, New York, pp 487–509

Aguilar-Ruiz J, Riquelme J, Toro M (2003) Evolutionary learning of hierarchical decision rules. IEEE Trans Syst Man Cybern B 33(2):324–331

Ahluwalia M, Bull L (1999) A genetic programming-based classifier system. In: Banzhaf et al. (eds) GECCO-99: Proceedings of the genetic and evolutionary computation conference. Morgan Kaufmann, San Francisco, pp 11–18

Andersen HC, Tsoi AC (1993) A constructive algorithm for the training of a multi-layer perceptron based on the genetic algorithm. Complex Syst, 7(4):249–268

Angeline PJ, Sauders GM, Pollack JB (1994) An evolutionary algorithm that constructs recurrent neural networks. IEEE Trans Neural Netw 5:54–65

Angelov P (2002) Evolving rule-based models: a tool for design of flexible adaptive systems. Studies in fuzziness and soft computing, vol 92. Springer, Heidelberg

Asuncion A, Newman DJ (2009) UCI machine learning repository. http://www.ics.uci.edu/~mlearn/ML Repository.html

Bacardit J, Stout M, Hirst JD and Krasnogor N (2008) Data mining in proteomics with learning classifier systems. In: Bull L, Bernadó Mansilla E, Holmes J (eds) Learning classifier systems in data mining. Springer, Berlin, pp 17–46

Bacardit J, Burke EK, Krasnogor N (2009a) Improving the scalability of rule-based evolutionary learning. Memetic Comput 1(1):55–57

Bacardit J, Stout M, Hirst JD, Valencia A, Smith RE, Krasnogor N (2009b) Automated alphabet reduction for protein datasets. BMC Bioinformatics. vol 10, 6

Bacardit J (2004) Pittsburgh genetic-based machine learning in the data mining era: representations, generalization, and run-time. PhD thesis, Universitat Ramon Llull, Barcelona, Spain

Bacardit J, Goldberg DE, Butz MV (2007) Improving the performance of a Pittsburgh learning classifier system using a default rule. In: Kovacs T, Llòra X, Takadama K, Lanzi PL, Stolzmann W, Wilson SW (eds) Learning Classifier Systems. International workshops, IWLCS 2003–2005, Revised selected papers, Lecture notes in computer science, vol 4399. Springer, Berlin, pp 291–307

Bacardit J, Krasnogor N (2008) Empirical evaluation of ensemble techniques for a Pittsburgh learning classifier system. In: Bacardit J, Bernadó-Mansilla E, Butz M, Kovacs T, Llorà X, Takadama K (eds) Learning Classifier Systems. 10th and 11th International Workshops (2006–2007), Lecture notes in computer science, vol 4998. Springer, Berlin, pp 255–268

Bagnall AJ, Zatuchna ZV (2005) On the classification of maze problems. In: Bull L, Kovacs T (eds) Applications of learning classifier systems. Springer, Berlin, pp 307–316

Banzhaf W, Daida J, Eiben AE, Garzon MH, Honavar V, Jakiela M, Smith RE (eds) (1999) GECCO-99: Proceedings of the genetic and evolutionary computation conference. Morgan Kaufmann, San Francisco, CA

Barry A (1996) Hierarchy formulation within classifiers system: a review. In: Goodman EG, Uskov VL, Punch WF (eds) Proceedings of the first international conference on evolutionary algorithms and their application EVCA'96. The Presidium of the Russian Academy of Sciences, Moscow, pp 195–211

Barry A (2000) XCS performance and population structure within multiple-step environments. PhD thesis, Queen's University Belfast, Belfast

Beielstein T, Markon S (2002) Threshold selection, hypothesis tests and DOE methods. In: 2002 congress on evolutionary computation. IEEE Press, Washington, DC, pp 777–782

Belew RK, McInerney J, Schraudolph NN (1992) Evolving networks: using the genetic algorithm with connectionistic learning. In: Langton CG, Taylor C, Farmer JD, Rasmussen S (eds) Proceedings of the 2nd conference on artificial life. Addison-Wesley, New York, pp 51–548

Bernadó E, Llorà X, Garrell JM (2002) XCS and GALE: a comparative study of two learning classifier systems on data mining. In: Lanzi PL, Stolzmann W, Wilson SW (eds) Advances in learning classifier systems, Lecture notes in artificial intelligence, vol 2321. Springer, Berlin, pp 115–132

Bernadó-Mansilla E, Garrell-Guiu JM (2003) Accuracy-based learning classifier systems: models, analysis and applications to classification tasks. Evolut Comput 11(3):209–238

Bernadó-Mansilla E, Ho TK (2005) Domain of competence of XCS classifier system in complexity measurement space. IEEE Trans Evolut Comput 9(1):82–104

Bonarini A (2000) An introduction to learning fuzzy classifier systems. In: Lanzi PL, Stolzmann W, Wilson SW (eds) Learning classifier systems: from foundations to applications, Lecture note in artificial intelligence, vol 1813. Springer, Berlin, pp 83–104

Bonelli P, Alexandre P (1991) An efficient classifier system and its experimental comparison with two representative learning methods on three medical domains. In: Booker LB, Belew RK (eds) Proceedings of the 14th international conference on genetic algorithms (ICGA'91). Morgan Kaufmann, San Francisco, CA, pp 288–295

Booker LB (1989) Triggered rule discovery in classifier systems. In: Schaffer JD (ed) Proceedings of the 3rd international conference on genetic algorithms (ICGA-89). Morgan Kaufmann, San Francisco, CA, pp 265–274

Booker LB (1991) Representing attribute-based concepts in a classifier system. In: Rawlins GJE (ed) Proceedings of the first workshop on foundations of genetic algorithms (FOGA91). Morgan Kaufmann, San Mateo, CA, pp 115–127

Booker LB (2005a) Adaptive value function approximations in classifier systems. In: GECCO '05: proceedings of the 2005 workshops on genetic and evolutionary computation. ACM, New York, pp 90–91

Booker LB (2005b) Approximating value functions in classifier systems. In: Bull L, Kovacs T (eds) Foundations of learning classifier systems (Studies in fuzziness and soft computing), Lecture notes in artificial intelligence, vol 183, Springer, Berlin, pp 45–61

Booker LB, Belew RK (eds) (1991) Proceedings of the 4th international conference on genetic algorithms (ICGA91). Morgan Kaufmann, San Francisco, CA

Bot MCJ, Langdon WB (2000) Application of genetic programming to induction of linear classification trees. In: Genetic programming: proceedings of the 3rd European conference (EuroGP 2000), Lecture notes in computer science, vol 1802, Springer, Berlin, pp 247–258

Breiman L (1996) Bagging predictors. Mach Learn 24 (2):123–140

Breiman L (1998) Arcing classifiers. Ann Stat 26(3): 801–845

Brown G (2010) Ensemble learning. In: Sammut C, Webb G (eds) Encyclopedia of machine learning. Springer, Berlin

Brown G, Wyatt J, Harris R, Yao X (2005) Diversity creation methods: A survey and categorisation. J Inform Fusion 6(1):5–20

Bull L (2009) On dynamical genetic programming: simple Boolean networks in learning classifier systems. IJPEDS 24(5):421–442

Bull L, Studley M, Bagnall T, Whittley I (2007) On the use of rule-sharing in learning classifier system ensembles. IEEE Trans Evolut Comput 11:496–502

Bull L (2005) Two simple learning classifier systems. In: Bull L, Kovacs T (eds) Foundations of learning classifier systems (Studies in fuzziness and soft computing), Lecture notes in artificial intelligence, vol 183, Springer, Berlin, pp 63–90

Bull L, O'Hara T (2002) Accuracy-based neuro and neuro-fuzzy classifier systems. In: Langdon WB, Cantú-Paz E, Mathias K, Roy R, Davis R, Poli R, Balakrishnan K, Honavar V, Rudolph G, Wegener J, Bull L, Potter MA, Schultz AC, Miller JF, Burke E, Jonoska N (eds) GECCO 2002: Proceedings of the genetic and evolutionary computation conference. Morgan Kaufmann, San Francisco, CA, pp 905–911

Burke EK, Hyde MR, Kendall G, Ochoa G, Ozcan E, Woodward JR (2009) Exploring hyper-heuristic methodologies with genetic programming. In: Mumford C, Jain L (eds) Collaborative computational intelligence. Springer, Berlin

Burke EK, Kendall G (2005) Introduction. In: Burke EK, Kendall G (eds) Search methodologies: introductory tutorials in optimization and decision support techniques. Springer, Berlin, pp 5–18

Burke EK, Kendall G, Newall J, Hart E, Russ P, Schulenburg S (2003) Hyper-heuristics: an emerging direction in modern search technology. In: Glover F,
Kochenberger G (eds) Handbook of meta-heuristics. Kluwer, Norwell, MA, pp 457–474

Butz MV, Kovacs T, Lanzi PL, Wilson SW (2004b) Toward a theory of generalization and learning in XCS. IEEE Trans Evolut Comput 8(1):8–46

Butz MV (2002a) An algorithmic description of ACS2. In: Lanzi PL, Stolzmann W, Wilson SW (eds) Learning classifier systems: from foundations to applications, Lecture notes in artificial intelligence, vol 2321. Springer, Berlin, pp 211–229

Butz MV (2002b) Anticipatory learning classifier systems. Kluwer, Norwell, MA

Butz MV, Goldberg DV, Stolzmann W (2000a) Introducing a genetic generalization pressure to the anticipatory classifier system – part 1: theoretical approach. In: Whitley D, Goldberg D, Cantú-Paz E, Spector L, Parmee I, Beyer H-G (eds) Proceedings of genetic and evolutionary computation conference (GECCO 2000). Morgun Kaufmann, San Francisco, CA, pp 34–41

Butz MV, Goldberg DE, Stolzmann W (2000b) Introducing a genetic generalization pressure to the anticipatory classifier system – part 2: performance analysis. In: Whitley D, Goldberg D, Cantú-Paz E, Spector L, Parmee I, Beyer H-G (eds) Proceedings of genetic and evolutionary computation conference (GECCO 2000). Morgun Kaufmann, San Francisco, CA, pp 42–49

Butz MV, Wilson SW (2001) An algorithmic description of XCS. In: Lanzi PL, Stolzmann W, Wilson SW (eds) Advances in learning classifier systems, Lecture notes in artificial intelligence, vol 1996. Springer, Berlin, pp 253–272

Butz MV (2005) Kernel-based, ellipsoidal conditions in the real-valued XCS classifier system. In: Beyer HG et al. (eds) Proceedings of the genetic and evolutionary computation conference (GECCO 2005). ACM, New York, pp 1835–1842

Butz MV (2006) Rule-based evolutionary online learning systems: a principled approach to LCS analysis and design. Studies in fuzziness and soft computing. Springer, Berlin

Butz MV, Goldberg DE, Lanzi PL (2004a) Bounding learning time in XCS. In Genetic and evolutionary computation (GECCO 2004), Lecture notes in computer science, vol 3103. Springer, Berlin, pp 739–750

Butz MV, Goldberg DE, Lanzi PL (2005a) Computational complexity of the XCS classifier system. In: Bull L, Kovacs T (eds) Foundations of learning classifier systems, Studies in fuzziness and soft computing, Lecture notes in artificial intelligence, vol 183. Springer, Berlin, pp 91–126

Butz MV, Goldberg DE, Lanzi PL (2005b) Gradient descent methods in learning classifier systems: improving XCS performance in multistep problems. IEEE Trans Evolut Comput 9(5):452–473

Butz MV, Goldberg DE, Lanzi PL, Sastry K (2007) Problem solution sustenance in XCS: Markov chain analysis of niche support distributions and the impact on computational complexity. GP and Evol Machines 8(1):5–37

Butz MV, Lanzi PL, Wilson SW (2006) Hyper-ellipsoidal conditions in XCS: rotation, linear approximation, and solution structure. In: Cattolico M (ed) Proceedings of the genetic and evolutionary computation conference (GECCO 2006). ACM, New York, pp 1457–1464

Butz MV, Pelikan M (2006) Studying XCS/BOA learning in Boolean functions: structure encoding and random Boolean functions. In: Cattolico M et al. (eds) Genetic and evolutionary computation conference, GECCO 2006. ACM, New York, pp 1449–1456

Butz MV, Pelikan M, Llorà X, Goldberg DE (2005) Extracted global structure makes local building block processing effective in XCS. In: Beyer HG, O'Reilly UM (eds) Proceedings of the genetic and evolutionary computation conference, GECCO 2005. ACM, New York, pp 655–662

Butz MV, Pelikan M, Llorà X, Goldberg DE (2006) Automated global structure extraction for effective local building block processing in XCS. Evolut Comput 14(3):345–380

Butz MV, Stalph P, Lanzi PL (2008) Self-adaptive mutation in XCSF. In GECCO '08: Proceedings of the 10th annual conference on genetic and evolutionary computation. ACM, New York, pp 1365–1372

Cantú-Paz E, Kamath C (2003) Inducing oblique decision trees with evolutionary algorithms. IEEE Trans Evolut Comput 7(1):54–68

Cantú-Paz E (2002) Feature subset selection by estimation of distribution algorithms. In: GECCO '02: Proceedings of the genetic and evolutionary computation conference. Morgan Kaufmann, San Francisco, CA, pp 303–310

Caruana R, Niculescu-Mizil A (2006) An empirical comparison of supervised learning algorithms. In: ICML '06: Proceedings of the 23rd international conference on machine learning. ACM, New York, pp 161–168

Casillas J, Carse B, Bull L (2007) Fuzzy-XCS: a Michigan genetic fuzzy system. IEEE Trans Fuzzy Syst 15: 536–550

Castilloa PA, Merelo JJ, Arenas MG, Romero G (2007) Comparing evolutionary hybrid systems for design and optimization of multilayer perceptron structure along training parameters. Inform Sciences, 177 (14):2884–2905

Chalmers D (1990) The evolution of learning: An experiment in genetic connectionism. In: Touretsky E (ed) Proceedings 1990 connectionist models summer school. Morgan Kaufmann, San Francisco, CA pp 81–90

Chandra A, Yao X (2006a) Ensemble learning using multi-objective evolutionary algorithms. J Math Model Algorithm 5(4):417–445

Chandra A, Yao X (2006b) Evolving hybrid ensembles of learning machines for better generalisation. Neurocomputing 69(7–9):686–700

Cho S, Cha K (1996) Evolution of neural net training set through addition of virtual samples. In: Proceedings of the 1996 IEEE international conference on evolutionary computation. IEEE Press, Washington DC, pp 685–688

Cho S-B (1999) Pattern recognition with neural networks combined by genetic algorithm. Fuzzy Set Syst 103:339–347

Cho S-B, Park C (2004) Speciated GA for optimal ensemble classifiers in DNA microarray classification. In: Congress on evolutionary computation (CEC 2004), vol 1. pp 590–597

Corcoran AL, Sen S (1994) Using real-valued genetic algorithms to evolve rule sets for classification. In: Proceedings of the IEEE conference on evolutionary computation. IEEE Press, Washington DC, pp 120–124

Cordón O, Herrera F, Hoffmann F, Magdalena L (2001) Genetic fuzzy systems. World Scientific, Singapore

Cribbs III HB, Smith RE (1996) Classifier system renaissance: new analogies, new directions. In: Koza JR, Goldberg DE, Fogel DB, Riolo RL (eds) Genetic programming 1996: proceedings of the first annual conference. MIT Press, Cambridge, MA, Stanford University, CA, USA, 28–31 July 1996. pp 547–552

Dam HH, Abbass HA, Lokan C, Yao X (2008) Neural-based learning classifier systems. IEEE Trans Knowl Data Eng 20(1):26–39

Dam HH, Abbass HA, Lokan C (2005) DXCS: an XCS system for distributed data mining. In: Beyer HG, O'Reilly UM (eds) Genetic and evolutionary computation conference, GECCO 2005. pp 1883–1890

Dasdan A, Oflazer K (1993) Genetic synthesis of unsupervised learning algorithms. Technical Report BU-CEIS-9306, Department of Computer Engineering and Information Science, Bilkent University, Ankara

De Jong KA, Spears WM (1991) Learning concept classification rules using genetic algorithms. In: Proceedings of the twelfth international conference on artificial intelligence IJCAI-91, vol 2. Morgan Kaufmann, pp 651–656

De Jong KA, Spears WM, Gordon DF (1993) Using genetic algorithms for concept learning. Mach Learn 3:161–188

Dietterich TG (1998) Machine-learning research: four current directions. AI Mag 18(4):97–136

Divina F, Keijzer M, Marchiori E (2002) Non-universal suffrage selection operators favor population diversity in genetic algorithms. In: Benelearn 2002: proceedings of the 12th Belgian-Dutch conference on machine learning (Technical report UU-CS-2002-046). pp 23–30

Divina F, Keijzer M, Marchiori E (2003) A method for handling numerical attributes in GA-based inductive concept learners. In: Proceedings of the genetic and evolutionary computation conference (GECCO-2003). Springer, Berlin, pp 898–908

Divina F, Marchiori E (2002) Evolutionary concept learning. In: Langdon WB, Cantú-Paz E, Mathias K, Roy R, Davis D, Poli R, Balakrishnan K, Honavar V, Rudolph G, Wegener J, Bull L, Potter MA, Schultz AC, Miller JF, Burke E, Jonoska N (eds) GECCO 2002: Proceedings of the Genetic and Evolutionary Computation Conference. Morgan Kaufmann, San Francisco, CA, New York, 9–13 July 2002. pp 343–350

Dixon PW, Corne D, Oates MJ (2002) A ruleset reduction algorithm for the XCS learning classifier system. In: Lanzi PL, Stolzmann W, Wilson SW (eds) Learning classifier systems, 5th international workshop (IWLCS 2002), Lecture notes in computer science, vol 2661. Springer, Berlin, pp 20–29

Donnart J-Y (1998) Cognitive architecture and adaptive properties of an motivationally autonomous animat. PhD thesis, Université Pierre et Marie Curie. Paris, France

Donnart J-Y, Meyer J-A (1996a) Hierarchical-map Building and Self-positioning with MonaLysa. Adapt Behav 5(1):29–74

Donnart J-Y, Meyer J-A (1996b) Learning reactive and planning rules in a motivationally autonomous animat. IEEE Trans Syst Man Cybern B Cybern 26 (3):381–395

Dorigo M, Colombetti M (1998) Robot shaping: an experiment in behavior engineering. MIT Press/Bradford Books, Cambridge, MA

Drugowitsch J, Barry A (2005) XCS with eligibility traces. In: Beyer H-G, O'Reilly U-M (eds) Genetic and evolutionary computation conference, GECCO 2005. ACM, New York, pp 1851–1858

Drugowitsch J (2008) Design and analysis of learning classifier systems: a probabilistic approach. Springer, Berlin

Drugowitsch J, Barry A (2007) A formal framework and extensions for function approximation in learning classifier systems. Mach Learn 70(1):45–88

Edakunni NE, Kovacs T, Brown G, Marshall JAR, Chandra A (2009) Modelling UCS as a mixture of experts. In: Proceedings of the 2009 Genetic and Evolutionary Computation Conference (GECCO'09). ACM, pp 1187–1994

Floreano D, Dürr P, Mattiussi C (2008) Neuroevolution: from architectures to learning. Evol Intell 1(1): 47–62

Folino G, Pizzuti C, Spezzano G (2003) Ensemble techniques for parallel genetic programming based classifiers. In: Proceedings of the sixth European conference on genetic programming (EuroGP'03), Lecture notes in computer science, vol 2610. Springer, Berlin, pp 59–69

Freitas AA (2002a) Data mining and knowledge discovery with evolutionary algorithms. Spinger, Berlin

Freitas AA (2002b) A survey of evolutionary algorithms for data mining and knowledge discovery. In: Ghosh A, Tsutsui S (eds) Advances in evolutionary computation. Springer, Berlin, pp 819–845

Freund Y, Schapire R (1996) Experiments with a new boosting algorithm. In: Proceedings of the international conference on machine learning (ICML'96), Bari, Italy, pp 148–156

Freund Y, Schapire R (1999) A short introduction to boosting. J Jpn Soc Artif Intell 14(5):771–780

Fürnkranz J (1998) Integrative windowing. J Artif Intell Res 8:129–164

Gagné C, Sebag M, Schoenauer M, Tomassini M (2007) Ensemble learning for free with evolutionary algorithms? In: GECCO '07: Proceedings of the 9th annual conference on genetic and evolutionary computation. ACM, New York, pp 1782–1789

Gathercole C, Ross P (1997) Tackling the Boolean even n parity problem with genetic programming and limited-error fitness. In: Koza JR, Deb K, Dorigo M, Fogel DB, Garzon M, Iba H, Riolo RL (eds) Genetic programming 1997: proceedings second annual conference. Morgan Kaufmann, San Francisco, CA, pp 119–127

Gérard P, Sigaud O (2003) Designing efficient exploration with MACS: Modules and function approximation. In: Cantú-Paz E, Foster JA, Deb K, Davis D, Roy R, O'Reilly U-M, Beyer H-G, Standish R, Kendall G, Wilson S, Harman M, Wegener J, Dasgupta D, Potter MA, Schultz AC, Dowsland K, Jonoska N, Miller J (eds) Genetic and evolutionary computation – GECCO-2003, Lecture notes in computer science, vol 2724. Springer, Berlin, pp 1882–1893

Gerard P, Stolzmann W, Sigaud O (2002) YACS, a new learning classifier system using anticipation. J Soft Comput 6(3–4):216–228

Geyer-Schulz A (1997) Fuzzy rule-based expert systems and genetic machine learning. Physica, Heidelberg

Giordana A, Neri F (1995) Search-intensive concept induction. Evolut Comput 3:375–416

Giordana A, Saitta L (1994) Learning disjunctive concepts by means of genetic algorithms. In: Proceedings of the international conference on machine learning, Brunswick, NJ, pp 96–104

Giraldez R, Aguilar-Ruiz J, Riquelme J (2003) Natural coding: a more efficient representation for evolutionary learning. In: Proceedings of the genetic and evolutionary computation conference (GECCO-2003). Springer, Berlin, pp 979–990

Giraud-Carrier C, Keller J (2002) Meta-learning. In: Meij J (ed) Dealing with the data flood. STT/Beweton, Hague, The Netherlands

Goldberg DE (1989) Genetic algorithms in search, optimization, and machine learning. Addison-Wesley, Reading, MA

Goldberg DE, Horn J, Deb K (1992) What makes a problem hard for a classifier system? In: Collected abstracts for the first international workshop on learning classifier system (IWLCS-92). (Also technical report 92007 Illinois Genetic Algorithms Laboratory, University of Illinois at Urbana-Champaign). Available from ENCORE (ftp://ftp.krl.caltech.edu/pub/EC/Welcome.html) in the section on Classifier Systems

Greene DP, Smith SF (1993) Competition-based induction of decision models from examples. Mach Learn 13:229–257

Greene DP, Smith SF (1994) Using coverage as a model building constraint in learning classifier systems. Evolut Comput 2(1):67–91

Greene DP, Smith SF (1987) A genetic system for learning models of consumer choice. In: Proceedings of the second international conference on genetic algorithms and their applications. Morgan Kaufmann, San Francisco, CA, pp 217–223

Greenyer A (2000) The use of a learning classifier system JXCS. In: van der Putten P, van Someren M (eds) CoIL challenge 2000: the insurance company case. Leiden Institute of Advanced Computer Science, June 2000. Technical report 2000–09

Gruau F (1995) Automatic definition of modular neural networks. Adapt Behav 3(2):151–183

Hanebeck D, Schmidt K (1996) Genetic optimization of fuzzy networks. Fuzzy set syst 79:59–68

Hansen LK, Salamon P (1990) Neural network ensembles. IEEE Trans Pattern Anal Mach Intell 12(10): 993–1001

Hart WE, Krasnogor N, Smith JE (eds) (2004) Special issue on memetic algorithms, evolutionary computation, vol 12, 3

Hart WE, Krasnogor N, Smith JE (eds) (2005) Recent advances in memetic algorithms, Studies in fuzziness and soft computing, vol 166. Springer, Berlin

He L, Wang KJ, Jin HZ, Li GB, Gao XZ (1999) The combination and prospects of neural networks, fuzzy logic and genetic algorithms. In: IEEE midnight-sun workshop on soft computing methods in industrial applications. IEEE, Washington, DC, pp 52–57

Heitkötter J, Beasley D (2001) The hitch-hiker's guide to evolutionary computation (FAQ for comp.ai.genetic). Accessed 28/2/09. http://www.aip.de/~ast/EvolCompFAQ/

Hekanaho J (1995) Symbiosis in multimodal concept learning. In: Proceedings of the 1995 international conference on machine learning (ICML'95). Morgan Kaufmann, San Francisco, pp 278–285

Herrera F (2008) Genetic fuzzy systems: taxonomy, current research trends and prospects. Evolut Intell 1(1):27–46

Holland JH (1976) Adaptation. In: Rosen R, Snell FM (eds) Progress in theoretical biology. Plenum, New York

Holland JH (1986) Escaping brittleness: the possibilities of general-purpose learning algorithms applied to parallel rule-based systems. In: Mitchell T, Michalski R, Carbonell J (eds) Machine learning, an artificial intelligence approach. vol. II, chap. 20. Morgan Kaufmann, San Francisco, CA, pp 593–623

Holland JH, Booker LB, Colombetti M, Dorigo M, Goldberg DE, Forrest S, Riolo RL, Smith RE, Lanzi PL, Stolzmann W, Wilson SW (2000) What is a learning classifier system? In: Lanzi PL, Stolzmann W, Wilson SW (eds) Learning classifier systems: from foundations to application, Lecture notes in artificial intelligence, vol 1813. Springer, Berlin, pp 3–32

Holland JH, Holyoak KJ, Nisbett RE, Thagard PR (1986) Induction: processes of inference, learning, and discovery. MIT Press, Cambridge, MA

Holland JH, Reitman JS (1978) Cognitive systems based on adaptive algorithms. In: Waterman DA, Hayes-Roth F (eds) Pattern-directed Inference Systems. Academic Press, New York. Reprinted in: Evolutionary Computation. The Fossil Record. Fogel DB (ed) (1998) IEEE Press, Washington DC. ISBN: 0-7803-3481-7

Homaifar A, Mccormick E (1995) Simultaneous design of membership functions and rule sets for fuzzy controllers using genetic algorithms. IEEE Trans Fuzzy Syst, 3(2):129–139

Howard D, Bull L (2008) On the effects of node duplication and connection-orientated constructivism in neural XCSF. In: Keijzer M et al. (eds) GECCO-2008: proceedings of the genetic and evolutionary computation conference. ACM, New York, pp 1977–1984

Howard D, Bull L, Lanzi PL (2008) Self-adaptive constructivism in neural XCS and XCSF. In: Keijzer M et al. (eds) GECCO-2008: proceedings of the genetic and evolutionary computation conference. ACM, New York, pp 1389–1396

Hu Y-J (1998) A genetic programming approach to constructive induction. In: Genetic programming 1998: proceedings of the 3rd annual conference. Morgan Kaufmann, San Francisco, CA, pp 146–151

Hurst J, Bull L (2003) Self-adaptation in classifier system controllers. Artifi Life Robot 5(2):109–119

Hurst J, Bull L (2004) A self-adaptive neural learning classifier system with constructivism for mobile robot control. In: Yao X et al. (eds) Parallel problem solving from nature (PPSN VIII), Lecture notes in computer science, vol 3242. Springer, Berlin, pp 942–951

Husbands P, Harvey I, Cliff D, Miller G (1994) The use of genetic algorithms for the development of sensori-motor control systems. In: Gaussier P, Nicoud J-D (eds) From perception to action. IEEE Press, Washington DC, pp 110–121

Iba H (1999) Bagging, boosting and bloating in genetic programming. In: Proceedings of the genetic and evolutionary computation conference (GECCO'99). Morgan Kaufmann, San Francisco, CA, pp 1053–1060

IEEE (2000) Proceedings of the 2000 congress on evolutionary computation (CEC'00). IEEE Press, Washington DC

Ishibuchi H, Nakashima T (2000) Multi-objective pattern and feature selection by a genetic algorithm. In: Proceedings of the 2000 genetic and evolutionary computation conference (GECCO'2000). Morgan Kaufmann, San Francisco, CA, pp 1069–1076

Islam MM, Yao X, Murase K (2003) A constructive algorithm for training cooperative neural network ensembles. IEEE Trans Neural Networ 14:820–834

Jain A, Zongker D (1997) Feature selection: evaluation, application and small sample performance. IEEE Trans. Pattern Anal Mach Intell 19(2):153–158

Janikow CZ (1991) Inductive learning of decision rules in attribute-based examples: a knowledge-intensive genetic algorithm approach. PhD thesis, University of North Carolina

Janikow CZ (1993) A knowledge-intensive genetic algorithm for supervised learning. Mach Learn 13:189–228

Jin Y, Sendhoff B (2004) Reducing fitness evaluations using clustering techniques and neural network ensembles. In: Genetic and evolutionary computation conference (GECCO–2004), Lecture notes in computer science, vol 3102. Springer, Berlin, pp 688–699

John G, Kohavi R, Phleger K (1994) Irrelevant features and the feature subset problem. In: Proceedings of the 11th international conference on machine learning. Morgan Kaufmann, San Francisco, CA, pp 121–129

Juan Liu J, Tin-Yau Kwok J (2000) An extended genetic rule induction algorithm. In: Proceedings of the 2000 congress on evolutionary computation (CEC'00). IEEE Press, Washington DC, pp 458–463

Kelly JD Jr, Davis L (1991) Hybridizing the genetic algorithm and the k nearest neighbors classification algorithm. In: Booker LB, Belew RK (eds) Proceedings of the 4th international conference on genetic algorithms (ICGA91). Morgan Kaufmann, San Francisco, CA, pp 377–383

Karr C (1991) Genetic algorithms for fuzzy controllers. AI Expert 6(2):26–33

Kasabov N (2007) Evolving connectionist systems: the knowledge engineering approach. Springer, Berlin

Keijzer M, Babovic V (2000) Genetic programming, ensemble methods, and the bias/variance/tradeoff – introductory investigation. In: Proceedings of the European conference on genetic programming (EuroGP'00), Lecture notes in computer science, vol 1802. Springer, Berlin, pp 76–90

Kitano H (1990) Designing neural networks by genetic algorithms using graph generation system. J Complex Syst 4:461–476

Kolman E, Margaliot M (2009) Knowledge-based neurocomputing: a fuzzy logic approach, Studies in fuzziness and soft computing, vol 234. Springer, Berlin

Kovacs T (1996) Evolving optimal populations with XCS classifier systems. Master's thesis, University of Birmingham, Birmingham, UK

Kovacs T (1997) XCS classifier system reliably evolves accurate, complete, and minimal representations for Boolean functions. In: Chawdhry PK, Roy R, Pant RK (eds) Soft computing in engineering design and manufacturing. Springer, London, pp 59–68 ftp://ftp.cs.bham.ac.uk/pub/authors/T. Kovacs/index.html

Kovacs T (2000) Strength or accuracy? Fitness calculation in learning classifier systems. In: Lanzi PL, Stolzmann W, Wilson SW (eds) Learning classifier systems: from foundations to applications, Lecture notes in artificial intelligence, vol 1813. Springer, Berlin, pp 143–160

Kovacs T (2004) Strength or accuracy: credit assignment in learning classifier systems. Springer, Berlin

Kovacs T (2009) A learning classifier systems bibliography. Department of Computer Science, University of Bristol. http://www.cs.bris.ac.uk/~kovacs/lcs/search.html

Kovacs T, Kerber M (2001) What makes a problem hard for XCS? In: Lanzi PL, Stolzmann W, Wilson SW (eds) Advances in learning classifier systems, Lecture notes in artificial intelligence, vol 1996. Springer, Berlin, pp 80–99

Kovacs T, Kerber M (2004) High classification accuracy does not imply effective genetic search. In: Deb K et al. (eds) Proceedings of the 2004 genetic and evolutionary computation conference (GECCO), Lecture notes in computer science, vol 3102. Springer, Berlin, pp 785–796

Koza JR (1992) Genetic programming: on the programming of computers by means of natural selection. MIT Press, Cambridge, MA

Koza JR (1994) Genetic Programming II. MIT Press

Krasnogor N (2002) Studies on the theory and design space of memetic algorithms. PhD thesis, University of the West of England

Krasnogor N, Smith JE (2005) A tutorial for competent memetic algorithms: model, taxonomy and design issues. IEEE Trans Evolut Comput 9(5): 474–488

Krasnogor N (2004) Self-generating metaheuristics in bioinformatics: the protein structure comparison case. GP and Evol Machines 5(2):181–201

Krasnogor N, Gustafson S (2004) A study on the use of self-generation in memetic algorithms. Natural Comput 3(1):53–76

Krawiec K (2002) Genetic programming-based construction of features for machine learning and knowledge discovery tasks. GP and Evol Machines 3(4):329–343

Krogh A, Vedelsby J (1995) Neural network ensembles, cross validation and active learning. NIPS 7:231–238

Kudo M, Skalansky J (2000) Comparison of algorithms that select features for pattern classifiers. Pattern Recogn 33:25–41

Kuncheva LI (2004) Combining pattern classifiers: methods and algorithms. Wiley, Hoboken

Kushchu I (2002) An evaluation of evolutionary generalization in genetic programming. Artif Intell Rev 18(1):3–14

Lam L, Suen CY (1995) Optimal combination of pattern classifiers. Pattern Recogn Lett 16:945–954 See Kuncheva (2004a) p.167

Landau S, Sigaud O, Schoenauer M (2005) ATNoSFERES revisited. In: Proceedings of the genetic and evolutionary computation conference GECCO-2005. ACM, New York, pp 1867–1874

Langdon W, Gustafson S, Koza J (2009) The genetic programming bibliography. http://www.cs.bham.ac.uk/~wbl/biblio/

Lanzi PL (1999a) Extending the representation of classifier conditions, part I: from binary to messy coding. In: Banzhaf W et al. (eds) GECCO-99: Proceedings of the genetic and evolutionary computation conference. Morgan Kaufmann, San Francisco, CA, pp 337–344

Lanzi PL (1999b) Extending the representation of classifier conditions part II: from messy coding to S-expressions. In: Banzhaf W et al. (eds) GECCO-99: Proceedings of the genetic and evolutionary computation conference. Morgan Kaufmann, San Francisco, CA, pp 345–352

Lanzi PL (2002a) Learning classifier systems from a reinforcement learning perspective. J Soft Comput 6(3–4):162–170

Lanzi PL (2001) Mining interesting knowledge from data with the XCS classifier system. In: Spector L, Goodman ED, Wu A, Langdon WB, Voigt H-M, Gen M, Sen S, Dorigo M, Pezeshk S, Garzon MH, Burke E (eds) Proceedings of the genetic and evolutionary computation conference (GECCO-2001). Morgan Kaufmann, San Francisco, CA, pp 958–965

Lanzi PL (2008) Learning classifier systems: then and now. Evolut Intell 1(1):63–82

Lanzi PL, Loiacono D, Wilson SW, Goldberg DE (2006a) Classifier prediction based on tile coding. In: Genetic and evolutionary computation – GECCO-2006. ACM, New York, pp 1497–1504

Lanzi PL, Loiacono D, Wilson SW, Goldberg DE (2006b) Prediction update algorithms for XCSF: RLS, Kalman filter and gain adaptation. In: Genetic and Evolutionary Computation – GECCO-2006. ACM, New York, pp 1505–1512

Lanzi PL, Loiacono D, Zanini M (2008) Evolving classifiers ensembles with heterogeneous predictors. In: Bacardit J, Bernadó-Mansilla E, Butz M, Kovacs T, Llorà X, Takadama K (eds) Learning classifier systems. 10th and 11th international workshops (2006–2007), Lecture notes in computer science, vol 4998. Springer, Berlin, pp 218–234

Lanzi PL, Riolo RL (2000) A roadmap to the last decade of learning classifier system research (from 1989 to 1999). In: Lanzi PL, Stolzmann W, Wilson SW (eds) Learning classifier systems: from foundations to applications, Lecture notes in artificial intelligence, vol 1813. Springer, Berlin, pp 33–62

Lanzi PL, Stolzmann W, Wilson SW (eds) (2000) Learning classifier systems: from foundations to applications, Lecture notes in artificial intelligence, vol 1813. Springer, Berlin

Lanzi PL, Stolzmann W, Wilson SW (eds) (2001) Advances in learning classifier systems, Lecture notes in artificial intelligence, vol 1996. Springer, Berlin

Lanzi PL, Stolzmann W, Wilson SW (eds) (2002) Advances in learning classifier systems, Lecture notes in artificial intelligence, vol 2321. Springer, Berlin

Lanzi PL, Butz MV, Goldberg DE (2007) Empirical analysis of generalization and learning in XCS with gradient descent. In: Lipson H (ed) Proceedings of the Genetic and evolutionary computation conference, GECCO 2007, vol 2. ACM, New York, pp 1814–1821

Lanzi PL, Loiacono D (2006) Standard and averaging reinforcement learning in XCS. In: Cattolico M (ed) Proceedings of the 8th annual conference on genetic and evolutionary computation, GECCO 2006. ACM, New York, pp 1480–1496

Lanzi PL, Loiacono D (2007) Classifier systems that compute action mappings. In: Lipson H (ed) Proceedings of the Genetic and evolutionary computation conference, GECCO 2007. ACM, New York, pp 1822–1829

Lanzi PL, Wilson SW (2006) Using convex hulls to represent classifier conditions. In: Cattolico M (ed) Proceedings of the genetic and evolutionary computation conference (GECCO 2006). ACM, New York, pp 1481–1488

Liangjie Z, Yanda L (1996) A new global optimizing algorithm for fuzzy neural networks. Int J Elect 80(3):393–403

Linkens DA, Nyongesa HO (1996) Learning systems in intelligent control: an appraisal of fuzzy, neural and genetic algorithm control applications. IEE Proc Contr Theo Appl 143(4):367–386

Liu Y, Yao X (1999) Ensemble learning via negative correlation. Neural Networ 12:1399–1404

Liu Y, Yao X, Higuchi T (2000) Evolutionary ensembles with negative correlation learning. IEEE Trans Evolut Comput 4(4):380–387

Llorà X (2002) Genetic based machine learning using fine-grained parallelism for data mining. PhD thesis, Enginyeria i Arquitectura La Salle. Ramon Llull University

Llorà X, Garrell JM (2001) Knowledge-independent data mining with fine-grained parallel evolutionary algorithms. In: Spector L, Goodman ED, Wu A, Langdon WB, Voigt H-M, Gen M, Sen S, Dorigo M, Pezeshk S, Garzon MH, Burke E (eds) Proceedings of the genetic and evolutionary computation conference (GECCO'2001). Morgan Kaufmann, San Francisco, CA, pp 461–468

Llorà X, Sastry K, Goldberg DE (2005a) Binary rule encoding schemes: a study using the compact classifier system. In: Rothlauf F (ed) Proceedings of the 2005 conference on genetic and evolutionary computation GECCO '05. ACM Press, New York, pp 88–89

Llorà X, Sastry K, Goldberg DE (2005b) The compact classifier system: scalability analysis and first results. In: Rothlauf F (ed) Proceedings of the IEEE congress on evolutionary computation, CEC 2005. IEEE, Press, Washington, DC, pp 596–603

Llorà X, Wilson SW (2004) Mixed decision trees: minimizing knowledge representation bias in LCS. In: Kalyanmoy Deb et al. (eds) Proceedings of the genetic and evolutionary computation conference (GECCO-2004), Lecture notes in computer science, Springer, Berlin, pp 797–809

Loiacono D, Marelli A, Lanzi PL (2007) Support vector regression for classifier prediction. In: GECCO '07: Proceedings of the 9th annual conference on genetic and evolutionary computation. ACM, Berlin, pp 1806–1813

Marmelstein RE, Lamont GB (1998) Pattern classification using a hybrid genetic algorithm – decision tree approach. In: Genetic programming 1998: proceedings of the 3rd annual conference (GP'98). Morgan Kaufmann, San Francisco, CA, pp 223–231

Marshall JAR, Kovacs T (2006) A representational ecology for learning classifier systems. In: Keijzer M et al. (ed) Proceedings of the 2006 genetic and evolutionary computation conference (GECCO 2006). ACM, New York, pp 1529–1536

Martin-Bautista MJ, Vila M-A (1999) A survey of genetic feature selection in mining issues.

In: Proceedings of the congress on evolutionary computation (CEC'99). IEEE Press, Washington, DC, pp 1314–1321

Meir R, Rätsch G (2003) An introduction to boosting and leveraging. In: Advanced lectures on machine learning. Springer, Berlin, pp 118–183

Mellor D (2005a) A first order logic classifier system. In: Rothlauf F (ed), GECCO '05: Proceedings of the 2005 conference on genetic and evolutionary computation. ACM, New York, pp 1819–1826

Mellor D (2005b) Policy transfer with a relational learning classifier system. In: GECCO Workshops 2005. ACM, New York, pp 82–84

Mellor D (2008) A learning classifier system approach to relational reinforcement learning. In: Bacardit J, Bernadó-Mansilla E, Butz M, Kovacs T, Llorà X, Takadama K (eds) Learning classifier systems. 10th and 11th international workshops (2006–2007), Lecture notes in computer science, vol 4998. Springer, New York, pp 169–188

Michalski RS, Mozetic I, Hong J, Lavrac N (1986) The AQ15 inductive learning system: an overview and experiments. Technical Report UIUCDCS-R-86-1260, University of Illinois

Miller GF, Todd PM, Hegde SU (1989) Designing neural networks using genetic algorithms. In: Schaffer JD (ed) Proceedings of the 3rd international conference genetic algorithms and their applications, Morgan Kaufmann, San Francisco, CA, pp 379–384

Mitra S, Hayashi Y (2000) Neurofuzzy rule generation: survey in soft computing framework. IEEE Trans Neural Networ 11(3):748–768

Morimoto T, Suzuki J, Hashimoto Y (1997) Optimization of a fuzzy controller for fruit storage using neural networks and genetic algorithms. Eng Appl Artif Intell 10(5):453–461

Nolfi S, Miglino O, Parisi D (1994) Phenotypic plasticity in evolving neural networks. In: Gaussier P, Nicoud J-D (eds) From perception to action. IEEE Press, Washington, DC, pp 146–157

O'Hara T, Bull L (2005) A memetic accuracy-based neural learning classifier system. In: Proceedings of the IEEE congress on evolutionary computation (CEC 2005). IEEE Press, Washington, DC, pp 2040–2045

Ong Y-S, Krasnogor N, Ishibuchi H (eds) (2007) Special issue on memetic algorithms, IEEE Transactions on Systems, Man and Cybernetics - Part B

Ong Y-S, Lim M-H, Neri F, Ishibuchi H (2009) Emerging trends in soft computing - memetic algorithms, Special Issue of Soft Computing. vol 13, 8–9

Ong YS, Lim MH, Zhu N, Wong KW (2006) Classification of adaptive memetic algorithms: A comparative study. IEEE Trans Syst Man Cybern B 36(1):141–152

Opitz D, Maclin R (1999) Popular ensemble methods: an empirical study. J Artif Intell Res 11:169–198

Opitz DW, Shavlik JW (1996) Generating accurate and diverse members of a neural-network ensemble. Advances in neural information processing systems, vol 8. Morgan Kaufmann, pp 535–541

Orriols-Puig A, Bernadó-Mansilla E (2006) Bounding XCS's parameters for unbalanced datasets. In: Keijzer M et al. (eds) Proceedings of the 2006 genetic and evolutionary computation conference (GECCO 2006). ACM, New York, pp 1561–1568

Orriols-Puig A, Casillas J, Bernadò-Mansilla E (2007a) Fuzzy-UCS: preliminary results. In: Lipson H (ed) Proceedings of the genetic and evolutionary computation conference, GECCO 2007. ACM, New York, pp 2871–2874

Orriols-Puig A, Goldberg DE, Sastry K, Bernadó-Mansilla E (2007b) Modeling XCS in class imbalances: population size and parameter settings. In: Lipson H et al. (eds) Genetic and evolutionary computation conference, GECCO 2007. ACM, New York, pp 1838–1845

Orriols-Puig A, Goldberg DE, Sastry K, Bernadó-Mansilla E (2007c) Modeling XCS in class imbalances: population size and parameter settings. In: Lipson H (eds) Proceedings of the genetic and evolutionary computation conference, GECCO 2007. ACM, New York, pp 1838–1845

Orriols-Puig A, Sastry K, Lanzi PL, Goldberg DE, Bernadò-Mansilla E (2007d) Modeling selection pressure in XCS for proportionate and tournament selection. In: Lipson H (ed) Proceedings of the genetic and evolutionary computation conference, GECCO 2007. ACM, New York, pp 1846–1853

Orriols-Puig A, Bernadó-Mansilla E (2008) Revisiting UCS: description, fitness sharing, and comparison with XCS. In: Bacardit J, Bernadó-Mansilla E, Butz M, Kovacs T, Llorà X, Takadama K (eds) Learning classifier systems. 10th and 11th international workshops (2006–2007), Lecture notes in computer science, vol 4998. Springer, Berlin, pp 96–111

Orriols-Puig A, Casillas J, Bernadó-Mansilla E (2008a) Evolving fuzzy rules with UCS: preliminary results. In: Bacardit J, Bernadó-Mansilla E, Butz M, Kovacs T, Llorà X, Takadama K (eds) Learning classifier systems. 10th and 11th international workshops (2006–2007), Lecture notes in computer science, vol 4998. Springer, Berlin, pp 57–76

Orriols-Puig A, Casillas J, Bernadó-Mansilla E (2008b) Genetic-based machine learning systems are competitive for pattern recognition. Evolut Intell 1(3):209–232

Pal S, Bhandari D (1994) Genetic algorithms with fuzzy fitness function for object extraction using cellular networks. Fuzzy Set Syst 65(2–3):129–139

Pappa GL, Freitas AA (2010) Automating the design of data mining algorithms. An evolutionary computation approach. Natural computing series. Springer

Paris G, Robilliard D, Fonlupt C (2001) Applying boosting techniques to genetic programming. In: Artificial evolution 2001, Lecture notes in computer science, vol 2310. Springer, Berlin, pp 267–278

Pereira FB, Costa E (2001) Understanding the role of learning in the evolution of busy beaver: a comparison between the Baldwin effect and Lamarckian strategy. In: Proceedings of the genetic and evolutionary computation conference (GECCO–2001). Morgan Kaufmann, San Francisco, pp 884–891

Perneel C, Themlin J-M (1995) Optimization of fuzzy expert systems using genetic algorithms and neural networks. IEEE Trans Fuzzy Syst 3 (3):301–312

Pham DT, Karaboga D (1991) Optimum design of fuzzy logic controllers using genetic algorithms. J Syst Eng 1:114–118

Poli R, Langdon WB, McPhee NF (2008) A field guide to genetic programming, freely available at http://www.gp-field-guide.org.uk.lulu.com

Punch WF, Goodman ED, Pei M, Chia-Shun L, Hovland P, Enbody R (1993) Further research on feature selection and classification using genetic algorithms. In: Forrest S (ed) Proceedings of the 5th international conference on genetic algorithms (ICGA93). Morgan Kaufmann, San Francisco, CA, pp 557–564

Radi A, Poli R (2003) Discovering efficient learning rules for feedforward neural networks using genetic programming. In: Abraham A, Jain L, Kacprzyk J (eds) Recent advances in intelligent paradigms and applications. Springer, Berlin, pp 133–159

Raymer ML, Punch WF, Goodman ED, Kuhn LA, Jain AK (2000) Dimensionality reduction using genetic algorithms. IEEE Trans Evolut Comput 4(2): 164–171

Reeves CR, Rowe JE (2002) Genetic algorithms – principles and perspectives. A guide to GA theory. Kluwer, Norwell

Riolo RL (1987) Bucket brigade performance: I. long sequences of classifiers. In: Grefenstette JJ (eds) Proceedings of the 2nd international conference on genetic algorithms (ICGA'87), Lawrence Erlbaum Associates, Cambridge, MA, pp 184–195

Rivest RL (1987) Learning decision lists. Mach Learn 2(3):229–246

Romaniuk S (1994) Towards minimal network architectures with evolutionary growth networks. In: Proceedings of the 1993 international joint conference on neural networks, vol 3. IEEE Press, Washington, DC, pp 1710–1713

Rouwhorst SE, Engelbrecht AP (2000) Searching the forest: using decision trees as building blocks for evolutionary search in classification databases. In: Proceedings of the 2000 congress on evolutionary computation (CEC00). IEEE Press, Washington, DC, pp 633–638

Rozenberg G, Bäck T, Kok J (eds) (2012) Handbook of natural computing. Springer, Berlin

Ruta D, Gabrys B (2001) Application of the evolutionary algorithms for classifier selection in multiple classifier systems with majority voting. In: Kittler J, Roli F (eds) Proceedings of the 2nd international workshop on multiple classifier systems, Lecture notes in computer science, vol 2096. Springer, Berlin, pp 399–408. See Kuncheva (2004a) p.321

Sánchez L, Couso I (2007) Advocating the use of imprecisely observed data in genetic fuzzy systems. IEEE Trans Fuzzy Syst 15(4):551–562

Sasaki T, Tokoro M (1997) Adaptation toward changing environments: why Darwinian in nature? In: Husbands P, Harvey I (eds) Proceedings of the 4th European conference on artificial life. MIT Press, Cambridge, MA, pp 145–153

Saxon S, Barry A (2000) XCS and the Monk's problems. In: Lanzi PL, Stolzmann W, Wilson SW (eds) Learning classifier systems: from foundations to applications, Lecture notes in artificial intelligence, vol 1813. Springer, Berlin, pp 223–242

Schaffer C (1994) A conservation law for generalization performance. In: Hirsh H, Cohen WW (eds) Machine learning: proceedings of the eleventh international conference. Morgan Kaufmann, San Francisco, CA, pp 259–265

Schaffer JD (ed) (1989) Proceedings of the 3rd international conference on genetic algorithms (ICGA-89), George Mason University, June 1989. Morgan Kaufmann, San Francisco, CA

Schmidhuber J (1987) Evolutionary principles in self-referential learning. (On learning how to learn: The meta-meta-... hook.). PhD thesis, Technische Universität München, Germany

Schuurmans D, Schaeffer J (1989) Representational difficulties with classifier systems. In: Schaffer JD (ed) Proceedings of the 3rd international conference on genetic algorithms (ICGA-89). Morgan Kaufmann, San Francisco, CA, pp 328–333

Sharkey AJC (1996) On combining artificial neural nets. Connection Sci 8(3–4):299–313

Sharpe PK, Glover RP (1999) Efficient GA based techniques for classification. Appl Intell 11:277–284

Sirlantzis K, Fairhurst MC, Hoque MS (2001) Genetic algorithms for multi-classifier system configuration: a case study in character recognition. In: Kittler J, Roli F (eds) Proceedings of the 2nd international workshop on multiple classifier systems, Lecture notes in computer science, vol 2096. Springer, Berlin, pp 99–108. See Kuncheva (2004a) p.321

Smith JE (2007) Coevolving memetic algorithms: a review and progress report. IEEE Trans Syst Man Cybern B Cybern 37(1):6–17

Smith MG, Bull L (2005) Genetic programming with a genetic algorithm for feature construction and selection. GP and Evol Machines 6(3):265–281

Smith RE (1992) A report on the first international workshop on learning classifier systems (IWLCS-92). NASA Johnson Space Center, Houston, Texas, Oct. 6–9. ftp://lumpi.informatik.uni-dortmund.de/pub/LCS/papers/lcs92.ps.gz or from ENCORE, The Electronic Appendix to the Hitch-Hiker's Guide to Evolutionary Computation (ftp://ftp.krl.caltech.edu/pub/EC/Welcome.html) in the section on Classifier Systems

Smith RE (1994) Memory exploitation in learning classifier systems. Evolut Comput 2(3):199–220

Smith RE, Cribbs HB (1994) Is a learning classifier system a type of neural network? Evolut Comput 2(1):19–36

Smith RE, Goldberg DE (1991) Variable default hierarchy separation in a classifier system. In: Rawlins GJE (ed) Proceedings of the first workshop on foundations of genetic algorithms. Morgan Kaufmann, San Mateo, pp 148–170

Smith RE, Cribbs III HB (1997) Combined biological paradigms. Robot Auton Syst 22(1):65–74

Song D, Heywood MI, Zincir-Heywood AN (2005) Training genetic programming on half a million patterns: an example from anomaly detection. IEEE Trans Evolut Comput 9(3):225–239

Srinivas N, Deb K (1994) Multi-objective function optimization using non-dominated sorting genetic algorithm. Evolut Comput 2(3):221–248

Stagge P (1998) Averaging efficiently in the presence of noise. In: Parallel problem solving from nature, vol 5. pp 188–197

Stolzmann W (1996) Learning classifier systems using the cognitive mechanism of anticipatory behavioral control, detailed version. In: Proceedings of the first European workshop on cognitive modelling. Berlin, TU, pp 82–89

Stone C, Bull L (2003) For real! XCS with continuous-valued inputs. Evolut Comput 11(3):298–336

Storn R, Price K (1996) Minimizing the real functions of the ICEC'96 contest by differential evolution. In: Proceedings of the IEEE international conference Evolutionary Computation. IEEE Press, Washington, DC, pp 842–844

Stout M, Bacardit J, Hirst JD, Krasnogor N (2008) Prediction of recursive convex hull class assignment for protein residues. Bioinformatics 24(7):916–923

Sutton RS (1986) Two problems with backpropagation and other steepest-descent learning procedures for networks. In: Proceedings of the 8th annual conference cognitive science society. Erlbaum, pp 823–831

Sziranyi T (1996) Robustness of cellular neural networks in image deblurring and texture segmentation. Int J Circuit Theory App 24(3):381–396

Tamaddoni-Nezhad A, Muggleton SH (2000) Searching the subsumption lattice by a genetic algorithm. In: Cussens J, Frisch A (eds) Proceedings of the 10th international conference on inductive logic programming. Springer, Berlin, pp 243–252

Tamaddoni-Nezhad A, Muggleton S (2003) A genetic algorithms approach to ILP. In: Inductive logic programming, Lecture notes in computer science, vol 2583. Springer, Berlin, pp 285–300

Tharakannel K, Goldberg D (2002) XCS with average reward criterion in multi-step environment. Technical report, Illinois Genetic Algorithms Laboratory, University of Illinois at Urbana-Champaign

Thompson S (1998) Pruning boosted classifiers with a real valued genetic algorithm. In: Research and development in expert systems XV – proceedings of ES'98. Springer, Berlin, pp 133–146

Thompson S (1999) Genetic algorithms as postprocessors for data mining. In: Data mining with evolutionary algorithms: research directions – papers from the AAAI workshop, Tech report WS–99–06. AAAI Press, Menlo Park, CA, pp 18–22

Thrift P (1991) Fuzzy logic synthesis with genetic algorithms. In: Booker LB, Belew RK (eds) Proceedings of 4th international conference on genetic algorithms (ICGA'91). Morgan Kaufmann, San Francisco, CA, pp 509–513

Tomlinson A (1999) Corporate classifier systems. PhD thesis, University of the West of England

Tomlinson A, Bull L (1998) A corporate classifier system. In: Eiben AE, Bäck T, Shoenauer M, Schwefel H-P (eds) Proceedings of the fifth international conference on parallel problem solving from nature – PPSN V, Lecture notes in computer science, vol 1498. Springer, Berlin, pp 550–559

Tomlinson A, Bull L (2002) An accuracy-based corporate classifier system. J Soft Comput 6(3–4):200–215

Tran TH, Sanza C, Duthen Y, Nguyen TD (2007) XCSF with computed continuous action. In: Genetic and evolutionary computation conference (GECCO 2007). ACM, New York, pp 1861–1869

Tumer K, Ghosh J (1996) Analysis of decision boundaries in linearly combined neural classifiers. Pattern Recogn 29(2):341–348

Turney P (1996) How to shift bias: lessons from the Baldwin effect. Evolut Comput 4(3):271–295

Valentini G, Masulli F (2002) Ensembles of learning machines. In: WIRN VIETRI 2002: Proceedings of the 13th Italian workshop on neural nets-revised papers. Springer, Berlin, pp 3–22

Valenzuela-Rendón M (1989) Two analysis tools to describe the operation of classifier systems. PhD thesis, University of Alabama. Also TCGA technical report 89005

Valenzuela-Rendón M (1991) The fuzzy classifier system: a classifier system for continuously varying variables. In: Booker LB, Belew RK (eds) Proceedings of the 4th international conference on genetic algorithms (ICGA'91). Morgan Kaufmann, San Francisco, CA, pp 346–353

Valenzuela-Rendón M (1998) Reinforcement learning in the fuzzy classifier system. Expert Syst Appl 14:237–247

Vallim R, Goldberg D, Llorà X, Duque T, Carvalho A (2003) A new approach for multi-label classification based on default hierarchies and organizational learning. In: Proceedings of the genetic and evolutionary computation conference, workshop sessions: learning classifier systems. ACM, New York, pp 2017–2022

Vanneschi L, Poli R (2012) Genetic programming: introduction, applications, theory and open issues. In: Rozenberg G, Bäck T, Kok J (eds) Handbook of natural computing. Springer, Berlin

Venturini G (1993) SIA: a supervised inductive algorithm with genetic search for learning attributes based concepts. In: Brazdil PB (ed) ECML-93 - Proceedings of the European conference on machine learning. Springer, Berlin, pp 280–296

Vilalta R, Drissi Y (2002) A perspective view and survey of meta-learning. Artif Intell Rev 18(2):77–95

Wada A, Takadama K, Shimohara K, Katai O (2005c) Learning classifier systems with convergence and generalization. In: Bull L, Kovacs T (eds) Foundations of learning classifier systems. Springer, Berlin, pp 285–304

Wada A, Takadama K, Shimohara K (2005a) Counter example for Q-bucket-brigade under prediction problem. In: GECCO Workshops 2005. ACM, New York, pp 94–99

Wada A, Takadama K, Shimohara K (2005b) Learning classifier system equivalent with reinforcement learning with function approximation. In: GECCO Workshops 2005. ACM, New York, pp 92–93

Wada A, Takadama K, Shimohara K (2007) Counter example for Q-bucket-brigade under prediction problem. In: Kovacs T, LLòra X, Takadama K, Lanzi PL, Stolzmann W, Wilson SW (eds) Learning classifier systems. International workshops, IWLCS 2003-2005, revised selected papers, Lecture notes in computer science, vol 4399. Springer, Berlin, pp 128–143

Whitley D, Goldberg D, Cantú-Paz E, Spector L, Parmee I, Beyer HG (eds) (2000) Proceedings of the genetic and evolutionary computation conference (GECCO-2000). Morgan Kaufmann, San Francisco, CA

Whiteson S, Stone P (2006) Evolutionary function approximation for reinforcement learning. J Mach Learn Res 7:877–917

Whitley D, Starkweather T, Bogart C (1990) Genetic algorithms and neural networks: optimizing connections and connectivity. Parallel Comput 14(3):347–361

Whitley D, Gordon VS, Mathias K (1994) Lamarckian evolution, the Baldwin effect and function optimization. In: Parallel problem solving from nature (PPSN-III). Springer, Berlin, pp 6–15

Wilcox JR (1995) Organizational learning within a learning classifier system. Master's thesis, University of Illinois. Also Technical Report No. 95003 IlliGAL

Wilson SW (2001a) Mining oblique data with XCS. In: Lanzi PL, Stolzmann W, Wilson SW (eds) Advances in learning classifier systems, third international workshop, IWLCS 2000, Lecture notes in computer science, vol 1996. Springer, Berlin, pp 158–176

Wilson SW (1989) Bid competition and specificity reconsidered. Complex Syst 2:705–723

Wilson SW (1994) ZCS: a zeroth level classifier system. Evolut Comput 2(1):1–18. http://prediction-dynamics.com/

Wilson SW (1995) Classifier fitness based on accuracy. Evolut Comput 3(2):149–175. http://prediction-dynamics.com/

Wilson SW (1998) Generalization in the XCS classifier system. In: Koza JR, Banzhaf W, Chellapilla K, Deb K, Dorigo M, Fogel DB, Garzon MH, Goldberg DE, Iba H, Riolo R (eds) Genetic programming 1998: proceedings of the third annual conference, Morgan Kaufmann, San Francisco, CA, pp 665–674. http://prediction-dynamics.com/

Wilson SW (1999) Get real! XCS with continuous-valued inputs. In: Booker L, Forrest S, Mitchell M, Riolo RL (eds) Festschrift in honor of John H. Holland. Center for the Study of Complex Systems. pp 111–121. http://prediction-dynamics.com/

Wilson SW (2000) Mining oblique data with XCS. In: Proceedings of the international workshop on learning classifier systems (IWLCS-2000), in the joint workshops of SAB 2000 and PPSN 2000. Extended abstract

Wilson SW (2001b) Function approximation with a classifier system. In: Spector L, Goodman ED, Wu A, Langdon WB, Voigt HM, Gen M, Sen S, Dorigo M, Pezeshk S, Garzon MH, Burke E (eds) Proceedings of the genetic and evolutionary computation conference (GECCO-2001). Morgan Kaufmann, San Francisco, CA, pp 974–981

Wilson SW (2002a) Classifiers that approximate functions. Natural Comput 1(2–3):211–234

Wilson SW (2002b) Compact rulesets from XCSI. In: Lanzi PL, Stolzmann W, Wilson SW (eds) Advances

in learning classifier systems, Lecture notes in artificial intelligence, vol 2321. Springer, Berlin, pp 196–208

Wilson SW (2007) Three architectures for continuous action. In: Kovacs T, LLòra X, Takadama K, Lanzi PL, Stolzmann W, Wilson SW (eds) Learning classifier systems. International workshops, IWLCS 2003-2005, revised selected papers, Lecture notes in computer science, vol 4399. Springer, Berlin, pp 239–257

Wilson SW (2008) Classifier conditions using gene expression programming. In: Bacardit J, Bernadó-Mansilla E, Butz M, Kovacs T, Llorà X, Takadama K (eds) Learning classifier systems. 10th and 11th international workshops (2006–2007), Lecture notes in computer science, vol 4998. Springer, Berlin, pp 206–217

Wilson SW, Goldberg DE (1989) A critical review of classifier systems. In: Schaffer JD (ed) Proceedings of the 3rd international conference on genetic algorithms. Morgan Kaufmann, San Francisco, CA, pp 244–255. http://prediction-dynamics.com/

Wolpert DH (1996) The lack of a priori distinctions between learning algorithms. Neural Comput 8(7):1341–1390

Wong ML, Leung KS (2000) Data mining using grammar based genetic programming and applications. Kluwer, Norwell

Woods K, Kegelmeyer W, Bowyer K (1997) Combination of multiple classifiers using local accuracy estimates. IEEE Trans Pattern Anal Mach Intell 19:405–410

Woodward JR (2003) GA or GP? That is not the question. In: Proceedings of the 2003 congress on evolutionary computation, CEC2003. IEEE Press, Washington DC, pp 1056–1063

Yamasaki K, Sekiguchi M (2000) Clear explanation of different adaptive behaviors between Darwinian population and Larmarckian population in changing environment. In: Proceedings of the fifth international symposium on artificial life and robotics. pp 120–123

Yao X (1999) Evolving artificial neural networks. Proc IEEE 87(9):1423–1447

Yao X, Islam MM (2008) Evolving artificial neural network ensembles. IEEE Comput Intell Mag 3(1):31–42

Yao X, Liu Y (1997) A new evolutionary system for evolving artificial neural networks. IEEE Trans Neural Networ 8:694–713

Yao X, Liu Y (1998) Making use of population information in evolutionary artificial neural networks. IEEE Trans Syst Man Cybern B 28(3):417–425

Zatuchna ZV (2005) AgentP: a learning classifier system with associative perception in maze environments. PhD thesis, University of East Anglia

Zatuchna ZV (2004) AgentP model: Learning Classifier System with Associative Perception. In 8th parallel problem solving from nature international conference (PPSN VIII). pp 1172–1182

Zhang B-T, Veenker G (1991) Neural networks that teach themselves through genetic discovery of novel examples. In: Proceedings 1991 IEEE international joint conference on neural networks (IJCNN'91) vol 1. IEEE Press, Washington DC, pp 690–695

31 Coevolutionary Principles

Elena Popovici[1] · Anthony Bucci[2] · R. Paul Wiegand[3] · Edwin D. de Jong[4]
[1]Icosystem Corporation, Cambridge, MA, USA
elena@icosystem.com
[2]Icosystem Corporation, Cambridge, MA, USA
anthony@icosystem.com
[3]Institute for Simulation and Training, University of Central Florida,
Orlando, FL, USA
wiegand@ist.ucf.edu
[4]Institute of Information and Computing Sciences, Utrecht University,
The Netherlands
dejong@cs.uu.nl

G. Rozenberg et al. (eds.), *Handbook of Natural Computing*, DOI 10.1007/978-3-540-92910-9_31,
© Springer-Verlag Berlin Heidelberg 2012

Abstract

Coevolutionary algorithms approach problems for which no function for evaluating potential solutions is present or known. Instead, algorithms rely on the aggregation of outcomes from interactions among evolving entities in order to make selection decisions. Given the lack of an explicit yardstick, understanding the dynamics of coevolutionary algorithms, judging whether a given algorithm is progressing, and designing effective new algorithms present unique challenges unlike those faced by optimization or evolutionary algorithms. The purpose of this chapter is to provide a foundational understanding of coevolutionary algorithms and to highlight critical theoretical and empirical work done over the last two decades. This chapter outlines the ends and means of coevolutionary algorithms: what they are meant to find, and how they should find it.

1   Introduction

The inspiration for coevolutionary algorithms (CoEAs) is the same as for traditional evolutionary algorithms (EAs): attempt to harness the Darwinian notions of *heredity* and *survival of the fittest* for simulation or problem-solving purposes. To put it simply, a *representation* is chosen to encode some aspects of potential solutions to a problem into individuals, those individuals are altered during search using genetic-like *variation* operators such as mutation and crossover, and search is directed by selecting better individuals as determined by a fitness *evaluation*. With any luck, the iteration of these steps will eventually lead to high-quality solutions to a problem, if problem solving is the aim, or to interesting or realistic system behavior.

Usually, EAs begin with a *fitness function*, which for the purposes of this chapter is a function of the form $f : G \to \mathbb{R}$ that assigns a real value to each possible genotype in G. (Note that in certain branches of biology, fitness refers to the number of offspring an individual receives; thus, the use of the term fitness function in this way, while common in the evolutionary computation literature, breaks somewhat with biological tradition. However, seemingly less loaded alternative terms such as *objective function* evoke multi-objective optimization, a topic that will also be discussed briefly. In order to avoid confusion with other uses of the term *objective*, the term fitness function is used throughout the chapter.) Given such a function, the fitness relationship between any two genotypes $g_1, g_2 \in G$ is clear: $f(g_1)$ is compared with $f(g_2)$ to see which is more fit. By contrast, CoEAs do not use such a direct metric of the fitness of individuals. Instead, two individuals are compared on the basis of their outcomes from interactions with other individuals. (For this reason, the fitness in a CoEA has been called *subjective fitness*, subject as it is to the changing population(s). The objectively given and unchanging fitness function used in typical EA applications is *objective*.) As should become apparent as this chapter unfolds, this seemingly small change to how evaluation is done creates a variety of fundamental differences from traditional evolutionary computation. Most importantly, the fitness ranking of two individuals can change over time, a phenomenon that cannot happen in an ordinary EA. Indeed, virtually all of the techniques and ideas presented in this chapter are affected one way or another by this fundamental question: How could one possibly build an effective algorithm when any time one believes A is better than B it is possible that later one will believe B is better than A?

This chapter is constructed to provide a foundational understanding of coevolutionary algorithms from a problem-solving point of view. Along the way, it will survey some of the many CoEAs that have been developed and the wide variety of domains to which they have been applied. Before going into detail, however, two coevolutionary algorithm schemes and problem classes are presented to give context for what follows.

1.1 Two Simple Coevolutionary Algorithms

In a *single population* CoEA (❯ *Algorithm 1*), individuals are evaluated by interacting with other individuals from that population. In a *multi-population* CoEA (❯ *Algorithm 2*), individuals in one population interact with individuals in one or several other populations.

Algorithm 1 SINGLE POPULATION CoEA

Initialize *population*
Select *evaluators* from *population*
Evaluate *individuals* from *population* by interacting with *evaluators*
while not done **do**
    Select *parents* from *population*
    Produce *children* from *parents* via variation
    Select *evaluators* from (*children*$^+$, *parents*)
    Evaluate *individuals* from *children* by interacting with *evaluators*
    Select survivors for next generation
end while
return solution

Algorithm 2 MULTI POPULATION CoEA

for each *pop* ∈ *populations* **do**
    Initialize *pop*
    Select *evaluators* from (*populations* − *pop*)
    Evaluate *individuals* from *pop* by interacting with *evaluators*
end for
while not done **do**
    for each *pop* ∈ *populations* **do**
        Select *parents* from *pop*
        Produce *children* from *parents* via variation
        Select *evaluators* from (*populations* − *pop*)
        Evaluate *individuals* from *children* by interacting with *evaluators*
        Select survivors for next generation
    end for
end while
return solution

There has been some amount of controversy over whether a single-population algorithm could be fairly called coevolutionary. After all, biologists use the term coevolution to refer to genetic influence of two species over each other. Since a population refers to a collection of members of a species, it seems that one must have two populations to have coevolution. A pragmatic approach to the term is taken here. Problems in which the ranking of two entities can change depending on the presence or absence of other entities cause a certain set of issues for population-based problem-solving algorithms, issues that do not arise with ordinary optimization or multi-objective optimization problems. While these issues manifest differently in single- and multi-population algorithms, they are nevertheless present and must be taken into account if one hopes to design a successful problem-solving algorithm. Thus, at the risk of breaking with the biological terminology, both single- and multi-population algorithms are referred to as coevolutionary.

While many CoEAs are variations of these two main frameworks, several details have been omitted, some of which will be discussed later in this chapter. For example, a common mechanism not addressed by those frameworks is for a CoEA (either single- or multi-population) to have an *archive* that serves to evaluate individuals or to store potential solutions and is updated from the main population(s). Also, multi-population CoEAs can perform simultaneous or concurrent evolutionary steps, while the example algorithm is sequential. Finally, these two algorithmic schemes do not specify which interactions occur or how the results of interaction outcomes are aggregated into an evaluation that can be used by a selection method. These points will be addressed in ❯ Sect. 2.

1.2 Problem Classes

As a consequence of subjective evaluation, the dynamics of coevolutionary algorithms can be frustratingly complex. One can view the interactions oneself as the basic unit of evaluation; in that view, the fitness landscape for an individual changes as a function of the content of the population(s). Intuitions about fitness landscapes from evolutionary computation do not easily map to this population-dependent, dynamic landscape situation. Nevertheless, coevolutionary algorithms ideally will leverage this mutability of the fitness landscape, adaptively focusing on relevant areas of a search space. This can be particularly helpful when problem spaces are very large or infinite. Additionally, some coevolutionary algorithms appear natural for domains that contain certain, known structure (Potter 1997; Stanley 2004) since search on smaller components in the larger structure can be emphasized. Most usefully, though, coevolution is appropriate for domains that have no intrinsic objective measure, which will be called *interactive domains*.

While the analogy with dynamic optimization (where the landscape also changes, but independently of the search process) may ground the reader, historically it has not generated insights into how to design successful coevolutionary algorithms. Instead, a fruitful approach has been to view interactive domains (and the problems that can be defined over them) as static, but structurally different and more complex than those of traditional optimization. This chapter is structured around this view.

Historically, the terms *cooperative* and *competitive* have been used to classify the domains to which coevolution is often applied. Indeed, game theory provides some guidance for making such distinctions. It can be argued that while such distinctions are relevant and at times useful for classifying the interactive domain over which an algorithm operates, they have not been appropriate for classifying *problems* or algorithms. The interactive domain, like the fitness function, simply gives values to interactions. The problem specifies what to find and the algorithm finds it. Chess is an interactive domain, finding a chess-playing strategy capable of grand master rating is a problem. Experience shows that details of the problem definition and algorithm design have much more impact on the overall behavior of a CoEA than details of the interactive domain.

Thus, problems are primarily divided into classes based on what constitutes a solution. Two types of problems are highlighted: *test-based* problems and *compositional* problems. A test-based problem is one in which the quality of a potential solution is determined by its performance when interacting with some set of tests. By contrast, in compositional problems the quality of a solution to the problem involves an interaction among many components that together might be thought of as a team or assembly.

1.3 Successful Applications of Coevolutionary Algorithms

In the subsequent sections interactive domains and coevolutionary algorithms will be described in detail; however, it is useful to have some context with which to begin the discussion. Below are three simple successful applications of coevolution to three different problems. They will be presented as examples for the ideas discussed above, but they were also chosen for their historical impact.

1.3.1 Single Population CoEA Applied to a Test-Based Problem

In single population CoEAs applied to test-based problems, individuals serve two roles: at times they are used as (components of) potential solutions, while at other times they are used as tests to provide evaluation information about other individuals. The problem of finding good game-playing strategies can be viewed as a test-based problem since strategies are tested by playing against other strategies. Chellapilla and Fogel's AI checkers player *Blondie24* is a popular and successful application of a CoEA to such a problem (Chellapilla and Fogel 1999). A strategy in Blondie24 employs a traditional minimax algorithm, but uses a neural network for board evaluation. Individuals in the CoEA are vectors of weights for a fixed-structure neural network, and are evaluated by playing against other individuals. Points are awarded based on the win/loss/draw record, and fairly standard evolutionary programming methods are used to select players, hence weight vectors, with better records. The CoEA employed for this problem was able to produce a checkers player that is competitive with the best existing human and AI players.

▶ *Interactive Domain:* Game of checkers; *Test:* Opponent playing strategy; *Potential Solution:* Playing strategy; *Problem:* Find a strategy that beats the most opponents.

1.3.2 Two-Population CoEA Applied to a Test-Based Problem

Hillis' coevolution of sorting networks and challenging data sets uses two populations (Hillis 1990). One population represents sorting networks (that is, arrangements of compare-and-swap circuits) while another represents unsorted data sets to test a network's sorting capability. The goal of the algorithm is to produce the smallest network possible that correctly sorts any given data set, but it does this while simultaneously honing ever more challenging and representative data sets. The technique produced a 61-comparator network, which is just one comparison larger than what was, at the time, the smallest-known network for a 16-input problem. A similar, non-coevolutionary technique described in that work was unable to produce networks smaller than 63 comparators.

▶ *Interactive Domain:* Running sorting network on data set; *Test:* Data set; *Potential Solution:* Sorting network; *Problem:* Find smallest, correct network.

1.3.3   Multi-population CoEA Applied to a Compositional Problem

There are a number of successful applications of CoEAs to compositional problems wherein problems are decomposed in some way and separate EAs are applied in parallel to the components, even though evaluation must involve some kind of aggregation or composition of components from the whole system. Perhaps the oldest, and still among the most successful, of these is Husbands and Mills' work on job-shop scheduling (Husbands and Mill 1991). In this work, individuals encode potential floor plans for managing jobs involving the processing of a particular widget constructed by the shop, and separate populations are used to optimize these process plans for each widget. Fitness, however, includes accounting for shared resources in the shop (time, space, etc.). There is also a population of arbitrators, agents that resolve conflicts between process plans for different widgets. Resulting schedulers can deal with a great deal of uncertainty on the job-shop floor.

While perhaps not as well known as the Blondie24 or sorting network examples, this work showcases the difference between a cooperative *domain* and compositional *problem*. In this case, the solution to the problem is a collection of floor plans (one for each component) and an arbitrator, which means the problem is compositional. However, the floor plans compete with one another in the evaluation function because they must share resources, so it is not strictly cooperative.

▶ *Interactive Domain:* Determination of complete job-shop schedule; *Components:* floor plans for different widgets, arbitrator for conflict resolution; *Potential Solution:* set of floor plans for each widget plus an arbitrator; *Problem:* Find an efficient and robust job-shop schedule.

1.4   Chapter Organization

History has shown that the naïve application of CoEAs to ill-understood domains is as likely to produce confusing and unsatisfactory results as to succeed. Therefore, at least from the perspective of problem solving, a careful algorithm design process is critical. This chapter is organized to emphasize this process: define the problem, define the relationship between the algorithm and the problem, implement the algorithm in a principled way, and analyze its behavior based on these principles.

The next section establishes clear definitions of domains, problems, and the types of solutions one expects to find. ❷ Section 3 describes how problems relate to evaluation and representation choices within a coevolutionary algorithm. This is followed by the presentation of analytical approaches currently available for understanding coevolutionary algorithm design, performance, and behavior. The concluding section provides some broader views of coevolutionary systems and outlines possible future directions of research and application.

2 Problems and Solutions

Coevolutionary algorithms are typically applied to interactive domains. Such domains generally lack an objective function giving a value to each potential solution. Rather, interactive domains encode the outcomes of interactions between two or more entities; depending on the domain, individual entities or the interaction itself may receive value as a result of an interaction. An algorithm must then decide how to use these outcomes to make decisions about which entities to promote in the next generation and which entities to demote or discard.

In a problem-solving context, an algorithm will need more than just an interactive domain. It will also require some way of deciding which entities in the domain are better than others. Coupled with information like that, an interactive domain becomes a co-search or co-optimization problem.

This section is concerned with detailing interactive domains, co-search, and co-optimization problems in the abstract, surveying some examples from both test-based and compositional problems, and beginning the discussion of what it means to extract a solution from such problems. Here, all dynamic or algorithmic considerations are left aside and instead the focus is on formalizing static definitions of problem classes to which CoEAs have been applied. Coevolutionary algorithms and their behavior on these problem classes are discussed in subsequent sections.

2.1 Interactive Domains

The formal notion of interactive domain can be defined as follows.

Definition 1 (Interactive Domain) An *interactive domain* consists of one or more functions, called *metrics*, of the form $p : X_1 \times X_2 \times \ldots \times X_n \to R$, where

- Each i with $1 \leq i \leq n$ is a *domain role*
- An element $x \in X_i$ is an *entity* (playing the domain role i)
- Each X_i is an *entity set* (for the domain role i)
- A tuple $(x_1, x_2, \ldots, x_n) \in X_1 \times \ldots \times X_n$ is an *interaction*
- The value $p(x_1, x_2, \ldots, x_n) \in R$ is an *outcome* (of the interaction)
- The ordered set R is the *outcome set* □

Some remarks about this definition are in order.

Interactions may be relatively simple, such as applying a sorting network to an input sequence; or highly complex, such as simulating the job-shop scheduling activities given some arbitrator and several floor plans. Further, they may be direct, such as, again, applying a sorting network to an input sequence; or indirect, via an *environment*, such as the board and dice in a checkers game or the shared resources in the scheduling example. We abstract away from such low level details and we are merely concerned with the outcome of the interaction.

The discussion of problem classes in ❯ Sect. 1.2 touched on the distinction between cooperative and competitive domains. Here, those terms will be linked directly to the interactive domain. *Cooperative domains* have n metrics p_i, one for each domain role. The metric p_i is interpreted as giving an outcome to the entities playing role i. In many examples, $p_i = p_j$ for all roles i and j, so that the metrics only differ by which entity receives the outcome (but see (Popovici 2006) for a more nuanced discussion of cooperative domains). *Competitive domains*, by contrast, typically have only two roles. The two corresponding metrics often obey $p_2 = -p_1$, making the domain equivalent to a zero-sum game. Naturally, there is an enormous space of interactive domains that do not fall into either the cooperative or competitive categories. Note also that the terms "cooperative coevolution" and "competitive coevolution" refer to coevolutionary algorithms operating on such domains.

While an interactive domain has n entity sets, two or more of these sets may be the same. The word *type* will be used to refer to the sets from which entities are drawn, independently of what domain roles the entities are playing. There are special cases in which all domain roles are played by entities of the same set and, furthermore, the outcomes a particular entity receives do not depend on which role it plays. These will be referred to as *role-symmetric* domains.

Example 1 (Rock–paper–scissors) The game rock–paper–scissors furnishes a particularly simple example of an interactive domain. This game is generally expressed with a payoff matrix

	Rock	Paper	Scissors
Rock	0	−1	0
Paper	1	0	−1
Scissors	−1	1	0

To view rock–paper–scissors as an interactive domain, observe

- There are two roles
- $X_1 = X_2 = \{rock, paper, scissors\}$ are the entity sets, so there is one type
- The matrix encodes a single metric p, whose value is given to the entity on the line. This domain is role symmetric; for instance, *rock* receives the same payoff versus *paper* regardless of whether it is playing the first role or the second.

Example 2 Here is another example coming from an abstract game

	t_1	t_2	t_3
s_1	1	1	0
s_2	0	1	2
s_3	2	1	1

As an interactive domain, note

- There are two roles
- $X_1 = \{s_1, s_2, s_3\}$; $X_2 = \{t_1, t_2, t_3\}$ are the entity sets (there are two types)
- The matrix encodes a single metric p. This domain is not role symmetric

For a large majority of interactive domains that have been studied in practice, the outcome set R is a subset of $\mathbb{R}$; however, that requirement is not strictly necessary (see (Bucci and Pollack 2002) for a discussion of ordered outcome sets that are not subsets of the reals).

Finally, note that this definition of interactive domain closely resembles the notion of the game defined by Ficici (2004). What we are calling entity is there called a behavior, reflecting that work's closer focus on agents selecting behaviors. What we are here calling an interaction is there called an event, and the notion of outcome corresponds to what Ficici calls a measurement. That work should be consulted for a more detailed treatment of interactive domains.

2.2 Co-search and Co-optimization Problems

A function $f : X \rightarrow \mathbb{R}$ gives values to elements of X. However, more information is needed to define a *problem* that can be solved. For instance, one may wish to find all the elements of X that maximize the value of f. Or, one may wish to find only one such element. Or, one may wish to find the elements that minimize f. Notice that for a fixed function f, there are many, distinct ways to decide what makes a "good" element of X. To give a well-defined search or optimization problem, one must specify not just the function f but also one way for deciding which elements of X are to be found.

By analogy, once an interactive domain is defined, one has a way of giving one or more values to interactions of entities, but does not yet have a well-defined problem to solve. The purpose of this section is to detail the kinds of problem that can be defined over interactive domains. These will be called *co-search problems* and *co-optimization problems*. First:

Definition 2 (Co-search Problem) Given an interactive domain, a *co-search problem* over it consists of

- A non-empty subset $I \subset (1, \dots, n)$ of the n domain roles
- A set $\mathscr{C}$ of *potential solutions* aggregated from entities of the domain roles in I
- A *solution concept* that specifies a partitioning of the potential solutions into (actual) solutions and non-solutions. This is represented as a subset $\mathscr{S} \subset \mathscr{C}$ of the potential solutions □

The last two components of the definition will be examined in more detail.

Potential Solutions: There are two critical points to emphasize:

- In many problems of practical interest, no single entity makes for a sufficient solution. Rock–paper–scissors stands as an intuitive example: none of the entities *rock, paper,* or *scissors* is, by itself, a reasonable strategy to play in this game. In such cases, potential solutions are *aggregated* from, meaning built or constructed out of, multiple entities.
- In some problems, one would like potential solutions to either lie in or be aggregated from entities in a single domain role, while in other problems, potential solutions should be aggregates of entities from several domain roles. The subset I is used to distinguish cases like these. If $I = (1)$, solutions are aggregated only from entities playing domain role 1, while if $I = (1, 2, \dots, n)$, solutions are aggregated from entities playing all n roles. One could, of course, have an I that is not all of $(1, \dots, n)$ but contains more than one domain role.

❯ Section 2.3 will give examples of what is meant by saying potential solutions are "aggregated from" entities. While a fully general formalization of this notion is beyond the scope of this chapter, in typical examples, potential solutions are sets of entities, mixtures of (distributions over) entities, tuples of entities, or perhaps combinations of these. (For instance, Nash equilibria are usually pairs of mixtures.) The notion of potential solution here is closely related to the idea of a *configuration* in (Ficici 2004), which should be consulted for more detail; Popovici (2006) discusses the idea of aggregating solutions in more detail as well.

Solution Concepts: The phrase "solution concept" originates in game theory (Osborne and Rubinstein 1994) and is largely synonymous with solution specification or solution definition. The phrase has been widely used in coevolution since being imported into the field by Ficici (2004). While the idea of a solution concept is presented in our abstract, as a subset of $\mathscr{C}$, in reality, solutions are identified using the metrics of the interactive domain. (What Ficici calls an *intensional* solution concept (Ficici 2004).) That is, it is typical to seek entities or combinations of entities that in some sense optimize the metrics of the interactive domain. ❯ Section 2.4 illustrates the point by describing several examples of solution concepts that have arisen in practice.

A *co-optimization problem* is closely related to a co-search problem. However, instead of a solution concept, a co-optimization problem specifies an order on the potential solutions and implicitly requires that solutions be maximal elements of the order. Specifically:

Definition 3 (Co-optimization Problem) Given an interactive domain, a *co-optimization problem* over it consists of

- A non-empty subset $I \subset (1, \ldots, n)$ of the n domain roles;
- A set $\mathscr{C}$ of *potential solutions* built from entities of the domain roles in I;
- An ordering (which may be a partial order or even a preorder) $\leq$ of $\mathscr{C}$ such that if $c_1, c_2 \in \mathscr{C}$ and $c_1 \leq c_2$, c_2 is interpreted as being no worse a solution than c_1. (Note that $\leq$ is an ordering on potential solutions, not on numbers. In defining it, it may be the case that lower values from the domain's metrics are better.) □

A co-optimization problem can be converted into a co-search problem by defining the solution concept $\mathscr{S} \subset \mathscr{C}$ to be the set of maximal elements of $\leq$. In that sense, co-optimization is a more refined notion than co-search.

As a final bit of terminology, recall that both co-search and co-optimization problems have a subset I of the domain roles that specifies which entities are aggregated into potential solutions. Let $\widehat{I}$ denote the complement of I in $(1, \ldots, n)$, so, for instance, if $n = 3$ and $I = (1, 2)$ then $\widehat{I} = (3)$. Then two *problem roles* can be distinguished:

- If $i \in I$, X_i is a *component role* and $x \in X_i$ are *component entities* or *components*
- if $i \in \widehat{I}$, X_i is a *test role* and $x \in X_i$ are *test entities* or *tests*

Note that $\widehat{I}$ may be empty, meaning there is no test role explicitly defined by the problem. However, I cannot be empty by definition, meaning there is always at least one set of entities playing the role of component.

To summarize some of the key concepts relating to entities that have been introduced in this section:

- An interactive domain defines two or more *domain roles* and entities that play each role.
- The notion of type is coarser than that of domain role: several domain roles might correspond to the same type of entity, but each type corresponds to at least one domain role.
- A co-search or co-optimization problem lies at a conceptually higher level, defined over an interactive domain.
- Co-search and co-optimization problems define two *problem roles*, components and tests.
- At the problem level, the component role corresponds to one or more lower-level domain roles. The test role may correspond to zero or more domain roles.
- Since types sit at the domain role level, the relationship between types and problem roles can be complicated. There may be components of different types, tests of different types, components and tests that both have the same type, etc.

❯ Section 1.2, besides distinguishing cooperative and competitive domains, also mentioned the distinction between compositional and test-based problems. Before elaborating on potential solutions in ❯ Sect. 2.3 and solution concepts in ❯ Sect. 2.4, we will define and illustrate these two important classes of co-search and co-optimization problems. Briefly, compositional and test-based problems correspond to the extremes where $|I| = n$ and $|I| = 1$, respectively. The middle ground between these extremes is virtually unexplored.

2.2.1 Compositional Problems

A *compositional problem* is a co-search or co-optimization problem in which all domain roles are used as components to build solutions. That is, $|I| = n$ and, conversely, $|\widehat{I}| = 0$. The idea is that each entity in each domain role is a component. To build potential solutions, one must use entities from each of the domain roles; one does not have a complete potential solution until one has used at least one component from each domain role. An intuitive example is a baseball team: one does not have a baseball team until one has a pitcher, a catcher, outfielders, etc.

Compositional problems have largely been explored under the rubric of cooperative coevolutionary algorithms (CCEAs). Besides theoretical works that study features of algorithm behavior on arbitrary or abstract test problems (Jansen and Wiegand 2003a, b, 2004), or empirical work of algorithm dynamics on simple test problems (Popovici and De Jong 2005a, c), work applying CCEAs to multivariate function optimization or multi-agent learning have also appeared. In multivariate function optimization, one treats each input variable as a distinct domain role, so that each entity in a particular domain role is a potential setting for one of the input variables (Potter and De Jong 1994, 2000; Panait et al. 2004). Multi-agent learning applications have treated each domain role as the space of possible behaviors or actions for agents that form a team performing a task together (Parker and Blumenthal 2003; Panait et al. 2006c).

2.2.2 Test-Based Problems

A *test-based problem* is a co-search or co-optimization problem in which $|I| = 1$; that is, in which a single domain role contains components, and all other domain roles are tests. Intuitively speaking, the test entities are used to probe or give information about the potential solutions that can be aggregated from the components.

Test-based problems are discussed as such in (De Jong and Pollack 2004) and analyzed in connection with multi-objective optimization in (De Jong and Bucci 2007); however, the idea is implicit in early work on Pareto coevolution (Noble and Watson 2001; Ficici and Pollack 2001) and is more explicit in later works (De Jong 2004a, c; Monroy et al. 2006). Bucci and Pollack (2002, 2003b) formally analyze test-based problems from the perspective of order-theory, while Bucci and Pollack (2005) and Panait and Luke (2006) treat compositional problems using ideas developed to study test-based problems. De Jong (2005), Jensen (2001), Branke and Rosenbusch (2008), Barbosa (1999), and Stuermer et al. (2009) approach two types of test-based problems not based on Pareto dominance.

In many of these works, the terms *candidate, candidate solution,* or *learner* are used to describe what is here described as a component. Likewise, the term *test* is used in those works to denote what is here called an entity playing the test role.

2.3 Potential Solutions in Co-search and Co-optimization Problems

❯ Section 2.2 defined a co-search problem as one that, given an interactive domain, specifies a solution concept as a subset of a set of *potential solutions* that are built from the interactive domain. This section is intended to detail, by example, several common extant ways of creating potential solutions from entities.

2.3.1 Single Entity

The simplest and most obvious set of potential solutions is a set of entities. Consider an interactive domain with n roles and a co-search problem with $I = (1)$, so that the component role is being played by the entities in domain role 1. Then, one set of potential solutions for this problem is the set X_1 of components itself. In this case, the set $\mathscr{C}$ is X_1, so that a solution concept $\mathscr{S}$ gives a subset of X_1.

Seminal work by Hillis (1990) and Sims (1994) both use single entities as potential solutions. Hillis' algorithm sought a single sorting network; there was no sense in which two or more sorting networks could be combined. Likewise, Sims' algorithm sought a morphology and behavior for a simulated creature; again, there was no mechanism for combining several creatures into a composite.

Stipulating that single entities serve as solutions is appropriate in certain domains, particularly those involving the design of a complex object (that is, when no clear way of combining entities together is present); or, those in which one expects that a single entity that solves the problem may exist. However, as noted above about rock–paper–scissors, many domains have no single entity that is adequate to solve the problem.

2.3.2 Sets

Consider, for concreteness, that we have an interactive domain with two roles and our co-search problem specifies that $I = (1)$. The entities in X_1 are components, while those in X_2 are tests. In this example, the set $\mathscr{C}$ is the powerset of X_1, $\mathscr{P}(X_1)$. A solution concept gives a subset of $\mathscr{P}(X_1)$, equivalently one or more subsets of X_1, where each subset is a solution.

Set-based notions of potential solution have been studied under the umbrella of Pareto coevolution (Ficici and Pollack 2001; Noble and Watson 2001). Treating each test as an objective to be optimized, the problem-solving goal is to find (an approximation of) the non-dominated front among the components. Therefore, a potential solution is a subset of X_1, in other words, an element of $\mathscr{C} = \mathscr{P}(X_1)$. The non-dominated front itself is the solution.

When solution concepts are further elaborated in ❷ Sect. 2.4, one will see the situation is more subtle; in fact, a set related to, but not quite, the non-dominated front is desirable. However, that discussion will be deferred until then.

2.3.3 Mixtures

Let's maintain the co-search problem of the previous example. However, let's say that a solution is not a subset of X_1, but a probability distribution over X_1. Following Vose (1999), denote the set of these by Λ^{X_1}. Then another choice for the set of potential solutions $\mathscr{C}$ is Λ^{X_1}.

Mixture-based solutions are most often discussed in the context of Nash equilibria; the Nash memory (Ficici and Pollack 2003) and parallel Nash memory (Oliehoek et al. 2006) are examples of algorithms that treat mixtures as potential solutions.

2.3.4 Compositions in General

Sets and mixtures are both examples of aggregates; that is, potential solutions created by putting together components drawn from one or more of the component roles defined in the problem. Sets and mixtures are particularly simple kinds of aggregates. More complicated compositions are conceivable.

Compositional coevolution in general aims to discover good assemblies. The originally stated aim of cooperative coevolutionary algorithms, from which the more general notion of compositional coevolution sprang, was to attack the problem of evolving complicated objects by explicitly breaking them into parts, evolving the parts separately, and then assembling the parts into a working whole (Potter and De Jong 2000). The nature of the "coadapted subcomponents" discussed in that work was not strictly defined; hence, the idea is that any subcomponent of any assembly (aggregate) was fair game for the method. Since then, compositional coevolution has been applied to coevolving teams of agents performing a task, for instance, where the composition here is of several different agents collaborating as a team.

In the language developed in this section, agents or coadapted subcomponents are domain entities playing the component role of a compositional problem. A collaboration among entities is an interaction, and the means for evaluating a team or assembly would constitute the metric of the problem. The potential solutions in these cases, then, are the possible tuples over the domain roles; that is, $\mathscr{C} = X_1 \times \cdots \times X_n$.

Though the translation into the notion of a co-search problem is less obvious, another important, nontrivial example of composition found in coevolutionary algorithms are in applications of neuro-evolutionary algorithms. The neuro-evolution through augmenting topologies (NEAT) algorithm (Stanley and Miikkulainen 2002a) has been applied to coevolve robot controllers in a simulated robot duel (Stanley and Miikkulainen 2002a, b, 2004) as well as players of a variant of the video game Pong (Monroy et al. 2006). A hallmark of the NEAT algorithm is the separation between components of neural networks, which consist of small assemblies of neurons and associated weighted synapses, and the topology of a complete network, which functions as a blueprint for how assemblies are put together. A second, important feature of NEAT emphasized in these works is the possibility for complexification or elaboration: since neural network topologies are not limited by NEAT, they can grow to arbitrary complexity, elaborating on previously evolved behaviors. In contradistinction to typical CCEAs, which prespecify the size and complexity of compositions, NEAT permits open-ended complexity increase. (And, one would hope, an increase in capability as well.)

The Symbiogenic Evolutionary Adaptation Model (SEAM) described by Watson (2002) includes a form of variational operator designed to explicitly compose evolving entities. Additionally, SEAM employs a group evolutionary mechanism of composition in conjunction with notions of test-based evolution by using Pareto dominance to determine whether variation is likely to result in productive compositions.

Finally, it is worth pointing out that in general game-theoretic analysis, Nash equilibria are pairs of mixtures of pure strategies. In many cases, only the first mixture in the pair is used to solve a problem; the other is there for testing purposes. However, should one actually need both mixtures in the pair, the set of potential solutions would be $\Lambda^{X_1} \times \Lambda^{X_2}$, which involves two levels of aggregating: first mixing (over X_1 and X_2), then pairing.

2.4 Solution Concepts

This section will detail several solution concepts for both compositional and test-based problems that have been used in coevolutionary algorithms. Recall that a solution concept specifies a subset of the potential solutions, $\mathscr{S} \subset \mathscr{C}$. Each example will detail the interactive domain, the co-search problem including the space of potential solutions, and how solutions are determined from the metric(s) of the domain. Solution concepts for compositional problems are presented first.

2.4.1 Compositional Solution Concepts

Ideal Team

Consider an interactive domain with n domain roles, entity sets $X_1, X_2, \ldots, X_n$ and one metric $p : X_1 \times \cdots \times X_n \to \mathbb{R}$. Consider also a co-search problem over this domain with potential solutions drawn from the set of tuples $\mathscr{C} = X_1 \times \cdots \times X_n$, so in fact this is a compositional problem. Observe there is a one-to-one correspondence between a potential solution and an interaction. The ideal team solution concept defines as solutions those potential solutions that maximize the value received from the metric

$$\mathscr{S} = \{\bar{x} \in \mathscr{C} \mid \forall \bar{x}' \in \mathscr{C}.p(\bar{x}) \leq p(\bar{x}') \Rightarrow p(\bar{x}) = p(\bar{x}')\} \tag{1}$$

where the shorthand $\bar{x}$ is used to denote an arbitrary tuple in $\mathscr{C}$. In multi-agent parlance, the teams for which no other teams perform better are ideal teams.

The ideal team is not the only relevant compositional solution concept, and the question has been posed to what extent cooperative coevolution algorithms converge to it (Wiegand 2004). Another solution concept of interest arising from that investigation is *maximizing robustness*. The idea here is that rather than maximizing the outcome of a single potential solution, a solution should be robust in the sense that a "small" change in one of the components results in "small" changes in the composite's value.

2.4.2   Test-Based Solution Concepts

Unless otherwise specified, the following examples of solution concepts for test-based problems are appropriate for problems of the following form:

- The interactive domain has two roles with entity sets X_1 and X_2.
- There is a single metric $p : X_1 \times X_2 \to \mathbb{R}$.
- The co-search problem specifies X_1 as the only component role and X_2 as the only test role.

What varies in these examples is the set of potential solutions and how the subset of solutions is defined. These will be detailed now.

Best Worst Case

Best worst case operates over single-entity potential solutions, meaning $\mathscr{C} = X_1$. This solution concept specifies as solutions those components that maximize the minimum possible outcome over interactions with all tests. (This version has also been called *maximin*. The corresponding *minimax* can similarly be defined.) That is,

$$\mathscr{S} = \left\{ x \in \mathscr{C} \mid \forall x' \in \mathscr{C}. \min_{t \in X_2} p(x, t) \le \min_{t \in X_2} p(x', t) \Rightarrow \min_{t \in X_2} p(x, t) = \min_{t \in X_2} p(x', t) \right\} \quad (2)$$

This criterion is appropriate in real-world domains where one needs to protect against the worst scenario possible.

Simultaneous Maximization of All Outcomes

Simultaneous maximization of all outcomes requires a solution to be a component that maximizes its outcome over all possible tests simultaneously. That is, $\mathscr{C} = X_1$ is a single entity set of potential solutions, and the solution concept is

$$\mathscr{S} = \{ x \in \mathscr{C} \mid \forall x' \in \mathscr{C} \, \forall t \in X_2. [p(x, t) \le p(x', t) \Rightarrow p(x, t) = p(x', t)] \} \quad (3)$$

This solution concept has a limited application scope, as for many problems, there does not exist a single potential solution that simultaneously maximizes the outcome against all possible tests.

Maximization of Expected Utility

Maximization of expected utility (MEU) is also relevant when $\mathscr{C} = X_1$. It specifies as solutions those components that maximize the expected score against a randomly selected opponent

$$\mathscr{S} = \{ x \in \mathscr{C} \mid \forall x' \in \mathscr{C}. \mathbf{E}(p(x, X_2)) \le \mathbf{E}(p(x', X_2)) \Rightarrow \mathbf{E}(p(x, X_2)) = \mathbf{E}(p(x', X_2)) \} \quad (4)$$

where one is abusing notation and treating X_2 as a uniformly distributed random variable ranging over the set of tests, so that $\mathbf{E}(p(x, X_2))$ is the expected value of $p(x, t)$ when t is selected uniformly randomly from X_2.

> ❯ Equation 4 is equivalent to

$$\mathscr{S} = \left\{ x \in \mathscr{C} \mid \forall x' \in \mathscr{C}. \sum_{t \in X_2} p(x, t) \le \sum_{t \in X_2} p(x', t) \Rightarrow \sum_{t \in X_2} p(x, t) = \sum_{t \in X_2} p(x', t) \right\} \quad (5)$$

when all the sums are defined. That is, maximizing expected utility is equivalent to maximizing the sum of outcome values over all tests when that sum is defined for all the components. Thus, MEU essentially assumes that all tests are of equal importance, which limits its generality.

Nash Equilibrium

The Nash equilibrium solution concept is inspired by game theory (Osborne and Rubinstein 1994) and operates over mixtures of components. When the problem is as before, $\mathscr{C}_1 = \Lambda^{X_1}$. However, deviating slightly from the previous examples, problems with $I = (1, 2)$ will also be considered, so that $\mathscr{C}_2 = \Lambda^{X_1} \times \Lambda^{X_2}$. That is, there are no tests, so in fact $\mathscr{C}_2$ defines a compositional problem. Furthermore, in both cases it will be assumed that there are two metrics, $p_1 : X_1 \times X_2 \to \mathbb{R}$ and $p_2 : X_1 \times X_2 \to \mathbb{R}$ interpreted as giving outcomes to entities in X_1 and X_2, respectively. In many domains such as those arising from zero sum games, $p_1(x, y) = -p_2(x, y)$ for all $x \in X_1$ and $y \in X_2$.

Consider the case $\mathscr{C}_2$ first. Let $\alpha \in \Lambda^{X_1}$ and $\beta \in \Lambda^{X_2}$ be two mixtures over X_1 and X_2, respectively. These can be written as formal sums

$$\alpha = \sum_{x \in X_1} \alpha_x \cdot x \tag{6}$$

and

$$\beta = \sum_{y \in X_2} \beta_y \cdot y \tag{7}$$

where α_x is the probability assigned to the component $x \in X_1$ by the mixture α, and β_y is the probability assigned to the test $y \in X_2$ by β and, since α and β are both distributions, $\sum_{x \in X_1} \alpha_x = 1$ and $\sum_{y \in X_2} \beta_y = 1$. Using this notation, define a function $\mathbf{E}_{p1} : \Lambda^{X_1} \times \Lambda^{X_2} \to \mathbb{R}$ such that for all $(\alpha, \beta) \in \Lambda^{X_1} \times \Lambda^{X_2}$

$$\mathbf{E}_{p1}(\alpha, \beta) = \sum_{\substack{x \in X_1 \\ y \in X_2}} \alpha_x \cdot \beta_y \cdot p_1(x, y) \tag{8}$$

$\mathbf{E}_{p1}(\alpha, \beta)$ is interpreted as giving the expected outcome that the mixture α receives when interacting with β. $\mathbf{E}_{p2}(\alpha, \beta)$ is defined similarly and gives the expected outcome to β.

A Nash equilibrium is a pair $(\alpha, \beta) \in \Lambda^{X_1} \times \Lambda^{X_2}$ such that neither α nor β can unilaterally change to some other mixture and receive a higher payoff. That is, (α, β) is a Nash equilibrium if the following two conditions hold:

- For all $\alpha' \in \Lambda^{X_1}$, $\mathbf{E}_{p1}(\alpha, \beta) \leq \mathbf{E}_{p1}(\alpha', \beta) \Rightarrow \mathbf{E}_{p1}(\alpha, \beta) = \mathbf{E}_{p1}(\alpha', \beta)$
- For all $\beta' \in \Lambda^{X_2}$, $\mathbf{E}_{p2}(\alpha, \beta) \leq \mathbf{E}_{p2}(\alpha, \beta') \Rightarrow \mathbf{E}_{p2}(\alpha, \beta) = \mathbf{E}_{p2}(\alpha, \beta')$

The Nash equilibrium solution concept for problems with $\mathscr{C}_2 = \Lambda^{X_1} \times \Lambda^{X_2}$ is then

$$\mathscr{S}_2 = \{(\alpha, \beta) \in \mathscr{C}_2 \mid (\alpha, \beta) \text{ is a Nash equilibrium}\} \tag{9}$$

For $\mathscr{C}_1 = \Lambda^{X_1}$, in other words problems in which one only cares about the mixture over components in X_1, the Nash equilibrium solution concept is

$$\mathscr{S}_1 = \pi_1(\mathscr{S}_2) \tag{10}$$

where π_1 projects a pair (α, β) onto the first coordinate, α.

An attractive feature of the Nash equilibrium as a solution concept is that Nash equilibria have certain "security" guarantees: The expected payoff to α in a Nash equilibrium (α, β) can be no lower than $\mathbf{E}_{p1}(\alpha, \beta)$, regardless of the strategy with which it interacts. In a game like {*rock, paper, scissors*}, for example, any individual strategy like *rock* will receive a -1 payoff against some opponent (if *paper* plays against *rock*, for example). The mixture $\frac{1}{3} \cdot rock + \frac{1}{3} \cdot paper + \frac{1}{3} \cdot scissors$, by contrast, will never receive an expected payoff lower than 0 against any opponent because $(\frac{1}{3} \cdot rock + \frac{1}{3} \cdot paper + \frac{1}{3} \cdot scissors, \frac{1}{3} \cdot rock + \frac{1}{3} \cdot paper + \frac{1}{3} \cdot scissors)$ is a Nash equilibrium for that game and $\mathbf{E}_{p1}(\frac{1}{3} \cdot rock + \frac{1}{3} \cdot paper + \frac{1}{3} \cdot scissors, \frac{1}{3} \cdot rock + \frac{1}{3} \cdot paper + \frac{1}{3} \cdot scissors) = 0$. Note that this is a kind of worst-case guarantee for the mixture α that is similar in spirit to that sought in the best worst case solution concept; the primary difference is that the best worst case seeks single components with such a guarantee, while Nash equilibrium seeks a mixture of many components.

Pareto Optimal Set

Multi-objective optimization extends traditional optimization through the introduction of multiple *objectives*. This may be viewed as the use of a function that is vector valued rather than scalar. In Pareto-based solution concepts, every possible test is viewed as an objective to be optimized.

Potential solutions for the Pareto optimal set solution concept are subsets of X_1; that is, $\mathscr{C} = \mathscr{P}(X_1)$. Here, the set of solutions will consist of a single member, the non-dominated front, which is defined

$$\mathscr{F} = \{x \in \mathscr{C} \mid \forall x' \in \mathscr{C}.[\forall t \in X_2.[p(x, t) \leq p(x', t)] \Rightarrow \forall t \in X_2.[p(x, t) = p(x', t)]]\} \quad (11)$$

Let the *Pareto covering* order be an order on X_1 defined as follows:

$$x \preceq x' \quad \text{if } \forall t \in X_2.p(x, t) \leq p(x', t) \quad (12)$$

for all x and x' in X_1. This definition essentially says that the outcome of x' against any test is at least as high as that of x. However, $\preceq$ is not a total order on X_1, because it is possible that $p(x, t) < p(x', t)$ while $p(x, t') > p(x', t')$ for two tests $t, t' \in X_2$. Furthermore, while $\preceq$ is reflexive and transitive, it need not be a partial order: two distinct components x and x' might receive precisely the same outcomes on all tests, so $x \preceq x'$ and $x' \preceq x$ both hold and anti-symmetry fails to hold.

The definition of $\mathscr{F}$ can be simplified using $\preceq$

$$\mathscr{F} = \{x \in \mathscr{C} \mid \forall x' \in \mathscr{C}.x \prec x' \Rightarrow x' \preceq x\} \quad (13)$$

The Pareto optimal set solution concept is defined as

$$\mathscr{S} = \{\mathscr{F}\} \quad (14)$$

Some remarks:

- There is a formal similarity between ❷ Eq. 3, the simultaneous maximization of all objectives, and the non-dominated front $\mathscr{F}$ defined in ❷ Eq. 11. They differ only in the placement of the quantifier $\forall t \in X_2$. However, while formally similar, these two solution concepts have significantly different interpretations: while ❷ Eq. 13 shows that $\mathscr{F}$ is the set of *maximal* elements of the order $\preceq$, in fact ❷ Eq. 3 is the set of *maxima* of that order, the difference being that maxima must be larger than all other components across all tests simultaneously. Thus, the non-dominated front puts weaker constraints on its members than the simultaneous maximization of all objectives, and one would expect the former to be a larger set of components.

- The Pareto optimal set solution concept has only one member, $\mathcal{F}$, which is itself a set of components. This definition is chosen for consistency with the other solution concepts; each solution concept is a set whose members are solutions. The non-dominated front forms a single, unique solution to a multi-objective problem. A limitation of the Pareto optimal set solution concept is that this set may be very large; while not being covered is a sensible minimum requirement for solutions, for certain problems it may be insufficiently weak, in that it can insufficiently narrow down the set of possible solutions.
- As noted, two components may receive precisely the same outcomes against all possible tests. Thus, while they may be distinct components in the set X_1, as far as performance against the tests in X_2 goes, they are indistinguishable.

Pareto Optimal [Minimal] Equivalence Set

The indistinguishability of some elements of the non-dominated front entails this set may be too large. One can partition this set in equivalence classes by this property, like so. Say that $x \sim x'$ if $x \preceq x' \wedge x' \preceq x$; that is, x and x' are equivalent if they receive the same outcomes against all tests in X_2. $\sim$ is an equivalence relation on X_1. (This is a simple, well-known consequence of $\preceq$ being a preorder.) It is reasonable to suppose that one only needs one representative of each equivalence class under $\sim$, and this intuition leads to two more solution concepts.

A Pareto optimal equivalence set is a set containing *at least one* element from each equivalence class of the full Pareto optimal set. Note that while there is a single Pareto optimal set, there may be a combinatorial number of Pareto optimal equivalence sets. A Pareto optimal minimal equivalence set contains *exactly one* component from each equivalence class of the full Pareto optimal set. There may be a combinatorial number of Pareto optimal minimal equivalence sets as well.

These two solution concepts typically define a smaller set of potential solutions than the Pareto optimal set. Yet, depending on the characteristics of the problem, such a set can still be very large.

3 Design of Coevolutionary Algorithms

As the previous sections have shown, there is a wide spectrum of problems in interactive domains. While throughout this chapter we discuss generic issues that any algorithm targeted at such co-search problems will have to address, in this section, the main focus is on issues that are specific to approaching these problems via coevolutionary algorithms. Like with any method, application of coevolutionary algorithms to a problem tends to go more smoothly and have more opportunity for success when algorithm engineers *first* go through the process of formalizing the domain and the problem, and *then* design the algorithm.

For domains that are inherently interactive, the formalizing process involves more rationalization than choosing. However, sometimes one may want to reformulate a noninteractive domain into an interactive one. The domains of problems featuring the ideal team solution concept are often the result of such a reformulation. In such cases, many different choices may exist as to how to decompose elements in the original domain, and they may have different effects on problem difficulty.

The design process entails finding useful and productive ways of mapping problem particulars into the algorithmic framework, and heuristic search tends to require some means of making qualitative comparisons and decisions in order to guide the path of the algorithm. There are principles behind, and effects from these decisions. In this section, the choices available to the algorithm designer are discussed, while the biases introduced by these mapping choices are discussed throughout ❿ Sect. 4.

Once the engineer identifies the types, domain roles, problem roles, and potential solution set, they must determine how a CoEA can be structured to represent them, and encode data structures within the algorithm to instantiate them in a way that is consistent with the problem's solution concept. Such decisions are considered to relate to *representation*. Additionally, one must decide how to explore the set of interactions, how outcomes from multiple interaction will be aggregated for selection purposes, and how interaction information is communicated throughout various parts of the algorithm. Such decisions are considered to relate to *evaluation*.

3.1 Representation

Successful design and application of any heuristic depends on a number of key representation decisions for how search knowledge is encoded and manipulated. In any evolutionary system, these choices involve questions about how aspects of potential solutions are represented genotypically, modified by genetic operators, and expressed phenotypically.

Coevolution brings additional subtleties to representation issues that are worth special consideration. For one, the relationship between basic terms such as *individuals*, *populations*, and *solutions* can be quite complicated in coevolution. Related to these terms are the foundational, domain-centric terms discussed above such as *test* and *component*, and a well-designed CoEA should consider how these problem-based concepts map to representations used by the search system. Additionally, many modern CoEAs make use of archival and memory mechanisms for a variety of purposes, and it is important to understand how they relate to the above notions. These notions will be discussed in turn and examples of common/possible mappings from problem to algorithm will be provided.

3.1.1 Individuals

In most traditional evolutionary systems, individuals represent potential solutions to some search problem, and the EA manipulates encoded potential solutions by modifying them using variational operators (mutation, crossover) and choosing the better from among them using selection operators (proportionate selection, rank selection, etc.). When optimizing some real-valued function, an individual might be encoded as a vector of real values to be used as arguments for the function, mutations could be accomplished by applying vectors of Gaussian noise, and selection might involve simply choosing those with the highest objective function value. Since the goal of the search is to find an optimal argument vector, one can typically equate potential solution and individual. Even when the EA is searching program spaces, such as in genetic programming, traditional algorithms still make use of individuals who are essentially procedural solutions to the problem.

In coevolution, individuals serve the same mechanistic purpose as in any evolutionary algorithm: they are the fundamental units being manipulated by the search operators themselves. That is, mutation and crossover modify individuals, and selection chooses among them.

Definition 4 (Individual) An *individual* is a representational unit that is subject to selection and variational operators. □

But individuals in a CoEA will not always represent potential solutions to the problem being solved. In coevolution, one must think of individuals in terms of how they relate to types, domain roles, and problem roles, *in addition to* traditional issues of how to encode problem aspects. For example, suppose an AI is being developed for controlling an airplane that generalizes over different flight conditions and this is formalized as follows: an interactive domain consists of airplane controllers interacting with different flight conditions/scenarios; the problem is test-based where the potential solutions are controllers and the flight scenarios are tests; and the solution concept is maximum expected utility. One might design a CoEA that maintains two kinds of individuals, corresponding to the two types of the domain, namely individuals representing controllers and individuals representing flight scenarios. But one still has to decide how controllers and scenarios will *actually be encoded*, which will involve additional choices. The consideration of how types, domain roles, and problem roles map to individuals is subtly different from traditional genotype/phenotype decisions.

Most often, individuals represent entities from the domain's types. Thus, for a domain with two symmetric roles, one may have a single kind of individual, and such individuals might participate in interactions in either of the domain's roles. In general, one will have at least as many kinds of individuals as types in the domain. From a problem-role perspective, individuals might represent components of a potential solution or tests helping to judge the quality of potential solutions. Hillis (1990) provides a less common example, where some of the individuals do not correspond to entities, but to sets of entities, namely, those playing the test role. The relationship between individual and solution will be revisited in ❷ Sect. 3.1.4 dedicated to solutions.

3.1.2 Populations

Just as with traditional EAs, it is useful to think of populations as simply collections of individuals. Most traditional EAs (excepting island models) have only a single, explicit population, while many CoEAs have multiple populations.

But even in traditional methods, the notion of a population can be a bit murky when dealing with mechanisms such as spatial models, niching, or restrictive mating (Horn 1995; Spears 1994) since stable subsets of individuals in the overall population can arise between which very little genetic material is exchanged. These subsets are sometimes referred to as "populations," and it is interesting to note that many of these are essentially coevolutionary in nature since fitness can be quite subjective (e.g., niching methods using fitness sharing).

Definition 5 (Population) A *population* is:

1. A set (or multiset) of individuals
2. A subset of individuals isolated from other individuals by some kind of barrier (e.g., inability to interbreed successfully, geographic, etc.) □

Designing effective CoEAs involves decisions about how aspects of a problem map to populations. The main purpose of the populations is to provide an exploration mechanism for the entity sets of the different types in the domain. Thus, a CoEA will generally maintain at least one population per type and most often a single population per type.

For example, for an ideal team problem in a domain with a fixed number of asymmetric roles (and therefore types), the typical CoEA will maintain one population for each. For a test-based problem in a domain with two asymmetric roles (and therefore types), the typical CoEA will also maintain one population for each type.

For domains with two symmetric roles (thus a single type), the problem definition itself may not distinguish between the component role, and the test role, and the algorithm designer can choose to do the same, by maintaining a single population, or to make the distinction explicit by maintaining two populations. A unique work by Potter (1997) features an ideal team problem in a domain with a variable (unbounded) number of symmetric roles (neural networks with variable number of neurons) approached via a CoEA with a dynamic number of populations, all containing individuals of the same type.

Populations may also be used to represent potential solutions, for example when these are aggregates of entities (and therefore individuals) from one type, as is the case with Pareto optimal solution concepts. (This is analogous to traditional multi-objective optimization, where the population can be used as an approximation of the Pareto-front.)

Finally, yet importantly, individuals in a population, by taking part in interactions, serve the purpose of evaluating individuals in the same or other populations. This is because while a problem definition may specify only some of the domain roles as contributing to the potential solution set, an algorithm needs some criteria based on which to select and promote individuals for the test roles as well.

3.1.3 Archives

Coevolutionary methods often employ another kind of collection, typically referred to as an *archive*. Archives span generations, and thus can be considered a kind of search memory. The purpose of the archives is to help the populations with, or release them from, some of their multiple duties. Archives can allow algorithm designers to separate exploration from evaluation and/or solution representation.

Definition 6 (Archive) An *archive* is a collection of individuals that spans multiple generations of a coevolutionary algorithm. □

Thus, archives often contain the end solution (Rosin and Belew 1997) or actually are the solution, mainly when potential solutions are aggregations of entities (De Jong 2004a, c, 2005; Oliehoek et al. 2006; Ficici and Pollack 2003). So an analogy in traditional evolutionary computing (EC) for this purpose of an archive is the *best-so-far* individual. But they need not only represent solutions, they can and typically do influence the search – a bit like elitism would in a traditional generational EA.

Archives can also serve for evaluation purposes. A simple example of this are *hall-of-fame* methods (Rosin and Belew 1997) where successful individuals discovered during search are added to a hall-of-fame and part of the fitness of new individuals comes from the outcome of interactions with some subset of individuals from that archive.

In domains with multiple types, there might be an archive for each type (in addition to a population for each type). Most archive-based CoEAs involve an UPDATE step where individuals in a current population are considered for inclusion in the corresponding archive, and individuals in that archive may be reorganized or removed. There is a certain appeal to this approach because there are often straightforward ways to explicitly implement a particular solution concept by crafting the correct update procedure for an archive.

3.1.4 Solutions

A potential solution can be many different things even as part of the problem specification, the latter being a higher-level notion defined over the low-level elements of an interactive domain. It should therefore not be surprising that, as already seen, a potential solution can map to different things in the algorithm. A potential solution may be a single individual, in which case it may be extracted out of

- Any population (e.g., if approaching a maximum expected utility problem in a domain with two symmetric roles via two populations)
- A specific population (e.g., if approaching a maximum expected utility problem in a domain with two asymmetric roles via two populations)
- An archive (e.g., if using a hall-of-fame approach)

A potential solution may also be a set of individuals, in which case it may be

- The entire contents or a subset of an archive or population (e.g., for CoEAs approaching Pareto optimal solution concepts)
- A collection of individuals, one from each archive/population (e.g., for CoEAs approaching the ideal team solution concept for asymmetric domains)

Further, there may be populations/archives in a CoEA that never contribute any individuals to a potential solution (e.g., populations/archives corresponding to test roles in a test-based problem that are nevertheless used during evaluation).

3.2 Evaluation

Some issues concerning evaluation are pertinent to both single- and multi-population CoEAs. Regardless of whether interactions occur between individuals in the same population or in different populations, decisions need to be made as to what *interactions* should be assessed and how the outcomes of those interactions should be *aggregated* to give individuals fitness. When using multiple populations, the additional issue of *communication* between these populations arises. Each of these three matters will be discussed in turn.

3.2.1 Interactions

The definition of interactions was discussed in the previous section. Here, the focus is on the *selection* of interactions (also referred to as the *interaction method* or, in cooperative domains, *collaboration method*). The simplest choice is to assess all interactions possible, given the individuals present in the system at evaluation time. This has been referred to as full mixing or complete mixing. While no additional decisions would have to be made, this choice has the disadvantage of being very expensive, as the time cost of the algorithm is counted in interactions assessed. To reduce cost, one must assess only some of all possible interactions. This immediately raises the question "which ones?"

There are two main approaches to choosing a subset of interactions: individual centric and population centric. With the individual-centric approach, one individual at a time is considered, a set of interactions is chosen for that individual to take part in, and after the interactions are assessed, the individual's fitness is computed. These interactions may be reused (but generally are not) for computing the fitness of the other individuals that took part in them. These other individuals have historically been called *collaborators* in cooperative domains and *opponents* in competitive domains. In this context, the phrase *sample size* denotes the number of interactions used per fitness evaluation. Note that if there is no reuse, the number of interactions an individual takes part in may be greater than the number of interactions used for its evaluation. Thus, while the sample size may be (and usually is) the same for all individuals, the number of interactions that individuals are involved in may vary.

A simple and quite common approach is to use a single interaction per fitness evaluation and have the other individuals in this interaction be the best from their respective populations. When solving ideal team problems, this has been termed *single-best collaboration method* by Wiegand (2004), while for domains considered competitive, it has been referred to as *last elite opponent (LEO) evaluation* (Sims 1994). More generally, it is called the *best-of-generation* interaction scheme.

With the population-centric approach, a *topology* of interactions is picked such that any individual in the population(s) is used at least once, these interactions are assessed and then fitness is computed for all individuals. Tournaments, such as single-elimination or round-robin, are the most common examples for single-population algorithms. Both for single- and for multi-population models, shuffle and pair (also called bipartite pairing) is a population-centric method that simplifies the reuse of assessed interactions (meant to reduce computational time). With the single-elimination tournament, different individuals will participate in a different number of interactions. An approach introduced by Jaśkowski et al. (2008) uses repeated tournaments not to compute fitness for individuals, but to directly select individuals (the winner from each tournament) for breeding.

Information available from previous evaluation rounds may be used to influence how interactions are chosen for the current evaluation round. We call this *fitness bias*. Usually, individual-centric approaches use such biases, while population-centric ones do not. Clearly, to be able to perform such biasing, the algorithm must store some information about previous evaluations. This information usually consists of individuals and their fitness, but can also be more complex in nature (e.g., include other properties of individuals or relationships between individuals). The amount of information can range from remembering the last best individual to remembering all interaction assessments ever performed. With generational CoEAs, intermediary approaches include saving one or a few (usually the best) individuals from each previous generation, saving all individuals in the previous generation, or other memory or archive method (Panait et al. 2006a; De Jong 2004a, c; Ficici and Pollack 2003; Oliehoek et al. 2006).

Additional reviews and comparisons of various methods for selecting interactions can be found in the literature (Angeline and Pollack 1993; Panait and Luke 2002; Sims 1994).

With individual-centric methods, some of the additional choices (in particular the sample size) may be dynamic (i.e., they vary at run-time as a function of time) (Panait and Luke 2005; Panait et al. 2006c) or adaptive (i.e., they vary at run-time as a function of the internal state of the algorithm, e.g., population diversity and operator success) (Panait and Luke 2006; Panait et al. 2006a; Panait 2006).

Additional choices must be made when using spatially embedded CoEAs. These tend to use individual-centric approaches and interactions are localized in space, namely, neighborhood based. The size of the neighborhood corresponds to the sample size, but the shape is an additional decision to be made. Fitness-biased selection of interactions can still be used (generally within the neighborhood).

3.2.2 Aggregation

If an individual participates in a single interaction, then it must obtain a value from it, and that value becomes the individual's fitness. But when an individual participates in multiple interactions and thus obtains multiple values, then a choice must be made about how to aggregate these values.

One approach is to input values from multiple interactions into some computations and output a *single value* per individual (to be used as fitness). The computations can simply use all values obtained by the individual and determine the *best*, the *worst*, or the *average* of those values. A comparison of these three methods can be found in (Wiegand et al. 2001). Other, more complex computations can take into account values from other individuals as well, as is the case for competitive fitness sharing (Rosin and Belew 1995, 1997) or the biasing technique introduced by Panait et al. (2003, 2004). The advantage of the single-value approach is that traditional parent-selection methods can then be used by the algorithm based on single-valued fitness. Unfortunately, different ways of computing a single value have different (and strong) biases, and it is not always straightforward to tell which bias is more or less helpful for the solution concept at hand. In particular, when dealing with the ideal team solution concept, the biases of the averaging method proved harmful (Wiegand 2004) while choosing the best is helpful (Panait 2006). Even so, there has been some amount of controversy over the utility of biases (Bucci and Pollack 2005; Panait and Luke 2006); Bucci (2007) goes so far as to argue that single-valued fitness assessments are to be blamed for a number of pathological algorithm dynamics.

The alternative is to have fitness be a *tuple* of values, usually the values obtained by the individual from multiple interactions. Selecting parents based on such fitnesses requires more specialized methods, in particular, ones akin to multi-objective EAs. The tuples of different individuals of the same role must be somehow comparable, which imposes constraints on the interactions that generated the values in the tuples. The advantage of this approach is that its bias may be more appropriate for solution concepts such as the Pareto optimal set or simultaneous maximization of all outcomes, and it is mostly algorithms targeted at these solution concepts that use tuple fitness (De Jong 2004a, c; Ficici and Pollack 2001). An example using tuple fitness for the ideal team solution concept can be found in (Bucci and Pollack 2005).

3.2.3 Communication

When using a CoEA with multiple populations, in order for individuals in one population to interact with individuals from other populations, the populations must have access to one another's contents. This is an issue of communication, and thus entails choices typical of any distributed system, such as coordination, flow, and frequency.

In terms of *coordination*, communication can be synchronous or asynchronous. In the *asynchronous* case, the populations evolve at their own pace and communicate with one another through shared memory. They decide independently when to write new information about their state to the shared memory and when to check the memory for new information about the other populations (which may or may not be available). Asynchronous CoEAs are uncommon. In the *synchronous* case, there is a centralized clock that dictates when the populations exchange information.

In terms of *flow* (in Wiegand's (2004) hierarchy of CoEA properties, flow is called *update timing*) the asynchronous model is always *parallel*, in the sense that at any point in time, there may be more than one population running. (When a parallel CoEA is run on a single processor, this translates into the fact that there is no guarantee about the order in which the populations run.) The synchronous model can be either parallel or sequential (also called serial). In the *parallel* case, all populations run simultaneously for a certain period of time (dictated by the central clock), after which they all pause and exchange (communicate) information and then they all continue. In the *sequential* case, at any point in time, there is a single population running and populations take turns in a round-robin fashion.

Frequency refers to the number of evaluation–selection–breeding cycles that a population goes through between two communication events. The frequency may be uniform across all populations or it may differ from one population to another.

4 Analysis of Coevolution

Particularly in the last decade, there has been a great deal of activity in analyzing coevolutionary algorithms and their use as co-optimizers. There are two major drivers of this field. One is the fact that most co-optimization problems pose difficulties not encountered in traditional optimization, namely, measuring (and achieving) algorithmic progress and performance. The other is that coevolutionary algorithms as dynamical systems exhibit behaviors not akin to traditional evolutionary algorithms.

The first half of this section clarifies the complex issues pertaining to interactive domains and co-optimization problems and places much extant research in the context of addressing such issues. The second half surveys the wide variety of analytical methods applied to understanding the search and optimization biases of coevolutionary algorithms, as well as the insights that were gained.

4.1 Challenges of Co-optimization: Progress and Performance

For traditional optimization, judging the success of any algorithm attempting to solve a given problem instance is a fairly straightforward matter. There may be different criteria of success, such as the highest quality potential solution(s) found with a given budget of function evaluations or, conversely, what evaluation budget is needed to find potential solutions whose quality is above a certain threshold. Such success metrics also allow for easy comparisons of algorithm performance, even when aggregated over classes of problems.

Additionally, even when concerned with the behavior of a single algorithm on a single problem instance, a straightforward notion of progress through time exists. With each new function evaluation, the algorithm can easily keep track of the best quality potential solution found so far. Even if the quality of the overall best is not known, the algorithm can detect when it gets closer to it. Moreover, should an algorithm actually discover a global optimum, whether or not it is recognized, there is no danger of "losing" it. The best-so-far quality metric is a monotonic function of time. (The potential solution with the optimal quality may change, should the algorithm be deliberately implemented to replace the current best-so-far potential solution with a new one of equal quality.)

All of this is largely possible because the quality of a potential solution can be obtained with a single function evaluation and thus immediately perceived by the algorithm, which in turn allows reliable comparison of any two potential solutions. As trivial as the above statements may appear, they are no longer a given in many co-optimization problems.

Consider first the one area that is most similar to traditional optimization: compositional problems using the ideal team solution concept. The unique feature of these problems is that the interaction set is actually the same as the potential solution set *and* the interaction function serves as the potential-solution quality function. Thus, the quality of a potential solution can be obtained with a single function evaluation and any two potential solutions can reliably be compared. This, of course, is not very surprising since many of these problems are actually obtained via decomposition from traditional optimization problems. Therefore, measuring performance and progress is not an issue.

Nevertheless, approaching these problems via coevolution raises new issues. For example, how does one assign credit to the components of potential solutions that are evolving in different populations? More generally, how do different algorithm design choices affect algorithm performance? Much research around this topic has shown that, even for these most traditional-optimization-similar problems, design heuristics for traditional evolutionary algorithms do not transfer as-is to coevolution (Popovici and De Jong 2004). Still, the ability to easily measure performance and progress greatly facilitates the study of coevolutionary algorithms for ideal team compositional problems.

Note that the equality of the potential solution set and the interaction set is not enough to guarantee ease of performance/progress assessment. Consider compositional problems using the pure Nash equilibria solution concept. This problem is not expressed as a co-optimization problem, but as a generic co-search problem. It is fairly obvious that the single evaluation of the interaction function for this potential solution is not enough to determine whether it constitutes a Nash equilibrium. To determine that, one would need to, in turn, fix all but one component of the potential solution and evaluate interactions corresponding to, in the worst case, the entire set of entities for that component. While this does not mean exploring the entire interaction set, it does mean fully exploring all the component sets, which is to be assumed intractable. A negative decision might be reached sooner than that, but to reach a positive decision exhaustive search must be performed.

For test-based problems, the potential solution set is always different from the interaction set and determining the quality of a potential solution requires evaluating interactions corresponding to entire entity sets (the test sets). Should this be tractable, the problem may be expressed as a traditional optimization problem. We assume that is not the case.

Consider a generic search algorithm that, at every step, evaluates a new interaction and outputs a potential solution. (Algorithms that may not be ready to output a potential solution after every evaluation can be interpreted as outputting their latest output again until ready.)

How can one judge whether the algorithm makes progress toward the solution to the entire problem? How can its output be compared with the output of a different algorithm that has evaluated the same number of interactions? Since determining the true quality of a potential solution is generally computationally intractable for real problems, one is left with attempting to compare different potential solutions based on incomplete information.

The above observations paint a rather bleak view with respect to determining whether an algorithm could tackle one of these problems. However, the matter is not hopeless, though more can be said and done for some problems than for others.

For example, for Pareto dominance solution concepts, it has been shown that for any given set of n component entities (components), an *ideal evaluation set* of at most $n^2 - n$ test entities (tests) exists that results in exactly the same dominance relations among the components as would result from using the set of all possible tests (De Jong and Pollack 2004). Of course, the guaranteed existence of this limited set of tests leaves open the question of how to identify them. However, an operational criterion is available to determine whether any given test should be included in such a limited test set. Specifically, the criterion of whether a test induces a *distinction* between two components, introduced by Ficici and Pollack (2001), can be used for this purpose; see (De Jong and Pollack 2004).

The vast majority of research on these issues has concentrated on progress of (co-optimization) algorithms in individual runs and assessing performance on test problems. Only recently have aggregate performance comparisons between algorithms started receiving attention.

With respect to progress in individual runs, one fact facilitates the analysis: the incomplete information that an algorithm sees, and based on which one should compare different potential solutions outputted by the algorithm, is increasing over time. The remarkable discovery of Ficici (2004) is that some problems (solution concepts) permit the construction of algorithms that can properly leverage this incomplete but increasing information to guarantee that outputted potential solutions have monotonically increasing *global/true* quality over time. Any such algorithm is said to guarantee *global monotonicity* and the said solution concepts are called *globally monotonic.*

Some problems of practical interest have solution concepts that are not globally monotonic. This makes it difficult or even impossible to devise algorithms that guarantee global monotonic progress as described above. Still, for some of these problems, algorithms can be built that guarantee a different form of monotonicity called *local monotonicity,* which entails that the quality of outputted potential solutions with respect to all information seen so far will increase with time.

Both globally and locally monotonic algorithms guarantee convergence to a globally optimal solution, if the interaction space is finite and any point in this space can be generated with nonzero probability at any time.

When neither global nor local monotonicity guarantees can be achieved, some surrogate techniques can be used to instrument what the algorithm is doing, though they do not inform about progress toward the problem-solving goal. Finally, while all of the above assume the intractability of computing global quality, one research approach is to study various co-optimization algorithms on test problems, for which global quality can actually be computed. This metric is not made available to the algorithm, but used externally for instrumentation, and the hope is that the observed results will transfer to real problems.

The remainder of this section reviews global and local monotonicity, surrogate measures of progress, test-bed analytical techniques and aggregate performance comparisons.

4.1.1 Monotonicity

Central to the work on monotonicity is the idea that the solution concept can be interpreted as more general than just a means of partitioning the entire set of potential solutions based on the entire set of interactions. Namely, it can be extended to apply to any context consisting of a subset of interactions and a corresponding subset of potential solutions (i.e., potential solutions that can be built with entities that participated in the given interactions). Thus, it can be viewed as a means for an algorithm to output a potential solution when queried, based on all or some of the entities it has generated and interactions it has assessed. (Conversely, any mechanism for selecting a potential solution to be outputted from the algorithm can be viewed as a solution concept.)

In general, it may be difficult for an algorithm to determine the output of the solution concept when applied to the entire history of entities discovered and interactions assessed if it has discarded some of the entities or some of the interaction outcomes. However, for some solution concepts this is possible. For example, for the best worst case solution concept, an algorithm needs to remember only the worst value seen for any potential solution and not all the interactions it took part in, the other entities involved in those interactions or their outcomes.

One question that arises is whether an algorithm able to report the application of the solution concept to its entire history (whether by means of never discarding any information or by cleverly storing parts or aggregates of this information) can guarantee that the *global* quality of the potential solutions it outputs will never decrease over time, regardless of the means of generating new interactions to be assessed. Ficici (2004) showed that this is possible for some solution concepts, which we call *globally monotonic*, but not for others. Out of the solution concepts presented in ❷ Sect. 2.4, the ones proven to be globally monotonic are the Nash equilibrium, the Pareto optimal minimal equivalence set (Ficici 2004), maximization of all outcomes, and ideal team (Service 2009). Solution concepts proven not to be globally monotonic are maximization of expected utility (Ficici 2004), best worst case (Service 2009; Popovici and De Jong 2009), and all Pareto dominance ones besides the minimal equivalence set (Ficici 2004).

For the globally monotonic solution concepts, the challenge from an algorithmic perspective is being able to report the application of the solution concept to the entire history while limiting the amount of history that must be stored. The algorithms mentioned below are in many cases able to discard information without jeopardizing the guarantee of global monotonicity. Yet, all of the globally monotonic algorithms discussed below require unlimited memory.

For the Pareto optimal minimal equivalence set solution concept, De Jong (2004a) introduced an algorithm called *IPCA* (Incremental Pareto-Coevolution Archive) that guarantees global monotonicity (Popovici and De Jong 2009) while discarding information along the way. The key to achieving that lies in a carefully designed archiving mechanism. A variant named the LAyered Pareto Coevolution Archive (LAPCA) (De Jong 2004c) addresses the unbounded memory issue by maintaining a limited number of non-dominated layers. While it no longer provides the monotonicity guarantee, this algorithm achieves good progress on test problems. Similarly, the *Nash memory* algorithm (Ficici and Pollack 2003; Ficici 2004), guarantees global monotonic progress for the solution concept of the Nash equilibrium in symmetric zero-sum domains while discarding information. As before, bounding the memory loses the theoretical monotonicity guarantee but still achieves good progress on test problems. Based on the Nash memory, the *parallel Nash memory* algorithm (Oliehoek et al. 2006) guarantees global monotonic progress for the solution concept of the Nash equilibrium in asymmetric games. For the maximization of all outcomes solution concept, the *covering competitive algorithm* of Rosin (1997) guarantees global monotonic progress while discarding information.

While the guarantee of increasing global quality may not be possible for solution concepts that are not globally monotonic, this does not mean that all hope is lost for problems with such solution concepts. For such problems it may still be possible to devise algorithms that guarantee that the quality of outputted potential solutions *with respect to the interactions seen so far* increases monotonically over time. Solution concepts with this property are called *locally monotonic*.

For example, while the maximization of expected utility (MEU) solution concept introduced in ❥ Sect. 2.4.2 is not globally monotonic, a variant of it, namely, maximum utility sum applied to domains with a positive-valued metric, allows algorithms guaranteeing locally monotonic progress (Popovici and De Jong 2009). As the name suggests, the MEU solution concept is an optimization criterion, where the function to be optimized is utility sum. When extended to any context consisting of a subset of interactions and a corresponding subset of potential solutions, the special property this function has (if the metric values are positive) is that the "local" value for a potential solution in some context can only increase when the context is expanded.

Based on this property, for maximum utility sum applied to domains with a binary-valued metric, De Jong (2005) introduced an algorithm called *MaxSolve* that is guaranteed to output the application of the solution concept to the entire history (and therefore an increasing value with respect to an increasing context), while discarding information along the way. This property, coupled with unbounded memory, allows *MaxSolve* to achieve local monotonicity.

As Popovici and De Jong (2009) showed, for the best worst case solution concept, even the local monotonicity guarantee cannot be fully achieved via algorithms that use the solution concept as an output mechanism (and probably not in general).

While monotonicity is an interesting property that has served in clarifying the behavior and goals of coevolutionary solution concepts and algorithms, it should be made clear that guarantees of monotonicity can have limited value for a practitioner. Three important things to keep in mind are that (a) the solution concept is most often a given, not a choice, and for some solution concepts there are no known algorithms with monotonicity guarantees; (b) all algorithms known to have monotonicity guarantees require unbounded memory; and (c) even if the quality of potential solutions can only improve over time, this provides no information about the *rate* at which progress will be achieved. For these reasons, an approach that may turn out to have more practical relevance is to consider the *performance* of coevolutionary algorithms. This topic is treated in ❥ Sect. 4.1.4.

4.1.2 Surrogate Progress Metrics

For problems where the above monotonicity guarantees are unobtainable, two questions remain. One is how to design algorithms for tackling these problems. Another is how to instrument any such algorithm. The former is still an open research question. Here, the latter is addressed from the perspective of coevolutionary algorithms. However, while these techniques have been introduced in the context of coevolution, some of them may be adapted to be applicable to generic co-optimization algorithms.

Many techniques have been introduced over the years, and while they cannot tell how the algorithm is doing with respect to the goal, they may still provide useful information about algorithm behavior. The techniques involve computing metrics that are external to the algorithm (i.e., they do not influence its behavior) but rely on its history. They may measure all potential solutions considered by the algorithm or only select those that the algorithm outputs. Measuring all can provide more information, but it may make it difficult to extract the important patterns within and may also pose a challenge for visualization. One must be aware that compressing information (e.g., by means of averaging) may actually eliminate important trends in the data. The trends to look for usually include increase, decrease, stagnation, noise, and repeated values.

The first such technique was introduced by Cliff and Miller (1995) in the form of CIAO plots. The acronym stands for "current individual ancestral opponent." For a two-population CoEA, a CIAO plot is a matrix in which rows represent generations of one population and columns represent generations of the other population. Every cell represents an interaction between the best individual in each population (as reported by the algorithm) at the corresponding generations. Thus, individuals from later generations of one population interact with individuals from early generations of the other population (ancestors). The cells are color-coded on a gray scale, and can be constructed to reflect success from the perspective of either population.

The "master tournament" metric of Floreano and Nolfi (1997) is basically a compression of the information in CIAO plots. Averages are taken along lines for one population and across columns for the other population.

Both these methods are computationally expensive, as they require the evaluation of n^2 interactions, where n is the number of generations. The master tournament metric makes it easier to identify "broadly successful" individuals, but the averaging it performs may obscure some circularities that can be observed using the CIAO plots. An in-depth critique of CIAO plots is provided in (Cartlidge and Bullock 2004b).

While these metrics were introduced in the context of domains with two asymmetric roles, they are easily applicable to domains with two symmetric roles, whether approached by single-population or two-population CoEAs.

In the context of a domain with two symmetric roles and a two-population CoEA, Stanley and Miikkulainen (2002b) introduced a less costly technique called "dominance tournament." At each generation, the best individual in each population is determined and the two of them are paired; out of their interaction, the more successful one is designated the generation champion. The dominance property is then defined as follows. The champion from the first generation is automatically considered dominant. In every subsequent generation, the champion is paired only with previous dominant champions and it is itself labeled dominant only if it is more successful than all of them. The method is easily applicable to domains with two symmetric roles and a single-population CoEA. The paper also suggests a (less straightforward) extension to some domains with two asymmetric roles.

CIAO, master tournament, and dominance tournament are all techniques that track only the best of generation individual. This was contrasted by Bader-Natal and Pollack (2004) who introduced a "population-differential" technique monitoring all individuals in each generation. Plots similar to the CIAO plots are produced, but now, each cell is an aggregation over the results of all pair-wise interactions between individuals in the two populations at the generations corresponding to that cell. This is clearly more expensive, and a number of memory policies are introduced for reducing time complexity.

The unifying idea of these techniques is to determine whether an algorithm is making at least some sort of "local" progress away from the starting point.

All the metrics described so far require performing additional evaluations, rather than using the ones already performed by the algorithm. Funes and Pollack (2000) introduced a technique that took the latter approach. Their metric, called "relative strength" is still subjective, as the value it returns for a particular individual depends on the current history of all interactions, and this history grows with time. The metric is based on paired comparison statistics and is applicable to domains with two symmetric roles.

4.1.3   Test-Bed Analysis

A means of judging both an algorithm's ability to make progress as well as its expected performance on some problem(s) is to test the algorithm on artificial domains for which true quality can be computed. The domains are constructed such that they exhibit properties thought to be present in real domains or such that they expose a range of different behaviors of the algorithm. The hope is that the results observed on the artificial domains will transfer to real domains.

A class of domains introduced to this effect are *number games* (Watson and Pollack 2001). These are domains with two roles, a common entity set consisting of real-valued vectors and a binary-valued interaction function (based on comparisons between vector values). The roles can be interpreted as symmetric or asymmetric. The solution concept is Pareto dominance. What makes these problems easier to analyze is the fact that the interaction function is designed such that a genotype can be used as-is to represent its own true quality. There are a handful of such interaction functions in the literature and many progress-focused algorithm comparisons have been performed using them as a test bed (De Jong 2004a, c, 2005, 2007; Bucci and Pollack 2003a; Funes and Pujals 2005).

Popovici and De Jong (2005b) introduced a set of four domains with two asymmetric roles, a common entity set consisting of single real values and a real-valued interaction function given in closed form. These domains were used to showcase the biases of symmetric last elite opponent evaluation, and how this technique generally does not optimize expected utility. The closed form of the interaction function allowed for easy computation of true quality.

A number of domains of the same kind (two asymmetric roles, real-valued entities, closed-form real-valued interaction function) have been introduced specifically for studying algorithms targeted at the best worse case solution concept (Jensen 2001; Branke and Rosenbusch 2008; Stuermer et al. 2009), which has been shown to be non-monotonic and difficult in terms of achieving even very weak forms of progress (Popovici and De Jong 2009).

Test-bed analysis has also been performed for algorithms targeted at the ideal team solution concept, yet with a different motivation (e.g., studying problem properties and algorithm design choices) since in this case assessing global quality is not an issue. Examples are given in ❯ Sect. 4.2.4.

4.1.4   Comparing Performance

As emphasized throughout the previous section, instrumenting an algorithm from a progress-making perspective or, indeed, designing an algorithm such that it guarantees progress relies rather heavily on the fact that the amount of information seen by the algorithm increases over time. How an algorithm uses this information can be key to instrumenting/achieving progress. Therefore, the theoretical work on monotonicity has so far had nothing to say about how to compare algorithms since there is no connection between the information that different algorithms see. This line of research provided no generic advice on how to design efficient algorithms.

While test-bed analysis is useful for determining whether an algorithm makes progress or not, when global quality can be computed, one can also use test-beds to compare algorithms based on the expected rate of progress or expected level of solution quality over multiple runs. This kind of performance analysis, especially when backed by dynamics analysis of the kind reviewed in ❷ Sect. 4.2.5, has provided insight into the biases of design choices by providing a means of comparison; however, it is unclear how well the results transfer to real-world problems, so the generality of such results is unclear.

From traditional optimization research, the famous no free lunch (NFL) theorem has approached the issue of performance generality (Wolpert and Macready 1997; Schumacher et al. 2001). This result has led to a welcome shift in EC research toward identifying classes of problems where NFL does not hold and designing algorithms targeted at a specific class (Igel and Toussaint 2003). This is of particular current interest in co-optimization/co-evolution (Wolpert and Macready 2005) since recent work states the opposite, namely, that free lunches do exist for certain solution concepts.

Wolpert and Macready extended their formalism of traditional optimization algorithms to co-optimization algorithms (albeit without using this terminology) as having two components: a search heuristic and an output selection function. The search heuristic takes as input the sequence of interactions assessed so far and returns a new interaction to be assessed. The output selection function takes the same input and returns a potential solution. The need to explicitly model the output selection function comes from the algorithms' lack of access to a global quality function: in traditional optimization, output selection is based on the quality values seen so far (usually picking the best of them); in co-optimization, what an algorithm sees are values of the interaction function, which provide incomplete information about global quality. The use of archives in CoEAs is often targeted at implementing the output selection function, thus relieving the populations of this duty so they can focus on the search component.

Wolpert and Macready showed that aggregate performance advantages (free lunches) can be obtained both via the output selection function and via the search heuristic. The former means that if the search heuristic is fixed, some output functions perform better than others. The optimal output function is the so-called Bayes output function, which outputs the potential solution with the highest estimated global quality, estimated over all possible problems consistent with the measurements of the interactions seen so far. The latter was showcased in the context of the best worst case solution concept: fixing the output selection function to Bayes, two specific search heuristics were shown to have different average performance.

This line of work was extended by Service and Tauritz (2008b), which showed that the same two algorithms from Wolpert and Macready (2005) exhibit different aggregate performance for the maximization of all outcomes solution concept.

Generally, such free lunches are believed to exist in co-optimization whenever the performance of an algorithm depends on information not available in the algorithm's trace of interactions; in other words, it explicitly depends on the interaction function (as opposed to implicitly, via the values in the trace). Therefore, the ideal team solution concept does not exhibit free lunches since the interaction set is actually the same as the potential solution set and the interaction function serves as the potential solution quality function.

The existence or lack of free lunches can be seen as a property of solution concepts, which has been named "bias" (Service 2009). Since monotonicity is also a solution concept property, the natural question that arises is how these two properties relate to one another. Service (2009) showed they are orthogonal to one another: solution concepts exist in all four classes combining monotonic/non-monotonic with biased/unbiased.

Especially for monotonic and biased solution concepts, the next natural question is whether the algorithms that can achieve monotonicity (known to exist) are also the best performing. This question was investigated by Popovici and De Jong (2009), who showed that the output selection function used by monotonic algorithms is not equivalent to the best-performing Bayes, thus exposing a potential trade-off in algorithm design between monotonicity and performance.

4.2 Analysis of CoEA Biases: A Survey

The previous half of this section discussed the difficulties of measuring and achieving algorithmic progress for co-optimization problems, and reviewed the state of the art in understanding and dealing with these issues. However, when applying coevolutionary algorithms to such problems, an additional concern is understanding and harnessing the biases these algorithms have as dynamical systems. This section surveys the analytical research that contributed such understanding.

4.2.1 Model Analysis

A common algorithm analysis method is to *model* an algorithm mathematically, then to study the properties of that model. The advantage is that certain mathematical facts can sometimes be precisely learned from these models, facts that help to understand system behavior in ways that are simply not possible via traditional approaches. The disadvantage is that the model is *not* a real algorithm, and it is not always clear what can be transferred from model to algorithm. This step is typically bridged by empirical study of some kind.

The simplest way to approach modeling evolutionary computation as a dynamical system is the so-called "infinite population model" (Vose 1999). One assumes that a population is infinitely large and focuses on how the distribution of genotypes within the population changes over time. Questions about the existence (and sometimes the location) of stable, attracting fixed points, or their nature and quality, can sometimes be answered by applying traditional dynamical systems methods to such a model.

Evolutionary game theory (EGT) (Hofbauer and Sigmund 1998; Weibull 1992; Maynard Smith 1982) is an appealing dynamical systems model of biological systems that is well-suited for studying coevolution. In EGT, interactions between genotypes are treated as a *stage game*, and the payoff from that game informs how the distribution is altered for the next play of the game (*replicator dynamics*). Because of its incorporation of game theory, a number of game-theoretic properties of such systems become useful points of consideration (e.g., Nash equilibria). Additionally, EGT has received a great deal of attention by the economics community (Friedman 1998). Consequently, quite a bit is already known about the dynamical properties of these mathematical models under different conditions.

For the most part, EGT analysis has focused on the dynamics of CoEAs under selection only; however, variational operators in CoEAs can be (and have been) modeled as is typically done in dynamical systems models of traditional EAs, by constructing a matrix that transforms the probability distribution resulting from the selection operator into one that reflects the properties of the population after crossover and mutation are applied, the so-called *mixing matrix* (Vose 1999). To specify this, one must be explicit about representation. Despite these simplifications, some interesting things are known about the system that inform our understanding of CoEAs.

In single population EGT models of CoEAs, for example, stable attracting *polymorphic* equilibria can exist when the underlying stage game is not a constant sum game. That is, aside from the stochastic effects of genetic drift, selection alone can drive the systems to states where the population is a mixture of different kinds of genotypes, and these stable, mixed strategies correspond with a Nash equilibrium of the stage game. Moreover, examining this system using other selection methods has uncovered some fascinating results. Common EC selection methods (e.g., truncation, (μ, λ), linear rank, Boltzmann) can sometimes lead to systems in which there are stable cycles and even chaotic orbits (Ficici and Pollack 2000b). Indeed, the dynamics of CoEAs can be much more complex than traditional EAs in that stable attractors of infinite population models can be quite unrelated to the specific values in the game payoff matrix itself. Empirical experiments on real algorithms verify that these fundamental dynamical pressures of the selection operator do, in fact, affect how real CoEAs perform on real problems.

Additionally, by considering certain game-theoretic properties of the stage game (a particular variation of the notion of constant-sum), one can show that the behavior of certain single population CoEAs using nonparametric selection operators will, in principle, behave dynamically like a comparable EA on some unknown fitness function (Luke and Wiegand 2003), while this cannot be said for other systems *even when the payoff is fully transitive*. This reinforces the idea that certain co-optimization problems are fundamentally different from traditional optimization problems.

EGT-based models have also been used to study two-population compositional CoEAs, showing that any pure Nash equilibrium of the underlying payoff game is a stable, attracting fixed point. This suggests selection alone can drive compositional CoEAs toward points that are globally very suboptimal in terms of their actual payoff value. Experimentation confirms this is so in real, analogous compositional CoEAs even to the extent that highly inferior local optima draw significantly more populations than do the true global optima. Indeed, complete mixing implies a solution concept wherein "optimality" is defined solely based on how individual strategies do on average over all possible partners. If one is using such methods to try to find a globally maximal payoff value, this can be seen as a kind of pathology (*relative generalization*). This dynamic may well be consistent with obtaining solutions that meet certain specific definitions of *robustness* (Wiegand and Potter 2006).

But most compositional CoEAs use best-of-generation interaction methods, not complete mixing, and Panait et al. (2006b) noted that these more common methods bias compositional coevolutionary search of projections of the payoff space to favor aggregate projections that suit identifying the global optimum. Further, by incorporating archival notions that help retain *informative* partners in terms of these projection estimates, the optimization potential of compositional CoEAs can be further improved. These analytical lessons have been applied to both real compositional CoEAs, as well as other RL-based co-optimization processes (Panait 2006). Additionally, Vo et al. (2009) showed a certain infinite population EGT model of two-population compositional CoEAs to be equivalent to the distribution model for univariate estimation-of-distribution algorithms, thus arguing for the potential of cross-fertilization between these two subdisciplines of evolutionary computation. This work also provides a generalization of a subset of compositional CoEA models to n-populations.

These studies have shown that interaction and aggregation have profound impacts on the dynamical influences of selection in compositional coevolution. More specifically, the complete mixing interaction pattern may be ill-suited for problems with an ideal team solution concept. Another, altogether different dynamical systems approach by Subbu and Sanderson (2004) provides convergence results for a particular class of distributed compositional CoEAs to ideal team style solutions under conditions where complete mixing is not even possible.

In addition to the infinite population models described above, several works have focused on finite populations. EGT itself has been applied to finite population situations (Maynard Smith 1982; Ficici and Pollack 2000a), but typically under very constrained circumstances (two-strategy games). Outside of using EGT-based methods, there have been attempts to model the dynamics of finite, two-population CoEAs using Markov models (Liekens et al. 2003). While these methods also restrict the size of the problem spaces for obvious reasons, they have important implications for infinite population models in the sense that they clearly demonstrate that the long-term behavior of these systems can differ quite significantly from infinite population models.

4.2.2 Runtime Analysis

EC has made good use of classical algorithm analysis methods to find bounds on the expected number of evaluations necessary before a global optimum has been obtained as a function of the size of the (typically, but not always, combinatorial) genotype space (Oliveto et al. 2007). These methods have the advantage of providing precise and correct performance bounds on real algorithms for real problems, but their chief disadvantage is that it is often difficult to gain crucial insight into the behavior of an algorithm. This concern is typically addressed by selecting problem classes that help identify important properties in the problem and falsify hypotheses regarding how EAs address them.

Indeed, runtime analysis has been quite useful in understanding how simple compositional coevolutionary algorithms approach problems with ideal team solution concepts. A common hypothesis about such algorithms is that they tend to perform better when the representation choices result in a linear separation of problem components with respect to the objective function. Runtime analysis of simple (1 + 1) (Jansen and Wiegand 2003a, b) and steady-state-like (Jansen and Wiegand 2004) variants of compositional CoEAs revealed this hypothesis to be conclusively false: though nonlinear relationships between components can have profound effects on the relative performance differences between a coevolutionary algorithm and a similarly structured EA, separability itself is neither a sufficient nor necessary property with which to predict problem difficulty (Jansen and Wiegand 2004). Additionally, there exist certain problem classes that essentially *cannot be solved* by a large class of compositional coevolutionary algorithms. Finally, this work suggests that compositional CoEAs optimize by leveraging the decomposition to partition the problem *while* increasing the focus of the explorative effects of the variational operators (so-called *partitioning and focusing* effect).

Runtime analysis of other CoEAs is more challenging and has not yet been done. Fortunately, recent research provides several needed pieces to overcome these challenges. First, there has been a great deal of recent progress in analyzing increasingly more complex multi-objective methods (Laumanns et al. 2004), and much of this may help study CoEAs that employ multi-objective mechanisms (e.g., archives). Second, the formalisms for explicit definition of solution concepts and the guarantee of monotonicity in certain concepts provide clear global goals needed for runtime analysis.

4.2.3 Probabilistic/Convergence Analysis

Another way to investigate algorithms theoretically, is to consider the dynamical behaviors of the real algorithms (as opposed to abstractly modeling them). Typically, such analyses are focused on providing convergence guarantees, but often careful consideration of problem and algorithm properties can provide greater insight.

Schmitt (2003a, b) provides a strong type of convergence analysis. For the maximization of all outcomes solution concept, the proposed algorithm is guaranteed not only to find a solution, but to have its populations converge to containing only genotypes corresponding to the solution(s). Markov chain analysis is used for the proofs. Bounds on convergence speed are not given; however, the work provides some useful general advice for how to apply rates of genetic operators to ensure such convergence.

Additionally, Funes and Pujals (2005) conduct a probability-theory-based investigation into the relationships between problem properties, algorithm properties, and system dynamics. This work focuses on test-based CoEAs, and demonstrates that the interplay of the considered properties cause trajectories through the domain-role sets to either make progress or resemble random-walk behaviors.

4.2.4 Empirical Black-Box Analysis

Analysis of coevolution can be approached empirically by viewing the system as a black box whose inputs are the algorithm and the problem, and the output is observed performance (e.g., quality of potential solution outputted given a certain amount of time). The algorithm and/or the problem are then varied and the output of the system re-observed, with the goal of determining the rules governing the dependency between inputs and outputs. Such studies are performed for problems with computable global quality, such as ideal team problems or testbeds of the kind reviewed in ❷ Sect. 4.1.3.

Two different approaches can be distinguished within the black-box analysis category. One approach focuses on *properties* of the algorithms, of the problems, or of both. When algorithm properties are involved, this approach has been referred to as component analysis

(Wiegand 2004). The other approach varies the algorithm and/or the problem and compares the resulting performance without a clear isolation of the properties responsible for the differences in performance.

This latter approach is reviewed first. Most works introducing new algorithms are of this kind, including some of the early research comparing CoEAs with traditional EAs, both for test-based problems (Hillis 1990; Angeline and Pollack 1993) and for compositional problems (Potter and De Jong 1994). Later on, the approach was used to compare CoEAs among themselves, usually "enhanced" algorithms with basic ones. Examples include works for test-based problems (Ficici and Pollack 2001, 2003; Ficici 2004; De Jong and Pollack 2004; De Jong 2004a, c, 2005) and compositional problems (Bucci and Pollack 2005; Panait et al. 2006a; Panait and Luke 2006; Panait 2006). Some of the algorithms introduced by these works are backed up by theoretical results (De Jong 2004a, 2005; De Jong and Pollack 2004).

Property-focused approaches address problem properties, algorithm properties, or the interactions between them. These cases are discussed in order.

Some of the CoEA-versus-EA comparisons were extended to determine the classes of problems for which one type of algorithm would provide performance advantages over the other. Such work was performed for compositional problems by Potter (1997) and Wiegand and Potter (2006). The scrutinized problem properties were *separability* (Wiegand 2004), *inter-agent epistasis* (Bull 1998), *dimensionality* (here dimensionality refers to the number of roles), *noise*, and *relative sizes of basins of attraction of local optima*.

For test-based problems, empirical black-box analysis was used to investigate problem properties such as *role asymmetry* (Olsson 2001), *intransitivity* (De Jong 2004b), and *dimensionality* (De Jong and Bucci 2006). (Here dimensionality refers to the number of underlying objectives implicitly defined by the test role(s).) These works introduced specialized algorithms intended to target the respective problem property.

Studies focusing only on algorithm properties investigated either the mechanism for *selecting interactions* (Panait et al. 2006c; Panait and Luke 2005; Parker and Blumenthal 2003; Blumenthal and Parker 2004) or the mechanism for *aggregating* the result of those interactions (Panait et al. 2003; Wiegand et al. 2001; Cartlidge and Bullock 2002). All but the last cited work were concerned with compositional problems.

Finally and perhaps most importantly, some black-box empirical studies analyzed the effects on performance of the interdependency between algorithm properties and problem properties. Most studied has been the mechanism for selecting interactions, especially in the context of compositional problems (Potter 1997; Wiegand et al. 2001, 2002; Wiegand 2004; Bull 1997, 2001). The launched hypothesis was that the performance effects of the interaction method are tightly related to the (previously mentioned) problem property separability and the kind of *cross-population epistasis* created when representation choices split up inseparable pieces of the problem. This dependency however proved not to be as straightforward as thought, as it was later shown in (Popovici and De Jong 2005a) via empirical dynamics analysis. For test-based problems, two different methods for selecting interactions were analyzed by Panait and Luke (2002), suggesting that the amount of noise in the problem affects their influence on performance.

For compositional problems, Panait et al. (2004) extended their previous work on the mechanism for aggregating results from multiple interactions by studying how its performance effects are affected by the problem's local optima and their basins of attraction. Wiegand and Sarma (2004) studied the potential benefits of *spatially distributed schemes for selecting interactions and/or selecting parents* on domains with role asymmetry.

Bull (1998) studied the effects of mutation and crossover in the context of "mixed-payoff" domains Kauffman and Johnson's (1991) NKC landscapes) and found them to be sensitive to inter-agent epistasis.

4.2.5 Empirical Dynamics Analysis

While black-box analysis can provide some heuristics for improving performance, it cannot explain the causes for the observed dependencies between the inputs and the outputs of a CoEC setup. To understand these causes, one needs to observe the system while running and track some of its time-varying properties. Instead, dynamics analysis is used to explain the connection between algorithm and problem properties on one side and performance on the other side. Individual studies connect different subsets of the pieces, analyzing:

- Dynamics in isolation (Cliff and Miller 1995; Stanley and Miikkulainen 2002b; Bader-Natal and Pollack 2004; Floreano and Nolfi 1997; Axelrod 1989; Cartlidge and Bullock 2004b);
- Problem properties and dynamics (Bull 2005a);
- Problem properties, algorithm properties, and dynamics (Bull 2005b; Kauffman and Johnson 1991);
- Dynamics and performance (Pagie and Mitchell 2002; Juillé and Pollack 1998; Ficici and Pollack 1998; Miller 1996; Funes and Pollack 2000; Paredis 1997);
- Problem properties, dynamics, and performance (Watson and Pollack 2001);
- Algorithm properties, dynamics, and performance (Williams and Mitchell 2005; Pagie and Hogeweg 2000; Rosin and Belew 1995, 1997; Cartlidge and Bullock 2003 2004a; Bucci and Pollack 2003a);
- How the combination of algorithm properties and problem properties influences dynamics and how dynamics influences performance (Popovici 2006).

These analyses used some of the progress metrics described in ❷ Sect. 4.1.2, and/or various techniques for tracking genotypic/phenotypic changes. Studies involving the latter were performed for domains and representations that are easily amenable to visualization.

Cliff and Miller (1995) were also the first to use techniques for tracking genotypic changes in order to analyze the run-time behavior (dynamics) of CoEAs. They introduced "elite bitmaps," "ancestral Hamming plots," and "consensus distance plots" as tools complementary to CIAO plots. They all work on binary representations. Elite bitmaps simply display the genotype of best-of-generation individuals next to each other in temporal sequence. Some additional processing can reveal interesting patterns. Ancestral Hamming plots display the Hamming distance between elite individuals from different generations. Consensus distance plots monitor the genotypic make up of the whole population through the distribution of Hamming distances from the genotype of each individual to the population's "consensus sequence." All three techniques are applicable to EAs in general, not just CoEAs.

Popovici (2006) extensively used best-of-generation trajectory plots for understanding the effects of various CoEA design choices and exposed best-response curves as a problem property that is a strong driver of coevolutionary algorithm behavior.

It is via the empirical dynamics analysis approach that most knowledge about the pros and cons of using CoEAs for co-optimization was generated. And just as traditional canonical evolutionary algorithms can be used for optimization, yet they are not intrinsically optimizers (De Jong 1992), canonical coevolutionary algorithms may be useful in co-optimization, yet they are not intrinsically co-optimizers.

5 The Future of Coevolution

The bulk of this chapter has focused on how coevolutionary algorithms, co-search problems, and interactive domains have been studied. This section is dedicated to the future of coevolution, including current lines of work, research questions, and results with intriguing but so-far-untapped potential. We end with some concluding remarks.

5.1 Open Issues

Recent research has helped establish a firm, formal foundation for coevolutionary problem solvers, and it has provided some principled ways to think about, design, and apply such systems. Nevertheless, many issues remain quite unresolved. These are grouped into two categories: those relating disparate theoretical analyses and those relating the analytical understanding of CoEAs to their practical application.

5.1.1 Relating Theories

Dynamical systems analysis of compositional coevolution has demonstrated that certain robustness properties in the problem can attract populations (Wiegand and Potter 2006). But there has yet to be any attempt to formalize this idea as a solution concept, and it remains to be seen whether such a concept is monotonic. Similarly, Lothar Schmitt's convergence analysis (Schmitt 2003b) provides useful feedback about how to manage the parameters of operators; however, it makes very specific assumptions about the solution concept, and no work has yet approached the task of generalizing his ideas to include CoEAs as applied to problems with different concepts of solution. Runtime analysis for compositional approaches has yielded precise bounds for those algorithms on certain optimization problems (Jansen and Wiegand 2004), and there has been a lot of success using similar methods to analyze the performance of EAs on multi-objective optimization problems. This suggests it should be possible to apply such analyses to CoEAs for test-based problems, due to their underlying multi-objective nature.

Though in a handbook of natural computing, this chapter has tried to present coevolutionary algorithms in the broader context of co-optimization. As pointed out in ❯ Sect. 4.1, some of the difficulties CoEAs face, such as the difficulty to monitor and achieve progress, are specific to the nature of certain co-optimization problems and not the evolutionary aspect of these algorithms. Any other algorithms attempting to solve such problems will have to deal with these issues. Formalizing co-optimization algorithms as having two components, a search heuristic and an output selection function, allows for easy transfer of results pertaining to monotonicity and performance to algorithms using non-evolutionary search heuristics, such as the ones introduced in (Service and Tauritz 2008a).

Last but not least, many research questions on how algorithms should be designed have been opened by the recent confluence of the two generic co-optimization theories concerning monotonicity and performance (Service 2009; Popovici and De Jong 2009). This is expounded below.

5.1.2 Relating Theory to Practice

As discussed earlier in the chapter, the notion of monotonic solution concept not only provides a way to characterize problems but also allows engineers to construct principled algorithms for which consistent progress toward a solution can be expected (Ficici 2004). Still, on a practical level, several questions remain. Many algorithms that implement monotonic solution concepts have unrealistic memory requirements. Researchers are only beginning to try to understand what the impact of practical limits on mechanisms such as archive size have on performance and theoretical guarantees about progress. Similarly, hidden in the free lunch proofs of Wolpert and Macready (2005) was advice on how to obtain some performance advantage, such as avoiding unnecessary evaluations, exploiting biases introduced by the solution concept, and using a Bayes output selection function. Still, determining how to use this advice in real algorithms for real problems is very much an open research issue. Further, the design choices required by monotonicity can conflict with those required for performance (Popovici and De Jong 2009) and additional research is needed to determine when/if this trade-off can be avoided or at least how to minimize it. Indeed, recent work relating monotonicity, solution concepts, and the NFL suggest this relationship could be quite complex (Service 2009; Popovici and De Jong 2009).

The same types of questions can be raised with respect to methods used for the discovery of informative dimensions of co-optimization problems: Can heuristic methods be developed that can approximate this search process in the general case, and what are the precise trade-offs of such an approximation? Finally, under certain conditions, it is theoretically possible for some CoEAs to prefer components considered later in the run, which in turn will tend to require larger supporting test sets (Bucci 2007). This implies the possibility of a kind of bloating that practitioners may need to address in practical application.

5.2 Discovery of Search Space Structure

De Jong and Pollack (2004) present empirical results suggesting that a Pareto coevolutionary algorithm could find what were dubbed the *underlying objectives* of a problem. These are hypothetical objectives that determine the performance of components without the need to have them interact with all possible tests. De Jong and Pollack (2004) apply a two-population Pareto coevolutionary algorithm, DELPHI, to instances of a class of abstract interactive domains. Figures 13 and 15 of that work suggest that evaluator individuals (what we have called tests) evolve in a way that tracks the underlying objectives of the domain. The results suggest that the algorithm is sensitive to the presence of underlying objectives even though it is not given explicit information about those objectives, exhibiting what has been called an "emergent geometric organization" of test individuals (Bucci 2007). Bucci and Pollack (2003a) make a similar observation, also empirical though using a different algorithm; Fig. 5 of that work suggests a similar sensitivity to underlying objectives. In both cases, clusters of individuals, rather than single individuals, move along or collect around the known objectives of the domain. The problems considered, namely, numbers games (Watson and Pollack 2001), were designed to have a known and controllable number of objectives, but the algorithms used in these two studies did not rely on that fact. The work therefore raises the question of whether underlying objectives exist in all domains, and whether algorithms can discover them.

A partial answer to this question is found in the notion of *coordinate system* defined in (Bucci et al. 2004). Coordinate systems, which were defined for a class of test-based problems (specifically, problems with a finite number of components and a finite number of binary-outcome tests), can be viewed as a formalization of the empirically observed underlying objectives of De Jong and Pollack (2004). To elaborate, a coordinate system consists of several *axes*. Each axis is a list of tests ordered in such a way that any component can be placed somewhere in the list. A component's placement is such that it does well against all tests before that spot and does poorly against all tests after it. For this reason, an axis can be viewed as measuring some aspect of a component's performance: A component that places high on an axis is "better than" one that sits lower in the sense that it does well against more tests (more of the tests present in the axis, not more tests of the domain as a whole). Formally, an axis corresponds to a numerical function over the components, in other words to an objective function. It can be proven (Bucci et al. 2004) that every domain of the considered class possesses at least one coordinate system, meaning it has a decomposition into a set of axes. In short, every such domain has some set of objective functions associated with it, one for each axis in a coordinate system for the problem.

Besides defining coordinate systems formally, Bucci et al. (2004) give an algorithm that finds a coordinate system for an interactive domain in polynomial time. The algorithm, though fast, is not guaranteed to produce the smallest-possible coordinate system for the domain, however. Finite domains must have a minimal coordinate system, but in general, even finite domains can have distinct coordinate systems of different sizes. The algorithm is not coevolutionary per se, as it examines the outcomes of tests on components. It is therefore applicable to the entire class of test-based problems and agnostic about the specific algorithm used to solve the problem.

The above work leaves open the question of whether underlying objectives or coordinate systems are simply mathematical curiosities with no practical or theoretical utility. The work of De Jong and Bucci (2007) addresses that question in two ways. First, it provides an exact coordinate system extraction algorithm that, unlike the approximate algorithm of Bucci et al. (2004), is slow but is guaranteed to return a minimal-sized coordinate system for a finite interactive domain. (Interactive domains can, in general, have more than one minimal-sized coordinate system. Hence *a* and not *the.*) Second, it applies the exact extraction algorithm to several small instances of the game of Nim, and observes that for all instances tested, the axes of extracted coordinate systems contain significant information about how a strategy performs at the game. Thus, far from being theoretical peculiarities, at least in this case (minimal) coordinate systems intuitively relate to real structure in the domain.

Given this last fact and that certain coevolutionary algorithms seem to be sensitive to the underlying dimensions of a domain, one is led to an intriguing possibility: that appropriately designed coevolutionary algorithms could discover and extract meaningful structure from interactive domains in the process of solving problems over them. Besides solving problems, algorithms might be able to simultaneously output useful knowledge about domains. For complex, ill-understood domains, the knowledge output might be at least as useful as a solution.

5.3 Open-Ended Evolution and Novelty

An important possibility that should not be lost to our solution-based, problem-oriented view is the idea that the goal of evolutionary systems need not necessarily be to find a particular solution (be it a set of points or otherwise) at all. In particular, in cases where a representation space is not explicitly limited, evolutionary systems can be seen as dynamical methods to explore "interesting" spaces. Here "interesting" might correspond with some problem-oriented view or relate to aspects of some simulation that involves the evolutionary process itself, but it may also simply correspond with notions of novelty.

In traditional, non-coevolutionary systems, an objective fitness measure in conjunction with selection methods will tend to drive the systems in particular directions. Explicit construction of fitness functions that encourage some direct notion of novelty can certainly be developed; however, coevolution offers some very natural and interesting possibilities here. Indeed, some of the earliest forms of artificial evolution involved game playing as means of exploring some of basic mechanisms of evolution and self-maintenance themselves (Barricelli 1963).

Many of the concepts discussed above (e.g., *informativeness* and *distinctions*) are very natural measures of novelty, and algorithms that literally function by moving in directions that work to maintain these are compelling for this purpose. Even more compelling is the idea just discussed above: Since CoEAs can be used to discover geometries of comparisons, to construct and report dimensions of informativeness in a space, they are very appealing mechanisms for novelty search in open-ended evolution because they may be capable of providing much more than new, unusual individuals – they may well be able to place such individuals in relationship with things already seen.

6 Concluding Remarks

The great promise of coevolutionary algorithms to find sorting networks, teams, 3-D creatures, game-playing strategies, or other complicated entities given only information about how these interact with other entities is far from realized. However, the theoretical tools and empirical methodologies presented in this chapter help in getting closer to that goal. One noteworthy outcome of theoretical investigations is that free lunch is possible: algorithms that provably perform better than others across all problem instances. This observation is relatively new and unexplored, but represents an exciting direction in coevolutionary algorithms research. It offers the possibility that a coevolutionary algorithm could not only find interesting or capable entities, but do so faster than random search would. We hope that the principled introduction to co-optimization and coevolutionary computation given in this chapter will contribute to the realization of these goals, by helping future efforts leverage the strong foundation laid by prior research.

References

Angeline PJ, Pollack JB (1993) Competitive environments evolve better solutions for complex tasks. In: Proceedings of the 5th international conference on genetic algorithms ICGA-1993. Urbana-Champaign, IL, pp 264–270

Axelrod R (1989) The evolution of strategies in the iterated prisoner's dilemma. In: Davis L (ed) Genetic algorithms and simulated annealing. Morgan Kaufmann, San Francisco, CA, pp 32–41

Bader-Natal A, Pollack JB (2004) A population-differential method of monitoring success and failure in coevolution. In: Proceedings of the genetic and evolutionary computation conference, GECCO-2004. Lecture notes in computer science, vol 3102. Springer, Berlin, pp 585–586

Barbosa H (1999) A coevolutionary genetic algorithm for constrained optimization. In: Proceedings of the congress on evolutionary computation, CEC 1999. IEEE Press, Washington, DC

Barricelli N (1963) Numerical testing of evolution theories. Part II. Preliminary tests of performance. Symbiogenesis and terrestrial life. Acta Biotheor 16(3–4):99–126

Blumenthal HJ, Parker GB (2004) Punctuated anytime learning for evolving multi-agent capture strategies. In: Proceedings of the congress on evolutionary computation, CEC 2004. IEEE Press, Washington, DC

Branke J, Rosenbusch J (2008) New approaches to coevolutionary worst-case optimization. In: Parallel problem solving from nature, PPSN-X, Lecture notes in computer science, vol 5199. Springer, Berlin, pp 144–153

Bucci A (2007) Emergent geometric organization and informative dimensions in coevolutionary algorithms. Ph.D. thesis, Michtom School of Computer Science, Brandeis University, Waltham, MA

Bucci A, Pollack JB (2002) Order-theoretic analysis of coevolution problems: coevolutionary statics. In: Langdon WB et al. (eds) Genetic and evolutionary computation conference workshop: understanding coevolution. Morgan Kaufmann, San Francisco, CA

Bucci A, Pollack JB (2003a) Focusing versus intransitivity: geometrical aspects of coevolution. In: Cantú-Paz E et al. (eds) Proceedings of the genetic and evolutionary computation conference, GECCO 2003. Springer, Berlin

Bucci A, Pollack JB (2003b) A mathematical framework for the study of coevolution. In: De Jong KA et al. (eds) Foundations of genetic algorithms workshop VII. Morgan Kaufmann, San Francisco, CA, pp 221–235

Bucci A, Pollack JB (2005) On identifying global optima in cooperative coevolution. In: Beyer HG et al. (eds) Proceedings of the genetic and evolutionary computation conference, GECCO 2005. ACM Press, New York

Bucci A, Pollack JB, De Jong ED (2004) Automated extraction of problem structure. In: Proceedings of the genetic and evolutionary computation conference, GECCO 2004, Lecture notes in computer science, vol 3102. Springer, Berlin, pp 501–512

Bull L (1997) Evolutionary computing in multi-agent environments: partners. In: Baeck T (ed) Proceedings of the 7th international conference on genetic algorithms. Morgan Kaufmann, San Francisco, CA, pp 370–377

Bull L (1998) Evolutionary computing in multi-agent environments: operators. In: Wagen D, Eiben AE (eds) Proceedings of the 7th international conference on evolutionary programming. Springer, Berlin, pp 43–52

Bull L (2001) On coevolutionary genetic algorithms. Soft Comput 5(3):201–207

Bull L (2005a) Coevolutionary species adaptation genetic algorithms: a continuing saga on coupled fitness landscapes. In: Capcarrere M et al. (eds) Proceedings of the 8th European conference on advances in artificial life, ECAL 2005. Springer, Berlin, pp 845–853

Bull L (2005b) Coevolutionary species adaptation genetic algorithms: growth and mutation on coupled fitness landscapes. In: Proceedings of the congress on evolutionary computation, CEC 2005. IEEE Press, Washington, DC

Cartlidge J, Bullock S (2002) Learning lessons from the common cold: how reducing parasite virulence improves coevolutionary optimization. In: Proceedings of the congress on evolutionary computation, CEC 2002. IEEE Press, Washington, DC, pp 1420–1425

Cartlidge J, Bullock S (2003) Caring versus sharing: how to maintain engagement and diversity in coevolving populations. In: Banzhaf W et al. (eds) Proceedings of the 7th European conference on advances in artificial life, ECAL 2003, Lecture notes in computer science, vol 2801. Springer, Berlin, pp 299–308

Cartlidge J, Bullock S (2004a) Combating coevolutionary disengagement by reducing parasite virulence. Evolut Comput 12(2):193–222

Cartlidge J, Bullock S (2004b) Unpicking tartan CIAO plots: understanding irregular co-evolutionary cycling. Adapt Behav 12(2):69–92

Chellapilla K, Fogel DB (1999) Evolving neural networks to play checkers without expert knowledge. IEEE Trans Neural Networks 10(6):1382–1391

Cliff D, Miller GF (1995) Tracking the red queen: measurements of adaptive progress in co-evolutionary simulations. In: Proceedings of the 3rd European conference on advances in artificial life, ECAL 1995. Lecture notes in computer science, vol 929. Springer, Berlin, pp 200–218

De Jong KA (1992) Genetic algorithms are not function optimizers. In: Whitley LD (ed) Foundations of genetic algorithms II. Morgan Kaufmann, San Francisco, CA, pp 5–17

De Jong ED (2004a) The incremental Pareto-coevolution archive. In: Proceedings of the genetic and evolutionary computation conference, GECCO 2004, Lecture notes in computer science, vol 3102. Springer, Berlin, pp 525–536

De Jong ED (2004b) Intransitivity in coevolution. In: Yao X et al. (eds) Parallel problem solving from nature, PPSN-VIII, Birmingham, UK, Lecture notes in computer science, vol 3242. Springer, Berlin, pp 843–851

De Jong ED (2004c) Towards a bounded Pareto-coevolution archive. In: Proceedings of the congress on evolutionary computation, CEC 2004. IEEE Press, Washington, DC, pp 2341–2348

De Jong ED (2005) The MaxSolve algorithm for coevolution. In: Beyer HG et al. (eds) Proceedings of the genetic and evolutionary computation conference, GECCO 2005. ACM Press, New York

De Jong ED (2007) Objective fitness correlation. In: Proceedings of the genetic and evolutionary computation conference, GECCO 2007. ACM Press, New York, pp 440–447

De Jong ED, Bucci A (2006) DECA: dimension extracting coevolutionary algorithm. In: Proceedings of the genetic and evolutionary computation conference, GECCO 2006. ACM Press, New York

De Jong ED, Bucci A (2007) Objective set compression: test-based problems and multiobjective optimization. In: Multiobjective problem solving from nature: from concepts to applications, Natural Computing Series. Springer, Berlin

De Jong ED, Pollack JB (2004) Ideal evaluation from coevolution. Evolut Comput 12(2):159–192

Ficici SG (2004) Solution concepts in coevolutionary algorithms. Ph.D. thesis, Department of Computer Science, Brandeis University, Waltham, MA

Ficici SG, Pollack JB (1998) Challenges in coevolutionary learning: arms-race dynamics, open-endedness, and mediocre stable states. In: Adami C et al. (eds) Artificial life VI proceedings. MIT Press, Cambridge, MA, pp 238–247

Ficici SG, Pollack JB (2000a) Effects of finite populations on evolutionary stable strategies. In: Whitley D et al. (eds) Proceedings of the genetic and evolutionary computation conference, GECCO 2000. Morgan Kaufmann, San Francisco, CA, pp 880–887

Ficici SG, Pollack JB (2000b) Game–theoretic investigation of selection methods used in evolutionary algorithms. In: Whitley D (ed) Proceedings of the 2000 congress on evolutionary computation, IEEE Press, Washington, DC, pp 880–887

Ficici SG, Pollack JB (2001) Pareto optimality in coevolutionary learning. In: Proceedings of the 6th European conference on advances in artificial life, ECAL 2001. Springer, London, pp 316–325

Ficici SG, Pollack JB (2003) A game-theoretic memory mechanism for coevolution. In: Cantú-Paz E et al. (eds) Genetic and evolutionary computation conference, GECCO 2003. Springer, Berlin, pp 286–297

Floreano D, Nolfi S (1997) God save the red queen! competition in co-evolutionary robotics. In: Koza JR et al.

(eds) Proceedings of the 2nd genetic programming conference, GP 1997. Morgan Kaufmann, San Francisco, CA, pp 398–406

Friedman D (1998) On economic applications of evolutionary game theory. J Evol Econ 8:15–43

Funes P, Pollack JB (2000) Measuring progress in coevolutionary competition. In: From animals to animats 6: Proceedings of the 6th international conference on simulation of adaptive behavior. MIT Press, Cambridge, MA, pp 450–459

Funes P, Pujals E (2005) Intransitivity revisited coevolutionary dynamics of numbers games. In: Proceedings of the conference on genetic and evolutionary computation, GECCO 2005. ACM Press, New York, pp 515–521

Hillis WD (1990) Co-evolving parasites improve simulated evolution as an optimization procedure. In: CNLS '89: Proceedings of the 9th international conference of the center for nonlinear studies on self-organizing, collective, and cooperative phenomena in natural and artificial computing networks on emergent computation. North-Holland Publishing Co., Amsterdam, pp 228–234

Hofbauer J, Sigmund K (1998) Evolutionary games and population dynamics. Cambridge University Press, Cambridge

Horn J (1995) The nature of niching: genetic algorithms and the evolution of optimal, cooperative populations. Ph.D. thesis, University of Illinois at Urbana-Champaign, Urbana-Champaign, IL

Husbands P, Mill F (1991) Simulated coevolution as the mechanism for emergent planning and scheduling. In: Belew R, Booker L (eds) Proceedings of the fourth international conference on genetic algorithms, Morgan Kaufmann, San Francisco, CA, pp 264–270

Igel C, Toussaint M (2003) On classes of functions for which no free lunch results hold. Inform Process Lett 86(6):317–321

Jansen T, Wiegand RP (2003a) Exploring the explorative advantage of the CC (1+1) EA. In: Proceedings of the 2003 genetic and evolutionary computation conference. Springer, Berlin

Jansen T, Wiegand RP (2003b) Sequential versus parallel cooperative coevolutionary (1+1) EAs. In: Proceedings of the congress on evolutionary computation, CEC 2003. IEEE Press, Washington, DC

Jansen T, Wiegand RP (2004) The cooperative coevolutionary (1+1) EA. Evolut Comput 12(4):405–434

Jaśkowski W, Wieloch B, Krawiec K (2008) Fitnessless coevolution. In: Proceedings of the genetic and evolutionary computation conference, GECCO 2008. ACM Press, New York, pp 355–365

Jensen MT (2001) Robust and flexible scheduling with evolutionary computation. Ph.D. thesis, Department of Computer Science, University of Aarhus, Denmark

Juillé H, Pollack JB (1998) Coevolving the ideal trainer: application to the discovery of cellular automata rules. In: Koza JR et al. (eds) Proceedings of the 3rd genetic programming conference, GP 1998. Morgan Kaufmann, San Francisco, CA, pp 519–527

Kauffman S, Johnson S (1991) Co-evolution to the edge of chaos: coupled fitness landscapes, poised states and co-evolutionary avalanches. In: Langton C et al. (eds) Artificial life II proceedings, vol 10. Addison-Wesley, Reading, MA, pp 325–369

Laumanns M, Thiele L, Zitzler E (2004) Running time analysis of multiobjective evolutionary algorithms on pseudo-Boolean functions. IEEE Trans Evolut Comput 8(2):170–182

Liekens A, Eikelder H, Hilbers P (2003) Finite population models of co-evolution and their application to haploidy versus diploidy. In: Cantú-Paz E et al. (eds) Proceedings of the genetic and evolutionary computation conference, GECCO 2003. Springer, Berlin, pp 344–355

Luke S, Wiegand RP (2003) Guaranteeing coevolutionary objective measures. In: De Jong KA et al. (eds) Foundations of genetic algorithms VII, Morgan Kaufmann, San Francisco, CA, pp 237–251

Maynard-Smith J (1982) Evolution and the theory of games. Cambridge University Press, Cambridge

Miller JH (1996) The coevolution of automata in the repeated prisoner's dilemma. J Econ Behav Organ 29(1):87–112

Monroy GA, Stanley KO, Miikkulainen R (2006) Coevolution of neural networks using a layered Pareto archive. In: Proceedings of the genetic and evolutionary computation conference, GECCO 2006. ACM Press, New York, pp 329–336

Noble J, Watson RA (2001) Pareto coevolution: using performance against coevolved opponents in a game as dimensions for Pareto selection. In: Spector L et al. (eds) Proceedings of the genetic and evolutionary computation conference, GECCO 2001. Morgan Kaufmann, San Francisco, CA, pp 493–500

Oliehoek FA, De Jong ED, Vlassis N (2006) The parallel Nash memory for asymmetric games. In: Proceedings of the genetic and evolutionary computation conference, GECCO 2006. ACM Press, New York, pp 337–344

Oliveto P, He J, Yao X (2007) Time complexity of evolutionary algorithms for combinatorial optimization: a decade of results. Int J Autom Comput 4(3): 281–293

Olsson B (2001) Co-evolutionary search in asymmetric spaces. Inform Sci 133(3–4):103–125

Osborne MJ, Rubinstein A (1994) A course in game theory. MIT Press, Cambridge, MA

Pagie L, Hogeweg P (2000) Information integration and red queen dynamics in coevolutionary optimization. In: Proceedings of the congress on evolutionary computation, CEC 2000. IEEE Press, Piscataway, NJ, pp 1260–1267

Pagie L, Mitchell M (2002) A comparison of evolutionary and coevolutionary search. Int J Comput Intell Appl 2(1):53–69

Panait L (2006) The analysis and design of concurrent learning algorithms for cooperative multiagent systems. Ph.D. thesis, George Mason University, Fairfax, VA

Panait L, Luke S (2002) A comparison of two competitive fitness functions. In: Langdon WB et al. (eds) Proceedings of the genetic and evolutionary computation conference, GECCO 2002. Morgan Kaufmann, San Francisco, CA, pp 503–511

Panait L, Luke S (2005) Time-dependent collaboration schemes for cooperative coevolutionary algorithms. In: AAAI fall symposium on coevolutionary and coadaptive systems. AAAI Press, Menlo Park, CA

Panait L, Luke S (2006) Selecting informative actions improves cooperative multiagent learning. In: Proceedings of the 5th international joint conference on autonomous agents and multi agent systems, AAMAS 2006. ACM Press, New York

Panait L, Wiegand RP, Luke S (2003) Improving coevolutionary search for optimal multiagent behaviors. In: Gottlob G, Walsh T (eds) Proceedings of the 18th international joint conference on artificial intelligence, IJCAI 2003. Morgan Kaufmann, San Francisco, CA, pp 653–658

Panait L, Wiegand RP, Luke S (2004) A sensitivity analysis of a cooperative coevolutionary algorithm biased for optimization. In: Proceedings of the genetic and evolutionary computation conference, GECCO 2004. Lecture notes in computer science, vol 3102. Springer, Berlin, pp 573–584

Panait L, Luke S, Harrison JF (2006a) Archive-based cooperative coevolutionary algorithms. In: Proceedings of the genetic and evolutionary computation conference, GECCO 2006. ACM Press, New York

Panait L, Luke S, Wiegand RP (2006b) Biasing coevolutionary search for optimal multiagent behaviors. IEEE Trans Evolut Comput 10(6):629–645

Panait L, Sullivan K, Luke S (2006c) Lenience towards teammates helps in cooperative multiagent learning. In: Proceedings of the 5th international joint conference on autonomous agents and multi agent systems, AAMAS 2006. ACM Press, New York

Paredis J (1997) Coevolving cellular automata: be aware of the red queen. In: Bäck T (ed) Proceedings of the 7th international conference on genetic algorithms, ICGA 1997. Morgan Kaufmann, San Francisco, CA

Parker GB, Blumenthal HJ (2003) Comparison of sample sizes for the co-evolution of cooperative agents. In: Proceedings of the congress on evolutionary computation, CEC 2003. IEEE Press, Washington, DC

Popovici E (2006) An analysis of two-population coevolutionary computation. Ph.D. thesis, George Mason University, Fairfax, VA

Popovici E, De Jong KA (2004) Understanding competitive co-evolutionary dynamics via fitness landscapes. In: Luke S (ed) AAAI fall symposium on artificial multiagent learning. AAAI Press, Menlo Park, CA

Popovici E, De Jong KA (2005a) A dynamical systems analysis of collaboration methods in cooperative co-evolution. In: AAAI fall symposium series co-evolution workshop. AAAI Press, Menlo Park, CA

Popovici E, De Jong KA (2005b) Relationships between internal and external metrics in co-evolution. In: Proceedings of the congress on evolutionary computation, CEC 2005. IEEE Press, Washington, DC

Popovici E, De Jong KA (2005c) Understanding cooperative co-evolutionary dynamics via simple fitness landscapes. In: Beyer HG et al. (eds) Proceedings of the genetic and evolutionary computation conference, GECCO 2005. ACM Press, New York

Popovici E, De Jong KA (2009) Monotonicity versus performance in co-optimization. In: Foundations of genetic algorithms X. ACM Press, New York

Potter M (1997) The design and analysis of a computational model of cooperative coevolution. Ph.D. thesis, Computer Science Department, George Mason University

Potter M, De Jong KA (1994) A cooperative coevolutionary approach to function optimization. In: Parallel problem solving from nature, PPSN-III, Jerusalem, Israel. Springer, Berlin, pp 249–257

Potter MA, De Jong KA (2000) Cooperative coevolution: an architecture for evolving coadapted subcomponents. Evolut Comput 8(1):1–29

Rosin CD (1997) Coevolutionary search among adversaries. Ph.D. thesis, University of California, San Diego, CA

Rosin CD, Belew RK (1995) Methods for competitive coevolution: finding opponents worth beating. In: Proceedings of the 6th international conference on genetic algorithms, ICGA 1995. Morgan Kaufmann, San Francisco, CA, pp 373–381

Rosin CD, Belew RK (1997) New methods for competitive coevolution. Evolut Comput 5(1):1–29

Schmitt LM (2003a) Coevolutionary convergence to global optima. In: Cantú-Paz E et al. (eds) Genetic and evolutionary computation conference, GECCO 2003. Springer, Berlin, pp 373–374

Schmitt LM (2003b) Theory of coevolutionary genetic algorithms. In: Guo M et al. (eds) International symposium on parallel and distributed processing and applications, ISPA 2003. Springer, Berlin, pp 285–293

Schumacher C, Vose M, Whitley L (2001) The no free lunch and description length. In: Proceedings of the genetic and evolutionary computation conference, GECCO 2001. Morgan Kaufmann, San Francisco, CA, pp 565–570

Service TC (2009) Unbiased coevolutionary solution concepts. In: Foundations of genetic algorithms X. ACM Press, New York

Service TC, Tauritz DR (2008a) Co-optimization algorithms. In: Keijzer M et al. (eds) Proceedings of the genetic and evolutionary computation conference, GECCO 2008. ACM Press, New York, pp 387–388

Service TC, Tauritz DR (2008b) A no-free-lunch framework for coevolution. In: Keijzer M et al. (eds) Proceedings of the genetic and evolutionary computation conference, GECCO 2008. ACM Press, New York, pp 371–378

Sims K (1994) Evolving 3D morphology and behaviour by competition. In: Brooks R, Maes P (eds) Artificial life IV proceedings. MIT Press, Cambridge, MA, pp 28–39

Spears W (1994) Simple subpopulation schemes. In: Proceedings of the 1994 evolutionary programming conference. World Scientific, Singapore

Stanley KO (2004) Efficient evolution of neural networks through complexification. Ph.D. thesis, The University of Texas at Austin, Austin, TX

Stanley KO, Miikkulainen R (2002a) Continual coevolution through complexification. In: Langdon WB et al. (eds) Proceedings of the genetic and evolutionary computation conference, GECCO 2002. Morgan Kaufmann, San Francisco, CA, pp 113–120

Stanley KO, Miikkulainen R (2002b) The dominance tournament method of monitoring progress in coevolution. In: Langdon WB et al. (eds) Proceedings of the genetic and evolutionary computation conference, GECCO 2002. Morgan Kaufmann, San Francisco, CA

Stanley KO, Miikkulainen R (2004) Competitive coevolution through evolutionary complexification. J Arti Intell Res 21:63–100

Stuermer P, Bucci A, Branke J, Funes P, Popovici E (2009) Analysis of coevolution for worst-case optimization. In: Proceedings of the genetic and evolutionary computation conference, GECCO 2009. ACM Press, New York

Subbu R, Sanderson A (2004) Modeling and convergence analysis of distributed coevolutionary algorithms. IEEE Trans Syst Man Cybern B Cybern 34(2):806–822

Vo C, Panait L, Luke S (2009) Cooperative coevolution and univariate estimation of distribution algorithms. In: Foundations of genetic algorithms X, ACM Press, New York

Vose M (1999) The simple genetic algorithm. MIT Press, Cambridge, MA

Watson RA, Pollack JB (2001) Coevolutionary dynamics in a minimal substrate. In: Spector L et al. (eds) Proceedings of the genetic and evolutionary computation conference, GECCO 2001. Morgan Kaufmann, San Francisco, CA, pp 702–709

Watson RA (2002) Compositional evolution: interdisciplinary investigations in evolvability, modularity, and symbiosis. Ph.D. thesis, Brandeis University, Waltham, Massachusetts

Weibull J (1992) Evolutionary game theory. MIT Press, Cambridge, MA

Wiegand RP (2004) An analysis of cooperative coevolutionary algorithms. Ph.D. thesis, George Mason University, Fairfax, VA

Wiegand RP, Potter M (2006) Robustness in cooperative coevolution. In: Proceedings of the genetic and evolutionary computation conference, GECCO 2006. ACM Press, New York

Wiegand RP, Sarma J (2004) Spatial embedding and loss of gradient in cooperative coevolutionary algorithms. In: Yao X et al. (eds) Parallel problem solving from nature, PPSN-VIII. Springer, Birmingham, UK, pp 912–921

Wiegand RP, Liles W, De Jong KA (2001) An empirical analysis of collaboration methods in cooperative coevolutionary algorithms. In: Spector L (ed) Proceedings of the genetic and evolutionary computation conference, GECCO 2001. Morgan Kaufmann, San Francisco, CA, pp 1235–1242, errata available at http://www.tesseract.org/paul/papers/gecco01-cca-errata.pdf

Wiegand RP, Liles WC, De Jong KA (2002) The effects of representational bias on collaboration methods in cooperative coevolution. In: Proceedings of the 7th conference on parallel problem solving from nature. Springer, Berlin, pp 257–268

Williams N, Mitchell M (2005) Investigating the success of spatial coevolution. In: Proceedings of the genetic and evolutionary computation conference, GECCO 2005. ACM Press, New York, pp 523–530

Wolpert D, Macready W (1997) No free lunch theorems for optimization. IEEE Trans Evolut Comput 1(1):67–82

Wolpert D, Macready W (2005) Coevolutionary free lunches. IEEE Trans Evolut Comput 9(6):721–735

32 Niching in Evolutionary Algorithms

Ofer M. Shir
Department of Chemistry, Princeton University, NJ, USA
oshir@princeton.edu

G. Rozenberg et al. (eds.), *Handbook of Natural Computing*, DOI 10.1007/978-3-540-92910-9_32,
© Springer-Verlag Berlin Heidelberg 2012

Abstract

Niching techniques are the extension of standard evolutionary algorithms (EAs) to multi-modal domains, in scenarios where the location of multiple optima is targeted and where EAs tend to lose population diversity and converge to a solitary basin of attraction. The development and investigation of EA niching methods have been carried out for several decades, primarily within the branches of genetic algorithms (GAs) and evolution strategies (ES). This research yielded altogether a long list of algorithmic approaches, some of which are bio-inspired by various concepts of organic speciation and ecological niches, while others are more computational-oriented. This chapter will lay the theoretical foundations for niching, from the perspectives of biology as well as optimization, provide a summary of the main contemporary niching techniques within EAs, and discuss the topic of experimental methodology for niching techniques. This will be accompanied by the discussion of specific case-studies, including the employment of the popular covariance matrix adaptation ES within a niching framework, the application to real-world problems, and the treatment of the so-called niche radius problem.

1   Introduction

Evolutionary Algorithms (EAs) have the tendency to lose diversity within their population of feasible solutions and to converge into a single solution (Bäck 1994, 1996; Mahfoud 1995a), even if the search landscape has multiple globally optimal solutions.

Niching methods, the extension of EAs to finding multiple optima in multimodal optimization within one population, address this issue by maintaining the diversity of certain properties within the population. Thus, they aim at obtaining parallel convergence into multiple attraction basins in the multimodal landscape within a single run.

The study of niching is challenging both from the theoretical point of view and from the practical point of view. The theoretical challenge is twofold – maintaining diversity within a population-based stochastic algorithm from the computational perspective, and also gaining an insight into *speciation* theory or *population genetics* from the evolutionary biology perspective. The practical aspect provides a real-world incentive for this problem – there is an increasing interest of the *applied optimization community* in providing decision makers with multiple solutions that ideally represent different conceptual designs, for single-criterion or multi-criterion search spaces (Avigad et al. 2004, 2005). The concept of *"going optimal"* is often extended nowadays into the aim of *"going multi optimal"*: **obtaining optimal results but also providing the decision maker with a variety of different options.** (The *Second Toyota Paradox* (Cristiano et al. 2001), which is often reviewed in management studies, promotes the consideration of multiple candidate solutions during the car production process: "Delaying decisions, communicating ambiguously, and pursuing an excessive number of prototypes, can produce better cars faster and cheaper").

Among the three conventional streams of EAs (Bäck 1996) – *genetic algorithms* (GAs) (Goldberg 1989), *evolution strategies* (ES) (Beyer and Schwefel 2002), and *evolutionary programming* (EP) (Fogel 1966) – niching methods have been studied in the past four decades almost exclusively within the framework of GAs. However, niching methods have been mostly a by-product of studying *population diversity*, and were hardly ever at the front of the evolutionary computation (EC) research as an independent subfield.

This chapter thus aims at achieving three main goals:

1. Laying the theoretical foundations for niching, both from the biological as well as the computational perspectives
2. Providing a summary of the main contemporary niching techniques within evolutionary algorithms
3. Discussing the topic of experimental methodology for niching techniques, and proposing a framework for it

We shall begin by providing the natural computing motivation for niching by means of the biological background for the evolutionary process of *speciation*, given in ❯ Sect. 2, where it will also be linked to the domain of function optimization. The important topic of *population diversity* within evolutionary algorithms is discussed in detail in ❯ Sect. 3, focusing on the two canonical streams of GAs and ES. This is followed by an overview of the existing niching techniques within EAs, in ❯ Sect. 4. A case study of a specific niching technique with the popular covariance matrix adaptation evolution strategy (CMA-ES) receives special attention by means of a detailed description in the end of that section. The topic of experimental methodology for research on niching is discussed in ❯ Sect. 5, where proposed synthetic test functions as well as recommended performance criteria are outlined. Niching with the CMA-ES is revisited in that section, as an experimental observation case study. ❯ Section 6 is dedicated to a crucial and challenging issue for niching methods, the so-called *niche radius problem*. Finally, ❯ Sect. 7 concludes this chapter and proposes directions for future research in this domain.

A Note on the Scope It should be stressed that there exist niching methods in other natural computing frameworks, such as in particle swarm optimization (PSO) or in ant colony optimization (ACO) (for general field reviews see Kennedy and Eberhart (2001) and Engelbrecht (2005) for PSO or Dorigo and Stützle (2004) and Blum (2005) for ACO). These frameworks exceed the scope of this chapter, which solely focuses on evolutionary algorithms. We refer the reader who wishes to learn about niching techniques in PSO to Brits et al. (2002) and Parrott and Li (2006), or to Angus (2006) in ACO.

A Note on Notation Upon consideration of optimization problems, this chapter will assume *minimization*, without loss of generality, unless specified otherwise.

2 Background: From Speciation to Ecological Optima

This section constitutes the introduction to niching, and thus covers a diverse set of interdisciplinary topics. It will review the *biological* elementary concepts that correspond to the core of niching methods, such as *population diversity* and *speciation*, while mainly relying on Freeman and Herron (2003). It will then make the linkage to computing, by shifting to the optimization arena and discussing the equivalent to ecological niches: *basins of attraction*.

A Note on Terminology A species is defined as the *smallest evolutionary independent unit*. The term *niche*, however, stems from ecology, and it has several different definitions. It is sometimes referred to as the collective environmental components which are favored by a specific species, but could also be considered as the ecosystem itself which hosts individuals of *various species*. Most definitions would typically also consider the *hosting capacity* of the niche, which refers to the limited available resources for sustaining life in its domain. In the context of function optimization, *niche* is associated with a *peak*, or a *basin of attraction*, whereas a *species* corresponds to the subpopulation of individuals occupying that *niche*.

2.1   Preliminary: Organic Evolution and Genetic Drift

Organic evolution can be broken down into four defining fundamental mechanisms: *natural selection, mutation, migration* or *gene flow*, and *genetic drift*. The latter, which essentially refers to *sampling errors in finite populations*, was overlooked by Darwin, who had not been familiar with Mendelian genetics, and thus did not discuss this effect in his *Origin of Species* (Darwin 1999). We assume that the reader is familiar with the first three fundamental elements, that is, selection–mutation–migration, and would like to elaborate on the drift effect, as it is the least intuitive mechanism among the four aforementioned evolutionary forces. In short, *genetic drift* (Fisher 1922; Wright 1931; Kimura 1983) is a stochastic process in which the diversity is lost in finite populations. A distribution of genetic properties is transferred to the following generation in a limited manner, due to the finite number of generated offspring, or equivalently, the limited statistical sampling of the distribution. As a result, the distribution is likely to approach an *equilibrium distribution*, for example, fixation of specific alleles when subject to equal fitness. This is why *genetic drift* is often considered as a *neutral effect*. The smaller the population, the faster and stronger this effect occurs. An analogy is occasionally drawn between genetic drift to *Brownian motion* of particles in mechanics.

The *genetic drift* effect had been originally recognized by Fisher (1922) (referred to as *random survival*), and was explicitly mentioned by Wright when studying Mendelian populations (Wright 1931). It was, however, revisited and given a new interpretation in the *Neutral Theory of Molecular Evolution* of Kimura (1983). The *neutral theory* suggested that the *random genetic drift* effect is the main driving force within molecular evolution, rather than the *nonrandom natural selection* mechanism. *Natural selection* as well as *genetic drift* are considered nowadays, by the contemporary evolutionary biology community, as the combined driving force of organic evolution. Moreover, the importance of the *neutral theory* is essentially in its being a **null hypothesis model** for the *natural selection theory*, by definition.

2.2   Organic Diversity

Diversity among individuals or populations in nature can be attributed to different evolutionary processes which occur at different levels. We distinguish here between variations that are observed within a single species to a *speciation* process, during which a new species arises, and both of them are reviewed shortly.

2.2.1   Variations Within a Species

Diversity of organisms within a single species stems from a variance at the genotypic level, referred to as *genetic diversity*, or from the existence of a spectrum of phenotypic realizations to a specific genotype. These effects are quantified and are usually associated with *genotypic variance* and *phenotypic variance*, respectively. Several hypotheses explaining *genetic diversity* have been proposed within the discipline of *population genetics*, including the *neutral evolution theory*. It should be noted that genetic diversity is typically considered to be advantageous for survival, as it may allow better adaptation of the population to environmental changes, such as climate variations, diseases, etc.

Phenotypic variance is measured on a continuous spectrum, also known as quantitative variation. Roughly speaking, the main sources of quantitative variations (Freeman and Herron 2003; Scheiner and Goodnight 1984) are outlined here:

1. Genes have *multiple loci*, and hence are mapped onto a large set of phenotypes.
2. *Environmental effects* have direct influence on natural selection; fitness is time-dependent, and thus phenotypic variations in the outcome of selection are expected to occur.
3. *Phenotypic plasticity* is the amount in which the *genotypic expression* vary in different environments. (Bradshaw (1965) gave the following qualitative definition to phenotypic plasticity: "The amount by which the expressions of individual characteristics of a genotype are changed by different environments is a measure of the plasticity of these characters.") And it is a direct source of variation at the phenotypic level.
4. The plastic response of the genotype to the environment, that is, the joint effect of genetic and environmental elements, also affects the selection of a specific phenotype, and thus can lead to variations. This effect is known as *genotype–environment interaction* ("G-by-E").

Thus, *quantitative variations* are mainly caused by genotypic and phenotypic realizations and their interaction with the environment. The ratio between *genetic variance* to total *phenotypic variance* is defined as *heritability* (Wright 1931).

2.2.2 Speciation

Speciation, on the other hand, is the process during which a new species arises. In this case, statistical disassociation, which is the trigger to speciation, originates from gradually decreasing physical linkage.

The essence of the speciation process is **lack of gene flow**, where physical isolation often plays the role of the barrier to gene flow. Lack of gene flow is only one of the necessary conditions for speciation. Another necessary condition for speciation to occur is that the reduction of gene flow will be followed by a phase of **genetic divergence**, by means of *mutation, selection,* or *drift*. Finally, the completion or elimination of divergence can be assessed via the so-called *secondary contact* phase: interbreeding between the parental populations would possibly fail (offspring is less fit), succeed (offspring is fitter), or have a neutral outcome (offspring has the same fitness). This would correspond, respectively, to increasing, decreasing, or stabilizing the differentiation between the two arising species. Note that the speciation can occur *de facto*, without the actual secondary contact taking place; the latter is for observational assessment purposes.

In organic evolution, four different levels of *speciation* are considered, corresponding to four levels of physical linkage between the subpopulations:

1. **Allopatric speciation** The split in the population occurs only due to complete geographical separation, for example, migration or mountain building. It results in two geographically isolated populations.

2. **Peripatric speciation** Species arise in small populations, which are not geographically separated but rather isolated in practice; the effect occurs mainly due to the *genetic drift* effect.

3. **Parapatric speciation** The geographical separation is limited, with a physical overlap between the two zones where the populations split from each other.

4. **Sympatric speciation** The two diverging populations coexist in the same zone, and thus the speciation is strictly non-geographical. This is observed in nature in parasite populations, that are located in the same zone, but associated with different plant or animal hosts (McPheron et al. 1988).

These four modes of speciation correspond to four levels of geographically decreasing linkages. Generally speaking, *statistical association* of genetic components in nature, such as *loci*, typically results from *physical linkage*. In this case, we claim that statistical disassociation, which is the trigger to speciation, originates from gradually decreasing physical linkage.

In summary, speciation typically occurs throughout three steps:

1. Geographic isolation or reduction of gene flow
2. Genetic divergence (mutation, selection, drift)
3. Secondary contact (observation/assessment)

2.3 "Ecological Optima": Basins of Attraction

We devote this section to the definition of basins of attraction. The task of defining a *generic basin of attraction* seems to be one of the most difficult problems in the field of *global optimization*, and there have only been few attempts to treat it theoretically. (Intuitively, and strictly metaphorically speaking, we may think of a *region of attraction* of $\mathbf{x}_L$ as the region, where if water is poured, it will reach $\mathbf{x}_L$. Accordingly, we may then think of the basin of $\mathbf{x}_L$ as the maximal region that will be covered when the cavity at $\mathbf{x}_L$ is filled to the lowest part of its rim (Törn and Zilinskas 1987).)

Rigorously, it is possible to define the basin by means of a *local optimizer*. In particular, consider a gradient descent algorithm starting from $\mathbf{x}_0$, which is characterized by the following dynamics:

$$\frac{d\mathbf{x}(t)}{dt} = -\nabla f(\mathbf{x}(t)) \tag{1}$$

with the initial condition $\mathbf{x}(0) = \mathbf{x}_0$. Now, consider the set of points for which the limit exists:

$$Y = \left\{ \mathbf{x} \in \mathbb{R}^n \middle| \mathbf{x}(0) = \mathbf{x} \wedge \mathbf{x}(t)|_{t \geq 0} \text{ satisfies Eq. 1} \wedge \lim_{t \to \infty} \mathbf{x}(t) \text{ exists} \right\} \tag{2}$$

Definition 1 The *region of attraction* $A(\mathbf{x}_L)$ of a local minimum, $\mathbf{x}_L$, is

$$A(\mathbf{x}_L) = \left\{ \mathbf{x} \in Y \middle| \mathbf{x}(0) = \mathbf{x} \wedge \mathbf{x}(t)|_{t \geq 0} \text{ satisfies Eq. 1} \wedge \lim_{t \to \infty} \mathbf{x}(t) = \mathbf{x}_L \right\} \tag{3}$$

The *basin* of $\mathbf{x}_L$ is the **maximal level set** that is fully contained in $A(\mathbf{x}_L)$.

In the case of several disconnected local minima with the same function value, it is possible to define the region of attraction as the union of the nonoverlapping connected sets.

2.4 Classification of Optima: The Practical Perspective

On a related note to the theoretical definition of the basin, the practical perspective for the *classification of optima shapes*, also referred to as global topology, is worth mentioning. This topic is strongly related to the emerging subfield of *robustness study* (see, e.g., Tsui 1992), which aims at attaining optima of high quality with large basins (i.e., low partial derivative values in the proximity of the peak). Moreover, yet visited from a different direction, another approach was introduced recently by Lunacek and Whitley for classifying different classes of multimodal landscapes with respect to algorithmic performance (Lunacek and Whitley 2006). The latter defines the *dispersion metric* of a landscape as the degree to which the local optima are globally clustered near one another. Landscapes with low dispersion have their best local optima clustered together in a single *funnel*. (We deliberately avoid the definition of a funnel as it is rather vague. We refer the reader to Doye et al. (2004).) This classification of low dispersion and high dispersion may be associated with the algorithmic trade-off between exploration of the landscape and exploitation of local structures. Upon considering landscapes with multiple funnels, a recent study (Lunacek et al. 2008) investigated the impact of a global structure of two uneven funnels on the evolutionary search. It concluded that EAs tend to converge into the larger funnel, even if it is of suboptimal quality, and thus put their effectiveness in global exploration in multi-funnel landscapes into question.

3 Population Diversity in Evolutionary Algorithms

The term *population diversity* is commonly used in the context of Evolutionary Algorithms, but it rarely refers to a rigorous definition. Essentially, it is associated both with *genetic diversity* as well as with *speciation* – the two different concepts from organic evolution that were discussed in the previous section. This is simply due to the fact that the differences between the two concepts do not have any practical effect on the evolutionary search nor on the goal of maintaining diversity among the evolving candidate solutions. In the well-known trade-off between *exploration* and *exploitation* of the landscape during a search, *maintaining population diversity* is a driving force in the *exploration front*, and thus it is an important component. However, among EC researchers, population diversity is primarily considered as a component due to play a role in the important exploration of the landscape for the sake of obtaining a single solution, while its role in obtaining multiple solutions is typically considered as a secondary one.

Mahfoud's Formalism

Mahfoud constructed a formalism for characterizing *population diversity* in the framework of Evolutionary Algorithms (see Mahfoud 1995a, pp. 50–59). Mahfoud's formal framework was based on the partitioning of the search space into equivalence classes (set to *minima* in the search landscape), a descriptive relation (typically, *genotypic* or *phenotypic* mappings), and the measurement of distance between the current distribution of subpopulations to some predefined *goal distribution*.

Let P be a discrete distribution describing the current partitioning of the population into subpopulations, and let Q be the goal distribution of the population with respect to the defined sites. The formalism focuses in defining the *directed divergence*, or distance of distribution P to distribution Q. Several well-known metrics follow this formalism by satisfying its various criteria, such as Shannon's information entropy, standard distance metrics, etc. (Mahfoud 1995a).

Diversity Loss

Subject to the complex dynamics of the various forces within an evolutionary algorithm, population diversity is typically lost, and the search is likely to converge into a single basin of attraction in the landscape. *Population diversity loss* within the population of solutions is the fundamental effect which niching methods aim to treat. In fact, from the historical perspective, the quest for diversity-promoting techniques was the main goal within the EC community for some time, and niching methods were merely obtained as *by-products*, so to speak, of that effort.

Due to the fundamental differences between GAs and ES, we choose to describe the effect of population diversity loss for each one of them separately.

3.1 Diversity Loss Within Genetic Algorithms

Mahfoud devoted a large part of his doctoral dissertation to studying population diversity within GAs (Mahfoud 1995a). He concluded that three main components can be attributed to the effect of population diversity loss within GAs.

Selection Pressure The traditional GA applies a probabilistic selection mechanism, namely, the *roulette-wheel selection* (RWS). This mechanism belongs to a broad set of selection mechanisms, which follow the fitness-proportionate selection principle. Selection pressure is thus associated with the first *moment of the selection operator*. It has been demonstrated by Mahfoud (1995a) that the selection pressure, or equivalently the nonzero expectation of the selection operator, prevents the algorithm from converging in parallel into more than a single attractor.

Selection Noise Selection noise is associated with the second *moment of the selection operator*, or its *variance*. Mahfoud (1995a) demonstrated that the high variance of the RWS, as well as of other selection mechanisms, is responsible for the fast convergence of a population into a single attractor, even when there exists a set of equally fit attractors. This effect can be considered as a *genetic drift* in its broad definition, that is, sampling error of a distribution.

Operator Disruption Evolutionary operators in general, and the *mutation* and *recombination* operators in particular, boost the evolution process toward exploration of the search space. In that sense, they have a constructive effect on the process, since they allow locating new and better solutions. However, their action also has a destructive effect. This is due to the fact that, by applying them, good solutions that have been previously located might be lost. Therefore, they may eliminate competition between highly fit individuals, and "assist" some of them to take over. The mutation operator usually has a small effect, since it acts in small steps – low mutation probability in the traditional GA, which means infrequent occurrence of bit flips. Thus, the mutation operator can be considered to have a negligible disruption. The recombination operator, on the other hand, has a more considerable effect. In the GA field, where the *crossover* operator is in use (single-point, two-point or *n*-point crossovers), the latter has been shown to have a disruptive nature by breaking desired patterns within the population (the well-known *schema theorem* discusses the schema disruption by the crossover operator and states that schemata with high defining length will most likely be disrupted by the crossover operator; see, for example, Goldberg (1989)).

Corollary 1 *The traditional GA, which employs the standard set of operators, is exposed to statistical as well as disruptive effects that are responsible for the loss of population diversity. This outcome is likely to occur due to the first and second moments of the RWS operator, as well as to the disruptive nature of the crossover operator. We conclude that the traditional GA is expected to lose diversity among its candidate solutions.*

3.2 Diversity Loss Within Evolution Strategies

The defining mechanism of ES is strongly dictated by the mutation operator as well as by the deterministic selection operator. As defining operators, they have a direct influence on the diversity property of the population. The recombination operator, nevertheless, does not play a critical role in the ES mechanism.

We attribute two main components to the *population diversity loss* within ES: fast *take-over*, which is associated with the *selection* operator, and *genetic drift* (or *neutrality* effect), which is associated both with the *selection* and the *recombination* operators.

3.2.1 Selective Pressure: Fast Take-Over

Evolution strategies have a strictly deterministic, rank-based approach to selection. In the two traditional approaches (Bäck 1996) – (μ, λ) and $(\mu + \lambda)$ – the best individuals are selected, implying, rather intuitively, high *selective pressure*. Due to the crucial role of the selection operator within the evolution process, its impact within the ES field has been widely investigated.

Goldberg and Deb introduced the important concept of *take-over time* (Deb and Goldberg 1989), which gives a quantitative description of selective pressure **with respect only to the selection operator.**

Definition 2 *The take-over time, τ^*, is the minimal number of generations until repeated application of the selection operator yields a uniform population filled with copies of the best individual.*

The selective pressure has been further investigated by Bäck (1994), who analyzed all the ES selection mechanisms also with respect to take-over times. He concluded that upon employing the typical selection mechanisms, very short *take-over times* are yielded. This result implies that ES are typically subject to high *selective pressures*.

3.2.2 ES Genetic Drift

We consider two different ES neutral effects that could be together ascribed as a general ES genetic drift: *recombination drift*, and *selection drift*. We argue that these two components are responsible for the loss of population diversity within ES.

Recombination Drift

Beyer explored extensively the so-called *mutation-induced speciation by recombination* (MISR) principle (see, e.g., Beyer 1999). According to this important principle, repeated application of the mutation operator, subject to a dominant recombination operator, would lead to a stable distribution of the population, which resembles a species or a cloud of individuals. When fitness-based selection is applied, this cloud is likely to move together toward fitter regions of the landscape. Furthermore, Beyer managed to prove analytically (Beyer 1999) that the MISR principle is indeed universal when finite populations are employed, subject to sampling-based recombination. The latter was achieved by analyzing the ES dynamics without fitness-based selection, deriving the expected population variance, and showing that it is reduced with random sampling in finite populations. This result was also corroborated by numerical simulations. This study provides one with an analytical result that a sampling-based recombination is subject to genetic drift, and leads to loss of population diversity.

Selection Drift

A recent study on the extinction of subpopulations on a simple *bimodal equi-fitness* model investigated the drift effect of the selection operator (Schönemann et al. 2004). It considered the application of *selection* on finite populations, when the fitness values of the different attractors were equal (i.e., eliminating the possibility of a *take-over effect*), and argued that a neutral effect (*drift*) would occur, pushing the population into a single attractor. The latter study, indeed, demonstrated this effect of *selection drift* in ES, which resulted in a convergence to an equilibrium distribution around a single attractor. It was also shown that the time of extinction increases proportionally with μ. The analysis was conducted by means of Markov chain models, supported by statistical simulations.

Corollary 2 *Evolution Strategies that employ finite populations are typically underposed to several effects that are responsible for the loss of population diversity. It has been shown that the standard selection mechanisms may lead to a fast take-over effect. In addition, we argued that both the* recombination *and the* selection *operators experience their own* drift *effects that lead to population diversity loss. We concluded that an evolution strategy with a small population is likely to encounter a rapid effect of diversity loss.*

4 Evolutionary Algorithms Niching Techniques

Despite the fact that the motivation for multimodal optimization is beyond doubt, and the biological inspiration is real, there is no unique definition of the goal statement for *niching techniques*. There have been several attempts to provide a proper definition and functional specification for niching; we review here some of them:

1. Mahfoud (1995a) chose to put emphasis on locating as well as maintaining good optima, and formulated the following:
   ▶ The litmus test for a niching method, therefore, will be whether it possesses the capability to find multiple, final solutions within a reasonable amount of time, and to maintain them for an extended period of time.

2. Beyer et al. (2002) put forward also the actual maintenance of population diversity:
   ▸ *Niching*: process of separation of individuals according to their states in the search space or maintenance of diversity by appropriate techniques, for example, local population models, fitness sharing, or distributed EA.
3. Preuss (2006) considered the two definitions mentioned above, and proposed a third:
   ▸ Niching in EAs is a two-step procedure that (a) concurrently or subsequently distributes individuals onto distinct basins of attraction and (b) facilitates approximation of the corresponding (local) optimizers.

We choose to adopt Preuss' mission statement and define **the challenge in niching as follows:**

▸ **Attaining the optimal interplay between partitioning the search space into niches occupied by stable subpopulations, by means of population diversity preservation, to exploiting the search in each niche by means of a highly efficient optimizer with local-search capabilities.**

Next, we shall provide an overview of existing niching techniques. Special attention will be given to a specific niching method which is based on a state-of-the-art evolution strategy, the so-called covariance matrix adaptation ES (CMA-ES). The latter is considered to be particularly efficient for high-dimensional continuous optimization, and will be described in greater detail.

4.1 GA Niching Methods

Niching methods within genetic algorithms have been studied during the past few decades, initially triggered by the necessity to promote *population diversity* within EAs. The research has yielded a variety of different methods, which are the vast majority of existing work on niching in general.

The remainder of this section will focus on GA niching techniques, by providing a short survey of the main known methods, with emphasis on the important concepts of *sharing* and *crowding*. This survey is mainly based on Mahfoud (1995a) and Singh and Deb (2006).

4.1.1 Fitness Sharing

The *sharing* concept was one of the pioneering niching approaches. It was first introduced by Holland (1975), and later implemented as a niching technique by Goldberg and Richardson (1987). This strong approach of **considering the fitness as a shared resource** has essentially become an important concept in the broad field of evolutionary algorithms, and laid the foundation for various successful niching techniques for multimodal function optimization, mainly within GAs. A short description of the *fitness sharing* mechanism follows.

The basic idea of *fitness sharing* is to consider the fitness of the landscape as a resource to be shared among the individuals, in order to decrease redundancy in the population. Given a similarity metric of the population, which can be *genotypic* or *phenotypic*, the *sharing function* is defined as follows:

$$sh(d_{i,j}) = \begin{cases} 1 - \left(\frac{d_{i,j}}{\rho}\right)^{\alpha_{sh}} & \text{if } d_{i,j} < \rho \\ 0 & \text{otherwise} \end{cases} \tag{4}$$

where $d_{i,j}$ is the distance between individuals i and j, ρ (traditionally noted as σ_{sh}) is the fixed radius of every niche, and $\alpha_{sh} \geq 1$ is a control parameter, typically set to 1. Using the *sharing function*, the *niche count* is given by

$$m_i = \sum_{j=1}^{N} sh(d_{i,j}) \tag{5}$$

where N is the number of individuals to be considered in the selection phase.

Let an individual raw fitness be denoted by f_i, then the *shared fitness* is defined by:

$$f_i^{sh} = \frac{f_i}{m_i} \tag{6}$$

assuming that the fitness is *strictly positive* and subject to *maximization*. The evaluation of the shared fitness is followed by the selection phase, which is typically based on the RWS operator (Goldberg 1989); The latter takes into consideration the shared fitness. Thus, the *sharing* mechanism practically penalizes individuals that have similar members within the population via their fitness, and by that it aims at reducing redundancy in the gene pool, especially around the peaks of the fitness landscape.

One important auxiliary component of this approach is the *niche radius*, ρ. Essentially, this approach makes a strong assumption concerning the fitness landscape, stating that the optima are far enough from one another with respect to the *niche radius*, which is estimated for the given problem and remains fixed during the course of evolution. Furthermore, it is important to note that the formulas for determining the value of ρ, which will be given in ❯ Sect. 4.3, are dependent on q, the number of peaks of the target function. Hence, a second assumption is that q can be estimated. In practice, an accurate estimation of the expected number of peaks q in a given domain may turn out to be extremely difficult. Moreover, peaks may vary in shape, and this would make the task of determining ρ rather complicated. The aforementioned assumptions pose the so-called *niche radius problem*, to be discussed in ❯ Sect. 6.

In the literature, several GA niching *sharing*-based techniques, which implement and extend the basic concept of sharing, can be found (Mahfoud 1995a; Goldberg 1987; Yin and Germany 1993; Jelasity 1998; Miller and Shaw 1996; Petrowski 1996; Cioppa et al. 2004). Furthermore, the concept of sharing was successfully extended to other "yields of interest," such as *concept sharing* (Avigad et al. 2004).

4.1.2 Dynamic Fitness Sharing

In order to improve the *sharing* mechanism, a dynamic approach was proposed. The *dynamic niche sharing* method (Miller and Shaw 1996), which extended the *fitness sharing* technique, aimed at dynamically recognizing the q peaks of the forming niches, and based on that information, classified the individuals as either members of one of the niches, or as members of the "non-peaks domain."

Explicitly, let us introduce the *dynamic niche count*:

$$m_i^{dyn} = \begin{cases} n_j & \text{if individual } i \text{ is within dynamic niche } j \\ m_i & \text{otherwise (non-peak individual)} \end{cases} \tag{7}$$

where n_j is the size of the jth dynamic niche (i.e., the number of individuals which were classified to niche j), and m_i is the standard *niche count*, as defined in ❯ Eq. 5.

The shared fitness is then defined as follows:

$$f_i^{\text{dyn}} = \frac{f_i}{m_i^{\text{dyn}}} \tag{8}$$

The identification of the dynamic niches can be carried out by means of a *greedy* approach, as proposed in Miller and Shaw (1996) as the dynamic peak identification (DPI) algorithm (see ❷ *Algorithm 1*). As in the original *fitness sharing* technique, the *shared fitness evaluation* is followed by the selection phase, typically implemented with the RWS operator. Thus, this technique does not fixate the peak individuals, but rather provides them with an advantage in the selection phase, which is probability-based within GAs.

4.1.3 Clearing

Another variation to the *fitness sharing* technique, called *clearing*, was introduced by Petrowski (1996) at the same time as the *dynamic fitness sharing* (Miller and Shaw 1996). The essence of this mechanism is the *"winner takes it all"* principle, and its idea is to designate a specific number of individuals per niche, referred to as *winners*, which could enjoy the resources of that niche. This is equivalent to the introduction of a "death penalty" to the *losers* of the niche, the individuals of each niche that lose the generational competition to the actual peak individuals. Following a *radius-based* procedure of identifying the winners and losers of each niche in each generation, the winners are assigned with their raw-fitness values, whereas all the other individuals are assigned with *zero* fitness (*maximization* was assumed). This is called the *clearing phase*. The selection phase, typically based on the RWS operator, considers, *de facto*, only the winners of the different niches. The allowed number of winners per niche, also referred to as the *niche capacity*, is a control parameter that reflects the degree of elitism. In any case, as in the previous techniques, the peak individuals are never fixated, and are subject to the probabilistic selection of the GA. This method was shown to outperform the *fitness sharing* technique on a specific set of low-dimensional test problems (Petrowski 1996).

Algorithm 1 Dynamic Peak Identification
input: population *Pop*, number of niches *q*, niche radius *ρ*
1: Sort *Pop* in increasing fitness order {*minimization*}
2: *i* := 1
3: *NumPeaks* := 0
4: *DPS* := ∅ {Set of peak elements in population}
5: **while** *NumPeaks* ≠ *q* and *i* ≤ *popSize* **do**
6:     **if** *Pop*[*i*] is not within sphere of radius *ρ* around peak in *DPS* **then**
7:        *DPS* := *DPS* ∪ {*Pop*[*i*]}
8:        *NumPeaks* := *NumPeaks* + 1
9:     **end if**
10: *i* := *i* + 1
11: **end while**
output: *DPS*

4.1.4 Crowding

Crowding was one of the pioneering methods in this field, as introduced by De Jong (1975). It considered, and to some extent generalized, *preselection schemes*, which had been investigated in the doctoral dissertation of Cavicchio (1970). The latter had showed that certain preselection schemes boosted the preservation of population diversity. The *crowding* approach aimed at reducing changes in the population distribution between generations, in order to prevent *premature convergence*, by means of *restricted replacement*. Next, we will describe the method in more detail.

Given the traditional GA, a proportion G of the population is selected in each generation via fitness-proportionate selection to undergo variations (i.e., *crossover* and *mutation*) – out of which a part is chosen to die and to be replaced by the new offspring. Each offspring finds the individuals it replaces by taking a random sample of CF (referred to as **crowding factor**) individuals from the population, and replacing the **most similar individual** from the sample. An appropriate *similarity metric* should be chosen.

The crucial point of this niching mechanism is the calculation of the so-called *crowding distance* **between parents and offspring,** in order to control the *change rate* between generations. A different use of the *crowding distance*, applied among individuals of the same generation and assigned with reversed ranking, is widely encountered in the context of evolutionary multiobjective optimization (EMOA) (Deb 2001; Coello Coello et al. 2007). In the joint context of niching and EMOA see also Deb's "omni-optimizer" (Deb and Tiwari 2005).

Mahfoud, who analyzed the *crowding* niching technique (1995a), concluded that it was subject to disruptive effects, mainly *drift*, which prevented it from maintaining more than two peaks. He then proposed a mechanism called *deterministic crowding*, as an improvement to the original *crowding* scheme. The proposed procedure applies variation operators to pairs of individuals in order to generate their offspring, which are all then evaluated with respect to the crowding distance, and undergo *replacement selection* (see ❷ *Algorithm 2*, which assumes *maximization*).

4.1.5 Clustering

The application of *clustering* for niching is very intuitive from the computational perspective, as well as straightforward in its implementation. Yin et al. (1993) proposed a clustering

---

Algorithm 2 Deterministic Crowding: Replacement Selection (*maximization*)
1: Select two parents, p_1 and p_2, randomly, without replacement
2: Generate two variations, c_1 and c_2
3: **if** $d(p_1, c_1) + d(p_2, c_2) \leq d(p_1, c_2) + d(p_2, c_1)$ **then**
4:      **if** $f(c_1) > f(p_1)$ **then** replace p_1 with c_1
5:      **if** $f(c_2) > f(p_2)$ **then** replace p_2 with c_2
6: **else**
7:      **if** $f(c_2) > f(p_1)$ **then** replace p_1 with c_2
8:      **if** $f(c_1) > f(p_2)$ **then** replace p_2 with c_1
9: **end if**

framework for niching with GAs, which we describe here briefly. A clustering algorithm, such as the *K-Means* algorithm (Haykin 1999), first partitions the population into niches, and then considers the *centroids*, or center points of mass, of the newly partitioned subpopulations.

Let d_{ic} denote the distance between individual i and its *centroid*, and let f_i denote the raw fitness of individual i. Assuming that there are n_c individuals in the niche of individual i, its fitness is defined as:

$$f_i^{\text{Clustering}} = \frac{f_i}{n_c \cdot (1 - (d_{ic}/2d_{\max})^\alpha)} \tag{9}$$

where $d_{\max}$ is the maximal distance allowed between an individual and its niche centroid, and α is a defining parameter. It should be noted that the clustering algorithm uses an additional parameter, $d_{\min}$, for determining the minimal distance allowed between centroids, playing an equivalent role to the *niche radius* ρ of the *sharing*-based schemes.

This method is often subject to criticism for its strong dependency on a relatively large number of parameters. However, this *clustering* technique has become a popular kernel for niching with EAs, and its application was reported in various studies (see, e.g., Schönemann et al. 2004, Hanagandi and Nikolaou 1998, Branke 2001, Gan and Warwick 2001, Aichholzer et al. 2000, Streichert et al. 2003, and Ando et al. 2005).

4.1.6 The Sequential Niche Technique

A straightforward approach of *iteration* can be used to sequentially locate multiple peaks in the landscape, by means of an *iterative local search* (Ramalhinho-Lourenco et al. 2000). This procedure is blind to any information gathered in previous searches, and sequentially restarts stochastic search processes, hoping to hit a different peak every run. Obviously, it is likely to encounter *redundancy*, and the number of expected iterations is then increased by a factor. A **redundancy factor** can be estimated if the peaks are of equal height (equi-fitness landscape), that is, the probability to converge into any of the q peaks is equal to $1/q$:

$$R = \sum_{i=1}^{q} \frac{1}{i}$$

For $q > 3$, this can be approximated by:

$$R \approx \gamma + \ln(q) \tag{10}$$

where $\gamma \approx 0.577$ is the Euler–Mascheroni constant. This *redundancy factor* remains reasonably low for any practical value of q, but is expected to considerably increase if all optima are not likely to be found equal.

On a related note, we would like to mention a multirestart with an increasing population size approach, that was developed with the CMA-ES algorithm (Auger and Hansen 2005a). The latter aims at attaining the global minimum, while possibly visiting local minima along the process and restarting the algorithm with a larger population size and a modified initial step-size. It is not defined as a niching technique and does not target optima other than the global minimum, but it can capture suboptimal minima during its search.

Beasley et al. extended the naive *iteration* approach, and developed the so-called *Sequential Niche* technique (Beasley et al. 1993). This method, in contrast to the other niching methods presented earlier, does not modify the genetic operators nor any characteristics of the traditional GA, but rather creates a general search framework suitable for locating multiple solutions. By means of this method the search process turns into a sequence of independent runs of the traditional GA, where the basic idea is to suppress the fitness function at the observed optimum that was obtained in each run, in order to prevent the search from revisiting that optimum.

In further detail, the traditional GA is run multiple times sequentially: given the best solution of each run, it is first stored as a possible final solution and, second, the fitness function is artificially suppressed in all the points within the neighborhood of that optimum up to a desired radius. This modification is done immediately after each run. Its purpose is to discourage the following runs from revisiting these optima, and by that to encourage the exploration of other areas of the search landscape – aiming at obtaining all its optima. It should be noted that each function modification might yield artificial discontinuities in the fitness landscape. This method focuses only on locating multiple optima of the given search problem, without considering the concepts of parallel evolution and formation of subpopulations. In that sense, it has been claimed that it could not be considered as a niching method, but rather as a modified iterated search.

4.1.7 The Islands Model

This is probably the most intuitive niching approach from the biological perspective, directly inspired by organic evolution. Also referred to as the *regional population model*, this approach (see, e.g., Grosso 1985; Adamidis 1994; Martin et al. 1997) simulates the evolution of subpopulations on remote computational units (independent processors), aiming at achieving a speciation effect by **monitoring the gene flow**. The population is divided into multiple subpopulations, which evolve independently for a fixed number of generations, called *isolation period*. This is followed by a phase of controlled gene flow, or *migration*, when a portion of each subpopulation migrates to other nodes.

The genetic diversity and the amount of information exchange between subpopulations are determined by the following parameters – the number of exchanged individuals, the *migration rate*, the selection method of the individuals for migration (uniformly at random, or elitist fitness-based approach), and the scheme of migration, for example, complete net topology, ring topology, or neighborhood topology.

4.1.8 Other GA-Based Methods

Tagging (see, e.g., Spears 1994 and Deb and Spears 1997) is a mechanism that aims at improving the distance-based methods of *fitness sharing* and *crowding*, by labeling individuals with tag-bits. Rather than carrying out distance calculations, the tag-bits are employed for identifying the subpopulations, enforcing *mating restrictions*, and then implementing the *fitness sharing* mechanism. An individual is classified to a subpopulation by its genetic inheritance, so to speak, which is subject to generational variations, rather than by its actual spatial state. This concept simplifies the classification process, and obviously reduces the computational costs per generation, and at the same time it introduces a new bio-inspired approach into niching: individuals belong to a species because their parents did, and not because they are currently adjacent to a "peak individual," for instance. This technique was shown in Spears (1994) to be a rather efficient implementation of the *sharing* concept.

A complex subpopulation differentiation model, the so-called **multinational evolutionary algorithm**, was presented by Ursem (1999). This original technique considers a world of *"nations," "governments,"* and *"politicians,"* with dynamics dictated by migration of individuals, merging of subpopulations, and selection. Additionally, it introduces a topology-based auxiliary mechanism of *sampling*, which detects whether feasible solutions share the same basin of attraction. Due to the *curse of dimensionality*, this sampling-based mechanism is expected to lose its efficiency in high-dimensional landscapes.

Stoean et al. (2005) constructed the so-called **elitist generational genetic chromodynamics algorithm**. The idea behind this radius-based technique was the definition of a *mating region*, a *replacement region*, and a *merging region* — with appropriate mating, replacement, and merging radii — which dictates the dynamics of the genetic operations.

4.1.9 Miscellaneous: Mating Schemes

It has been observed that once the niche formation process starts, that is, when the population converges into the multiple basins of the landscape, crossbreeding between different niches is likely to fail in producing good offspring. In biological terms, this is the elimination of the divergence, by means of *hybridization*, in the **secondary contact phase**, as discussed in ❯ Sect. 2.2.

Deb and Goldberg (1989) proposed a so-called *mating restriction scheme*, which poses a limitation on the choice of partners in the reproduction phase and prevents recombination between competing niches. They employed a distance measure, subject to a distance threshold, which was set to the niche radius, and showed that it could be used to improve the *fitness sharing* algorithm.

Mahfoud (1995a) proved that the mating restriction scheme of Deb and Goldberg was not sufficient, *per se*, in maintaining the population diversity in GA niching. A different approach of Smith and Bonacina (2003), however, considered an evolutionary computation multi-agent system, as opposed to the traditional *centralized* EA, and did manage to show that the same mating restriction scheme in an agent-based framework was capable of maintaining diversity and converging with stability into the desired peaks.

From the biological perspective, the mating restriction scheme is obviously equivalent to keeping the geographical isolation, or the barrier to gene flow, in order to allow the completion of the speciation phase. As discussed earlier, the geographical element in organic evolution is the crucial component which creates the conditions for speciation, and it is not surprising that artificial niching techniques choose to enforce it, by means of mechanisms such as the niche radius or the mating restriction scheme.

4.2 ES Niching Methods

Researchers in the field of evolution strategies initially showed no particular interest in the topic of niching, leaving it essentially for genetic algorithms. An exception would be the employment of island models. Generally speaking, classical niching schemes such as *fitness sharing*, which redefine the selection mechanism, are likely to interfere with the core of evolution strategies – the *self-adaptation mechanism* – and thus doomed to fail in a straightforward implementation. Any manipulation of the fitness value is usually not suitable for evolution strategies, as in the case of constraints handling: death penalty is typically the chosen approach for a violation of a constraint in ES, rather than a continuous penalty as used in other EAs, in order to avoid the introduction of disruptive effects to the self-adaptation mechanism (see, e.g., Coello Coello 1999; Kramer and Schwefel 2006). Therefore, niching with evolution strategies would have to be addressed from a different direction. Moreover, the different nature of the ES dynamics, throughout the *deterministic selection* and the *mutation operator*, suggests that an alternative treatment is required here.

There are several, relatively new, niching methods that have been proposed within ES, mostly clustering-based (Schönemann et al. 2004; Aichholzer et al. 2000; Streichert et al. 2003). In addition, niching was also introduced to the mixed-integer ES framework (Li et al. 2008). A different approach, based on derandomized evolution strategies (DES) (for the latter see, e.g., Ostermeier et al. (1993, 1994) and Hansen et al. (1995)), was presented by Shir and Bäck (2008). One of its variants, which employs the popular covariance matrix adaptation evolution strategy (CMA-ES), will receive special attention here, and will be set as a detailed case study of niching techniques, to be outlined in the following section.

4.2.1 Case Study: Niching with CMA-ES

A niching framework for $(1 \overset{+}{,} \lambda)$, derandomized ES (DES) kernels subject to a fixed niche radius has been introduced recently (see, e.g., Shir and Bäck 2008). Following the *mission statement* presented earlier, the aim was the construction of a generic niching framework, which offers the combination of population diversity preservation and local search capabilities. Thus, DES were considered as an excellent choice for that purpose, as EAs with local search characteristics. Furthermore, DES typically employ small populations, which was shown to be a potential advantage for a niching technique, as it can boost the speciation effect (Shir 2008).

The Covariance Matrix Adaptation Evolution Strategy (CMA-ES)

The CMA-ES (Hansen and Ostermeier 2001), is a DES variant that has been successful in treating correlations among object variables by efficiently learning matching mutation distributions. Explicitly, given an initial search point $\mathbf{x}^{(0)}$, λ offspring are generated by means of normally distributed variations:

$$\mathbf{x}^{(g+1)} \sim \mathcal{N}\left(\langle \mathbf{x} \rangle_W^{(g)}, \sigma^{(g)^2} \mathbf{C}^{(g)}\right) \qquad (11)$$

Here, $\mathcal{N}(\mathbf{m}, \mathbf{C})$ denotes a normally distributed random vector with mean $\mathbf{m}$ and a covariance matrix $\mathbf{C}$. The best μ search points out of these λ offspring undergo weighted recombination and become the parent of the next generation, denoted by $\langle \mathbf{x} \rangle_W$. The covariance matrix $\mathbf{C}$ is initialized as the *unity matrix* and is learned during the course of evolution, based on cumulative information of successful past mutations (the *evolution path*). The global step-size, $\sigma^{(g)}$, is updated based on information extracted from *principal component analysis* of $\mathbf{C}^{(g)}$ (the *conjugate evolution path*). For more details, we refer the reader to Hansen and Ostermeier (2001).

An elitist sibling to the CMA comma strategy was also introduced (Igel et al. 2006), based upon the classical $(1+1)$-ES (Bäck 1996).

A Detailed Description of the Algorithm

This niching technique is based upon interacting search processes, which simultaneously perform a derandomized $(1,\lambda)$ or $(1+\lambda)$ search in different locations of the space. In case of multimodal landscapes these search processes are meant to explore different attractor basins of local optima.

An important point in this approach is to strictly enforce the fixed allocation of the population resources, that is, number of offspring per niche. The idea is thus to prevent a scenario of a take-over, where a subpopulation located at a fitter optimum can generate more offspring. The biological idea behind this fixed allocation of resources stems from the concept of limited *hosting capacities* of given ecological niches, as previously discussed.

The *speciation interaction* occurs every generation when all the offspring are considered together to become niches' representatives for the following iteration, or simply the next search points, based on the rank of their fitness and their location with respect to higher-ranked individuals. The focus here is on a simple framework without recombination ($\mu = 1$).

Given q, the estimated/expected number of peaks, $q + p$ "D-sets" are initialized, where a D-set is defined as the collection of all the dynamically adapted strategy as well as decision parameters of the CMA algorithm, which uniquely define the search at a given point of time. (A D-set originally referred to the *derandomized* set of strategy parameters. When the CMA kernel is in use, it is sometimes referred to in the literature as a CMA-set.) These parameters are the current search point, the covariance matrix, the step-size, as well as other auxiliary parameters. At every point in time the algorithm stores exactly $q + p$ D-sets, which are associated with $q + p$ search points: q for the peaks and p for the "non-peaks domain." The $(q + 1)th \ldots (q + p)th$ D-sets are associated with individuals, which are randomly re-generated every *epoch*, that is, a cycle of κ generations, as potential candidates for niche formation. This is basically a *quasi-restart* mechanism, which allows new niches to form dynamically. Setting the value of p should reflect the trade-off between applying a wide restart approach for exploring further the search space to exploiting computational resources for the existing niches. In any case, due to the *curse of dimensionality*, p loses its significance as the dimension of the search space increases.

Until the stopping criterion is met, the following procedure takes place. Each search point samples λ offspring, based on its evolving D-set. After the fitness evaluation of the new $\lambda \cdot (q + p)$ individuals, the classification into niches of the entire population is obtained in a *greedy* fashion, by means of the DPI routine (Miller and Shaw 1996) (❷ *Algorithm 1*). The latter is based on the fixed niche radius ρ. The peaks then become the new search points, while their D-sets are inherited from their parents and updated according to the CMA defining equations.

A pseudocode for a single iteration in this *niching routine* is presented as ❷ *Algorithm 3*.

Natural Interpretation: Alpha-Males Competition

We would like to point out the nature of the subpopulations dynamics in the aforementioned niching scheme. Due to the *greedy* classification to niches, which is carried out every generation, some niches can merge in principle, while all the individuals, except for the *peak individual*, die out in practice. Following the posed principle of fixed resources per niche, only the peak individual will be sampled λ times in the following generation. In socio-biological terms, the peak individual could be then associated with an **alpha-male**, which wins the local competition and gets all the sexual resources of its ecological niche. The algorithm as a whole can be thus considered as a competition between $q + p$ alpha-males, each of which is fighting for one of the available q "computational resources," after winning its local competition at the "ecological optimum" site. The domination battles take place locally every cycle, as dictated by the DPI scheme. An elitist-CMA kernel will allow aging alpha-males to keep participating in the ongoing competitions, whereas a comma strategy will force their replacement by fresh

---

Algorithm 3 (1 + λ)-CMA-ES Niching with Fixed Niche Radius (A Single Iteration)

1: **for** $i = 1...(q + p)$ search points **do**
2:   Generate λ samples based on the D-set of i
3: **end for**
4: Evaluate fitness of the population
5: Compute the Dynamic Peak Set with the DPI Algorithm
6: **for all** elements of *DPS* **do**
7:   Set peak as a search point
8:   Inherit the D-set and update it respectively
9: **end for**
10: **if** N_{DPS}=size of *DPS* $< q$ **then**
11:   Generate $q - N_{DPS}$ new search points, reset D-sets
12: **end if**
13: **if** *gen* mod $\kappa \equiv 0$ **then**
14:   Reset the $(q + 1)th...(q + p)th$ search points
15: **end if**

---

blood of new alpha-males. At the global level, the value of p determines the selection pressure of alpha-males; setting $p = 0$ will then eliminate global competition and will grant automatically λ offspring to each alpha-male in the following generation, respectively.

4.3   Niche-Radius Calculation

The original formula for ρ for *phenotypic sharing* in GAs was derived by Deb and Goldberg (1989). Analogously, by considering the decision parameters as the decoded parameter space of the GA, the same formula can be applied, using the Euclidean metric, to ES. Given q, the number of target peaks in the solution space, every niche is considered to be surrounded by an n-dimensional hypersphere with radius ρ, which occupies $\frac{1}{q}$ of the entire volume of the space. The volume of the hypersphere which contains the entire space is

$$V = cr^n \tag{12}$$

where c is a constant, given explicitly by:

$$c = \frac{\pi^{\frac{n}{2}}}{\Gamma(\frac{n}{2} + 1)} \tag{13}$$

with $\Gamma(n)$ as the Gamma function. Given lower and upper bound values $x_{k,min}$, $x_{k,max}$ of each coordinate in the decision parameters space, r is defined as follows:

$$r = \frac{1}{2}\sqrt{\sum_{k=1}^{n}(x_{k,max} - x_{k,min})^2} \tag{14}$$

If we divide the volume into q parts, we may write

$$c\rho^n = \frac{1}{q}cr^n \tag{15}$$

which yields

$$\rho = \frac{r}{\sqrt[n]{q}} \qquad (16)$$

Hence, by applying this niche radius approach, two assumptions are made:

1. The number of target peaks, q, is given or can be estimated.
2. All peaks are at least at distance 2ρ from each other, where ρ is the fixed radius of every niche.

5 Experimental Methodology

Since the topic of niching has not drawn considerable attention from the mainstream EC community, there have been no constructive attempts to generate a generalized experimental framework, for testing niching methods to be agreed upon. This section will focus on that, and propose an experimental methodology for EA niching algorithms. It will present a suite of synthetic test functions, discuss possible performance criteria, and conclude by revisiting the niching-CMA algorithm and some of its experimental observations.

In the broad context of function optimization, EA niching techniques are obviously within the general framework of stochastic algorithms, and as such should be treated carefully upon reporting their experiments. We recommend following Bartz-Beielstein (2006) when conducting experiments, and especially following the 7-points scheme of Preuss (2007) when reporting them.

5.1 Multimodal Test Functions

The choice of a numerical testbed for evaluating the performance of search or optimization methods is certainly one of the core issues among the scholars in the community of algorithms and operations research.

In a benchmark article, Whitley et al. (1996) criticized the commonly tested artificial landscapes in the evolutionary algorithms community, and offered general guidelines for constructing test problems. A remarkable effort was made almost a decade after that document, when a large group of scholars in the EC community joined their efforts and compiled an agreed test suite of single-objective artificial landscapes (Suganthan et al. 2005), to be tested in an open performance competition reported at the 2005 IEEE Congress on Evolutionary Computation (CEC) (Auger and Hansen 2005b). The latter also included multimodal functions.

The issue of developing a multimodal test suite received even less attention, likely due to historical reasons. Since multimodal domains were mainly treated by GA-based niching methods, their corresponding test suites were limited to low-dimensional continuous landscapes, typically with two decision parameters to be optimized ($n = 2$) (see, e.g., Goldberg and Richardson 1987; Mahfoud 1995a).

When compiling our proposed test suite, we aim at following Whitley's guidelines, and to include some traditional GA-niching test functions as well as functions from the 2005 CEC inventory (Suganthan et al. 2005). Some of the landscapes have symmetric or equal distributions of minima, and some do not. Some of the functions are *separable*, that is, they can be optimized by solving n 1-dimensional problems separately (Whitley et al. 1996), while some of them are non-separable.

◘ **Table 1**

Test functions to be *minimized* and initialization domains. For some of the non-separable functions, we apply translation and rotation: $y = \mathcal{O}\,(x - r)$ where $\mathcal{O}$ is an orthogonal rotation matrix, and r is a shifting vector

Separable:			
Name	**Function**	**Init**	**Niches**
$\mathcal{M}$	$\mathcal{M}(\mathbf{x}) = -\frac{1}{n}\sum_{i=1}^{n} \sin^{\alpha}(5\pi x_i)$	$[0,1]^n$	100
$\mathcal{A}$ [Ackley]	$\mathcal{A}(\mathbf{x}) = -c_1 \cdot \exp\left(-c_2\sqrt{\frac{1}{n}\sum_{i=1}^{n} x_i^2}\right)$ $- \exp\left(\frac{1}{n}\sum_{i=1}^{n} \cos(c_3 x_i)\right) + c_1 + e$	$[-10,10]^n$	$2n+1$
$\mathcal{L}$	$\mathcal{L}(\mathbf{x}) = -\prod_{i=1}^{n} \sin^{k}(l_1\pi x_i + l_2) \cdot \exp\left(-l_3\left(\frac{x_i-l_4}{l_5}\right)^2\right)$	$[0,1]^n$	$n+1$
$\mathcal{R}$ [Rastrigin]	$\mathcal{R}(\mathbf{x}) = 10n + \sum_{i=1}^{n}\left(x_i^2 - 10\cos(2\pi x_i)\right)$	$[-1,5]^n$	$n+1$
$\mathcal{G}$ [Griewank]	$\mathcal{G}(\mathbf{x}) = 1 + \sum_{i=1}^{n}\frac{x_i^2}{4000} - \prod_{i=1}^{n}\cos\left(\frac{x_i}{\sqrt{i}}\right)$	$[-10,10]^n$	5
$\mathcal{S}$ [Shekel]	$\mathcal{S}(\mathbf{x}) = -\sum_{i=1}^{10}\frac{1}{k_i(\mathbf{x}-a_i)(\mathbf{x}-a_i)^T+c_i}$	$[0,10]^n$	8
$\mathcal{V}$ [Vincent]	$\mathcal{V}(\mathbf{x}) = -\frac{1}{n}\sum_{i=1}^{n}\sin(10 \cdot \log(x_i))$	$[0.25,10]^n$	50
Non-separable:			
Name	**Function**	**Init**	**Niches**
$\mathcal{F}$ [Fletcher-Powell]	$\mathcal{F}(\mathbf{x}) = \sum_{i=1}^{n}(A_i - B_i)^2$ $A_i = \sum_{j=1}^{n}\left(a_{ij} \cdot \sin(\alpha_j) + b_{ij} \cdot \cos(\alpha_j)\right)$ $B_i = \sum_{j=1}^{n}\left(a_{ij} \cdot \sin(x_j) + b_{ij} \cdot \cos(x_j)\right)$ $a_{ij}, b_{ij} \in [-100,100]; \quad \alpha \in [-\pi,\pi]^n$	$[-\pi,\pi]^n$	10
$\mathcal{R}_{SR}$ [S.R. Rastrigin]	$\mathcal{R}_{SR}(\mathbf{x}) = 10n + \sum_{i=1}^{n}\left(y_i^2 - 10\cos(2\pi y_i)\right)$	$[-5,5]^n$	$n+1$
$\mathcal{G}_{SR}$ [S.R. Griewank]	$\mathcal{G}_{SR}(\mathbf{x}) = 1 + \sum_{i=1}^{n}\frac{y_i^2}{4000} - \prod_{i=1}^{n}\cos\left(\frac{y_i}{\sqrt{i}}\right)$	$[0,600]^n$	5

We propose for consideration a set of multimodal test functions, as indicated in ❷ *Table 1*. The table summarizes the proposed unconstrained multimodal test functions as well as their initialization intervals and the number of target niches. Next, we provide the reader with an elaborate description of the test functions, corresponding to the notation of ❷ *Table 1*:

- $\mathcal{M}$ is a basic hypergrid multimodal function with uniformly distributed minima of equal function value of -1. It is meant to test the stability of a particularly large number of niches: in the interval $[0,1]^n$ it has 5^n minima. We set $\alpha = 6$.
- The well-known Ackley function has one global minimum, regardless of its dimension n, which is surrounded isotropically by $2n$ local minima in the first hypersphere, followed by an exponentially increasing number of minima in successive hyperspheres. Ackley's function has been widely investigated in the context of *evolutionary computation* (see, e.g., Bäck 1996). We set $c_1 = 20$, $c_2 = 0.2$, and $c_3 = 2\pi$.

- $\mathscr{L}$ – also known as *F2*, as originally introduced by Goldberg and Richardson (1987) – is a sinusoid trapped in an exponential envelope. The parameter k determines the sharpness of the peaks in the function landscape; We set it to $k = 6$. $\mathscr{L}$ has one global minimum, regardless of n and k. It has been a popular test function for GA niching methods. We set $l_1 = 5.1$, $l_2 = 0.5$, $l_3 = 4 \cdot \ln(2)$, $l_4 = 0.0667$ and $l_5 = 0.64$.
- The Rastrigin function (Törn and Zilinskas 1987) has one global minimum, surrounded by a large number of local minima arranged in a lattice configuration.
  We also propose its shifted-rotated variant (Suganthan et al. 2005), with a linear transformation matrix of condition number 2 as the rotation operator.
- The Griewank function (Törn and Zilinskas 1987) has its global minimum ($f^* = 0$) at the origin, with several thousand local minima in the area of interest. There are four suboptimal minima $f \approx 0.0074$ with $\mathbf{x}^* \approx \left(\pm\pi, \pm\pi\sqrt{2}, 0, 0, 0, \ldots 0\right)$.
  We also propose its shifted-rotated variant (Suganthan et al. 2005), with a linear transformation matrix of condition number 3 as the rotation operator.
- The Vincent function is a sine function with a decreasing frequency. It has 6^n global minima in the interval $[0.25, 10]^n$.
- The Shekel function, suggested by Törn and Zilinskas (1987), introduces a landscape with a dramatically uneven spread of minima. It has one global minimum, and seven ordered local minima.
- The function after Fletcher and Powell (Bäck 1996) is a non-separable *nonlinear parameter estimation problem*, which has a nonuniform distribution of 2^n minima. It has non-isotropic attractor basins.

5.2 Performance Criteria

The performance criteria of niching methods include numerous possible attractive options and, generally speaking, vary in different experimental reports. The common ground is typically the examination of the population in its final stage, while aiming to observe the algorithm's ability to capture as well as *maintain* peaks of high-quality. Stability of good niches is thus considered as one of the implicit criteria.

Studies of traditional GA niching methods had been strongly interested in the distribution of the final population compared to a goal-distribution, as formalized by Mahfoud (1995a). While Mahfoud's formalism introduced a generic theoretical tool, being derived from information theory, other studies considered, *de facto*, specific performance calculations. For example, a very popular niching performance measurement, which satisfies Mahfoud formalism's criteria, is the *Chi-square-like performance statistic* (see, e.g., Deb and Goldberg 1989). The latter estimates the deviation of the actual distribution of individuals N_i from an ideal distribution (characterized by mean μ_i and variance σ_i^2) in all the $i = 1 \ldots q + 1$ subspaces (q peak subspaces and the non-peak subspace):

$$\chi^2 = \sqrt{\sum_{i=1}^{q+1} \left(\frac{N_i - \mu_i}{\sigma_i}\right)^2} \tag{17}$$

where the ideal-distribution characteristic values are derived per function.

Most of the existing studies have focused on the ability to identify global as well as local optima, and to converge in these directions through time, with no particular interest in the distribution of the population. One possible performance criterion is often defined as the *success-rate* of the niching process, which refers to the percentage of optima attained by the end of the run with respect to the target peaks, as defined *a priori*. Furthermore, as has been employed in earlier studies of GA niching (Miller and Shaw 1996), another performance criterion is called the *maximum peak ratio* (MPR) statistic. This metric measures the quality as well as the number of optima given as a final result by the evolutionary algorithm. Explicitly, assuming a *minimization problem*, given the fitness values of the subpopulations in the final population $\left\{\tilde{f}_i\right\}_{i=1}^{q}$, and the fitness values of the real optima of the objective function $\left\{\hat{\mathcal{F}}_i\right\}_{i=1}^{q}$, the *maximum peak ratio* is defined as follows:

$$\text{MPR} = \frac{\sum_{i=1}^{q}\hat{\mathcal{F}}_i}{\sum_{i=1}^{q}\tilde{f}_i} \tag{18}$$

where all values are assumed to be *strictly positive*. If this is not the case in the original parametrization of the landscape, the latter should be scaled accordingly with an additive constant for the sake of this calculation. Also, given a maximization problem, the MPR is defined as the sum of the obtained optima divided by the sum of the real optima. A drawback of this performance metric is that the real optima need to be known *a priori*. However, for many artificial test problems these can be derived analytically, or tight numerical approximations to them are available. We adopt the MPR performance criterion and recommend it for reporting experimental results of EA niching techniques.

5.2.1 Another Perspective: MPR Versus Time

Although the MPR metric was originally derived to be analyzed by means of its saturation value in order to examine the niches' stability, a new perspective was introduced by Shir and Bäck (2005a). That study investigated the MPR as a function of time, focusing on the early stages of the run, in addition to the saturation value. It was shown experimentally that the time-dependent MPR data fits a theoretical function, *the logistic curve*:

$$y(t) = \frac{a}{1 + \exp\{c(t - T)\}} \tag{19}$$

where a is the saturation value of the curve, T is its time shift, and c (in this context always negative) determines the shape of the exponential rise. This equation, known as the *logistic equation*, describes many processes in nature. All those processes share the same pattern of behavior: growth with *acceleration*, followed by *deceleration* and then a *saturation* phase. In the context of evolutionary niching methods, it was argued (Shir and Bäck 2005a) that the logistic parameters should be interpreted in the following way: T as the *learning period* of the algorithm, and the absolute value of c as its *niching formation acceleration*. a is clearly the *MPR* saturation value.

5.3 Experimental Observation Examples: Niching-CMA Revisited

We revisit the niching-CMA algorithm with three examples of experimental observations.

5.3.1 MPR Time-Dependent Analysis: Reported Observations

The MPR time-dependent analysis was applied in Shir and Bäck (2005a) to two ES-based niching techniques: niching with the CMA-ES and niching with the Standard-ES according to the Schwefel-approach (Shir and Bäck 2005b). In short, the latter method applies the same niching framework as the niching-CMA (❷ Sect. 4.2.1) except for one conceptual difference: It employs a (μ, λ) strategy in each niche, subject to *restricted mating*. Otherwise, it applies the standard ES operators (Bäck 1996).

We outline some of the conclusions of that study:

1. The ***niching formation acceleration***, expressed as the absolute value of c, had larger values for the CMA-ES kernel for all the observed test cases. That implied stronger niching acceleration and faster convergence.
2. A trend concerning the absolute value of c as a function of the dimensionality was observed: The higher the dimensionality of the search space, the lower the absolute value of c, that is, the slower the niching process.
3. The ***learning period***, expressed as the value of T in the curve fitting, has negative as well as positive values. Negative values mean that the niches' formation process, expressed as the exponential rise of the MPR, started immediately from generation zero.
4. The averaged ***saturation value*** a, that is, the MPR saturation value, was larger in all of the test cases for the CMA-ES mechanism. In that respect, the CMA kernel outperformed the standard-ES on the tested landscapes.

That study concluded with the claim that there was a clear *trade-off*: Either a long learning period followed by a high niching acceleration (CMA-ES), or a short learning period followed by a low niching acceleration (Standard-ES). A hypothesis concerning the existence of a general trade-off between the learning period T and the niching acceleration, c, was numerically assessed in a following study (Shir and Bäck 2008). It was shown that this trade-off stands for various DES niching variants on two synthetic landscapes (one separable, one non-separable) in a large spectrum of search space dimensions.

5.3.2 Performance on the Synthetic Multimodal Test-Suite

The proposed CMA-ES niching framework has been successfully applied to a suite of synthetic multimodal *high-dimensional* continuous landscapes, as reported by Shir and Bäck (2008), which in principle followed the recommended test functions proposed earlier in this chapter. That study addressed various *research questions* concerning the generalized DES niching framework, such as *which DES variant captures and maintains most desired optima* (i.e., best saturation MPR values), *what are the differences in the niching formation acceleration among the DES variants*, and others. The CMA-ES kernels were observed to perform very well among the DES variants, with the $(1+10)$ kernel performing best, while typically obtaining most of the desired basins of attraction of the various landscapes at different dimensions. Furthermore, upon carrying out behavioral analysis of the simulations, e.g., MPR analysis, some characteristic patterns for the different algorithmic kernels were revealed. For instance, it was observed that the elitist-CMA has consistently the lowest niching acceleration. A straightforward and rather intuitive explanation for the excellent behavior of the elitist-CMA variant would be its tendency to maintain convergence in any basin of attraction, versus a higher probability for the comma strategy to escape them. Moreover, another argument for the advantage of an elitist strategy for niching was suggested. The niching problem can be considered as an optimization task with constraints, that is, the formation of niches that restricts competing niches and their optimization routines from exploring the search space freely. It has been suggested in previous studies (see, e.g., Kramer and Schwefel 2006) that ES self-adaptation in constrained problems would tend to fail with a comma strategy, and thus an elitist strategy is preferable for such problems.

We choose to omit here specific details concerning the numerical simulations, and refer the reader to Shir and Bäck (2008) and Shir (2008).

5.3.3 Real-World Application: Quantum Control

As was, furthermore, reported by Shir and Bäck (2008), the proposed DES niching framework was successfully applied to a real-world landscape from the field of Quantum Control. (Symbolically, this interdisciplinary study forms a *closed natural computing circle*, where biologically oriented investigation of organic evolution and speciation helps to develop methods for solving applications in physics in general, and in quantum control in particular. By our reckoning, this symbolism is even further strengthened upon considering the stochastic nature of evolutionary algorithms. This process can be thus considered as throwing dice in order to solve quantum mechanics, sometimes referred to as the *science of dice*.) Namely, the dynamic molecular alignment problem (for a review see Shir et al. 2008). In short, the goal of this application is the maximization of the alignment of diatomic molecules after the interaction with an electric field arising from a laser source. The black-box noise-free simulator, which is designed in a lab-oriented fashion, provides a reliable physics simulation with a duration of 35 s per trial solution. The 80 object variables constitute a parametrized curve, to be learned, that undergoes spectral transformation for the construction of the electric field. This challenging high-dimensional application required the definition of a tailor-made diversity measure, due to certain invariance properties of the control function that stem from the spectral transformations. The resulting metric employed was the Euclidean distance in the second derivative space of the control function. DES niching variants were shown to perform well, and to obtain different pulse shapes of high quality, representing different conceptual designs. Also in this case, the elitist-CMA kernel was observed to perform best. While referring the reader to Shir and Bäck (2008) and Shir (2008) for more details on this particular study, we would like to conclude with a visualization of the resulting niching process. Three laser-pulse niches, obtained in a typical elitist-CMA run, are plotted for illustration in ❯ *Fig. 1* (left column): The thick line describes the attained pulse shape, while the thin line is the molecular reaction, also referred to as the *revival structure*. The right column provides the equivalent quantum pictures, in terms of population of the energy levels: The reader can observe three different modes of quantum energy-level population for the different attained niches.

6 The Niche-Radius Problem

While the motivation and usefulness of niching cast no doubt, the relaxation of assumptions and limitations concerning the hypothetical landscape is much needed if niching methods are

◘ **Fig. 1**

Experimental results for niching-CMA with a fixed niche-radius on the real-world problem of dynamic molecular alignment (quantum control). *Left column*: *alignment* and *revival-structure* of the three niches obtained by the (1+10)-CMA. *Thin red line*: alignment; *thick black line*: intensity of the laser pulse. *Right column*: Quantum picture of the solutions; a *Fourier transform* applied to the revival structures of the optimal solutions (the thin red alignment curves). The values are log scaled, and represent how high the rotational levels of the molecules are populated as a function of time. Note that the quality of the laser pulse cannot be measured in those plots.

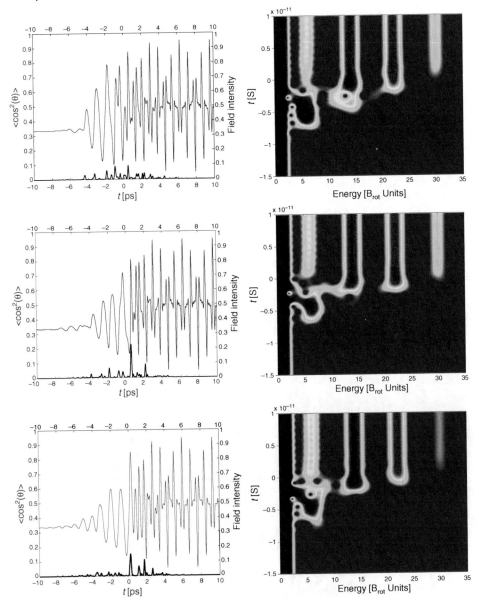

to be valid in a broader range of applications. In particular, consider the assumptions made in ❯ Sect. 4.3 concerning the fitness landscape, stating that the optima are far enough from one another with respect to the *niche radius*, which is estimated for the given problem and remains fixed during the course of evolution. Most prominently, the niche radius is used in the *sharing function*, which penalizes fitness values of individuals whose distance to other individuals is below that threshold (❯ Sect. 4.1.1). Obviously, there are landscapes for which this assumption is not applicable, and where this approach is most likely to fail (see ❯ *Figs. 2* and ❯ *3* for illustrations). This topic is directly linked to the task of defining a generic basin of attraction, which was discussed in ❯ Sect. 2.3.

6.1 Treating the Niche Radius Problem: Existing Work

There were several GA-oriented studies which addressed this so-called *niche radius problem*, aiming to relax the assumption specified earlier, or even to drop it completely. Jelasity (1998) suggested a cooling-based mechanism for the niche-radius, also known as the UEGO, which adapts the global radius as a function of time during the course of evolution. Gan and Warwick (2001) introduced the so-called dynamic niche clustering, to overcome the radius problem by using a clustering mechanism. A complex subpopulation differentiation model, the so-called multinational evolutionary algorithm, was presented by Ursem (1999). It introduces a topology-based auxiliary mechanism of sampling, which detects whether feasible solutions share the same basin of attraction. A recent study by Stoean et al. (2007) considered

■ Fig. 2

The Shekel function (see, e.g., Törn and Zilinskas 1987) in a 2D decision space: introducing a dramatically uneven spread of optima.

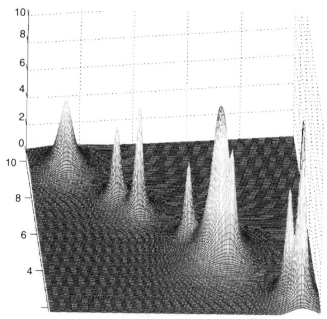

◻ Fig. 3
The Vincent function in a 2D decision space: a sine function with a decreasing frequency.

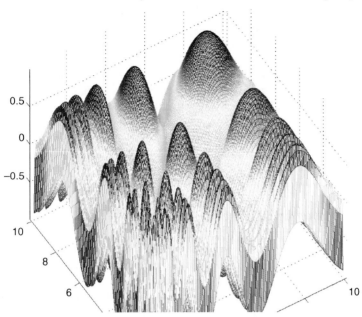

the hybridization of the latter with a radius-based niching method proposed in Stoean et al. (2005). An iterative statistical-based approach was introduced (Cioppa et al. 2004) for learning the optimal niche radius, without *a priori* knowledge of the landscape. It considers the *fitness sharing* strategy, and optimizes it as a function of the population size and the niche radius, without relaxing the landscape assumption specified earlier – that is, the niches are eventually obtained using a single fixed niche radius. Finally, two ES-based niche-radius adaptation niching algorithms were proposed recently (Shir and Bäck 2006; van der Goes 2008). They both considered individual niche radii for the evolving population members, based upon original learning schemes. While Shir and Bäck (2006) relied on coupling the niche radius to the niche's global step size in combination with a secondary selection scheme, van der Goes et al. (2008) introduced a pure self-adaptive niche-radius approach, independent of any coupling to strategy parameters. The latter considered inner and outer niche counts, and applied the original sharing function for the sake of niches classification. Both approaches were shown to successfully tackle landscapes with challenging distributions as well as shapes of attraction basins.

6.2 Learning Niche Shapes: Exploiting CMA Information

As an extension to the previously discussed niching-CMA case study (◐ Sect. 4.2.1), a self-adaptive niche-shaping approach was derived in a recent study (Shir et al. 2010). The latter introduced the *Mahalanobis distance metric* into the niching mechanism, aiming to allow a more accurate spatial classification of niches by considering rotatable ellipsoid shapes as classification regions. The shapes of these ellipsoids were obtained from the covariance matrix of the evolving multivariate normal distribution adapted by the CMA mutation scheme, replacing the default classification by means of Euclidean hyperspheres. This approach was implemented in a straightforward manner using the Mahalanobis metric as a replacement for the Euclidean, due to the fact that the former metric, the hyperspheres of which are ellipsoids when viewed in the Euclidean space, is parametrized by a covariance matrix that is obtained without additional cost from the CMA. The proposed approach was tested on high-dimensional artificial landscapes at several levels of difficulty, and was shown to be robust and to achieve satisfying results. ❷ *Figure 4* provides an illustration for the numerical results of this self-adaptive approach. It presents a snapshot gallery of the niching algorithm with the elitist-CMA kernel, employing the Mahalanobis distance, performing on the 2-dimensional Fletcher-Powell landscape.

7 Discussion and Outlook

Niching studies, following somehow various *mission statements*, introduce a large variety of approaches, some of which are more biologically inspired, whereas others are multimodal-optimization oriented. In both cases, those techniques were mainly tested on *low-dimensional synthetic landscapes*, and the application of these methods to real-world landscapes was hardly ever reported to date. We claim that niching methods should be implemented also for attaining multiple solutions in high-dimensional real-world problems, serving decision makers by providing them with the choice of optimal solutions, and representing well evolutionary algorithms in multimodal domains. By our reckoning, the *multimodal front* of real-world applications, that is, multimodal real-world problems, which pose the demand for multiple optimal solutions, should also enjoy the powerful capabilities of evolutionary algorithms, as other fronts do, for example, multi-objective or constrained domains.

On a different note, Preuss, in an important paper (Preuss 2006), raised the question: *"Under what conditions can niching techniques be faster than iterated local search algorithms?"* Upon considering a simplified model, and assuming the existence of an efficient basin identification method, he managed to show that it does pay off to employ EA niching techniques on landscapes whose basins of attraction vary significantly in size. However, the original question in its general form remained open. Mahfoud (1995b) drew a comparison of *parallel* versus *sequential* niching methods, while considering *fitness sharing, deterministic crowding, sequential niching*, and *parallel hillclimbing*. Generally speaking, he concluded that parallel niching GAs outperform parallel hillclimbers on a hard set of problems, and that *sequential niching* is always outperformed by the parallel approaches.

Obviously, there is *no free lunch*, and there is no best technique, especially in the domain of multimodal search spaces. In this respect, *local search* capabilities should not be underestimated, and *population diversity preservers* should not be overestimated. We claim that like any other complex component in organic as well as artificial systems, the success of niching is about the subtle interplay between the different, sometime conflicting, driving effects.

In this chapter, we have introduced *niching* as an algorithmic framework following the evolutionary process of *speciation*. Upon providing the practical motivation for such a framework, in terms of conceptual designs for better decision making, we outlined in detail the essential biological background, as well as the computational perspective of multimodal function

◻ **Fig. 4**

A snapshot gallery: the adaptation of the *classification-ellipses*, subject to the Mahalanobis metric with the updating covariance matrix, in the elitist-CMA strategy for a 2D Fletcher-Powell problem. Images are taken in the box $[-\pi, \pi]^2$. Contours of the landscape are given as the background, where the *X*s indicate the real optima, the dots are the evolving individuals, and the ellipses are plotted centered about the peak individual. A snapshot is taken every four generations (i.e., every 160 function evaluations), as indicated by the counter.

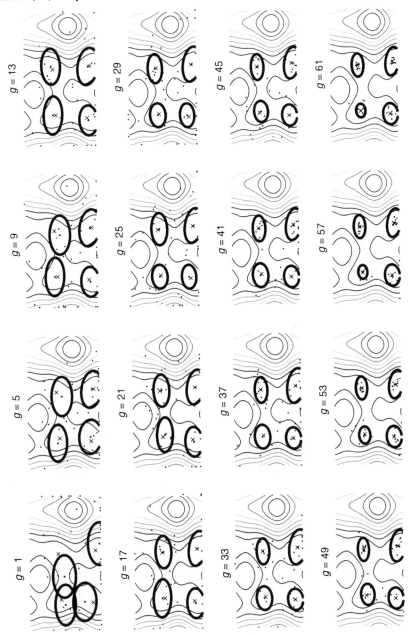

optimization. This was followed by a survey of existing EA niching techniques, mainly from the GA field, and a detailed description of a specific case-study technique, based on the CMA-ES. We highlighted the *natural computing* aspects of these techniques, especially the biologically oriented components in each scheme, e.g., hosting capacity, gene flow, alpha-males, etc. We then proposed an experimental framework for testing niching methods, and presented some experimental observations of the CMA-ES case-study technique, including results from a real-world problem. We would like to use this opportunity to encourage other scholars to apply the proposed experimental framework, and at the same time to consider ways to join forces for the construction of a general framework to be agreed upon. Finally, we discussed the important topic of the niche-radius problem, gave an overview of existing methods to treat it, and revisited the CMA-ES case study for an extended self-adaptive scheme, which employs the Mahalanobis metric for obtaining niches with more complex geometrical shapes. We conclude that even though the assumptions made for radius-based niching techniques are problematic, there are good proposed solutions that treat the problem at different levels.

We would like to propose possible directions for future research in this domain:

- Transferring existing niching algorithms into additional real-world applications in general, and into experimental optimization in particular
- Proceeding with the effort to tackle the niche radius problem, in order to develop state-of-the-art niching techniques, which are not subject to the niche radius assumptions
- Extending the study of niching to environments with uncertainty
- Developing theoretical frameworks for the investigation of niching, for example, by means of simplified modeling

References

Adamidis P (1994) Review of parallel genetic algorithms bibliography. Tech. rep., Automation and Robotics Lab., Dept. of Electrical and Computer Eng., Aristotle University of Thessaloniki, Greece

Aichholzer O, Aurenhammer F, Brandtstätter B, Ebner T, Krasser H, Magele C (2000) Niching evolution strategy with cluster algorithms. In: Proceedings of the 9th biennial IEEE conference on electromagnetic field computations. IEEE Press, New York

Ando S, Sakuma J, Kobayashi S (2005) Adaptive isolation model using data clustering for multimodal function optimization. In: Proceedings of the 2005 conference on genetic and evolutionary computation, GECCO 2005. ACM, New York, pp 1417–1424

Angus D (2006) Niching for population-based ant colony optimization. In: Second international conference on e-science and grid technologies (e-science 2006), December 4–6, 2006, Amsterdam, The Netherlands, IEEE Computer Society, p 115

Auger A, Hansen N (2005a) A restart CMA evolution strategy with increasing population size. In: Proceedings of the 2005 congress on evolutionary computation CEC 2005. IEEE Press, Piscataway, NJ, pp 1769–1776

Auger A, Hansen N (2005b) Performance evaluation of an advanced local search evolutionary algorithm. In: Proceedings of the 2005 congress on evolutionary computation CEC 2005. IEEE Press, Piscataway, NJ, pp 1777–1784

Avigad G, Moshaiov A, Brauner N (2004) Concept-based interactive brainstorming in engineering design. J Adv Comput Intell Intell Informatics 8(5): 454–459

Avigad G, Moshaiov A, Brauner N (2005) Interactive concept-based search using MOEA: the hierarchical preferences case. Int J Comput Intell 2 (3):182–191

Bäck T (1994) Selective pressure in evolutionary algorithms: a characterization of selection mechanisms. In: Michalewicz Z, Schaffer JD, Schwefel HP, Fogel DB, Kitano H (eds) Proceedings of the first IEEE conference on evolutionary computation (ICEC'94). Orlando FL. IEEE Press, Piscataway, NJ, pp 57–62

Bäck T (1996) Evolutionary algorithms in theory and practice. Oxford University Press, New York

Bartz-Beielstein T (2006) Experimental research in evolutionary computation – the new experimentalism. Natural computing series. Springer, Berlin

Beasley D, Bull DR, Martin RR (1993) A sequential niche technique for multimodal function optimization. Evolut Comput 1(2):101–125

Beyer HG (1999) On the dynamics of GAs without selection. In: Banzhaf W, Reeves C (eds) Foundations of genetic algorithms 5. Morgan Kaufmann, San Francisco, CA, pp 5–26

Beyer HG, Schwefel HP (2002) Evolution strategies a comprehensive introduction. Nat Comput Int J 1(1):3–52

Beyer HG, Brucherseifer E, Jakob W, Pohlheim H, Sendhoff B, To TB (2002) Evolutionary algorithms - terms and definitions. http://ls11-www.cs.uni-dortmund.de/people/beyer/EA-glossary/

Blum C (2005) Ant colony optimization: introduction and recent trends. Phys Life Rev 2:353–373

Bradshaw A (1965) Evolutionary significance of phenotypic plasticity in plants. Adv Genet 13:115–155

Branke J (2001) Evolutionary optimization in dynamic environments. Kluwer, Norwell, MA

Brits R, Engelbrecht AP, Bergh FVD (2002) A niching particle swarm optimizer. In: The fourth Asia-Pacific conference on simulated evolution and learning (SEAL2002). Singapore, pp 692–696

Cavicchio D (1970) Adaptive search using simulated evolution. Ph.D. thesis, University of Michigan, Ann Arbor, MI

Cioppa AD, Stefano CD, Marcelli A (2004) On the role of population size and niche radius in fitness sharing. IEEE Trans Evolut Comput 8(6):580–592

Coello Coello CA, Lamont GB, Van Veldhuizen DA (2007) Evolutionary algorithms for solving multi-objective problems. Springer, Berlin

Coello Coello CA (1999) A survey of constraint handling techniques used with evolutionary algorithms. Tech. Rep. Lania-RI-99-04, Laboratorio Nacional de Informática Avanzada. Xalapa, Veracruz, México

Cristiano JJ, White CC, Liker JK (2001) Application of multiattribute decision analysis to quality function deployment for target setting. IEEE Trans Syst Man Cybern Part C 31(3):366–382

Darwin CR (1999) The origin of species: by means of natural selection or the preservation of favoured races in the struggle for life. Bantam Classics, New York

De Jong KA (1975) An analysis of the behavior of a class of genetic adaptive systems. Ph.D. thesis, University of Michigan, Ann Arbor, MI

Deb K (2001) Multi-objective optimization using evolutionary algorithms. Wiley, New York

Deb K, Goldberg DE (1989) An investigation of niche and species formation in genetic function optimization. In: Proceedings of the third international conference on genetic algorithms. Morgan Kaufmann, San Francisco, CA, pp 42–50

Deb K, Spears WM (1997) Speciation methods. In: Bäck T, Fogel D, Michalewicz Z (eds) The handbook of evolutionary computation. IOP Publishing and Oxford University Press, Bristol

Deb K, Tiwari S (2005) Omni-optimizer: a procedure for single and multi-objective optimization. In: Evolutionary multi-criterion optimization, third international conference, EMO 2005, Lecture notes in computer science, vol 3410. Springer, Guanajuato, Mexico, pp 47–61

Dorigo M, Stützle T (2004) Ant colony optimization. MIT Press, Cambridge, MA

Doye J, Leary R, Locatelli M, Schoen F (2004) Global optimization of Morse clusters by potential energy transformations. INFORMS J Comput 16(4): 371–379

Engelbrecht A (2005) Fundamentals of computational swarm intelligence. New York

Fisher RA (1922) Darwinian evolution of mutations. Eugen Rev 14:31–34

Fogel LJ (1966) Artificial intelligence through simulated evolution. Wiley, New York

Freeman S, Herron JC (2003) Evolutionary analysis. Benjamin Cummings, 3rd edn. Redwood City, CA

Gan J, Warwick K (2001) Dynamic niche clustering: a fuzzy variable radius niching technique for multi-modal optimisation in GAs. In: Proceedings of the 2001 congress on evolutionary computation CEC2001, IEEE Press, COEX, World Trade Center, 159 Samseong-dong. Gangnam-gu, Seoul, Korea, pp 215–222

van der Goes V, Shir OM, Bäck T (2008) Niche radius adaptation with asymmetric sharing. In: Parallel problem solving from nature – PPSN X, Lecture notes in computer science, vol 5199. Springer, pp 195–204

Goldberg DE (1989) Genetic algorithms in search, optimization, and machine learning. Addison-Wesley, Reading, MA

Goldberg DE, Richardson J (1987) Genetic algorithms with sharing for multimodal function optimization. In: Proceedings of the second international conference on genetic algorithms and their application. Lawrence Erlbaum, Mahwah, NJ, pp 41–49

Grosso PB (1985) Computer simulations of genetic adaptation: parallel subcomponent interaction in a multilocus model. Ph.D. thesis, University of Michigan, Ann Arbor, MI

Hanagandi V, Nikolaou M (1998) A hybrid approach to global optimization using a clustering algorithm in a genetic search framework. Comput Chem Eng 22(12):1913–1925

Hansen N, Ostermeier A (2001) Completely derandomized self-adaptation in evolution strategies. Evolut Comput 9(2):159–195

Hansen N, Ostermeier A, Gawelczyk A (1995) On the adaptation of arbitrary normal mutation distributions in evolution strategies: the generating set adaptation. In: Proceedings of the sixth international conference on genetic algorithms (ICGA6). Morgan Kaufmann, San Francisco, CA, pp 57–64

Haykin S (1999) Neural networks: a comprehensive foundation, 2nd edn. Prentice Hall, NJ, USA

Holland JH (1975) Adaptation in natural and artificial systems. The University of Michigan Press, Ann Arbor, MI

Igel C, Suttorp T, Hansen N (2006) A computational efficient covariance matrix update and a (1+1)-CMA for evolution strategies. In: Proceedings of the genetic and evolutionary computation conference, GECCO 2006. ACM, New York, pp 453–460

Jelasity M (1998) UEGO, an abstract niching technique for global optimization. In: Parallel problem solving from nature - PPSN V, Lecture notes in computer science, vol 1498. Springer, Amsterdam, pp 378–387

Kennedy J, Eberhart R (2001) Swarm intelligence. Morgan Kaufmann, San Francisco, CA

Kimura M (1983) The neutral theory of molecular evolution. Cambridge University Press, Cambridge

Kramer O, Schwefel HP (2006) On three new approaches to handle constraints within evolution strategies. Nat Comput Int J 5(4):363–385

Li R, Eggermont J, Shir OM, Emmerich M, Bäck T, Dijkstra J, Reiber J (2008) Mixed-integer evolution strategies with dynamic niching. In: Parallel problem solving from nature - PPSN X, Lecture notes in computer science, vol 5199. Springer, pp 246–255

Lunacek M, Whitley D (2006) The dispersion metric and the CMA evolution strategy. In: Proceedings of the genetic and evolutionary computation conference, GECCO 2006. ACM, New York, pp 477–484

Lunacek M, Whitley D, Sutton A (2008) The impact of global structure on search. In: Parallel problem solving from nature - PPSN X, Lecture notes in computer science, vol 5199. Springer, pp 498–507

Mahfoud SW (1995a) Niching methods for genetic algorithms. Ph.D. thesis, University of Illinois at Urbana Champaign, IL

Mahfoud SW (1995b) A comparison of parallel and sequential niching methods. In: Eshelman L (ed) Proceedings of the sixth international conference on genetic algorithms. Morgan Kaufmann, San Francisco, CA, pp 136–143

Martin W, Lienig J, Cohoon J (1997) Island (migration) models: evolutionary algorithms based on punctuated equilibria. In: Bäck T, Fogel DB, Michalewicz Z (eds) Handbook of evolutionary computation. Oxford University Press, New York, and Institute of Physics, Bristol, pp C6.3:1–16

McPheron BA, Smith DC, Berlocher SH (1988) Genetic differences between host races of Rhagoletis pomonella. Nature 336:64–66

Miller B, Shaw M (1996) Genetic algorithms with dynamic niche sharing for multimodal function optimization. In: Proceedings of the 1996 IEEE international conference on evolutionary computation (ICEC'96). New York, pp 786–791

Ostermeier A, Gawelczyk A, Hansen N (1993) A derandomized approach to self adaptation of evolution strategies. Tech. rep., TU Berlin

Ostermeier A, Gawelczyk A, Hansen N (1994) Step-size adaptation based on non-local use of selection information. In: Parallel problem solving from nature - PPSN III, Lecture notes in computer science, vol 866. Springer, Berlin, pp 189–198

Parrott D, Li X (2006) Locating and tracking multiple dynamic optima by a particle swarm model using speciation. Evolut Comput, IEEE Trans 10(4):440–458, doi: 10.1109/TEVC.2005.859468

Petrowski A (1996) A clearing procedure as a niching method for genetic algorithms. In: Proceedings of the 1996 IEEE international conference on evolutionary computation (ICEC'96). New York, pp 798–803

Preuss M (2006) Niching prospects. In: Proceedings of the international conference on bioinspired optimization methods and their applications, BIOMA 2006. Jožef Stefan Institute, Slovenia, pp 25–34

Preuss M (2007) Reporting on experiments in evolutionary computation. Tech. Rep. CI-221/07, University of Dortmund, SFB 531

Ramalhinho-Lourenco H, Martin OC, Stützle T (2000) Iterated local search. Economics Working Papers 513, Department of Economics and Business, Universitat Pompeu Fabra

Scheiner SM, Goodnight CJ (1984) The comparison of phenotypic plasticity and genetic variation in populations of the grass Danthonia spicata. Evolution 38(4):845–855

Schönemann L, Emmerich M, Preuss M (2004) On the extinction of sub-populations on multimodal landscapes. In: Proceedings of the international conference on bioinspired optimization methods and their applications, BIOMA 2004. Jožef Stefan Institute, Slovenia, pp 31–40

Shir OM (2008) Niching in derandomized evolution strategies and its applications in quantum control. Ph.D. thesis, Leiden University, The Netherlands

Shir OM, Bäck T (2005a) Dynamic niching in evolution strategies with covariance matrix adaptation. In: Proceedings of the 2005 congress on evolutionary computation CEC-2005. IEEE Press, Piscataway, NJ, pp 2584–2591

Shir OM, Bäck T (2005b) Niching in evolution strategies. Tech. Rep. TR-2005-01, LIACS, Leiden University

Shir OM, Bäck T (2006) Niche radius adaptation in the CMA-ES niching algorithm. In: Parallel problem solving from nature - PPSN IX, Lecture notes in computer science, vol 4193. Springer, pp 142–151

Shir OM, Bäck T (2008) Niching with derandomized evolution strategies in artificial and real-world landscapes. Nat Comput Int J (2008), doi: 10.1007/s11047-007-9065-5

Shir OM, Beltrani V, Bäck T, Rabitz H, Vrakking MJ (2008) On the diversity of multiple optimal controls for quantum systems. J Phys B At Mol Opt Phys 41(7):(2008). doi: 10.1088/0953-4075/41/7/074021

Shir OM, Emmerich M, Bäck T (2010) Adaptive niche-radii and niche-shapes approaches for niching with the CMA-ES. Evolut Comput 18(1):97–126. doi: 10.1162/evco.2010.18.1.18104

Singh G, Deb K (2006) Comparison of multi-modal optimization algorithms based on evolutionary algorithms. In: Proceedings of the 2006 annual conference on genetic and evolutionary computation, GECCO 2006. ACM Press, New York, pp 1305–1312

Smith RE, Bonacina C (2003) Mating restriction and niching pressure: results from agents and implications for general EC. In: Proceedings of the 2003 conference on genetic and evolutionary computation, GECCO 2003, Lecture notes on computer science, vol 2724. Springer, Chicago, IL, pp 1382–1393

Spears WM (1994) Simple subpopulation schemes. In: Proceedings of the 3rd annual conference on evolutionary programming, World Scientific. San Diego, CA, Singapore, pp 296–307

Stoean C, Preuss M, Gorunescu R, Dumitrescu D (2005) Elitist generational genetic chromodynamics – a new radii-based evolutionary algorithm for multimodal optimization. In: Proceedings of the 2005 congress on evolutionary computation (CEC'05). IEEE Press, Piscataway NJ, pp 1839–1846

Stoean C, Preuss M, Stoean R, Dumitrescu D (2007) Disburdening the species conservation evolutionary algorithm of arguing with radii. In: Proceedings of the genetic and evolutionary computation conference, GECCO 2007. ACM Press, New York, pp 1420–1427

Streichert F, Stein G, Ulmer H, Zell A (2003) A clustering based niching EA for multimodal search spaces. In: Proceedings of the international conference evolution artificielle, Lecture notes in computer science, vol 2936. Springer, Heidelberg, Berlin, pp 293–304

Suganthan PN, Hansen N, Liang JJ, Deb K, Chen YP, Auger A, Tiwari S (2005) Problem definitions and evaluation criteria for the CEC 2005 special session on real-parameter optimization. Tech. rep., Nanyang Technological University, Singapore

Törn A, Zilinskas A (1987) Global optimization, Lecture notes in computer science, vol 350. Springer, Berlin

Tsui K (1992) An overview of Taguchi method and newly developed statistical methods for robust design. IIE Trans 24:44–57

Ursem RK (1999) Multinational evolutionary algorithms. In: Proceedings of the 1999 congress on evolutionary computation (CEC 1999). IEEE Press, Piscataway NJ, pp 1633–1640

Whitley D, Mathias KE, Rana SB, Dzubera J (1996) Evaluating evolutionary algorithms. Artif Intell 85(1–2):245–276

Wright S (1931) Evolution in Mendelian populations. Genetics 16:97–159

Yin X, Germany N (1993) A fast genetic algorithm with sharing using cluster analysis methods in multimodal function optimization. In: Proceedings of the international conference on artificial neural nets and genetic algorithms, Innsbruck. Austria, 1993, Springer, pp 450–457

Printed by Publishers' Graphics LLC